Study Guide and Stu
Solutions Manual

LIAL • MILLER • HUNGERFORD

Mathematics with Applications

FIFTH EDITION

Prepared with the assistance of

Louis Hoezle
Bucks County Community College

Gerald Krusinski
College of DuPage

August Zarcone
College of DuPage

HarperCollins*Publishers*

Study Guide and Student's Solutions Manual to accompany Lial•Miller•Hungerford Mathematics with Applications, Fifth Edition, by Louise Hoezle et al.

ISBN: 0-673-46278-1
ι 91 92 93 94 9 8 7 6 5 4 3 2

PREFACE

This book provides solutions for selected exercises in Mathematics With Applications, Fifth Edition, by Margaret L. Lial, Charles D. Miller, and Thomas W. Hungerford.

The solutions should be used as an aid as you work to master your course work. Try to solve the exercises that your instructor assigns before you refer to the solutions in this book. Then, if you have difficulty, read these solutions to guide you in solving the exercises. The solutions have been written to follow the methods presented in the textbook.

You may find that some solutions in a set are given in greater detail than others. Thus, if you cannot find an explanation for a difficulty that you encountered in one exercise, you may find the explanation in a solution for a similar exercise elsewhere in the set.

Solutions that require answer graphs will refer you to the answer section of your textbook; these graphs are not included in this book.

In addition to solutions, this book contains, in the preliminary pages, a list of exercises from the chapter review section of each chapter. These exercises, as a group, can be used as a chapter test.

The following people have made valuable contributions to this book: Marjorie Seachrist, editor; Fenner and Associates, typists; and Carmen Eldersveld, proofreader.

CHAPTER TEST PROBLEMS

The following list indicates those exercises from the chapter review exercise sets that can be considered as tests for each chapter. Working the listed exercises for a chapter will enable you to gauge the extent of your proficiency with the concepts and skills in that chapter.

Chapter 1 Review Exercises

3, 5, 20, 31, 36, 40, 49, 53, 57, 67, 73, 77, 85, 89, 103, 107, 115, 119, 128, 133, 136, 137, 143, 151, 156, 157, 163, 169, 181, 187, 199, 205, 210

Chapter 2 Review Exercises

5, 13, 17, 25, 27, 31, 35, 39, 41, 42, 43, 47, 50, 61, 63, 69

Chapter 3 Review Exercises

1, 3, 7, 13, 19, 27, 29, 35, 43, 47, 51

Chapter 4 Review Exercises

3, 5, 7, 10, 11, 15, 19, 27, 31, 35, 43, 53, 55, 59, 63

Chapter 5 Review Exercises

4, 7, 13, 21, 27, 37, 49, 53, 55, 61, 63, 69

Chapter 6 Review Exercises

8, 11, 15, 19, 27, 35, 38, 39, 45, 51, 53, 63, 71, 75, 85, 87

Chapter 7 Review Exercises

3, 9, 19, 25, 29, 31, 35, 39, 41, 43, 49

Chapter 8 Review Exercises

3, 7, 13, 17, 21, 33, 35, 37, 43, 45, 52, 55, 58, 75, 77, 79, 85

Chapter 9 Review Exercises

3, 4, 13, 19, 25, 31, 37, 41, 47, 49

Chapter 10 Review Exercises

1, 5, 7, 9, 11, 13, 21, 25, 29, 37

Chapter 11 Review Exercises

5, 11, 13, 15, 25, 27, 31, 39, 43, 45, 49, 55, 63, 71, 75, 77, 79, 89, 97, 98

Chapter 12 Review Exercises

3, 5, 15, 21, 25, 32, 33, 43, 47, 49

Chapter 13 Review Exercises

3, 11, 19, 21, 25, 29, 35, 43, 45, 49, 55, 61, 63, 71, 75

Chapter 14 Review Exercises

3, 7, 13, 19, 21, 23, 27, 31, 35, 39

CONTENTS

8 SETS AND PROBABILITY

9 FURTHER TOPICS IN PROBABILITY

CHAPTER 1 FUNDAMENTALS OF ALGEBRA

Section 1.1

1. 6: Natural (or counting) number, whole number, integer, rational number, real number.

3. −7: Integer, rational number, real number.

5. $\frac{1}{2}$: Rational number, real number.

7. $\sqrt{7}$: Irrational number, real number.

9. π: Irrational number, real number.

11. Every integer is a rational number. True, since every integer can be written as the ratio of the integer and 1.

    For example, $5 = \frac{5}{1}$.

13. Every whole number is an integer. True, since the integers are composed of the whole numbers and 0 and the opposite of the whole numbers.

15. There is a natural number that is not a whole number. False. The whole numbers are composed of the natural numbers and 0.

17. Every natural number is a rational number. True. Every natural number can be written as the ratio of the natural number and 1.

19. $8 \cdot 9 = 9 \cdot 8$
Commutative property of multiplication.

21. $3 + (-3) = (-3) + 3$
Commutative property of addition.

23. $-7 + 0 = -7$
Additive identity property.

25. $[9 (-3)] \cdot 2 = 9[(-3) \cdot 2]$
Associative property of multiplication.

27. $x(y + 2) = xy + 2x$
Distributive and commutative properties.

29. $8(4 + 2) = (2 + 4)8$
Commutative properties of addition and multiplication.

See the graphs for Exercises 31-47 in the answers at the back of the textbook.

31. All integers x such that $-5 < x < 5$. Plot points for every integer between but not including −5 and 5. Thus, points will be located at −4, −3, −2, −1, 0, 1, 2, 3, 4.

33. All whole numbers x such that $x \leq 3$.
Plot points for every whole number less than or equal to 3. Thus, points will be located at 3, 2, 1, and 0.

35. All natural numbers x such that $-1 < x < 5$.
Plot points for every natural number between but not including −1 and 5. Thus, points will be located at 1, 2, 3, 4. Note −1 and 0 are not included since they are not natural numbers.

37. $(-\infty, 4)$. Start at 4 and draw a line to the left. Use an open circle at 4 to show that 4 is not part of the graph.

39. $[6, \infty)$. Start at 6 and draw a line to the right. Use a solid circle at 6 to show that 6 is part of the graph.

41. $(-2, \infty)$. Start at -2 and draw a line to the right. Use an open circle to show that -2 is not part of the graph.

43. $(-5, -3)$. Start at -5 and draw a line to the right ending at -3. Use open circles at both -5 and -3 to show that neither is part of thee graph.

45. $[-3, 6]$. Start at -3 and draw a line to the right ending at 6. Use solid circles at both -3 and 6 to show that both are part of the graph.

47. $(1, 6]$. Start at 1 and draw a line to the right ending at 6. Use an open circle at 1 to show it is not part of the graph and a solid circle at 6 to show it is part of the graph.

49. $9 - 4^2 - (-12)$
$= 9 - 16 - (-12)$
$= 9 - 16 + 12$
$= -7 + 12$
$= 5$

51. $-2(9 - 8) + (-7)(2)^3$
$= -2(1) + (-7)(2)^3$
$= -2(1) + (-7)(8)$
$= -2 + (-56)$
$= -58$

53. $(4 - 2^3)(-2 + \sqrt{25})$
$= (4 - 8)(-2 + 5)$
$= (-4)(3)$
$= -12$

55. $\dfrac{-9.23(5.87) + 6.993}{1.225(-8.601) - (148)(.0723)}$

$= \dfrac{-54.1801 + 6.993}{-10.536225 - 10.7004}$

$= \dfrac{-47.1871}{-21.236625}$

$= -2.221967945$
$= 2.22$ (rounded to the nearest hundredth)

57. $-3(p + 5q)$
$= -3[-2 + 5(4)]$ *Let $p = -2$, $q = 4$*
$= -3(-2 + 20)$
$= -3(18)$
$= -54$

59. $\dfrac{q + r}{q + p}$

$= \dfrac{4 + (-5)}{4 + (-2)}$ *Let $q = 4$, $p = -2$, $r = -5$*

$= \dfrac{-1}{2}$

$= -\dfrac{1}{2}$

61. $\dfrac{\frac{q}{4} - \frac{r}{5}}{\frac{p}{2} + \frac{q}{2}}$

$= \dfrac{\frac{4}{4} - \frac{-5}{5}}{\frac{-2}{2} + \frac{4}{2}}$ *Let $p = -2$, $q = 4$, $4 = -5$*

$= \dfrac{1 - (-1)}{-1 + 2}$

$= \dfrac{1 + 1}{-1 + 2}$

$= \dfrac{2}{1}$

$= 2$

63. $-|-4| = -(4) = -4$

65. $|6 - 4| = |2| = 2$

67. $-|12 + (-8)| = -|12 - 8|$
$= -|4| = -4$

69. $|8 - (-9)| = |8 + 9| = |17| = 17$

71. $|8| - |-4| = 8 - (4) = 4$

73. $-|-4| - |-1 - 14| = -(4) - |-15|$
$= -(4) - (15)$
$= -19$

75. $|5|$ ____ $|-5|$
$|5| = 5$
$|-5| = 5$
Since $5 = 5$, $|5| = |-5|$.

77. $-|7|$ ____ $|7|$
$-|7| = -(7) = -7$
$|7| = 7$
Since $-7 < 7$, $-|7| < |7|$.

79. $|10 - 3|$ ____ $|3 - 10|$
$|10 - 3| = |7| = 7$
$|3 - 10| = |-7| = 7$
Since $7 = 7$, $|10 - 3| = |3 - 10|$.

81. $|1 - 4|$ ____ $|4 - 1|$
$|1 - 4| = |-3| = 3$
$|4 - 1| = |3| = 3$
Since $3 = 3$, $|1 - 4| = |4 - 1|$.

83. $|-2 + 8|$ ____ $|2 - 8|$
$|-2 + 8| = |6| = 6$
$|2 - 8| = |-6| = 6$
Since $6 = 6$, $|-2 + 8| = |2 - 8|$.

85. $|3| \cdot |-5|$ ____ $|3(-5)|$
$|3| \cdot |-5| = 3 \cdot 5 = 15$
$|3(-5)| = |-15| = 15$
Since $15 = 15$, $|3| \cdot |-5| = |3(-5)|$.

87. $|3 - 2|$ ____ $|3| - |2|$
$|3 - 2| = |1| = 1$
$|3| - |2| = 3 - 2 = 1$
Since $1 = 1$, $|3 - 2| = |3| - |2|$.

89. Yes.
If both a and b are positive, then $a + b$ is positive, and so $|a + b| = a + b$. Likewise, $|a| + |b| = a + b$. If both a and b are negative, then $(a + b)$ is negative.
Consequently, $|a + b| = -(a + b)$
$= (-a) + (-b) = |a| + |b|$.

91. Yes.
$|a - b| = |-1\ (-a + b)|$
$= |-1\ (b - a)| = |-1|\ |b - a|$
$= 1 \cdot |b - a| = |b - a|$

93. Let x = the percent of additional market share.
$$x \geq 4$$

95. Let x = the percent of revenue from razors and blades.
$$x \geq 32$$

97. Let x = the amount in millions Gillette will spend on advertising.
$$x \leq 110$$

99. The highest possible average status $[(x + y)/2]$ would occur when $x = 100$ and $y = 100$, or
$$\frac{100 + 100}{2} = \frac{200}{2} = 100.$$
The lowest possible average status would occur when $x = 0$ and $y = 0$, or
$$\frac{0 + 0}{2} = \frac{0}{2} = 0.$$

101. For Jolene Rizzo, $x = 56$ and $y = 78$.

average status:

$$\frac{56 + 78}{2} = \frac{134}{2} = 67$$

status incongruity:

$$\left|\frac{56 - 78}{2}\right| = \left|\frac{-22}{2}\right| = |-11| = 11.$$

Section 1.2

1. $4x - 1 = 15$

$4x - 1 + 1 = 15 + 1$ *Add 1 to both sides*

$4x = 16$

$\frac{1}{4}(4x) = \frac{1}{4}(16)$ *Multiply both sides by 1/4*

$1x = 4$

$x = 4$

3. $.2m - .5 = .1m + 7$

$.2m - .5 + .5 = .1m + .7 + .5$ *Add .5 to both sides*

$.2m = .1m + 1.2$

$.2m + (-.1m) = .1m + 1.2 + (-.1m)$ *Add -.1m to both sides*

$.1m = 1.2$

$\frac{1}{.1}(.1m) = \frac{1}{.1}(1.2)$ *Multiply both sides by 1/.1*

$1m = 12$

$m = 12$

5. $\frac{5}{6}k - 2k + \frac{1}{3} = \frac{2}{3}$ *Common demonimator is 6*

$6(\frac{5}{6}k - 2k + \frac{1}{3}) = 6(\frac{2}{3})$ *Multiply both sides by 6*

$6(\frac{5}{6}k) - 6(2k) + 6(\frac{1}{3}) = 6(\frac{2}{3})$

$5k - 12k + 2 = 4$

$-7k + 2 = 4$ *Distributive property*

$-7k + 2 + (-2) = 4 + (-2)$ *Add -2 to both sides*

$-7k = 2$

$(-\frac{1}{7})(-7k) = (-\frac{1}{7})(2)$ *Multiply both sides by (-1/7)*

$k = -\frac{2}{7}$

7. $3r + 2 - 5(r + 1) = 6r + 4$

$3r + 2 - 5r - 5 = 6r + 4$

$-2r - 3 = 6r + 4$

$-2r - 3 + 2r = 6r + 4 + 2r$

$-3 = 8r + 4$

$-3 + (-4) = 8r + 4 + (-4)$

$-7 = 8r$

$\frac{1}{8}(-7) = \frac{1}{8}(8r)$

$-\frac{7}{8} = r$

9. $\frac{3x}{5} - \frac{4}{5}(x + 1) = 2 - \frac{3}{10}(3x - 4)$

Multiply both sides by the common denominator, 10.

$10(\frac{3x}{5}) - 10(\frac{4}{5})(x + 1)$

$= (10)(2) - (10)(\frac{3}{10})(3x - 4)$

$2(3x) - 8(x + 1) = 20 - 3(3x - 4)$

$6x - 8x - 8 = 20 - 9x + 12$

$-2x - 8 = 32 - 9x$

$-2x + 9x = 32 + 8$ *Add 9x*

$7x = 40$

$\frac{1}{7}(7x) = \frac{1}{7}(40)$

$x = \frac{40}{7}$

11. $\frac{3x - 2}{7} = \frac{x + 2}{5}$

$35\left(\frac{3x - 2}{7}\right) = 35\left(\frac{x + 2}{5}\right)$ *Multiply by 35*

$5(3x - 2) = 7(x + 2)$

$15x - 10 = 7x + 14$

$15x - 7x = 10 + 14$

$8x = 24$

$\frac{1}{8}(8x) = \frac{1}{8}(24)$

$x = 3$

13. $\frac{x}{3} - 7 = 6 - \frac{3x}{4}$

$12(\frac{x}{3} - 7) = 12(6 - \frac{3x}{4})$

Multiply by 12

$12(\frac{x}{3}) - (12)(7) = (12)(6) - (12)(\frac{3x}{4})$

$4x - 84 = 72 - 3(3x)$

$4x - 84 = 72 - 9x$

$4x + 9x = 84 + 72$

$13x = 156$

$\frac{1}{13}(13x) = \frac{1}{13}(156)$

$x = 12$

15. $\frac{1}{4p} + \frac{2}{p} = 3$

$4p(\frac{1}{4p} + \frac{2}{p}) = 4p(3)$

$4p(\frac{1}{4}p) + 4p(\frac{2}{p}) = 4p(3)$

$1 + 8 = 12p$

$9 = 12p$

$\frac{1}{12}(9) = \frac{1}{12}(12p)$

$\frac{9}{12} = p$

$\frac{3}{4} = p$

17. $\frac{m}{2} - \frac{1}{m} = \frac{6m + 5}{12}$

$12m(\frac{m}{2} - \frac{1}{m}) = 12m\left(\frac{6m + 5}{12}\right)$

$(12m)(\frac{m}{2}) - (12m)(\frac{1}{m}) = m(6m) + m(5)$

$6m^2 - 12 = 6m^2 + 5m$

$-12 = 5m$

$-12 = 5m$

$\frac{1}{5}(-12) = \frac{1}{5}(5m)$

$-\frac{12}{5} = m$

19. $\frac{2r}{r - 1} = 5 + \frac{2}{r - 1}$

Multiply by $r - 1$, assuming $r - 1 \neq 0$.

$(r - 1)\left(\frac{2r}{r - 1}\right) = (r - 1)(5) + (r - 1)\left(\frac{2}{r - 1}\right)$

$2r = 5(r - 1) + 2$

$2r = 5r - 5 + 2$

$2r = 5r - 3$

$3 = 5r - 2r$

$3 = 3r$

$1 = r$

Recall the assumption that $r - 1 \neq 0$, or $r \neq 1$. Substituting 1 for r gives denominators of 0. Since division by 0 is not defined, there is no solution.

21. $\frac{4}{x - 3} - \frac{8}{2x + 5} + \frac{3}{x - 3} = 0$

$(x - 3)(2x + 5)\left(\frac{4}{x - 3}\right)$

$- (x - 3)(2x + 5)\left(\frac{8}{2x + 5}\right)$

$+ (x - 3)(2x + 5)\left(\frac{3}{x - 3}\right)$

$= (x - 3)(2x + 5)0$

$4(2x-5) - 8(x - 3) + 3(2x + 5) = 0$

$8x + 20 - 8x + 24 + 6x + 15 = 0$

$6x + 59 = 0$

$6x = -59$

$x = -\frac{59}{6}$

23. $\frac{3}{2m + 4} = \frac{1}{m + 2} - 2$

$\frac{3}{2(m + 2)} = \frac{1}{m + 2} - 2$

$2(m + 2)\left(\frac{3}{2(m + 2)}\right)$

$= 2(m + 2)\left(\frac{1}{m + 2}\right) - 2(m + 2)(2)$

$3 = 2 - 4(m + 2)$

$3 = 2 - 4m - 8$

$3 = -6 - 4m$

$3 + 6 = -4m$

$9 = -4m$

$\frac{9}{-4} = m$

$-\frac{9}{4} = m$

25. $2(x - a) + b = 3x + a$

$2x - 2a + b = 3x + a$

$-2a + b - a = 3x - 2x$

$-3a + b = x$

27. $ax + b = 3(x - a)$

$ax + b = 3x - 3a$

$b + 3a = 3x - ax$

$b + 3a = x(3 - a)$

$\frac{1}{3 - a}(b + 3a) = \frac{1}{3 - a}[x(3 - a)]$

$\frac{b + 3a}{3 - a} = x$

29. $x = a^2x - ax + 3a - 3$

$3 - 3a = a^2x - ax - x$

$3 - 3a = x(a^2 - a - 1)$

$\frac{3 - 3a}{a^2 - a - 1} = x$

31. $a^2x + 3x = 2a^2$

$x(a^2 + 3) = 2a^2$

Multiply by $1/(a^2 = 3)$.

$x = \frac{2a^2}{a^2 + 3}$

33. $PV = k$ for V

$\frac{1}{P}(PV) = \frac{1}{P}(k)$

$V = \frac{k}{P}$

35. $V = V_0 + gt$ for g

$V - V_0 = gt$

$\frac{V - V_0}{t} = \frac{gt}{t}$

$\frac{V - V_0}{t} = g$

37. $A = \frac{1}{2}(B + b)h$ for B

$A = \frac{1}{2}(Bh + \frac{1}{2}bh$

$2A = Bh + bh$ *Multiply by 2*

$2A - bh = Bh$

$\frac{2A - bh}{h} = \frac{Bh}{h}$ *Multiply by 1/h*

$\frac{2A - bh}{h} = B$

$\frac{2A}{h} - \frac{bh}{h} = B$

$\frac{2a}{h} - b = B$

39. $\frac{1}{R} = \frac{1}{r_1} + \frac{1}{r_2}$ for R

Multiply by common denominator Rr_1r_2.

$Rr_1r_2\left(\frac{1}{R}\right) = Rr_1r_2\left(\frac{1}{r_1}\right) + Rr_1r_2\left(\frac{1}{r_2}\right)$

$r_1r_2 = Rr_2 + Rr_1$

$r_1r_2 = R(r_1 + r_2)$

Multiply by $1/(r_1 + r_2)$.

$\frac{r_1r_2}{r_1 + r_2} = R$

41. $9.06x + 3.59(8x - 5) = 12.07x + .5612$

$9.06x + 28.72x - 17.95 = 12.07x + .5612$

$9.06x + 28.72 - 12.07x = 17.95 + .5612$

$25.71x = 18.5112$

$x = \frac{18.5112}{25.71}$

$x = .72$

43. $\frac{2.5x - 7.8}{3.2} + \frac{1.2x + 11.5}{5.8} = 6$

Multiply by (3.2)(5.8).

$5.8(2.5x - 7.8) + 3.2(1.2x + 11.5)$

$= (3.2)(5.8)(6)$

$14.5x - 45.24 + 3.84x + 36.8 = 111.36$

$14.5x + 3.84x = 111.36 + 45.24 - 36.8$

$18.34x = 119.80$

$x = 6.53$

45. $\frac{2.63r - 8.99}{1.25} - \frac{3.90r - 1.77}{2.45} = r$

Multiply by (1.25)(2.45).

$(2.45)(2.63r - 8.99)$

$- (1.25)(3.90r - 1.77)$

$= (1.25)(2.45)r$

$35r - 22.0255 - 4.875r + 2.2125$

$= 3.0625r$

$6.4435r - 4.875r - 3.0625r$

$= 22.0255 - 2.2125$

$-1.494r = 19.813$

$r = -13.26$

47. $F = \frac{9}{5}C + 32$

$F = \frac{9}{5}(20) + 32$

$F = 36 + 32$

$F = 68°$

49. $C = \frac{5(F - 32)}{9} = \frac{5(59 - 32)}{9} = 15°$

51. $C = \frac{5(100 - 32)}{9} = 37.8°$

53. $F = \frac{9}{5}(40) + 32 = 72 + 32 = 104°$

55. $A = \frac{24f}{b(p + 1)} = \frac{24(800)}{(4000)(36 + 1)}$

$A = \frac{19200}{148000} = .130$

$A = 13\%$

57. $A = \frac{24f}{b(p + 1)}$

Solve for f:

$Ab(p + 1) = 24f$

$\frac{Ab(p + 1)}{24} = f$

So $f = \frac{(.08)(2000)(36 + 1)}{24}$

$f = \frac{5920}{24} = \$247$

59. $A = \frac{24f}{b(p + 1)}$

Solve for b.

$Ab(p + 1) = 24f$

$b = \frac{24f}{A(p + 1)}$

$b = \frac{24(370)}{(.06)(36 + 1)} = \frac{8880}{2.22}$

$b = \$4000$

61. $u = f \cdot \frac{n(n + 1)}{q(q + 1)} = 800 \cdot \frac{18(19)}{36(37)}$

$= \$205.41.$

63. $u = f \cdot \frac{n(n + 1)}{q(q + 1)} = 950 \cdot \frac{(6)(7)}{(24)(25)}$

$= \$66.50$

65. Let x = the shortest side.
Then $2x$ = the second side
$x + 7$ = the third side.

$$x + 2x + x + 7 = 27$$
$$4x + 7 = 27$$
$$4x = 20$$
$$x = 5$$

So the shortest side is 5 centimeters.

67. Let x = amount invested at 6%.
Then $20,000 - x$ = amount invested at 8%.

$$.06x + .08(20,000 - x) = 1360$$
$$.06x + 1600 - .08x = 1360$$
$$-.02x = 1360 - 1600$$
$$-.02x = -240$$
$$x = 12,000$$

The amount invested at 8% is 20,000 - 12,000, or $8000.

69. Taxes = .30($100,000) = $30,000
Amount to be invested
= $100,000 - $30,000 = $70,000
Let x = amount invested at 6%.
Then $70,000 - x$
= amount invested at 8 1/2%.

$$.06x + .085(70,000 - x) = 5450$$
$$.06x + 5950 - .085X = 5450$$
$$-.025x = 5450 - 5950$$
$$-.025x = -500$$
$$x = 20,000$$

So $20,000 is invested at 6%.

71. Let x = price of first plot.
Then $120,000 - x$ = price of second plot.
$.15x$ = profit from first plot.
$-.10(120,000 - x)$
= loss from second plot.

$$.15x - .10(120,000 - x) = 5500$$
$$.15x - 12,000 + .10x = 5500$$
$$.25x = 17500$$
$$x = 70,000$$

She paid $70,000 for the first plot and 120,000 - 70,000, or $50,000 for the second plot.

73. Let x = the number of pounds of peanuts.
Since 10 = the number of pounds of cashews
8 = the number of pounds of hazelnuts,
then $x + 10 + 8$ = the total number of nuts.

$$1(x) + 4(10) + 3(8) = 2.50(x + 10 + 8)$$
$$x + 40 + 24 = 2.5x + 45$$
$$x - 2.5x = 45 - 40 - 24$$
$$-1.5x = -19$$
$$15x = 190 \quad \textit{Multiply by 10}$$
$$x = \frac{190}{15} = \frac{38}{3}$$

So 38/3 pounds are needed.

75. x = the number of liters of 94 octane gas
200 = the number of liters of 99 octane gas
$200 + x$ = the number of liters of 97 octane gas

$$94x + 99(200) = 97(200 + x)$$
$$94x + 19,800 = 19,400 + 97x$$
$$19,800 - 19,400 = 97x - 94x$$
$$400 = 3x$$
$$\frac{400}{3} = x$$

So 400/3 liters of 94 octane gas are needed.

Section 1.3

See the graphs for Exercises 1-31 in the answers at the back of the textbook.

1.
$$6x \le -18$$
$$\frac{6x}{6} \le \frac{-18}{6}$$
$$x \le -3$$

The solution is $(-\infty, -3]$.

3. $-3p < 18$

$\frac{-3p}{-3} > \frac{18}{-3}$ *Reverse the < symbol*

$p > -6$

The solution is $(-6, \infty)$.

5. $-9a < 0$

$\frac{-9a}{-9} > \frac{0}{-9}$

$a > 0$

The solution is $(0, \infty)$.

7. $2x + 1 \le 9$

$2x + 1 - 1 \le 9 - 1$

$2x \le 8$

$\frac{2x}{2} \le \frac{8}{2}$

$x \le 4$

The solution is $(-\infty, 4]$.

9. $-3p - 2 \ge 1$

$-3p - 2 + 2 \ge 1 + 2$

$-3p \ge 3$

$\frac{-3p}{-3} \le \frac{3}{-3}$

$p \le -1$

The solution is $(-\infty, -1]$.

11. $6k - 4 < 3k - 1$

$6k - 3k < 4 - 1$

$3k < 3$

$\frac{3k}{3} < \frac{3}{3}$

$k < 1$

The solution is $(-\infty, 1)$.

13. $m - (4 + 2m) + 3 < 2m + 2$

$m - 4 - 2m + 3 < 2m + 2$

$-1 - m < 2m + 2$

$-m - 2m < 2 + 1$

$-3m < 3$

$\frac{-3m}{-3} > \frac{3}{-3}$

$m > -1$

The solution is $(-1, \infty)$.

15. $-2(3y - 8) \ge 5(4y - 2)$

$-6y + 16 \ge 20y - 10$

$16 + 10 \ge 20y + 6y$

$26 \ge 26y$

$1 \ge y$

or $y \le 1$

The solution is $(-\infty, 1]$.

17. $3p - 1 < 6p + 2(p - 1)$

$3p - 1 < 6p + 2p - 2$

$-1 + 2 < 6p + 2p - 3p$

$1 < 5p$

$\frac{1}{5} < p$

or $p > \frac{1}{5}$

The solution is $(1/5, \infty)$.

19. $-7 < y - 2 < 4$

$-7 + 2 < y < 4 + 2$

$-5 < y < 6$

The solution is $(-5, 6)$.

21. $8 \le 3r + 1 \le 13$

$8 - 1 \le 3r \le 13 - 1$

$7 \le 3r \le 12$

$\frac{7}{3} \le r \le 4$

The solution is $[7/3, 4]$.

23. $-4 \le \frac{2k - 1}{3} \le 2$

$-4(3) \le 3\left(\frac{2k - 1}{3}\right) \le 2(3)$

$-12 \le 2k - 1 \le 6$

$-12 + 1 \le 2k \le 6 + 1$

$-11 \le 2k \le 7$

$-\frac{11}{2} \le k \le \frac{7}{2}$

The solution is $[-11/2, 7/2]$.

25. $z + 1 \le 2$ or $z - 5 \ge 1$

$z \le 2 - 1$ or $z \ge 1 - 1 + 5$

$z \le 1$ or $z \ge 6$

The solution is all numbers in $(-\infty, 1]$ or $[6, \infty)$.

27. $6m + 4 \ge 4 + m$ or $2m + 6 < -2 -2m$

$6m - m \ge 4 - 4$ or $2m + 2m < -2 -6$

$5m \ge 0$ or $4m < -8$

$\frac{5m}{5} \ge \frac{0}{5}$ or $\frac{4m}{4} < \frac{-8}{4}$

$m \ge 0$ or $m < -2$

The solution is all numbers in $(-\infty, -2)$ or $[0, \infty)$.

29. $\frac{3}{2}b - 1 < 4$ or $\frac{3}{4}b + \frac{1}{3} > \frac{19}{3}$

$\frac{3}{2}b < 4 + 2$ or $\frac{3}{4}b > \frac{19}{3} - \frac{1}{3}$

$\frac{3}{2}b < 6$ or $\frac{3}{4}b > \frac{18}{3}$

$\frac{2}{3}(\frac{3}{2}b) < \frac{2}{3}(6)$

or $\frac{4}{3}(\frac{3}{4}b) > \frac{4}{3}(\frac{18}{3})$

$b < 4$ or $b > 8$

The solution is all numbers in $(-\infty, 4)$ or $(8, \infty)$.

31. $\frac{3}{5}(2p + 3) \ge \frac{1}{10}(5p + 1)$

$10 \cdot \frac{3}{5}(2p + 3) \ge 10 \cdot \frac{1}{10}(5p + 1)$

$6(2p + 3) \ge (5p + 1)$

$12p + 18 \ge 5p + 1$

$12p - 5p \ge -18 + 1$

$7p \ge -17$

$p \ge -\frac{17}{7}$

The solution is $[-17/7, \infty)$.

33. $7.6092k \ge 2.28276$

$\frac{7.6092k}{7.6092} \ge \frac{2.28276}{7.6092}$

$k \ge .3$

The solution is $[.3, \infty)$.

35. $8.0413z - 9.7268$

$< 1.7251z - .25250$

$8.0413z - 1.7251z$

$< -.25250 + 9.7268$

$6.3162z < 9.4743$

$z < \frac{9.4743}{6.3162}$

$z < 1.5$

The solution is $(-\infty, 1.5)$.

37. $-(1.42m + 7.63) + 3(3.7m - 1.12)$

$\le 4.81m - 8.555$

$-1.42m - 7.63 + 11.1m - 3.36$

$\le 4.81m - 8.555$

$-1.42m + 11.1m - 4.81m$

$\le -8.555 + 7.63 + 3.36$

$4.87m \le 2.435$

$m \le \frac{2.435}{4.87}$

$m \le .5$

The solution is $(-\infty, .5]$.

39. Let x = the number.

Then 5x = five times the number.

$-8 < 5x < 6$

$-\frac{8}{5} < x < \frac{6}{5}$

The number is between $-\frac{8}{5}$ and $\frac{6}{5}$.

41. Let x = the number. Then 3x = three times the number. 5 + 3x = 5 added to three times the number.

$5 + 3x \ge 11$

$3x \ge 11 - 5$

$3x \ge 6$

$x \ge 2$

The number is equal to or greater than 2.

43. Let x = the number.

Then $\frac{1}{3}x$ = one-third the number.

$2 + \frac{1}{3}x$ = one-third of the number added to 2.

$$2 + \frac{1}{3}x \ge 8$$
$$\frac{1}{3}x \ge 8 - 2$$
$$\frac{1}{3}x \ge 6$$
$$3\left(\frac{1}{3}x\right) \ge 3(6)$$
$$x \ge 18$$

The number is at least 18.

45. Let x = the number of points she can get on the final.

Then 970 + x = her total number of points.

.81(1300) = the number of points needed to get a B.

$$970 + x \ge .81(1300)$$
$$970 + x \ge 1053$$
$$x \ge 83$$

Her lowest score could be 83 points.

47. Let x = the number of summers he must work.

$$1610\ x \ge 6440$$
$$x \ge 4$$

He must work 4 summers.

49. To at least break even, R ≥ C.

So $60x \ge 50x + 5000$

$$60x - 50x \ge 5000$$
$$10x \ge 5000$$
$$x \ge 500$$

The number of units of wire must be in the interval $[500, \infty)$.

51.

$$R \ge C$$
$$105x \ge 85x + 900$$
$$105x - 85x \ge 900$$
$$20x \ge 900$$
$$x \ge \frac{900}{20}$$
$$x \ge 45$$

x must be in the interval $[45, \infty)$.

53.

$$R \ge C$$
$$900x \ge 1000x + 5000$$
$$900x - 1000x \ge 5000$$
$$-100x \ge 5000$$
$$x \le \frac{5000}{-100}$$ *Reverse ≥ symbol.*
$$x \le -50$$

This is impossible so the product cannot break even. That is, for positive values of x, C is always greater than R, so it is impossible to make a profit.

Section 1.4

1. $|a - 2| = 1$

$a - 2 = 1$ or $a - 2 = -1$

$a = 1 + 2$ $\quad$ $a = -1 + 2$

$a = 3$ or $a = 1$

3. $|3m - 1| = 2$

$3m - 1 = 2$ or $3m - 1 = -2$

$3m = 2 + 1$ $\quad$ $3m = -2 + 1$

$3m = 3$ $\quad$ $3m = -1$

$m = 1$ or $m = -\frac{1}{3}$

5. $|5 - 3x| = 3$

$5 - 3x = 3$ or $5 - 3x = -3$

$-3x = 3 - 5$ $\quad$ $-3x = -3 - 5$

$-3x = -2$ $\quad$ $-3x = -8$

$x = \frac{-2}{-3}$ $\quad$ $x = \frac{-8}{-3}$

$x = \frac{2}{3}$ or $x = \frac{8}{3}$

7. $\left|\frac{z - 4}{2}\right| = 5$

$\frac{z - 4}{2} = 5$

$z - 4 = 10$

$z = 10 + 4$

$z = 14$

or $\frac{z - 4}{2} = -5$

$z - 4 = -10$

$z = -10 + 4$

$z = -6$

9. $\left|\frac{5}{r - 3}\right| = 10$

$\frac{5}{r - 3} = 10$

$5 = 10(r - 3)$

$5 = 10r - 30$

$5 + 30 = 10r$

$35 = 10r$

$\frac{35}{10} = r$

$\frac{7}{2} = r$

or $\frac{5}{r - 3} = -10$

$5 = -10(r - 3)$

$5 = -10r + 30$

$5 - 30 = -10r$

$-25 = -10r$

$\frac{-25}{-10} = r$

$\frac{5}{2} = r$

11. $\left|\frac{6y + 1}{y - 1}\right| = 3$

$\frac{6y + 1}{y - 1} = 3$

$6y + 1 = 3(y - 1)$

$6y + 1 = 3y - 3$

$6y - 3y = -3 - 1$

$3y = -4$

$y = \frac{-4}{3}$

$y = -\frac{4}{3}$

or $\frac{6y + 1}{y - 1} = -3$

$6y + 1 = -3(y - 1)$

$6y + 1 = -3y + 3$

$6y + 3y = 3 - 1$

$9y = 2$

$y = \frac{2}{9}$

13. $|2k - 3| = |5k + 4|$

$2k - 3 = 5k + 4$

$2k - 5k = 4 + 3$

$-3k = 7$

$k = \frac{7}{-3}$

$k = -\frac{7}{3}$

or $2k - 3 = -(5k + 4)$

$2k - 3 = -5k - 4$

$2k + 5k = -4 + 3$

$7k = -1$

$k = -\frac{1}{7}$

15. $|4 - 3y| = |7 + 2y|$

$4 - 3y = 7 + 2y$

$-3y - 2y = 7 - 4$

$-5y = 3$

$y = \frac{3}{-5}$

$y = -\frac{3}{5}$

or $4 - 3y = -(7 + 2y)$

$$4 - 3y = -7 - 2y$$

$$-3y + 2y = -7 - 4$$

$$-y = -11$$

$$y = \frac{-11}{-1}$$

$$y = 11$$

See the graphs for Exercises 17-31 in the answers at the back of the textbook.

17. $|x| \le 3$

$$-3 \le x \le 3$$

The solution is all numbers in $[-3, 3]$.

19. $|m| > 1$

$$m > 1 \text{ or } m < -1$$

The solution is all numbers in $(-\infty, -1)$ or $(1, \infty)$.

21. $|a| < -2$

Since the absolute value of a number is never negative, the inequality has no solution.

23. $|x| - 3 \le 7$

$$|x| \le 7 + 3$$

$$|x| \le 10$$

$$-10 \le x \le 10$$

The solution is all numbers in $[-10, 10]$.

25. $|2x + 5| < 3$

$$-3 < 2x + 5 < 3$$

$$-3 - 5 < 2x < 3 - 5$$

$$-8 < 2x < -2$$

$$-4 < x < -1$$

The solution is all numbers in $(-4, -1)$.

27. $|3m - 2| > 4$

$$3m - 2 > 4 \quad \text{or} \quad 3m - 2 < -4$$

$$3m > 6 \qquad 3m < -2$$

$$m > 2 \quad \text{or} \quad m < -\frac{2}{3}$$

The solution is all numbers in $(-\infty, -2/3)$ or $(2, \infty)$.

29. $|3z + 1| \ge 7$

$$3z + 1 \ge 7 \quad \text{or} \quad 3z + 1 \le -7$$

$$3z \ge 7 - 1 \qquad 3z \le -7 - 1$$

$$3z \ge 6 \qquad 3z \le -8$$

$$z \ge 2 \quad \text{or} \quad z \le -\frac{8}{3}$$

The solution is all numbers in $(-\infty, -8/3]$ or $[2, \infty)$.

31. $\left|5x + \frac{1}{2}\right| - 2 < 5$

$$\left|5x + \frac{1}{2}\right| < 7$$

$$-7 < 5x + \frac{1}{2} < 7$$

$$-7 - \frac{1}{2} < 5x < 7 - \frac{1}{2}$$

$$\frac{-15}{2} < 5x < \frac{13}{2}$$

$$\frac{-15}{2} \cdot \frac{1}{5} < x < \frac{13}{2} \cdot \frac{1}{5}$$

$$-\frac{3}{2} < x < \frac{13}{10}$$

The solution is all numbers in $(-3/2, 13/10)$.

33. No more than means "less than or equal to."

$$|x| \le 6$$

$$-6 \le x \le 6$$

The number is any number between -6 and 6, inclusive.

35. Let x = the number.
Then $4x$ = four times the number.
$6 + 4x$ = six added to four times the number.
So $|6 + 4x| \le 1$
$-1 \le 6 + 4x \le 1$
$-1 - 6 \le 4x \le 1 - 6$
$-7 \le 4x \le -5$
$-\frac{7}{4} \le x \le -\frac{5}{4}$
The number is any number between $-7/4$ and $-5/4$, inclusive.

37. Let x = the number.
Then $x + 2$ = the sum of the number and 2.
$|x + 2|$ = absolute value of the sum.
Thus $|x + 2| - 8 \ge 4$
$|x + 2| \ge 12$
$x + 2 \le -12$ or $x + 2 \ge 12$
$x \le -14$ or $x \ge 10$
The number is any number less than or equal to -14 or greater than or equal to 10.

39. If x is within 4 units of 2, then the distance from x to 2 is less than or equal to 4.
So $|x - 2| \le 4$.

41. If z is no less than 2 units from 12, then the distance from z to 12 is greater than or equal to 2.
So $|z - 12| \ge 2$.

43. If k is 6 units from 9, then the distance from 9 to k is exactly 6 units.
So $|k - 9| = 6$.

45. If $|x - 2| \le .0004$,
then $|y - 7| \le .00001$.

Section 1.5

1. $5^4 = 5\cdot5\cdot5\cdot5 = 625$

3. $\left(\frac{4}{5}\right)^3 = \frac{4}{5}\cdot\frac{4}{5}\cdot\frac{4}{5} = \frac{64}{125}$

5. $-5^2 = -(5 \cdot 5) = -25$

7. $(-2)^6 = (-2)(-2)(-2)(-2)(-2)(-2) = 64$

9. $5\cdot2^3 = 5\cdot2\cdot2\cdot2 = 5\cdot8 = 40$

11. $-2\cdot3^4 = -2(3\cdot3\cdot3\cdot3) = -2(81) = -162$

13. $2^4\cdot2^3 = 2^{4+3} = 2^7$

15. $(-5)^2\cdot(-5)^5 = (-5)^{2+5} = (-5)^7$ or -5^7

17. $(-3)^5\cdot3^4 = -(3^5)\cdot3^4 = -(3^{5+4}) = -3^9$

19. $(2z)^5\cdot(2z)^6 = (2z)^{5+6} = (2z)^{11}$

21. $(8m + 9) + (6m - 3)$
$= (8m + 6m) + (9 - 3)$
$= 14m + 6$

23. $(2x^2 - 6x + 11) + (-3x^2 + 7x - 2)$
$= (2x^2 - 3x^2) + (-6x + 7x) + (11 - 2)$
$= -x^2 + x + 9$.

25. $(-4y^2 - 3y + 8) - (2y^2 - 6y - 2)$
$= -4y^2 - 3y + 8 - 2y^2 + 6y + 2$
$= (-4y^2 - 2y^2) + (-3y + 6y) + (8 + 2)$
$= -6y^2 + 3y + 10$

27. $(2x^3 - 2x^2 + 4x - 3) - (2x^3 + 8x^2 - 1)$
$= 2x^3 - 2x^2 + 4x - 3 - 2x^3 - 8x^2 + 1$
$= (2x^3 - 2x^3) + (-2x^2 - 8x^2)$
$+ (4x) + (-3 + 1)$
$= -10x^2 + 4x - 2$

29. $(.613x^2 - 4.215x + .892)$
$- .47(2x^2 - 3x + 5)$
$= .613x^2 - 4.215x + .892$
$- .94x^2 + 1.41x - 2.35$
$= (.613x^2 - .94x^2) + (-4.215x + 1.41x)$
$+ (.892 - 2.35)$
$= -.327x^2 - 2.805x - 1.458$

31. $3p(2p - 5) = 3p(2p) - (3p)(5)$
$= 6p^2 - 15p$

33. $-9m(2m^2 + 3m - 1)$
$= (-9m)(2m^2) + (-9m)(3m) + (-9m)(-1)$
$= -18m^3 - 27m^2 + 9m$

35. $(3z + 5)(4z^2 - 2z - 11)$
$= (3z + 5)(4z^2) + (3z + 5)(-2z)$
$+ (3z + 5)(1)$
$= 12z^3 + 20z^2 - 6z^2 - 10z + 3z + 5$
$= 12z^3 + 14z^2 - 7z + 5$

37. $(6k - 1)(2k - 3)$
$= (6k)(2k - 3)$ *Distributive*
$+ (-1)(2k - 3)$ *property*
$= (6k)(2k) + (6k)(-3)$ *Distributive*
$+ (-1)(2k) + (-1)(-3)$ *property*
$= 12k^2 - 18k - 2k + 3$
$= 12k^2 - 20k + 3$

39. $(3y + 5)(2y - 1)$
$= 3y(2y) + 3y(-1)$
$+ 5(2y) + 5(-1)$ *FOIL*
$= 6y^2 - 3y + 10y - 5$
$= 6y^2 + 7y - 5$

41. $(5r - 3s)(5r + 4s)$
$= 25r^2 + 20rs - 15rs - 12s^2$ *FOIL*
$= 25r^2 + 5rs - 12s^2$

43. $(.012x - .17)(.3x + .54)$
$= (.012x)(.3x) + (.012x)(.54)$
$+ (-.17)(.3x) + (-.17)(.54)$
$= .0036x^2 + .00648x - .051x - .0918$
$= .0036x^2 - .04452x - .0918$

45. $2p - 3[4p - (3p + 1)]$
$= 2p - 3(4p - 3p - 1)$
$= 2p - 3(p - 1)$
$= 2p - 3p + 3$
$= -p + 3$

47. $(3x - 1)(x + 2) - (2x + 5)^2$
$= (3x^2 + 5x - 2) - (4x^2 + 20x + 25)$
$= 3x^2 + 5x - 2 - 4x^2 - 20x - 25$
$= 3x^2 - 4x^2 + 5x - 20x - 2 - 25$
$= -x^2 - 15x - 27$

49. $P = R - C$
$= (5x^3 - 3x + 1) - (4x^2 + 5x)$
$= 5x^3 - 3x + 1 - 4x^2 - 5x$
$= 5x^3 - 4x^2 - 8x + 1$

51. $P = R - C$
$= (2x^2 - 4x + 50) - (x^2 + 3x + 10)$
$= 2x^2 - 4x + 50 - x^2 - 3x - 10$
$= 2x^2 - x^2 - 4x - 3x + 50 - 10$
$= x^2 - 7x + 40$

Section 1.6

1. $4z + 4 = 4 \cdot z + 4 \cdot 1$
$= 4(z + 1)$

3. $8x + 6y + 4z = 2 \cdot 4x + 2 \cdot 3y + 2 \cdot 2z$
$= 2(4x + 3y + 2z)$

5. $m^3 - 9m^2 + 6m = m \cdot m^2 - m \cdot 9m + m \cdot 6$
$= m(m^2 - 9m + 6)$

7. $8a^3 - 16a^2 + 24a$
$= 8a \cdot a^2 + 8a(-2a) + 8a(3)$
$= 8a(a^2 - 2a + 3)$

9. $25p^4 - 20p^3q + 100p^2q^2$
$= 5p^2(5p^2 - 4pq + 20q^2)$

11. $2(5x - 1)^2 + 8(5x - 1)^3$
$= 2(5x -1)^2 \cdot 1 + 2(5x - 1)^2 \cdot 4(5x - 1)$
$= 2(5x - 1)^2 [1 + 4(5x - 1)]$
$= 2(5x - 1)^2 (1 + 20x - 4)$
$= 2(5x - 1)^2 (20x - 3)$

13. $9(x - 4)^5 - (x - 4)^3$
$= (x - 4)^3 [9(x - 4)^2 - 1]$
$= (x - 4)^3 [9(x^2 - 8x + 16) - 1]$
$= (x - 4)^3 (9x^2 - 72x + 144 - 1)$
$= (x - 4)^3 (9x^2 - 72x + 143)$

15. $x^2 + 4x - 5 = (x + \quad)(x - \quad)$
$= (x + 5)(x - 1)$ *Look for factors of -5 with a sum of 4*

17. $6a^2 - 48a - 120 = 6(a^2 - 8a - 20)$
$= 6(a - 10)(a + 2)$

19. $x^2 - 64 = x^2 - (8)^2 = (x + 8)(x - 8)$

21. $3m^3 + 12m^2 + 9m = 3m(m^2 + 4m + 3)$
$= 3m(m + 3)(m + 1)$

23. $b^2 - 8b + 7 = (b - 7)(b - 1)$

25. $m^2 - 6mn + 9n^2$
$= (m)^2 - 2(m)(3n) + (3n)^2$
$= (m - 3n)^2$

27. $3p^2 - 7p + 10$
We need factors of $3 \cdot 10$ or 30 with a sum of -7. There are no such numbers: Therefore, the polynomial cannot be factored.

29. $9m^2 - 25 = (3m)^2 - (5)^2$
$= (3m + 5)(3m - 5)$

31. $a^2 + 4ab + 5b^2$
We are looking for two positive numbers that are factors of 5 and have the sum, 4. Since there are none, the polynomial cannot be factored.

33. $2x^2 - 5x - 3 = (2x + 1)(x - 3)$

35. $3k^2 + 2k - 8 = (3k - 4)(k + 2)$

37. $21m^2 + 13mn + 2n^2 = (7m + 2n)(3m + n)$

39. $121a^2 - 100 = (11a)^2 - (10)^2$
$= (11a + 10)(11a - 10)$

41. $5a^2 - 7ab - 6b^2 = (5a + 3b)(a - 2b)$

43. $y^2 - 4yz - 21z^2 = (y - 7z)(y + 3z)$

45. $9x^2 + 64$ *Sum of two squares, not difference*
The polynomial cannot be factored.

47. $z^2 + 14zy + 49y^2$
$= (z)^2 + 2(z)(7y) + (7y)^2 = (z + 7y)^2$

49. $24a^4 + 10a^3b - 4a^2b^2$
$= 2a^2(12a^2 + 5ab - 2b^2)$
$= 2a^2(4a - b)(3a + 2b)$

51. $6x^2 + x - 1 = (3x - 1)(2x + 1)$

53. $y^2 + 10y + 25 - z^2$
$= (y^2 + 10y + 25) - z^2$
$= (y + 5)^2 - z^2$
$= [(y + 5) + z][(y + 5) - z]$
$= (y + 5 + z)(y + 5 - z)$

55. $m^2 - n^2 + 2n - 1$
$= m^2 - (n^2 - 2n + 1)$
$= m^2 - (n - 1)^2$
$= [m + (n - 1)][m - (n - 1)]$
$= (m + n - 1)(m - n + 1)$

57. $3x^4(x^2 + 9)^3 + 2x^2(x^2 + 9)^4$

$= x^2(x^2 + 9)^3 [3x^2 + 2(x^2 + 9)]$

$= x^2(x^2 + 9)^3 (3x^2 + 2x^2 + 18)$

$= x^2(x^2 + 9)^3 (5x^2 + 18)$

59. $a^3 - 216 = a^3 - (6)^3$

$= (a - 6)(a^2 + 6a + 36)$

61. $8r^3 - 27s^3 = (2r)^3 - (3s)^3$

$= (2r - 3s)[(2r)^2 + (2r)(3s) + (3s)^2]$

$= (2r - 3s)(4r^2 + 6rs + 9s^2)$

63. $64m^3 + 125 = (4m)^3 + (5)^3$

$= (4m + 5)[(4m)^2 - (4m)(5) + (5)^2]$

$= (4m + 5)(16m^2 - 20m + 25)$

65. $1000y^3 - z^3 = (10y)^3 - (z)^3$

$=(10y - z)[(10y)^2 + (10y)(z) + (z)^2]$

$=(10y - z)(100y^2 + 10yz + z^2)$

Section 1.7

1. $\frac{6m}{24} = \frac{6 \cdot m}{6 \cdot 4} = \frac{m}{4}$

3. $\frac{7z^2}{14z} = \frac{7z \cdot z}{7z \cdot 2} = \frac{z}{2}$

5. $\frac{25p^3}{10p^2} = \frac{5p^2(5p)}{5p^2 \cdot 2} = \frac{5p}{2}$

7. $\frac{8k + 16}{9k + 18} = \frac{8(k + 2)}{9(k + 2)} = \frac{8}{9}$

9. $\frac{3(t + 5)}{(t + 5)(t - 3)} = \frac{3}{t - 3}$

11. $\frac{8x^2 + 16x}{4x^2} = \frac{8x(x + 2)}{4x^2} = \frac{2(x + 2)}{x}$

13. $\frac{m^2 - 4m + 4}{m^2 + m - 6} = \frac{(m - 2)(m - 2)}{(m + 3)(m - 2)} = \frac{m - 2}{m + 3}$

15. $\frac{x^2 + 3x - 4}{x^2 - 1} = \frac{(x + 4)(x - 1)}{(x + 1)(x - 1)} = \frac{x + 4}{x + 1}$

17. $\frac{8m^2 + 6m - 9}{16m^2 - 9} = \frac{(4m - 3)(2m + 3)}{(4m + 3)(4m - 3)}$

$= \frac{2m + 3}{4m + 3}$

19. $\frac{9k^2}{25} \cdot \frac{5}{3k} = \frac{45k^2}{75k} = \frac{(15k)(3k)}{(15k)(5)} = \frac{3k}{5}$

21. $\frac{15p^3}{9p^2} \div \frac{6p}{10p^2} = \frac{15p^3}{9p^2} \cdot \frac{10p^2}{6p} = \frac{105p^5}{54p^3}$

$= \frac{(6p^3)(25p^2)}{(6p^3)(9)} = \frac{25p^3}{9}$

23. $\frac{a + b}{2p} \cdot \frac{12}{5(a + b)} = \frac{12(a + b)}{10p(a + b)}$

$= \frac{12}{10p} = \frac{2 \cdot 6}{2 \cdot 5p} = \frac{6}{5p}$

25. $\frac{2k + 8}{6} \div \frac{3k + 12}{2} = \frac{2k + 8}{6} \cdot \frac{2}{3k + 12}$

$= \frac{2(k + 4)}{6} \cdot \frac{2}{3(k + 4)} = \frac{4(k + 4)}{18(k + 4)}$

$= \frac{4}{18} = \frac{2}{9}$

27. $\frac{9y - 18}{6y + 12} \cdot \frac{3y + 6}{15y - 30}$

$= \frac{9(y - 2)}{6(y + 2)} \cdot \frac{3(y + 2)}{15(y - 2)}$

$= \frac{27(y - 2)(y + 2)}{90(y + 2)(y - 2)} = \frac{27}{90} = \frac{3}{10}$

29. $\dfrac{4a + 12}{2a - 10} \div \dfrac{a^2 - 9}{a^2 - a - 20}$

$= \dfrac{4a + 12}{2a - 10} \cdot \dfrac{a^2 - a - 20}{a^2 - 9}$

$= \dfrac{4(a + 3)}{2(a - 5)} \cdot \dfrac{(a - 5)(a + 4)}{(a + 3)(a - 3)}$

$= \dfrac{4(a + 3)(a - 5)(a + 4)}{2(a - 5)(a + 3)(a - 3)}$

$= \dfrac{2(a + 4)}{a - 3}$

31. $\dfrac{k^2 - k - 6}{k^2 + k - 12} \cdot \dfrac{k^2 + 3k - 4}{k^2 + 2k - 3}$

$= \dfrac{(k - 3)(k + 2)}{(k + 4)(k - 3)} \cdot \dfrac{(k + 4)(k - 1)}{(k + 3)(k - 1)}$

$= \dfrac{(k - 3)(k + 2)(k + 4)(k - 1)}{(k + 4)(k - 3)(k + 3)(k - 1)} = \dfrac{k + 2}{k + 3}$

33. $\dfrac{m^2 + 3m + 2}{m^2 + 5m + 4} \div \dfrac{m^2 + 5m + 6}{m^2 + 10m + 24}$

$= \dfrac{m^2 + 2m + 2}{m^2 + 5m + 4} \cdot \dfrac{m^2 + 10m + 24}{m^2 + 5m + 6}$

$= \dfrac{(m + 1)(m + 2)}{(m + 1)(m + 4)} \cdot \dfrac{(m + 6)(m + 4)}{(m + 3)(m + 2)}$

$= \dfrac{m + 6}{m + 3}$

35. $\dfrac{2m^2 - 5m - 12}{m^2 - 10m + 24} \div \dfrac{4m^2 - 9}{m^2 - 9m + 18}$

$= \dfrac{2m^2 - 5m - 12}{m^2 - 10m + 24} \cdot \dfrac{m^2 - 9m + 18}{4m^2 - 9}$

$= \dfrac{(2m + 3)(m - 4)}{(m - 6)(m - 4)} \cdot \dfrac{(m - 6)(m - 3)}{(2m + 3)(2m - 3)}$

$= \dfrac{m - 3}{2m - 3}$

37. $\dfrac{2}{3y} - \dfrac{1}{4y} = \dfrac{4 \cdot 2}{4 \cdot 3y} - \dfrac{3 \cdot 1}{3 \cdot 4y}$

$= \dfrac{8}{12y} - \dfrac{3}{12y} = \dfrac{8 - 3}{12y} = \dfrac{5}{12y}$

39. $\dfrac{a + 1}{2} - \dfrac{a - 1}{2} = \dfrac{(a + 1) - (a - 1)}{2}$

$= \dfrac{a + 1 - a + 1}{2} = \dfrac{2}{2} = 1.$

41. $\dfrac{3}{p} + \dfrac{1}{2} = \dfrac{2 \cdot 3}{2 \cdot p} + \dfrac{p \cdot 1}{p \cdot 2}$

$= \dfrac{6}{2p} + \dfrac{p}{2p} = \dfrac{6 + p}{2p}$

43. $\dfrac{2}{y} - \dfrac{1}{4} = \dfrac{4 \cdot 2}{4 \cdot y} - \dfrac{y \cdot 1}{y \cdot 4}$

$= \dfrac{8}{4y} - \dfrac{y}{4y} = \dfrac{8 - y}{4y}$

45. $\dfrac{1}{6m} + \dfrac{2}{5m} + \dfrac{4}{m} = \dfrac{5 \cdot 1}{5 \cdot 6m} + \dfrac{6 \cdot 2}{6 \cdot 5m} + \dfrac{30 \cdot 4}{30 \cdot m}$

$= \dfrac{5}{30m} + \dfrac{12}{30m} + \dfrac{120}{30m}$

$= \dfrac{5 + 12 + 120}{30m} = \dfrac{137}{30m}$

47. $\dfrac{1}{m - 1} + \dfrac{2}{m} = \dfrac{m \cdot 1}{m \cdot (m - 1)} + \dfrac{(m - 1) \cdot 2}{(m - 1) \cdot m}$

$= \dfrac{m}{m(m - 1)} + \dfrac{2(m - 1)}{m(m - 1)} = \dfrac{m + 2(m - 1)}{m(m - 1)}$

$= \dfrac{m + 2m - 2}{m(m - 1)} = \dfrac{3m - 2}{m(m - 1)}$

49. $\dfrac{8}{3(a - 1)} + \dfrac{2}{a - 1}$

$= \dfrac{8}{3(a - 1)} + \dfrac{3 \cdot 2}{3(a - 1)}$

$= \dfrac{8}{3(a - 1)} + \dfrac{6}{3(a - 1)}$

$= \dfrac{8 + 6}{3(a - 1)} = \dfrac{14}{3(a - 1)}$

51. $\dfrac{2}{5(k - 2)} + \dfrac{3}{4(k - 2)}$

$= \dfrac{8}{20(k - 2)} + \dfrac{15}{20(k - 2)}$

$= \dfrac{8 + 15}{20(k - 2)} = \dfrac{23}{20(k - 2)}$

53. $\frac{2}{x^2 - 2x - 3} + \frac{5}{x^2 - x - 6}$

$= \frac{2}{(x - 3)(x + 1)} + \frac{5}{(x - 3)(x + 2)}$

$= \frac{2(x + 2)}{(x - 3)(x + 1)(x + 2)}$

$+ \frac{5(x + 1)}{(x - 3)(x + 2)(x + 1)}$

$= \frac{2(x + 2) + 5(x + 1)}{(x - 3)(x + 2)(x + 1)}$

$= \frac{2x + 4 + 5x + 5}{(x - 3)(x + 1)(x + 2)}$

$= \frac{7x + 9}{(x - 3)(x + 1)(x + 2)}$

55. $\frac{2y}{y^2 + 7y + 12} - \frac{y}{y^2 + 5y + 6}$

$= \frac{2y}{(y + 4)(y + 3)} - \frac{y}{(y + 3)(y + 2)}$

$= \frac{2y(y + 2)}{(y + 4)(y + 3)(y + 2)}$

$- \frac{y(y + 4)}{(y + 4)(y + 3)(y + 2)}$

$= \frac{2y(y + 2) - y(y + 4)}{(y + 4)(y + 3)(y + 2)}$

$= \frac{2y^2 + 4y - y^2 - 4y}{(y + 4)(y + 3)(y + 2)}$

$= \frac{y^2}{(y + 4)(y + 3)(y + 2)}$

57. $\frac{3k}{2k^2 + 3k - 2} - \frac{2k}{2k^2 - 7k + 3}$

$= \frac{3k}{(2k - 1)(k + 2)} - \frac{2k}{(2k - 1)(k - 3)}$

$= \frac{3k(k - 3)}{(2k - 1)(k + 2)(k - 3)}$

$- \frac{2k(k + 2)}{(2k - 1)(k + 2)(k - 3)}$

$= \frac{3k(k - 3) - 2k(k + 2)}{(2k - 1)(k + 2)(k - 3)}$

$= \frac{3k^2 - 9k - 2k^2 - 4k}{(2k - 1)(k + 2)(k - 3)}$

$= \frac{k^2 - 13k}{(2k - 1)(k + 2)(k - 3)}$

$= \frac{k(k - 13)}{(2k - 1)(k + 2)(k - 3)}$

59. $\frac{1 + \frac{1}{x}}{1 - \frac{1}{x}} = \frac{\frac{x}{x} + \frac{1}{x}}{\frac{x}{x} - \frac{1}{x}}$

$= \frac{\frac{x + 1}{x}}{\frac{x - 1}{x}} = \frac{x + 1}{x} \div \frac{x - 1}{x}$

$= \frac{x + 1}{x} \cdot \frac{x}{x - 1} = \frac{x + 1}{x - 1}$

61. $\frac{\frac{1}{x + h} - \frac{1}{x}}{h} = \frac{\frac{x - (x + h)}{x(x + h)}}{h}$

$= \frac{\frac{x - x - h}{x(x + h)}}{h} = \frac{\frac{-h}{x(x + h)}}{h}$

$= \frac{-h}{x(x + h)} \div h = \frac{-h}{x(x + h)} \cdot \frac{1}{h}$

$= \frac{-1}{x(x + h)}$

63. $\frac{1 + \frac{1}{1 - b}}{1 - \frac{1}{1 + b}} = \frac{\frac{1 - b + 1}{1 - b}}{\frac{1 + b - 1}{1 + b}}$

$= \frac{\frac{2 - b}{1 - b}}{\frac{b}{1 + b}} = \frac{2 - b}{1 - b} \div \frac{b}{1 + b}$

$= \frac{2 - b}{1 - b} \cdot \frac{1 + b}{b}$

$= \frac{(2 - b)(1 + b)}{b(1 - b)} = \frac{2 + b - b^2}{b - b^2}$

Section 1.8

1. $7^3 = 7 \cdot 7 \cdot 7 = 343$

3. $8^{-1} = \frac{1}{8^1} = \frac{1}{8}$

5. $2^{-3} = \frac{1}{2^3} = \frac{1}{8}$

7. $5^{-1} = \frac{1}{5}$

9. $8^{-3} = \frac{1}{8^3} = \frac{1}{8 \cdot 8 \cdot 8} = \frac{1}{512}$

11. $\left(\frac{1}{2}\right)^{-3} = \left(\frac{2}{1}\right)^3 = 2^3 = 8$

13. $\left(\frac{2}{7}\right)^{-2} = \left(\frac{7}{2}\right)^2 = \frac{7^2}{2^2} = \frac{49}{4}$

15. $\frac{3^{-4}}{3^2} = 3^{-4+2} = 3^{-6} = \frac{1}{3^6}$

17. $\frac{2^{-5}}{2^{-2}} = 2^{-5-(-2)} = 2^{-5+2} = 2^{-3} = \frac{1}{2^3}$

19. $\frac{6^{-1}}{6} = 6^{-1-1} = 6^{-2} = \frac{1}{6^2}$

21. $4^{-3} \cdot 4^6 = 4^{-3+6} = 4^3$

23. $7^{-5} \cdot 7^{-2} = 7^{-5-2} = 7^{-7} = \frac{1}{7^7}$

25. $\frac{8^9 \cdot 8^{-7}}{8^{-3}} = \frac{8^{9-7}}{8^{-3}} = \frac{8^2}{8^{-3}} = 8^{2-(-3)} = 8^{2+3} = 8^5$

27. $\frac{10^8 \cdot 10^{-10}}{10^4 \cdot 10^2} = \frac{10^{8-10}}{10^{4+2}}$

$= \frac{10^{-2}}{10^6} = 10^{-2-6} = 10^{-8} = \frac{1}{10^8}$

29. $\left(\frac{5^{-6} \cdot 5^3}{5^{-2}}\right)^{-1} = \left(\frac{5^{-6+3}}{5^{-2}}\right)^{-1}$

$= \left(\frac{5^{-3}}{5^{-2}}\right)^{-1} = (5^{-3-(-2)})^{-1}$

$= (5^{-3+2})^{-1} = (5^{-1})^{-1}$

$= 5^1 = 5$

31. $\frac{x^4 \cdot x^3}{x^5} = \frac{x^7}{x^5} = x^{7-5} = x^2$

33. $\frac{(4k^{-1})^2}{2k^{-5}} = \frac{16k^{-2}}{2k^{-5}} = 8k^{-2-(5)} = 8k^{-2+5}$

$= 8k^3$

35. $(a^4b^{-2})^{-5} = (s^4)^{-5} \cdot (b^{-2})^{-5} = a^{-20}b^{10}$

$= \frac{1}{a^{20}} \cdot b^{10} = \frac{b^{10}}{a^{20}}$

37. $(2p^{-1})^3 \cdot (5p^2)^{-2} = 2^3(p^{-1})^3(5)^{-2}(p^2)^{-2}$

$= 2^3(p^{-3})\left(\frac{1}{5^2}\right)(p^{-4})$

$= 2^3\left(\frac{1}{p^3}\right)\left(\frac{1}{5^2}\right)\left(\frac{1}{p^4}\right)$

$= \frac{2^3}{5^2p^7}$

39. $\frac{7^{-1} \cdot 7r^{-3}}{7^2 \cdot (r^{-2})^2} = \frac{7^{-1+1} \cdot r^{-3}}{7^2 \cdot r^{-4}} = \frac{7^0 \cdot r^{-3}}{7^2 \cdot r^{-4}}$

$= 7^{0-2} \cdot r^{-3-(-4)} = 7^{-2}r^{-3+4} = 7^{-2}r^1$

$= \frac{1}{7^2} \cdot r^1$

$= \frac{r}{7^2} = \frac{r}{49}$

41. $\frac{6k^{-4} \cdot (3k^{-1})^{-2}}{2^3k^2} = \frac{6k^{-4} \cdot 3^{-2}k^2}{2^3k^2}$

$= \frac{6 \cdot 3^{-2} k^{-4+2}}{2^3k^2} = \frac{6k^{-2}}{3^2 2^3 k^2}$

$= \frac{6k^{-2-2}}{8 \cdot 9} = \frac{6k^{-4}}{72} = \frac{1}{12k^4}$

43. $\frac{(2x)^{-2}(x^{-1}y)^{-3}}{(xy)^{-2}y^2} = \frac{2^{-2}x^{-2} \cdot x^3y^{-3}}{x^{-2}y^{-2}y^2}$

$= \frac{2^{-2}x^1y^{-3}}{x^{-2}y^0} = \frac{x^1x^2}{2^2y^3y^0} = \frac{x^3}{2^2y^3} = \frac{x^3}{4y^3}$

45. $(m^4p^{-2})^2(m^{-1}p^2) = (m^8p^{-4})(m^{-1}p^2)$

$= m^{8-1}p^{-4+2} = m^7p^{-2} = \frac{m^7}{p^2}$

47. $(5a^2b^{-3})^{-1} \cdot (3^{-1}a^{-2}b^2)^{-2}$

$= (5^{-1}a^{-2}b^3)(3^2a^4b^{-4})$

$= 5^{-1} \cdot 3^2 \cdot a^{-2+4} \cdot b^{3-4}$

$= 5^{-1} \cdot 3^2 \cdot a^2 \cdot b^{-1}$

$= \frac{1}{5} \cdot 3^2 \cdot a^2 \cdot \frac{1}{b^1} = \frac{3^2a^2}{5b} = \frac{9a^2}{5b}$

Let $a = 2$, $b = -3$, and $c = 0$ in Exercises 49-59.

49. $a^3 + b = (2)^3 + (-3) = 8 - 3 = 5$

51. $-b^2 + 3(c + 5) = -(-3)^2 + 3(0 + 5)$

$= -9 + 3(5)$

$= -9 + 15 = 6$

53. $a^7b^9c^5 = (2)^7(-3)^9(0)^5 = (2^7)(-3)^9 \cdot 0 = 0$

55. $a^b + b^a = (2)^{-3} + (-3)^2$

$= \frac{1}{2^3} + 9 = \frac{1}{8} + 9$

$= \frac{1}{8} + \frac{72}{8} = \frac{73}{8}$

57. $a^{-1} + b^{-1} = (2)^{-1} + (-3)^{-1} = \frac{1}{2} + \frac{1}{-3}$

$= \frac{1}{2} - \frac{1}{3} = \frac{3}{6} - \frac{2}{6} = \frac{1}{6}$

59. $a^{-b} + b^{-c} = 2^{-(-3)} + (-3)^{-0} = 2^3 + (-3)^0$

$= 2^3 + 1 = 8 + 1 = 9$

61. $2^{-1} - 3^{-1} = \frac{1}{2} - \frac{1}{3} = \frac{3}{6} - \frac{2}{6} = \frac{1}{6}$

63. $(6^{-1} + 2^{-1})^{-1} = \left(\frac{1}{6} + \frac{1}{2}\right)^{-1} = \left(\frac{1}{6} + \frac{3}{6}\right)^{-1}$

$= \left(\frac{4}{6}\right)^{-1} = \left(\frac{2}{3}\right)^{-1} = \frac{3}{2}$

65. $\frac{3^{-2} - 4^{-1}}{4^{-1}} = \frac{\frac{1}{3^2} - \frac{1}{4}}{\frac{1}{4}}$

$= \frac{\frac{1}{9} - \frac{1}{4}}{\frac{1}{4}} = \frac{\frac{4}{36} - \frac{9}{36}}{\frac{1}{4}}$

$= \frac{\frac{-5}{36}}{\frac{1}{4}} = \frac{-5}{36} \cdot \frac{4}{1} = -\frac{5}{9}$

67. $\frac{a^{-1} + b^{-1}}{(ab)^{-1}} = \frac{\frac{1}{a} + \frac{1}{b}}{\frac{1}{ab}} = \frac{\frac{b}{ab} + \frac{a}{ab}}{\frac{1}{ab}}$

$= \frac{\frac{b + a}{ab}}{\frac{1}{ab}} = \frac{b + a}{ab} \cdot \frac{ab}{1} = b + a$

69. $(a + b)^{-1}(a^{-1} + b^{-1}) = \left(\frac{1}{a + b}\right)\left(\frac{1}{a} + \frac{1}{b}\right)$

$= \left(\frac{1}{a + b}\right)\left(\frac{b + a}{ab}\right) = \frac{b + a}{(a + b)ab} = \frac{1}{ab}$

Section 1.9

1. $81^{1/2} = 9$ since $9^2 = 81$.

3. $27^{1/3} = 3$ since $3^3 = 27$.

5. $8^{2/3} = (8^{1/3})^2 = 2^2 = 4$.

7. $(1000)^{2/3} = (1000^{1/3})^2 = 10^2 = 100$.

9. $-125^{2/3} = -(125^{1/3})^2 = -(5^2) = -25$

11. $\left(\frac{4}{9}\right)^{1/2} = \frac{2}{3}$ since $\left(\frac{2}{3}\right)^2 = \frac{4}{9}$.

13. $\left(\frac{64}{27}\right)^{1/3} = \frac{4}{3}$ since $\left(\frac{4}{3}\right)^3 = \frac{64}{9}$.

15. $(16)^{-5/4} = \frac{1}{(16)^{5/4}} = \frac{1}{(16^{1/4})^5} = \frac{1}{2^5} = \frac{1}{32}$

17. $\left(\frac{27}{64}\right)^{-1/3} = \left(\frac{64}{27}\right)^{1/3} = \frac{4}{3}$

19. $2^{-1} + 4^{-1} = \frac{1}{2} + \frac{1}{4} = \frac{2}{4} + \frac{1}{4} = \frac{3}{4}$

21. $(3^2 + 4^2)^{1/2} = (9 + 16)^{1/2} = (25)^{1/2} = 5$

23. $2^{1/2} \cdot 2^{3/2} = 2^{1/2+3/2} = 2^{4/2} = 2^2 = 4$

25. $27^{2/3} \cdot 27^{-1/3} = 27^{1/3} = (3^3)^{1/3} = 3$

27. $\frac{4^{2/3}\ 4^{5/3}}{4^{1/3}} = 4^{2/3+5/3-1/3} = 4^{6/3} = 4^2 = 16$

29. $\frac{6^{2/5} \cdot 6^{-4/5}}{6^2 \cdot 6^{-1/5}} = \frac{6^{2/5-4/5}}{6^{2-1/5}} = \frac{6^{-2/5}}{6^{9/5}} = 6^{-2/5-9/5}$

$= 6^{-11/5} = \frac{1}{6^{11/5}}$

31. $\frac{4^{-2/3}\ 4^{1/5}}{4^{5/3}} = \frac{4^{-10/15} \cdot 4^{3/15}}{4^{25/15}}$

$= \frac{4^{-7/15}}{4^{25/15}} = 4^{-32/15} = \frac{1}{4^{32/15}} = \frac{1}{2^{64/15}}$

33. $(2p)^{1/2} \cdot (2p^3)^{1/3} = 2^{1/2}\ p^{1/2} \cdot 2^{1/3}\ (p3)^{1/3}$

$= 2^{1/2}\ 2^{1/2}\ p^{1/2}\ p^1 = 2^{3/6}\ 2^{2/6}\ p^{1/2}\ p^{2/2}$

$= 2^{5/6}\ p^{3/2}$

35. $\frac{(mn)^{1/5}\ (m^2n)^{-2/5}}{m^{1/3}n} = \frac{m^{1/5}\ n^{1/5}\ m^{-4/5}\ n^{-2/5}}{m^{1/3}n}$

$= \frac{m^{-3/5}\ n^{-1/5}}{m^{1/3}n} = m^{-3/5-1/3}n^{n-1/5-1}$

$= m^{-9/15-5/15}n^{-1/5-5/5}$

$= m^{-14/15}n^{-6/5} = \frac{1}{m^{14/15}\ n^{6/5}}$

37. $x^{3/4}(x^2 - 3x^3) = x^{3/4}\ x^2 - 3x^{3/4}\ x^3$

$= x^{3/4+2} - 3x^{3/4+3} = x^{3/4+8/4} - 3x^{3/4+12/4}$

$= x^{11/4} - 3x^{15/4}$

39. $2z^{1/2}(3z^{-1/2} + z^{1/2})$

$= 6z^{1/2-1/2} + 2z^{1/2+1/2}$

$= 6z^0 + 2z^1 = 6 + 2z$

41. $(m + 2)^{1/2}\ [4(m + 2)^{-1/2} - 3\ (m + 2)^{3/2}]$

$= 4(m + 2)^{1/2-1/2} - 3(m + 2)^{1/2+3/2}$

$= 4(m + 2)^0 - 3(m + 2)^2$

$= 4 - 3(m^2 + 4m + 4)$

$= 4 - 3m^2 - 12m - 12 = -3m^2 - 12m - 8$

43. $\sqrt[3]{125} = \sqrt[3]{5^3} = (5^3)^{1/3} = 5$

45. $\sqrt[4]{1296} = \sqrt[4]{6^4} = 6$

47. $\sqrt[5]{-3125} = \sqrt[5]{(-5)^5} = -5$

49. $\sqrt{50} = \sqrt{25 \cdot 2} = \sqrt{25}\sqrt{2} = 5\sqrt{2}$

51. $-\sqrt[4]{32} = -\sqrt[4]{16 \cdot 2} = -\sqrt[4]{16}\,\sqrt[4]{2} = -2\sqrt[4]{2}$

53. $-\sqrt{\dfrac{9}{5}} = \dfrac{-\sqrt{9}}{\sqrt{5}} = \dfrac{-3}{\sqrt{5}} \cdot \dfrac{\sqrt{5}}{\sqrt{5}} = -\dfrac{3\sqrt{5}}{5}$

55. $-\sqrt[3]{\dfrac{3}{2}} = \dfrac{-\sqrt[3]{3}}{\sqrt[3]{2}} \cdot \dfrac{\sqrt[3]{2^2}}{\sqrt[3]{2^2}} = \dfrac{-\sqrt[3]{3 \cdot 4}}{\sqrt[3]{8}} = -\dfrac{\sqrt[3]{12}}{2}$

57. $\sqrt[4]{\dfrac{3}{2}} = \dfrac{\sqrt[4]{3}}{\sqrt[4]{2}} \cdot \dfrac{\sqrt[4]{2^3}}{\sqrt[4]{2^3}} = \dfrac{\sqrt[4]{3 \cdot 8}}{\sqrt[4]{2^4}} = \dfrac{\sqrt[4]{24}}{2}$

59. $\sqrt{24 \cdot 3^2 \cdot 2^4} = \sqrt{24 \cdot 9 \cdot 16}$

$= \sqrt{6 \cdot 4 \cdot 9 \cdot 16} = \sqrt{6}\,\sqrt{4}\,\sqrt{9}\,\sqrt{16}$

$= \sqrt{6}\,(2 \cdot 3 \cdot 4) = \sqrt{0}\,(24) = 24\sqrt{6}$

61. $\sqrt{8z^5x^8} = \sqrt{4 \cdot 2 \cdot z^4 \cdot z \cdot x^8}$

$= \sqrt{4z^4x^8}\,\sqrt{2z} = 2z^2x^4\,\sqrt{2z}$

63. $\sqrt{m^2n^7p^8} = \sqrt{m^2 \cdot n^6 \cdot n \cdot p^8}$

$= \sqrt{m^2n^6p^8}\,\sqrt{n} = mn^3p^4\sqrt{n}$

65. $\sqrt{\dfrac{2}{3x}} = \dfrac{\sqrt{2}}{\sqrt{3x}} \cdot \dfrac{\sqrt{3x}}{\sqrt{3x}} = \dfrac{\sqrt{6x}}{3x}$

67. $\sqrt{\dfrac{x^5y^3}{3^2}} = \sqrt{\dfrac{x^4y^2}{3^2}}\,\sqrt{xy} = \dfrac{x^2y}{3}\,\sqrt{xy}$

69. $\sqrt[3]{\sqrt{4}} = \sqrt[3 \cdot 2]{4} = \sqrt[6]{4} = \sqrt[6]{2^2}$

$= (2^2)^{1/6} = 2^{1/3} = \sqrt[3]{2}$

71. $\sqrt[6]{\sqrt[3]{x}} = \sqrt[18]{x}$

73. $4\sqrt{3} - 5\sqrt{12} + 3\sqrt{75}$

$= 4\sqrt{3} - 5(2\sqrt{3}) + 3(5\sqrt{3})$

$= 4\sqrt{3} - 10\sqrt{3} + 15\sqrt{3}$

$= (4 - 10 + 15)\sqrt{3} = 9\sqrt{3}$

75. $\sqrt{50} - 8\sqrt{8} + 4\sqrt{18} = 5\sqrt{2} - 8(2\sqrt{2}) + 4(3\sqrt{2})$

$= 5\sqrt{2} - 16\sqrt{2} + 12\sqrt{2}$

$= (5 - 16 + 12)\sqrt{2} = \sqrt{2}$

77. $3\sqrt{28}p - 4\sqrt{63p} + \sqrt{112p}$

$= 3(2\sqrt{7p}) - 4(3\sqrt{7p}) + 4\sqrt{7p}$

$= 6\sqrt{7p} - 12\sqrt{7p} + 4\sqrt{7p}$

$= (6 - 12 + 4)\sqrt{7p} = -2\sqrt{7p}$

79. $3\sqrt[3]{16} - 4\sqrt[3]{2} = 3(\sqrt[3]{8}\,\sqrt[3]{2}) - 4\sqrt[3]{2}$

$= 3(2\sqrt[3]{2}) - 4\sqrt[3]{2} = 6\sqrt[3]{2} - 4\sqrt[3]{2} = 2\sqrt[3]{2}$

81. $2\sqrt[3]{3} + 4\sqrt[3]{24} - \sqrt[3]{81}$

$= 2\sqrt[3]{3} + 4(2\sqrt[3]{3}) - (3\sqrt[3]{3})$

$= 2\sqrt[3]{3} + 8\sqrt[3]{3} - 3\sqrt[3]{3} = 7\sqrt[3]{3}$

83. $(\sqrt{2} + 3)(\sqrt{2} - 3)$

$= (\sqrt{2})^2 - (3)^2 = 2 - 9 = -7$

85. $(\sqrt[3]{11} - 1)(\sqrt[3]{11^2} + \sqrt[3]{11} + 1)$

$= \sqrt[3]{11^3} + \sqrt[3]{11^2} + \sqrt[3]{11} - \sqrt[3]{11^2} - \sqrt[3]{11} - 1$

$= \sqrt[3]{11^3} - 1 = 11 - 1 = 10$

87. $(3\sqrt{2} + \sqrt{3})(2\sqrt{3} - \sqrt{2})$

$= 6\sqrt{6} - 3\sqrt{4} + 2\sqrt{9} - \sqrt{6}$

$= 6\sqrt{6} - 3(2) + 2(3) - \sqrt{6}$

$= 6\sqrt{6} - 6 + 6 - \sqrt{6} = 5\sqrt{6}$

89. $\dfrac{3}{1 - \sqrt{2}} \cdot \dfrac{1 + \sqrt{2}}{1 + \sqrt{2}} = \dfrac{3(1 + \sqrt{2})}{(1)^2 - (\sqrt{2})^2}$

$= \dfrac{3(1 + \sqrt{2})}{1 - 2} = \dfrac{3(1 + \sqrt{2})}{-1}$

$= -3(1 + \sqrt{2}) = -3 - 3\sqrt{2}$

91. $\frac{4 - \sqrt{2}}{2 - \sqrt{2}} \cdot \frac{2 + \sqrt{2}}{2 + \sqrt{2}} = \frac{8 + 4\sqrt{2} - 2\sqrt{2} - (\sqrt{2})^2}{(2)^2 - (\sqrt{2})^2}$

$= \frac{8 + 4\sqrt{2} - 2\sqrt{2} - 2}{4 - 2}$

$= \frac{6 + 2\sqrt{2}}{2} = \frac{2(3 + \sqrt{2})}{2}$

$= 3 + \sqrt{2}$

93. $\frac{p}{\sqrt{p} + 2} \cdot \frac{\sqrt{p} - 2}{\sqrt{p} - 2} = \frac{p(\sqrt{p} - 2)}{p - 4}$

95. $\frac{a}{\sqrt{a^2 - 4}} + \frac{3\sqrt{a^2 - 4}}{a}$

$= \frac{a^2}{a\sqrt{a^2 - 4}} + \frac{3\sqrt{a^2 - 4}\sqrt{a^2 - 4}}{a\sqrt{a^2 - 4}}$

$= \frac{a^2 + 3(a^2 - 4)}{a\sqrt{a^2 - 4}}$

$= \frac{a^2 + 3a^2 - 12}{a\sqrt{a^2 - 4}} = \frac{4a^2 - 12}{a\sqrt{a^2 - 4}} \cdot \frac{\sqrt{a^2 - 4}}{\sqrt{a^2 - 4}}$

$= \frac{(4a^2 - 12)\sqrt{a^2 - 4}}{a(a^2 - 4)}$

97. $p = 2x^{1/2} + 3x^{2/3}$ *Let x = 64*

$p = 2(64)^{1/2} + 3(64)^{2/3} = 2(8)\ 3(4)^2$

$= 16 + 3(16)$

$= 16 + 48 = 64$

The price is $64.

99. $\text{amount for large state} = \left(\frac{E_{large}}{E_{small}}\right)^{3/2} \times \text{amount for a small state}$

$= (\frac{48}{3})^{3/2}(1{,}000{,}000)$

$= (16)^{3/2}(1{,}000{,}000)$

$= (64)(1{,}000{,}000)$

$= \$64{,}000{,}000$

$64,000,000 should be spent in the large state.

101. $D = 1.22x^{1/2} = 1.22(5000)^{1/2}$

$= 1.22(70.7) = 86.2$

D is about 86 miles.

103. $D = 1.22x^{1/2} = 1.22(30{,}000)^{1/2}$

$= 1.22(173.2) = 211.3$

D is about 211 miles.

105. $S = 28.6A^{.32} = 28.6(1)^{.32}$

$= 28.6(1) = 28.6 \approx 29$

107. $S = 28.6A^{.32} = 28.6(300)^{.32}$

$= 28.6(6.204) = 177.4 \approx 177$

Section 1.10

1. $(y - 5)(y + 4) = 0$

$y - 5 = 0$ or $y + 4 = 0$

$y = 5$ or $y = -4$

The solutions are 5 and -4.

3. $x^2 + 5x + 6 = 0$

$(x + 2)(x + 3) = 0$

$x + 2 = 0$ or $x + 3 = 0$

$x = -2$ or $x = -3$

The solutions are -2 and -3.

5. $x^2 = 3 + 2x$

$x^2 - 2x - 3 = 0$

$(x - 3)(x + 1) = 0$

$x - 3 = 0$ or $x + 1 = 0$

$x = 3$ or $x = -1$

The solutions are 3 and -1.

7. $m^2 + 16 = 8m$

$m^2 - 8m + 16 = 0$

$(m - 4)^2 = 0$

$m - 4 = 0$

$m = 4$

The only solution is 4.

9. $2k^2 - k = 10$

$2k^2 - k - 10 = 0$

$(2k - 5)(k + 2) = 0$

$2k - 5 = 0$ or $k + 2 = 0$

$2k = 5$ $\quad k = -2$

$k = \frac{5}{2}$ or $k = -2$

The solutions are 5/2 and -2.

11. $6x^2 - 5x = 4$

$6x^2 - 5x - 4 = 0$

$(3x - 4)(2x + 1) = 0$

$3x - 4 = 0$ or $2x = -1$

$3x = 4$ $\quad x = -\frac{1}{2}$

$x = \frac{4}{3}$ or $x = -\frac{1}{2}$

The solutions are 4/3 and -1/2.

13. $m(m - 7) = -10$

$m^2 - 7m + 10 = 0$

$(m - 5)(m - 2) = 0$

$m - 5 = 0$ or $m - 2 = 0$

$m = 5$ or $m = 2$

The solutions are 5 and 2.

15. $9x^2 - 16 = 0$

$(3x + 4)(3x - 4) = 0$

$3x + 4 = 0$ or $3x - 4 = 0$

$3x = -4$ $\quad 3x = 4$

$x = -\frac{4}{3}$ or $x = \frac{4}{3}$

The solutions are -4/3 and 4/3.

17. $16x^2 - 16x = 0$

$16x(x - 1) = 0$

$16x = 0$ or $x - 1 = 0$

$x = 0$ or $x = 1$

The solutions are 0 and 1.

19. $x^2 = 29$

$x = \sqrt{29}$ or $x = -\sqrt{29}$

The solutions are $\sqrt{29}$ and $-\sqrt{29}$.

21. $(m - 3)^2 = 5$

$m - 3 = \sqrt{5}$ or $m - 3 = -\sqrt{5}$

$m = 3 + \sqrt{5}$ or $m = 3 - \sqrt{5}$

The solutions are $3 + \sqrt{5}$ and $3 - \sqrt{5}$.

23. $(3k - 1)^2 = 19$

$3k - 1 = \sqrt{19}$ or $3k - 1 = -\sqrt{19}$

$3k = 1 + \sqrt{19}$ $\quad 3k = 1 - \sqrt{19}$

$k = \frac{1 + \sqrt{19}}{3}$ or $k = \frac{1 - \sqrt{19}}{3}$

The solutions are $\frac{1 + \sqrt{19}}{3}$ and $\frac{1 - \sqrt{19}}{3}$.

25. $q^2 + 2q = 8$

$q^2 + 2q + 1 = 8 + 1$

$(q + 1)^2 = 9$

$q + 1 = 3$ $\quad qr + 1 = -3$

$q = 2$ or $q = -4$

The solutions are 2 and -4.

27. $4z^2 - 4z = 1$

$z^2 - z = \frac{1}{4}$

$z^2 - z + \frac{1}{4} = \frac{1}{4} + \frac{1}{4}$

$(z - \frac{1}{2})^2 = \frac{1}{2}$

$z - \frac{1}{2} = \sqrt{\frac{1}{2}}$ or $z - \frac{1}{2} = -\sqrt{\frac{1}{2}}$

$z = \frac{1}{2} + \sqrt{\frac{1}{2}}$ $\quad z = \frac{1}{2} - \sqrt{\frac{1}{2}}$

$z = \frac{1}{2} + \frac{1}{\sqrt{2}}$ $\quad z = \frac{1}{2} - \frac{1}{\sqrt{2}}$

$z = \frac{1}{2} + \frac{\sqrt{2}}{2}$ $\quad z = \frac{1}{2} - \frac{\sqrt{2}}{2}$

$z = \frac{1 + \sqrt{2}}{2}$ or $z = \frac{1 - \sqrt{2}}{2}$

The solutions are $\frac{1 + \sqrt{2}}{2}$ and $\frac{1 - \sqrt{2}}{2}$.

29. $3x^2 - 5x + 1 = 0$

$x = \frac{-(-5) \pm \sqrt{(-5)^2 - 4(3)(1)}}{2(3)}$

$= \frac{5 \pm \sqrt{13}}{6}$

$x = \frac{5 + \sqrt{13}}{6} = \frac{5 + 3.606}{6}$

$= \frac{8.606}{6} = 1.434$ or

$x = \frac{5 - \sqrt{13}}{6} = \frac{5 - 3.606}{6}$

$= \frac{1.394}{6} = .232$

The solutions are $\frac{5 + \sqrt{13}}{6} \approx 1.434$

and $\frac{5 - \sqrt{13}}{6} \approx .232$.

31. $2m^2 = m + 4$

$2m^2 - m - 4 = 0$

$m = \frac{-(-1) \pm \sqrt{(-1)^2 - 4(2)(-4)}}{2(2)}$

$= \frac{1 \pm \sqrt{33}}{4}$

$m = \frac{1 + \sqrt{33}}{4} = \frac{1 + 5.745}{4}$

$= 1.686$ or

$m = \frac{1 - \sqrt{33}}{4} = \frac{1 + 5.745}{4}$

$= -1.186$

The solutions are $\frac{1 + \sqrt{33}}{4} \approx 1.686$

and $\frac{1 - \sqrt{33}}{4} \approx -1.186$.

33. $k^2 - 10k = -20$

$k^2 - 10k + 20 = 0$

$k = \frac{-(-10) \pm \sqrt{(-10)^2 - 4(1)(20)}}{2(1)}$

$= \frac{10 \pm \sqrt{20}}{2} = \frac{10 \pm 2\sqrt{5}}{2}$

$k = 5 \pm \sqrt{5}$

$k = 5 + \sqrt{5} = 5 + 2.236 = 7.236$ or

$k = 5 - \sqrt{5} = 5 - 2.236 = 2.764$

The solutions are $5 + \sqrt{5} \approx 7.236$

and $5 - \sqrt{5} \approx 2.764$.

35. $2x^2 + 12x + 5 = 0$

$x = \frac{-12 \pm \sqrt{(12)^2 - 4(2)(5)}}{2(2)}$

$= \frac{-12 \pm \sqrt{104}}{4} = \frac{-12 \pm 2\sqrt{26}}{4}$

$x = \frac{-6 \pm \sqrt{26}}{2}$

$x = \frac{-6 + \sqrt{26}}{2} = \frac{-6 + 5.099}{2} = -.450$ or

$x = \frac{-6 - \sqrt{26}}{2} = \frac{-6 + 5.099}{2} = -5.550$

The solutions are $\frac{-6 + \sqrt{26}}{2} \approx -.450$

and $\frac{-6 - \sqrt{26}}{2} \approx -5.550$.

37. $2r^2 - 7r + 5 = 0$

$r = \frac{-(-7) \pm \sqrt{49 - 4(2)(5)}}{2(2)} = \frac{7 \pm \sqrt{9}}{4}$

$= \frac{7 \pm 3}{4}$

$r = \frac{7 + 3}{4} = \frac{10}{4} = \frac{5}{2}$ or

$r = \frac{7 - 3}{4} = \frac{4}{4} = 1$

The solutions are $\frac{5}{2}$ and 1.

39. $6k^2 - 11k + 4 = 0$

$$k = \frac{-(-11) \pm \sqrt{121 - 4(6)(4)}}{2(6)}$$

$$= \frac{11 \pm \sqrt{25}}{12} = \frac{11 \pm 5}{12}$$

$$k = \frac{11 + 5}{12} = \frac{16}{12} = \frac{4}{3} \text{ or}$$

$$k = \frac{11 - 5}{12} = \frac{6}{12} = \frac{1}{2}$$

The solutions are $\frac{4}{3}$ and $\frac{1}{2}$.

41. $x^2 + 3x = 10$

$x^2 + 3x - 10 = 0$

$$x = \frac{-3 \pm \sqrt{9 - 4(-10)}}{2}$$

$$= \frac{-3 \pm \sqrt{49}}{2} = \frac{-3 \pm 7}{2}$$

$$x = \frac{-3 + 7}{2} = \frac{4}{2} = 2 \text{ or}$$

$$x = \frac{-3 - 7}{2} = \frac{-10}{2} = -5$$

The solutions are 2 and -5.

43. $2x^2 - 7x + 30 = 0$

$$x = \frac{-(-7) \pm \sqrt{49 - 4(2)(30)}}{2(2)}$$

$$= \frac{7 \pm \sqrt{-191}}{4}$$

Since $\sqrt{-191}$ is not a real number, there are no real solutions.

45. $5m^2 + 5m = 0$

$$m = \frac{-5 \pm \sqrt{25-4(5)(0)}}{2(5)} = \frac{-5 \pm 5}{10}$$

$$m = \frac{-5 + 5}{10} = 0 \text{ or}$$

$$m = \frac{-5-5}{10} = \frac{-10}{10} = -1$$

The solutions are 0 and -1.

47. $m^4 - 3m^2 - 10 = 0$

$(m^2 - 5)(m^2 + 2) = 0$

$m^2 - 5 = 0$ or $m^2 = -2$

$m^2 = 5$ No real

$m = \pm\sqrt{5}$ solutions.

The solutions are $\sqrt{5}$ and $-\sqrt{5}$.

49. $6y^4 = y^2 + 15$

$6y^4 - y^2 - 15 = 0$

Let $y^2 = w$ so $y^4 = w^2$.

$6w^2 - w - 15 = 0$

$(3w - 5)(2w + 3) = 0$

$3w - 5 = 0$ or $2w + 3 = 0$

$3w = 5$ $2w = -3$

$w = \frac{5}{3}$ or $w = -\frac{3}{2}$

So $y^2 = \frac{5}{3}$ or $y^2 = -\sqrt{\frac{3}{2}}$

$y = \pm\frac{\sqrt{5}}{\sqrt{3}} \cdot \frac{\sqrt{3}}{\sqrt{3}}$ *This equation has no real solution.*

$y = \pm\frac{\sqrt{15}}{3}$

The solutions are $\frac{\sqrt{15}}{3}$ and $-\frac{\sqrt{15}}{3}$.

51. $b^4 - 3b^2 - 5 = 0$

Let $b^2 = w$ so $b^4 = w^2$.

$w^2 - 3w - 5 = 0$

$$w = \frac{3 \pm \sqrt{9-4(1)(-5)}}{2(1)} = \frac{3 \pm \sqrt{29}}{2}$$

So $b^2 = \dfrac{3 \pm \sqrt{29}}{2}$

$b = \pm\sqrt{\dfrac{3 \pm \sqrt{29}}{2}}$ but $\pm\sqrt{\dfrac{3 - \sqrt{29}}{2}}$ are not real numbers.

$$b = \frac{\pm\sqrt{3 + \sqrt{29}}}{\sqrt{2}} \cdot \frac{\sqrt{2}}{\sqrt{2}} = \frac{\pm\sqrt{2(3 + \sqrt{29})}}{2}$$

$$b = \frac{\pm\sqrt{6 + 2\sqrt{29}}}{2}$$

The solutions are $\dfrac{\sqrt{6 + 2\sqrt{29}}}{2}$ and $-\dfrac{\sqrt{6 + 2\sqrt{29}}}{2}$.

53. $6(p - 3)^2 + 5(p - 3) - 6 = 0$

Let $u = p - 3$ so $p = u + 3$.

$6u^2 + 5u - 6 = 0$

$(3u - 2)(2u + 3) = 0$

$3u - 2 = 0$ or $2u + 3 = 0$

$3u = 2$ $\quad 2u = -3$

$u = \dfrac{2}{3}$ or $u = -\dfrac{3}{2}$

Thus, $p = u + 3 = \dfrac{2}{3} + 3 = \dfrac{11}{3}$

or $p = u + 3 = \dfrac{-3}{2} + 3 = \dfrac{3}{2}$

The solutions are 11/3 and 3/2.

55. $2(p + 5)^2 - 3(p + 5) = 20$

Let $u = p + 5$ so $p = u - 5$

$2u^2 - 3u - 20 = 0$

$(2u + 5)(u - 4) = 0$

$2u + 5 = 0$ or $u - 4 = 0$

$u = \dfrac{-5}{2}$ or $u = 4$

Then $p = u - 5$ so

$p = \dfrac{-5}{2} - 5$ or $p = 4 - 5$

$p = \dfrac{-15}{2}$ or $p = -1$

The solutions are -15/2 and -1.

57. $1 + \dfrac{7}{2a} = \dfrac{15}{2a^2}$

$2a^2 + 7a = 15$

$2a^2 + 7a - 15 = 0$

$(2a - 3)(a + 5) = 0$

$2a - 3 = 0$ or $a + 5 = 0$

$2a = 3$ or $a = -5$

$a = 3/2$

The solutions are 3/2 and -5.

59. $\dfrac{-2}{3z^2} + \dfrac{1}{3} + \dfrac{8}{3z} = 0$

$-2 + z^2 + 8z = 0$

$z^2 + 8z - 2 = 0$

$$z = \frac{-8 \pm \sqrt{64 - 4(-2)}}{2(1)} = \frac{-8 \pm \sqrt{72}}{2}$$

$$z = \frac{-8 \pm 6\sqrt{2}}{2} = \frac{2(-4 \pm 3\sqrt{2})}{2} = -4 \pm 3\sqrt{2}$$

The solutions are $-4 + 3\sqrt{2}$ and $-4 - 3\sqrt{2}$.

61.(a) $\dfrac{6}{r} - \dfrac{5}{r - 2} - 1$

$$= \frac{6(r - 2)}{r(r - 2)} - \frac{5r}{r(r - 2)} - \frac{r(r - 2)}{r(r - 2)}$$

$$= \frac{6(r - 2) - 5r - r(r - 2)}{r(r - 2)}$$

$$= \frac{6r - 12 - 5r - r^2 + 2r}{r(r - 2)}$$

$$= \frac{-r^2 + 3r - 12}{r(r - 2)}$$

61.(b) $\frac{6}{r} = \frac{5}{r-2} - 1$

$$\frac{6}{r} - \frac{5}{r-2} + 1 = 0$$

$$\frac{6(r-2) - 5r + r(r-2)}{r(r-2)} = 0$$

$$\frac{6r - 12 - 5r + r^2 - 2r}{r(r-2)} = 0$$

$$\frac{r^2 - r - 12}{r(r-2)} = 0$$

$$r^2 - r - 12 = 0$$

$$(r-4)(r+3) = 0$$

$r - 4 = 0$ or $r + 3 = 0$

$r = 4$ or $r = -3$

The solutions are 4 and -3.

63.(a) $\frac{1}{2k} + \frac{3}{k-1} - 5$

$$= \frac{k-1}{2k(k-1)} + \frac{3(2k)}{2k(k-1)} - \frac{5(2k)(k-1)}{2k(k-1)}$$

$$= \frac{k - 1 + 3(2k) - 5(2k)(k-1)}{2k(k-1)}$$

$$= \frac{k - 1 + 6k - 10k^2 + 10k}{2k(k-1)}$$

$$= \frac{-10k^2 + 17k - 1}{2k(k-1)}$$

(b) $\frac{1}{2k} + \frac{3}{k-1} = -5$

$$\frac{1}{2k} + \frac{3}{k-1} + 5 = 0$$

$$\frac{k - 1 + 3(2k) + 5(2k)(k-1)}{2k(k-1)} = 0$$

$$\frac{k - 1 + 6k + 10k^2 - 10k}{2k(k-1)} = 0$$

$$\frac{10k^2 - 3k - 1}{2k(k-1)} = 0$$

$$10k^2 - 3k - 1 = 0$$

$$(5k+1)(2k-1) = 0$$

$5k + 1 = 0$ or $2k - 1 = 0$

$5k = -1$ or $2k = 1$

$k = -\frac{1}{5}$ or $k = \frac{1}{2}$

The solutions are -1/5 and 1/2.

65. Let x = the width.

Then $2x + 200$ = the length.

The area is the width times the length.

$40{,}000 = x(2x+200)$

$40{,}000 = 2x^2 + 200x$

$0 = x^2 + 100x - 20{,}000$

$0 = (x + 200)(x - 100)$

$x + 200 = 0$ or $x - 100 = 0$

$x = -200$ or $x = 100$

Discard -200 since width must be positive.

So the width is 100 yards, and the length is 2(100) + 200 = 400 yards.

67. Let x = the width of the picture.

Then $x + 3$ = the length of the picture.

$$x(x+3) = 70$$

$$x^2 + 3x - 70 = 0$$

$$(x+10)(x-7) = 0$$

$x = -10$ or $x = 7$

Discard -10 since width must be positive so the width is 7 inches and the length is 7 + 3 or 10 inches. The frame extends 1.5 inches on each side so the outside dimensions are 7 + 1.5 + 1.5 or 10 inches by 10 + 1.5 + 1.5 or 13 inches.

69. $C = 20x + 60$

$R = x^2 - 8x$

Find x if

$C = R.$

$20x + 60 = x^2 - 8x$

$0 = x^2 - 28x - 60$

$0 = (x - 30)(x + 2)$

$x - 30 = 0$ or $x + 2 = 0$

$x = 30$ or $x = -2$

Discard -2 because the numbers of bicycles must be positive. The break-even value is 30 bicycles.

71. (a) $h = 64t - 16t^2$

$64 = 64t - 16t^2$ *Let h = 64*

$16t^2 - 64t + 64 = 0$

$t^2 - 4t + 4 = 0$

$(t - 2)^2 = 0$

$t - 2 = 0$

$t = 2$

The time is 2 seconds at 64 feet.

(b) $h = 64t - 16t^2$

$28 = 64t - 16t^2$ *Let h = 28*

$16t^2 - 64t + 28 = 0$

$4t^2 - 16t + 7 = 0$

$(2t - 7)(2t - 1) = 0$

$2t - 7 = 0$ or $2t - 1 = 0$

$t = 7/2$ or $t = 1/2$

The time is either 1/2 second or 7/2 seconds at 28 feet.

(c) It reaches the given height twice--once on the way up and once on the way down. (Note that it reaches 64 feet only once because that is the maximum height.)

73. $S = \frac{1}{2}gt^2$ for t

$2S = gt^2$

$\frac{2S}{g} = t^2$

$\frac{\sqrt{2S}}{g} = t$ *All variables positive*

$\frac{\sqrt{2Sg}}{g} = t$

75. $L = \frac{d^4k}{h^2}$

$Lh^2 = d^4k$

$h^2 = \frac{d^4k}{L}$

$h = \frac{\sqrt{d^4k}}{L} \cdot \frac{\sqrt{L}}{\sqrt{L}} = \frac{\sqrt{d^4kL}}{L}$

$h = \frac{d^2\sqrt{kL}}{L}$

77. $S = S_0 + gt^2 + k$ for t

$S - S_0 - k = gt^2$

$\frac{S - S_0 - k}{g} = t^2$

$\frac{\sqrt{S - S_0 - k}}{g} = t$

$\frac{\sqrt{g(S - S_0 - k)}}{g} = t$

79. $S = 2\pi rh + 2\pi r^2$ for r

Write as a quadratic equation in r.

$(s\pi)r^2 + (2\pi h)r - S = 0$

Find r using the quadratic formula.

$$r = \frac{-2\pi h \pm \sqrt{4\pi^2h^2 - 4(2\pi)(-s)}}{2(2\pi)}$$

$$r = \frac{-2\pi h \pm \sqrt{4\pi^2h^2 + 8\pi s}}{4\pi}$$

$$r = \frac{-2\pi h \pm 2\sqrt{\pi^2h^2 + 2\pi s}}{4\pi}$$

$$r = \frac{2(-\pi h \pm \sqrt{\pi^2h^2 + 2\pi s})}{4\pi}$$

$$r = \frac{-\pi h \pm \sqrt{\pi^2h^2 + 2\pi s}}{2\pi}$$

Section 1.11

See the graphs for Exercises 1-17 in the answers at the back of the textbook.

1. $(m + 2)(m - 4) < 0$

Solve $(m + 2)(m - 4) = 0$:

$m = -2$ or $m = 4$.

These numbers divide the number line into three regions, shown as follows:

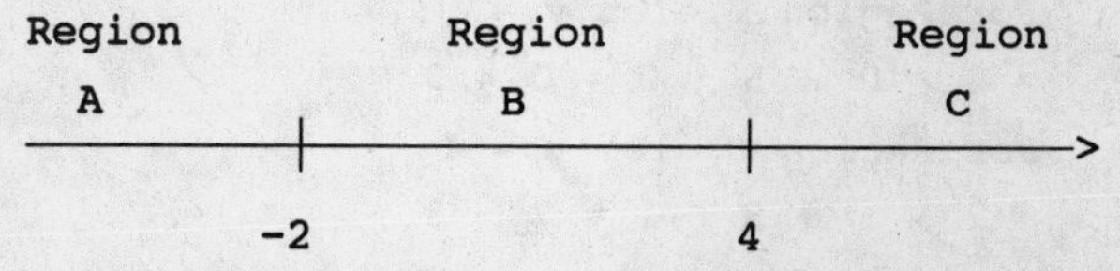

Test a value of m from each region to find the sign of $(m + 2)(m - 4)$ in the region.

For Region A, let $m = -2$:

$(-3 + 2)(-2 - 4) = 8 > 0$

For Region B, let $m = 0$:

$(0 + 2)(0 - 4) = -8 < 0$

For Region C, let $m = 5$:

$(5 + 2)(5 - 4) = 7 > 0$

The only region where the expression is negative is Region B, so the solution is $(-2, 4)$.

3. $(t + 6)(t - 1) \geq 0$

Solve $(t + 6)(t - 1) = 0$:

$t = -6$ or $t = 1$ *These solutions are also solutions of the given inequality.*

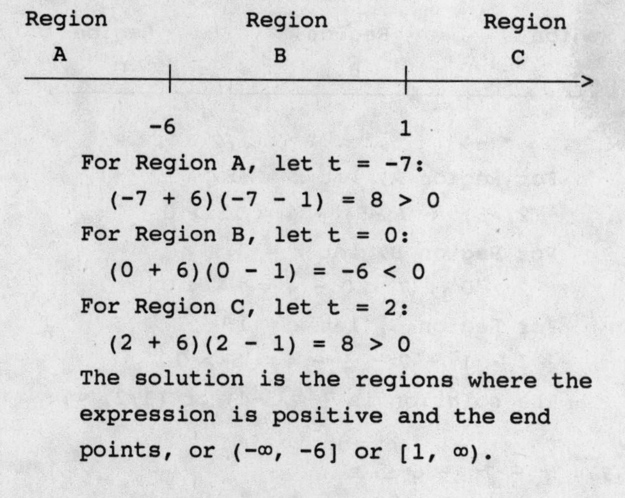

For Region A, let $t = -7$:

$(-7 + 6)(-7 - 1) = 8 > 0$

For Region B, let $t = 0$:

$(0 + 6)(0 - 1) = -6 < 0$

For Region C, let $t = 2$:

$(2 + 6)(2 - 1) = 8 > 0$

The solution is the regions where the expression is positive and the end points, or $(-\infty, -6]$ or $[1, \infty)$.

5. $y^2 - 3y + 2 < 0$

Solve $y^2 - 3y + 2 = 0$.

$(y - 2)(y - 1) = 0$

$y = 2$ or $y = 1$

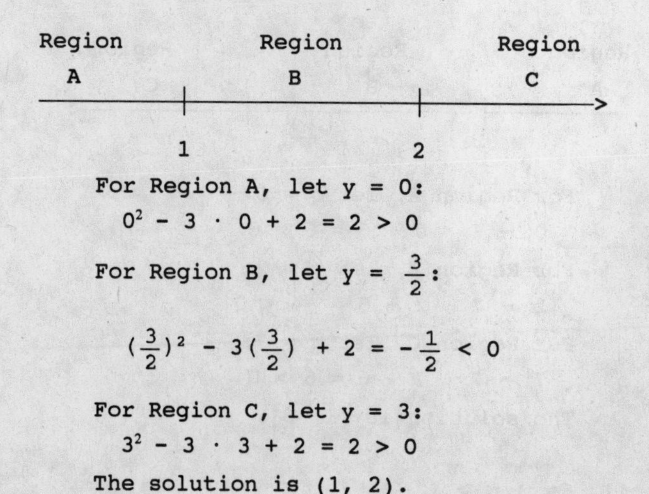

For Region A, let $y = 0$:

$0^2 - 3 \cdot 0 + 2 = 2 > 0$

For Region B, let $y = \frac{3}{2}$:

$(\frac{3}{2})^2 - 3(\frac{3}{2}) + 2 = -\frac{1}{2} < 0$

For Region C, let $y = 3$:

$3^2 - 3 \cdot 3 + 2 = 2 > 0$

The solution is $(1, 2)$.

7. $2k^2 + 7k - 4 > 0$

Solve $2k^2 + 7k - 4 = 0$.

$(2k - 1)(k + 4) = 0$

$k = \frac{1}{2}$ or $k = -4$

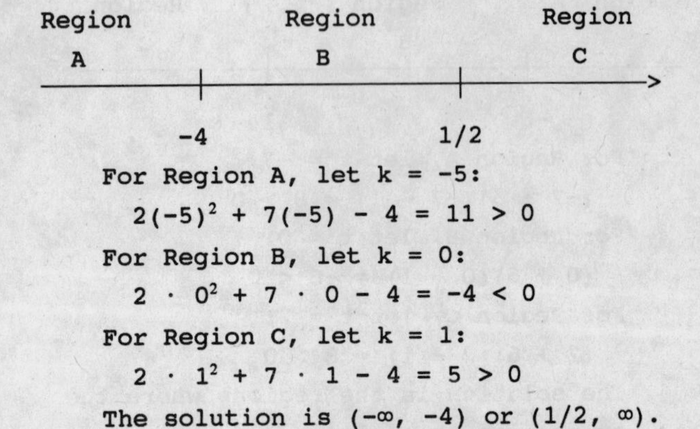

For Region A, let $k = -5$:

$2(-5)^2 + 7(-5) - 4 = 11 > 0$

For Region B, let $k = 0$:

$2 \cdot 0^2 + 7 \cdot 0 - 4 = -4 < 0$

For Region C, let $k = 1$:

$2 \cdot 1^2 + 7 \cdot 1 - 4 = 5 > 0$

The solution is $(-\infty, -4)$ or $(1/2, \infty)$.

9. $q^2 - 7q + 6 \le 0$

Solve $q^2 - 7q + 6 = 0$.

$(q - 1)(q - 6) = 0$

$q = 1$ or $q = 6$ *These are solutions of the inequality.*

Region A | Region B | Region C

1 | 6

For Region A, let $q = 0$:

$0^2 - 7 \cdot 0 + 6 = 6 > 0$

For Region B, let $q = 2$:

$2^2 - 7 \cdot 2 + 6 = -4 < 0$

For Region C, let $q = 7$:

$7^2 - 7 \cdot 7 + 6 = 6 > 0$

The solution is $[1, 6]$.

11. $6m^2 + m > 1$

Solve $6m^2 + m = 1$.

$6m^2 + m - 1 = 0$

$(2m + 1)(3m - 1) = 0$

$m = -\frac{1}{2}$ or $m = \frac{1}{3}$

Region A | Region B | Region C

-1/2 | 1/3

For Region A, let $m = -1$:

$6(-1)^2 + (-1) = 5 > 1$

For Region B, let $m = 0$:

$6 \cdot 0^2 + 0 = 0 < 1$

For Region C, let $m = 1$

$6 \cdot 1^2 + 1 = 7 > 1$

The solution is $(-\infty, -1/2)$ or $(1/3, \infty)$.

13. $2y^2 + 5y \le 3$

Solve $2y^2 + 5y = 3$.

$2y^2 + 5y - 3 = 0$

$(y + 3)(2y - 1) = 0$

$y = -3$ or $y = \frac{1}{2}$ *These are solutions of the inequality.*

Region A | Region B | Region C

-3 | 1/2

For Region A, let $y = -4$:

$2(-4)^2 + 5(-4) = 12 > 3$

For Region B, let $y = 0$:

$2 \cdot 0^2 + 5 \cdot 0 = 0 < 3$

For Region C, let $y = 1$:

$2 \cdot 1^2 + 5 \cdot 1 = 7 > 3$

The solution is $[-3, 1/2]$.

15. $x^2 \le 25$

Solve $x^2 = 25$.

$x = \pm 5$ *These are solutions of the inequality.*

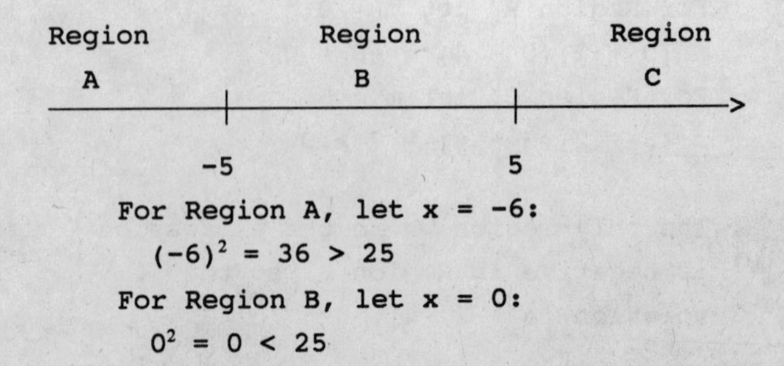

For Region A, let $x = -6$:

$(-6)^2 = 36 > 25$

For Region B, let $x = 0$:

$0^2 = 0 < 25$

For Region C, let $x = 6$:

$6^2 = 36 > 25$

The solution is $[-5, 5]$.

17. $p^2 - 16p > 0$

Solve $p^2 - 16p = 0$.

$p(p-16) = 0$

$p = 0$ or $p = 16$

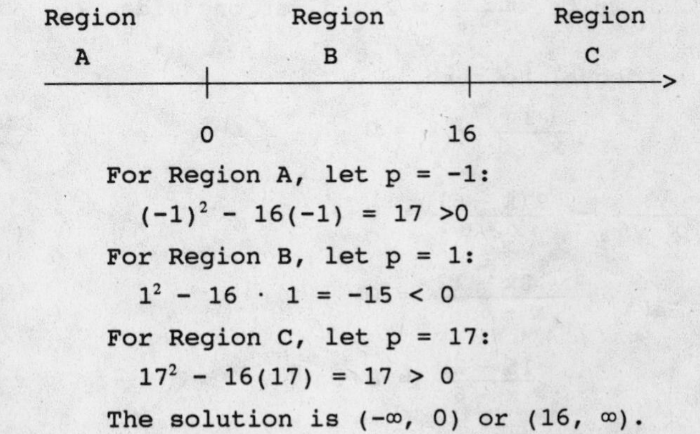

For Region A, let $p = -1$:

$(-1)^2 - 16(-1) = 17 > 0$

For Region B, let $p = 1$:

$1^2 - 16 \cdot 1 = -15 < 0$

For Region C, let $p = 17$:

$17^2 - 16(17) = 17 > 0$

The solution is $(-\infty, 0)$ or $(16, \infty)$.

19. $t^3 - 16t \le 0$

Solve $t^3 - 16t = 0$.

$t(t^2 - 16) = 0$

$t(t + 4)(t - 4) = 0$

$t = 0$ or $t = -4$ or $t = 4$ *These are solutions of the inequality.*

Region A | Region B | Region C | Region D

Number line: -4, 0, 4

For Region A, let $t = -5$:

$(-5)^3 - 16(-5) = -45 < 0$

For Region B, let $t = -1$:

$(-1)^3 - 16(-1) = 17 > 0$

For Region C, let $t = 1$:

$1^3 - 16 \cdot 1 = -15 < 0$

For Region D, let $t = 5$:

$5^3 - 16 \cdot 5 = 45 > 0$

The solution is $(-\infty, -4]$ or $[0, 4]$.

21. $2r^3 - 98r > 0$

Solve $2r^3 - 98r = 0$

$2r(r^2 - 49) = 0$

$2r(r + 7)(r - 7) = 0$

$r = 0$ or $r = -7$ or $r = 7$

Region A | Region B | Region C | Region D

Number line: -7, 0, 7

For Region A, let $r = -8$:

$2(-8)^3 - 98(-8) = -240 < 0$

For Region B, let $4 = -1$:

$2(-1)^3 - 98(-1) = 96 > 0$

For Region C, let $r = 1$:

$2 \cdot 1^3 - 98 \cdot 1 = -96 < 0$

For Region D, let $r = 8$:

$2 \cdot 8^3 - 98 \cdot 8 = 240 > 0$

The solution is $(-7, 0)$ or $(7, \infty)$.

23. $4m^3 + 7m^2 - 2m > 0$

Solve $4m^3 + 7m^2 - 2m = 0$

$m(4m^2 + 7m - 2) = 0$

$m(4m - 1)(m + 2) = 0$

$m = 0$ or $m = \frac{1}{4}$ or $m = -2$

Region A | Region B | Region C | Region D

Number line: -2, 0, $1/4$

For Region A, let $m = -3$:

$4(-3)^3 + 7(-3)^2 - 2(-3) = -39 < 0$

For Region B, let $m = -1$:

$4(-1)^3 + 7(-1)^2 - 2(-1) = 5 > 0$

For Region C, let $m = .1$:

$4(.1)^3 + 7(.1)^2 - 2(.1) = -.126 < 0$

For Region D, let $m = 1$:

$4 \cdot 1^3 + 7 \cdot 1^2 - 2 \cdot 1 = 9 > 0$

The solution is $(-2, 0)$ or $(1/4, \infty)$.

25. $\frac{m - 3}{m + 5} \le 0$

Write $\frac{m - 3}{m + 5} = 0$ to decide when the quotient can change sign. The quotient changes sign only when:

$m - 3 = 0$ or $m + 5 = 0$

$m = 3$ or $m = -5$.

3 is a solution of the inequality but the inequality is undefined when $m = -5$.

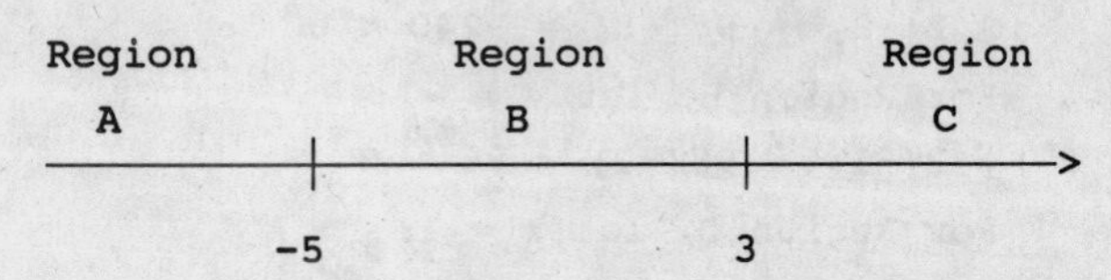

For Region A, let $m = -6$: $\frac{-6 - 3}{-6 + 5} = 9 > 0$

For Region B, let $m = 0$: $\frac{0 - 3}{0 + 5} = -\frac{3}{5} < 0$

For Region C, let $m = 4$: $\frac{4 - 3}{4 + 5} = \frac{1}{9} > 0$

The solution is $(-5, 3]$.

27. $\frac{p - 1}{p + 2} > 1$

$\frac{p - 1}{p + 2} = 1$

$\frac{p - 1}{p + 2} - 1 = 0$

$\frac{p - 1}{p + 2} - \frac{p - 2}{p + 2} = 0$

$\frac{1}{p + 2} = 0$

This equation has no solution but the quotient changes sign when

$p + 2 = 0$

$p = -2$.

Region A | Region B

-2

For Region A, let $k = -3$:

$\frac{-3 - 1}{-3 + 2} = 4 > 1$

For Region B, let $k = 0$

$\frac{0 - 1}{0 + 2} = -\frac{1}{2} < 1$

The solution is $(-\infty, -2)$.

29. $\frac{3}{x - 6} \le 2$

Write $\frac{3}{x - 6} = 2$ and get one side equal to zero.

$\frac{3}{x - 6} - 2 = 0$

$\frac{3}{x - 6} - \frac{2(x - 6)}{x - 6} = 0$

$\frac{3 - 2x + 12}{x - 6} = 0$

$\frac{15 - 2x}{x - 6} = 0$

The quotient changes sign when

$15 - 2x = 0$ or $x - 6 = 0$

$x = \frac{15}{2}$ or $x = 6$.

15/2 is a solution of the inequality but the inequality is undefined when $x = 6$.

Region A | Region B | Region C

6 | 15/2

For Region A, let $x = 0$:

$\frac{3}{0 - 6} = -\frac{1}{2} < 2$

For Region B, let $x = 7$:

$\frac{3}{7 - 6} = 3 > 2$

For Region C, let $x = 8$:

$\frac{3}{8 - 6} = \frac{3}{2} < 2$

The solution is $(-\infty, 6)$ or $[15/2, \infty)$.

31. $\frac{2y + 3}{y - 5} \le 1$

$\frac{2y + 3}{y - 5} = 1$

$\frac{2y + 3}{y - 5} - 1 = 0$

$\frac{2y + 3}{y - 5} - \frac{y - 5}{y - 5} = 0$

$\frac{y + 8}{y - 5} = 0$

The quotient changes sign when

$y + 8 = 0$ or $y - 5 = 0$

$y = -8$ or $y = 5$.

-8 is a solution of the inequality but the inequality is undefined when $y = 5$.

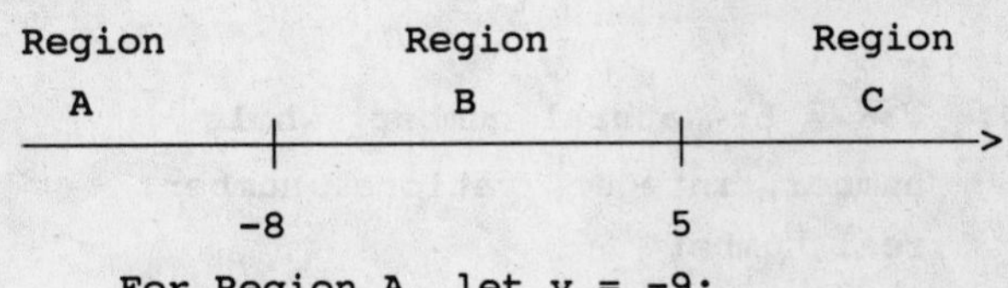

For Region A, let $y = -9$:

$\frac{2(-9) + 3}{-9 - 5} = \frac{15}{14} > 1$

For Region B, let $y = 0$:

$\frac{2 \cdot 0 + 3}{0 - 5} = -\frac{3}{5} < 1$

For Region C, let $y = 6$:

$\frac{2 \cdot 6 + 3}{6 - 5} = 15 > 1$

The solution is $[-8, 5)$.

33. $\frac{7}{k + 2} \ge \frac{1}{k + 2}$

$\frac{7}{k + 2} = \frac{1}{k + 2}$

$\frac{7}{k + 2} - \frac{1}{k + 2} = 0$

$\frac{6}{k + 2} = 0$

The quotient changes sign when $k + 2 = 0$ or $k = -2$, but the inequality is undefined when $k = 2$.

Region A   Region B

-2

For Region A, let $k = -3$:

$\frac{7}{-3 + 2} = -7$ and $\frac{1}{-3 + 2} = -1$

Since $-7 < -1$, -3 is not a solution of the inequality.

For Region B, let $k = 0$:

$\frac{7}{0 + 2} = \frac{7}{2}$ and $\frac{1}{0 + 2} = \frac{1}{2}$

Since $\frac{7}{2} > \frac{1}{2}$, 0 is a solution of the inequality.

The solution is $(-2, \infty)$.

35. $p = 4t^2 - 29t + 30$

When $t = 8$,

$p = 4(8)^2 - 29(8) + 30$

$p = 256 - 232 + 30$

$p = 54$.

Her total profit is \$54 after 8 months.

37. $p = 4t^2 - 29t + 30$

Let $p = 0$.

$0 = 4t^2 - 29t + 30$

$0 = (4t - 5)(t - 6)$

$4t - 5 = 0$ or $t - 6 = 0$

$4t = 5$ or $t = 6$

$t = \frac{5}{4}$

The times when she has just broken even is at $\frac{5}{4}$ months and 6 months.

39. $p = 3x^2 - 35x + 50$

The company makes a profit when $3x^2 - 35x + 50 > 0$.

Solve $3x^2 - 35x + 50 = 3$

$(3x - 5)(x - 10) = 0$

$x = \frac{5}{3}$ or $x = 10$

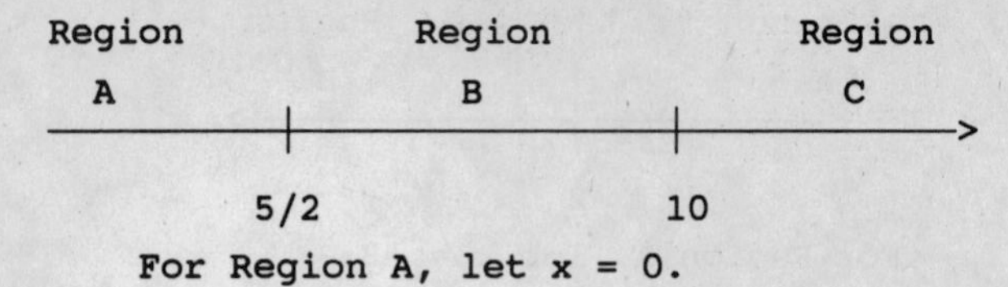

For Region A, let $x = 0$.

$3 \cdot 0^2 - 35 \cdot 0 + 50 = 50 > 0$

For Region B, let $x = 2$.

$3 \cdot 2^2 - 35 \cdot 2 + 50 = -8 < 0$

For Region C, let $x = 11$.

$3 \cdot 11^2 - 35 \cdot 11 + 50 = 28 > 0$

The company makes a profit when the amount spent on advertising in hundreds of dollars is in the intervals (0,5/3) or (10, ∞), that is, $0 < x < 5/3$ and $x > 10$ hundres of dollars.

41. $s = 220t - 16t^2$

Find t when $s \geq 624$.

$220t - 16t^2 \geq 624$

Solve $220t - 16t^2 = 624$

$16t^2 - 220t + 624 = 0$

$4t^2 - 55t + 156 = 0$

$(t - 4)(4t - 39) = 0$

$t = 4$ or $t = \frac{39}{4} = 9\frac{3}{4}$

These values are solutions of the inequality.

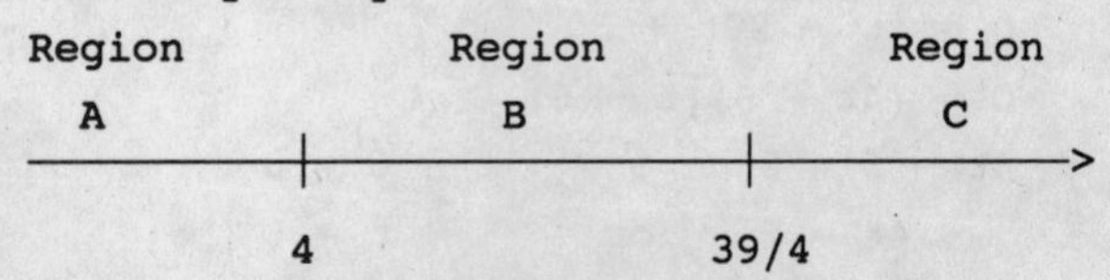

For Region A, let $t = 1$:

$220 \cdot 1 - 16 \cdot 1^2 = 204 < 624$

For Region B, let $5 = 5$:

$220 \cdot 5 - 16 \cdot 5^2 = 700 > 624$

For Region C, let $t = 10$:

$220 \cdot 10 - 16 \cdot 10^2 = 600 < 624$

So the projectile is at least 624 feet above the ground when the time in seconds is in the interval [4, 39/4], that is, $4 \leq t \leq 9.75$ seconds.

Chapter 1 Review Exercises

For Exercises 1-5, the list of numbers is $-12, -6, -9/10, -\sqrt{7}, -4, 0, 1/8, \pi/4, 6, \sqrt{11}$.

1. Only 6 is a natural number.

3. Since $-\sqrt{4} = -2$, $-12, -6, -\sqrt{4}, 0$, and 6 are integers.

5. $-\sqrt{7}$, $\pi/4$, $\sqrt{11}$ are irrational numbers.

7. 22: Natural number, whole number, integer, rational number, real number.

9. $\sqrt{36} = 6$: Natural number, whole number, integer, rational number, real number

11. -1: Integer, rational number, real number

13. $\sqrt{15}$: Irrational number, real number

15. $\frac{3\pi}{4}$: Irrational number, real number

17. $-7, -3, -2, 0, \pi, 8$ are in order. (π is about 3.14.)

19. $|6 - 4| = 2$, $-|-2| - 2$,

$|8 + 1| = |9| = 9$,

$-|3 - (-2)| = -|3 + 2| = -|5| = -5$

Since $-5, -2, 2, 9$ are in order, then $-|3 - (-2)|$, $-|-2|$, $|6 - 4|$, $|8 + 1|$ are in order.

21. $-|-6| + |3| = -6 + 3 = -3$

23. $7 - |-8| = 7 - 8 = -1$

See the graphs for Exercises 25 and 27 in the answers at the back of the textbook.

25. $x \geq -3$
Start at -3 and draw a line to the right. Use a solid circle at -3 to show that -3 is a part of the graph.

27. $x < -2$
Start at -2 and draw a line to the left. Use an open circle at -2 to show that -2 is not a part of the graph.

29. $(-6 + 2 \cdot 5)(-2) = (-6 + 10)(-2)$
$= 4(-2) = -8$

31. $\frac{-8 + (-6)(-3) + 9}{6 - (-2)} = \frac{-6}{8} = -\frac{3}{4}$

33. $4k - 11 + 3k = 2k - 1$
$7k - 11 = 2k - 1$
$5k = 10$
$k = 2$

35. $2m + 7 = 3m + 1$
$7 = m + 1$
$6 = m$

37. $5y - 2(y + 4) = 3(2y + 1)$
$5y - 2y - 8 = 6y + 3$
$3y - 8 = 6y + 3$
$-11 = 3y$
$y = -\frac{11}{3}$

39. $\frac{p}{2} - \frac{3p}{4} = 8 + \frac{p}{3}$
$6p - 9p = 96 + 4p$ *Multiply all terms by 12*
$-3p = 96 + 4p$
$p = -\frac{96}{7}$

41. $\frac{2z}{5} - \frac{4z - 3}{10} = \frac{-z + 1}{10}$
$4z - (4z - 3) = -z + 1$ *Multiply by 10*
$3 = -z + 1$
$z = -2$

43. $\frac{2m}{m - 3} = \frac{6}{m - 3} + 4$
$2m = 6 + 4(m - 3)$ *Multiply by $m - 3$, assuming $m - 3 \neq 0$, or $m \neq 3$.*
$2m = 6 + 4m - 12$
$2m = 4m - 6$
$6 = 2m$
$3 = m$ but $m \neq 3$, so there is no solution.

45. $9px - 2 = x$ for x
$9px - x = 2$
$x(9p - 1) = 2$
$x = \frac{2}{9p - 1}$

47. $3(x + 2b) + a = 2x - 6$ for x
$3x + 6b + a = 2x - 6$
$3x - 2x = -6b - a - 6$
$x = -6b - a - 6$

49. $\frac{x}{m - 2} = kx - 3$ for x
$x = (kx - 3)(m - 2)$
$x = kmx - 2kx - 3m + 6$
$x - kmx + 2kx = 6 - 3m$
$x(1 - km + 2k) = 6 - 3m$
$x = \frac{6 - 3m}{1 + 2k - km}$

51. Let x = the original price.
Then $.15x$ = the amount taken off in the sale.

$$x - .15x = 425$$
$$.85x = 425$$
$$x = 500$$

The original price is $500.

53. Let x = the amount invested at 8%.
Then $100{,}000 - x$ = the amount invested at 5%.
$.08x$ = the interest from 8% investment
$.05(100{,}000 - x)$ = the interest from 5% investment
6800 = the total interest.

$$.08x + .05(100{,}000 - x) = 6800$$
$$.08x + 5000 - .05x = 6800$$
$$.03x = 1800$$
$$x = 60{,}000$$
$$100{,}000 - x = 40{,}000$$

Invest $60,000 at 8% and $40,000 at 5%.

55.
$$-9x < 4x + 7$$
$$-13x < 7 \qquad \textit{Subtract } 4x$$
$$x > -\frac{7}{13} \qquad \textit{Divide by } -13$$

The solution is $(-\frac{7}{13}, \infty)$.

57.
$$-5z - 4 \geq 3(2z - 5)$$
$$-5z - 4 \geq 6z - 15$$
$$11 \geq 11z$$
$$1 \geq z$$

The solution is $(-\infty, 1]$.

59.
$$3r - 4 + r > 2(r - 1)$$
$$4r - 4 > 2r - 2$$
$$2r > 2$$
$$r > 1$$

The solution is $(1, \infty)$.

61.
$$5 \leq 2x - 3 \leq 7$$
$$8 \leq 2x \leq 10$$
$$4 \leq x \leq 5$$

The solution is $[4, 5]$.

63.
$$|a + 4| = 7$$
$$a + 4 = 7 \text{ or } a + 4 = -7$$
$$a = 3 \text{ or } a = -11$$

65.
$$\frac{|r - 5|}{3} = 6$$
$$|r - 5| = 18$$
$$r - 5 = 18 \quad \text{or} \quad r - 5 = -18$$
$$r = 23 \quad \text{or} \quad r = -13$$

67.
$$\left|\frac{8r - 1}{2}\right| = 7$$
$$\frac{8r - 1}{2} = 7 \quad \text{or} \quad \frac{8r - 1}{2} = -7$$
$$8r - 1 = 14 \qquad 8r - 1 = -14$$
$$8r = 15 \qquad 8r = -13$$
$$r = \frac{15}{8} \quad \text{or} \quad r = -\frac{13}{8}$$

69.
$$|5r - 1| = |2r + 3|$$
$$5r - 1 = 2r + 3 \quad \text{or} \quad 5r - 1 = -(2r + 3)$$
$$5r - 2r = 3 + 1 \qquad 5r - 1 = -2r - 3$$
$$5r + 2r = -3 + 1$$
$$3r = 4 \qquad 7r = -2$$
$$r = \frac{4}{3} \quad \text{or} \quad r = -\frac{2}{7}$$

71. $|m| \leq 7$ means
$-7 \leq m \leq 7$.
The solution is all numbers in $[-7, 7]$.

73. $|p| > 3$ means
$p < -3$ or $p > 3$.
The solution is all numbers in $(-\infty, -3)$ or $(3, \infty)$.

75. $|b| \le -1$

Since the absolute value of a number is always positive and no positive number is less than any negative number, the inequality has no solution.

77. $|7k - 3| < 5$ means

$$-5 < 7k - 3 < 5$$
$$-5 + 3 < 7k < 5 + 3$$
$$-2 < 7k < 8$$
$$-\frac{2}{7} < k < \frac{8}{7}.$$

The solution is all numbers in $(-2/7, 8/7)$.

79. $|3r + 7| > 5$ means

$$3r + 7 < -5 \quad \text{or} \quad 3r + 7 > 5$$
$$3r < -12 \qquad 3r > -2$$
$$r < -4 \quad \text{or} \quad r > -\frac{2}{3}.$$

The solution is all numbers in $(-\infty, -4)$ or $(-2/3, \infty)$.

81. $(-9m^2 + 11m - 7) + (2m^2 - 12m + 4)$
$= -9m^2 + 2m^2 + 11m - 12m - 7 + 4$
$= -7m^2 - m - 3$

83. $(5r^4 - 6r^2 + 2r) - (-3r^4 + 2r^2 - 9r)$
$= (5r^4 - 6r^2 + 2r) + (3r^4 - 2r^2 + 9r)$
$= 5r^4 + 3r^4 - 6r^2 - 2r^2 + 2r + 9r$
$= 8r^4 - 8r^2 + 11r$

85. $-(r^5 + 2r^4 + 8r^2) + 3(r^4 + 9r^2)$
$= -r^5 - 2r^4 - 8r^2 + 3r^4 + 27r^2$
$= -r^5 + r^4 + 19r^2$

87. $(8y - 7)(2y + 7)$
$= 16y^2 + 56y - 14y - 49$
$= 16y^2 + 42y - 49$

89. $(7z + 10y)(3z - 5y)$
$= 21z^2 - 35yz + 30yz - 50y^2$
$= 21z^2 - 5yz - 50y^2$

91. $(3k - 5m)^2$
$= (3k)^2 - 2(3k)(5m) + (5m)^2$
$= 9k^2 - 30mk + 25m^2$

93. $(5x - 2)^3$
$= (5x - 2)(5x - 2)(5x - 2)$
$= (25x^2 - 20x + 4)(5x - 2)$
$= 125x^3 - 150x^2 + 60x - 8$

95. $(3w - 2)(5w^2 - 4w + 1)$
$= 15w^3 - 12w^2 + 3w - 10w^2 + 8w - 2$
$= 15w^3 - 22w^2 + 11w - 2$

97. $7z^2 - 9z^3 + z$
$= z(7z - 9z^2 + 1)$

99. $12p^5 - 8p^4 + 20p^3$
$= 4p^3(3p^2 - 2p + 5)$

101. $r^2 + rp - 42p^2$
$= (r + 7p)(r - 6p)$

103. $6m^2 - 13m - 5$
$= (3m + 1)(2m - 5)$

105. $3m^2 - 8m - 35$
$= (3m + 7)(m - 5)$

107. $114p^2 - 169q^2$
$= (12p)^2 - (13q)^2$
$= (12p - 13q)(12p + 13q)$

109. $8y^3 - 1$
$= (2y)^3 - 1^3$
$= (2y - 1)(4y^2 + 2y + 1)$

111. $\frac{6p}{7} \cdot \frac{28p}{15} = \frac{2 \cdot 3p \cdot 4 \cdot 7p}{7 \cdot 3 \cdot 5} = \frac{8p^2}{5}$

113. $\frac{m^2 - 2m - 3}{m(m - 3)} \div \frac{m + 1}{4m}$

$= \frac{(m - 3)(m + 1)4m}{m(m - 3)(m + 1)} = 4$

115. $\frac{k^8 + k}{8k^3} \cdot \frac{4}{k^2 - 1} = \frac{k(k+1)(4)}{8k^3(k-1)(k+1)}$

$= \frac{1}{2k^2(k-1)}$

117. $\frac{5}{2r} + \frac{6}{r} = \frac{5}{2r} + \frac{12}{2r} = \frac{17}{2r}$

119. $\frac{8}{r-1} - \frac{3}{r} = \frac{8r - 3(r-1)}{(r-1)r}$

$= \frac{5r+3}{r(r-1)}$

121. $\frac{\frac{1}{p} + \frac{1}{q}}{1 - \frac{1}{pq}} = \frac{pq\left(\frac{1}{p} + \frac{1}{q}\right)}{pq\left(1 - \frac{1}{pq}\right)}$

$= \frac{q+p}{pq-1}$

123. $4^{-2} = \frac{1}{4^2} = \frac{1}{16}$

125. $5^{-3} = \frac{1}{5^3} = \frac{1}{125}$

127. $\left(\frac{3}{4}\right)^{-3} = \left(\frac{4}{3}\right)^3 = \frac{4^3}{3^3} = \frac{64}{27}$

129. $6^4 \cdot 6^{-3} = 6^{4-3} = 6^1 = 6$

131. $\frac{8^{-5}}{8^{-3}} = 8^{-5-(-3)}$

$= 8^{-5+3} = 8^{-2} = \frac{1}{8^2} = \frac{1}{64}$

133. $\frac{9^4 \cdot 9^{-5}}{(9^{-2})^2} = \frac{9^4 9^{-5}}{9^{-4}}$

$= 9^{4-5-(-4)} = 9^3 = 729$

135. $4^{-1} + 2^{-1} = \frac{1}{4} + \frac{1}{2}$

$= \frac{1}{4} + \frac{2}{4} = \frac{3}{4}$

137. $125^{2/3} = (125^{1/3})^2 = 5^2 = 25$

139. $9^{-5/2} = \frac{1}{9^{5/2}} = \frac{1}{(9^{1/2})^5} = \frac{1}{3^5}$

141. $\frac{5^{1/3}\, 5^{1/2}}{5^{3/2}} = 5^{1/3 + 1/2 - 3/2}$

$= 5^{2/6 + 3/6 - 9/6}$

$= 5^{-4/6} = 5^{-2/3}$

$= \frac{1}{5^{2/3}}$

143. $(3a^2)^{1/2} \cdot (3^2a)^{3/2} = 3^{1/2}a \cdot 3^3a^{3/2}$

$= 3^{1/2+3}a^{1+3/2}$

$= 3^{7/2}a^{5/2}$

145. $\sqrt[3]{27} = \sqrt[3]{3^3} = 3$

147. $\sqrt[5]{-32} = \sqrt[5]{(-2)^5} = -2$

149. $\sqrt{24} = \sqrt{4 \cdot 6} = 2\sqrt{6}$

151. $\sqrt[3]{54p^3q^5} = \sqrt[3]{27 \cdot 2p^3q^3q^2}$

$= 3pq\sqrt[3]{2q^2}$

153. $\sqrt{\frac{5n^2}{6m}} = \frac{n\sqrt{5}}{\sqrt{6m}} \cdot \frac{\sqrt{6m}}{\sqrt{6m}}$

$= \frac{n\sqrt{30m}}{6m}$

155. $\sqrt[3]{\sqrt{5}} = \sqrt[3 \cdot 2]{5} = \sqrt[6]{5}$

157. $2\sqrt{3} - 5\sqrt{12} = 2\sqrt{3} - 5\sqrt{4 \cdot 3}$
$= 2\sqrt{3} - 5 \cdot 2\sqrt{3}$
$= 2\sqrt{3} - 10\sqrt{3}$
$= -8\sqrt{3}$

159. $5\sqrt[3]{16r^2s^3} - s\sqrt[3]{54r^2}$
$= 5\sqrt[3]{2 \cdot 2^3r^2s^3} - s\sqrt[3]{2 \cdot 3^3r^2}$
$= 5 \cdot 2s\sqrt[3]{2r^2} - 3s\sqrt[3]{2r^2}$
$= 10s\sqrt[3]{2r^2} - 3s\sqrt[3]{2r^2}$
$= 7s\sqrt[3]{2r^2}$

161. $(\sqrt{5} - 1)(\sqrt{5} + 1) = (\sqrt{5})^2 - 1^2$
$= 5 - 1 = 4$

163. $(2\sqrt{5} - \sqrt{3})(\sqrt{5} + 2\sqrt{3})$
$= 2 \cdot 5 + 4\sqrt{15} - \sqrt{15} - 2 \cdot 3$
$= 10 - 6 + 3\sqrt{15}$
$= 4 + 3\sqrt{15}$

165. $\dfrac{\sqrt{2}}{1 + \sqrt{3}} = \dfrac{\sqrt{2}(1 - \sqrt{3})}{(1 + \sqrt{3})(1 - \sqrt{3})}$

$= \dfrac{\sqrt{2} - \sqrt{6}}{1 - 3} = \dfrac{\sqrt{6} - \sqrt{2}}{2}$

167. $x^2 = 7$
$x = \pm\sqrt{7}$
The solutions are $\sqrt{7}$ and $-\sqrt{7}$.

169. $(b + 7)^2 = 5$
$b + 7 = \pm\sqrt{5}$
$b = -7 \pm \sqrt{5}$
The solutions are $-7 + \sqrt{5}$ and $-7 - \sqrt{5}$.

171. $x^2 - 4x + 3 = 0$
$(x - 3)(x - 1) = 0$
$x - 3 = 0$ or $x - 1 = 0$
$x = 3$ or $x = 1$
The solutions are 3 and 1.

173. $2p^2 + 3p = 2$
$2p^2 + 3p - 2 = 0$
$(2p - 1)(p + 2) = 0$
$2p - 1 = 0$ or $p + 2 = 0$
$p = \dfrac{1}{2}$ or $p = -2$

The solutions are 1/2 and -2.

175. $x^2 - 2x = 2$
$x^2 - 2x - 2 = 0$

$x = \dfrac{2 \pm \sqrt{2^2 - 4(-2)}}{2}$

$x = \dfrac{2 \pm \sqrt{12}}{2} = \dfrac{2 \pm 2\sqrt{3}}{2}$

$x = 1 \pm \sqrt{3}$

The solutions are $1 + \sqrt{3}$ and $1 - \sqrt{3}$.

177. $2m^2 - 12m = 11$
$2m^2 - 12m - 11 = 0$

$m = \dfrac{12 \pm \sqrt{144 - 4(2)(-11)}}{2(2)}$

$m = \dfrac{12 \pm \sqrt{232}}{4}$

$m = \dfrac{12 \pm 2\sqrt{58}}{4}$

$m = \dfrac{6 \pm \sqrt{58}}{2}$

The solutions are $\dfrac{6 + \sqrt{58}}{2}$ and $\dfrac{6 - \sqrt{58}}{2}$.

179. $2a^2 + a - 15 = 0$
$(2a - 5)(a + 3) = 0$
$2a - 5 = 0$ or $a + 3 = 0$
$a = \dfrac{5}{2}$ or $a = -3$

The solutions are 5/2 and -3.

181. $2q^2 - 11q = 2/$

$2q^2 - 11q - 21 = 0$

$(2q + 3)(q - 7) = 0$

$2q + 3 = 0$ or $q - 7 = 0$

$q = -\frac{3}{2}$ or $q = 7$

The solutions are -3/2 and 7.

183. $6k^4 + k^2 = 1$

$6k^4 + k^2 - 1 = 0$

Let $p = k^2$ so $p^2 = k^4$.

$6p^2 + p - 1 = 0$

$(3p - 1)(2p + 1) = 0$

$3p - 1 = 0$ or $2p + 1 = 0$

$p = \frac{1}{3}$ or $p = -\frac{1}{2}$

If $p = \frac{1}{3}$, $k^2 = \frac{1}{3}$

$$k = \pm\sqrt{\frac{1}{3}}$$

$$= \pm\frac{\sqrt{3}}{3}$$

If $p = -\frac{1}{2}$, $k^2 = -\frac{1}{2}$ has no real number solution.

The solutions are $\sqrt{3}/3$ and $-\sqrt{3}/3$.

185. $2x^4 = 7x^2 + 15$

$2x^4 - 7x^2 + 15 = 0$

Let $p = x^2$, so $p^2 = x^4$.

$2p^2 - 7p + 15 = 0$

$(2p + 3)(p - 5) = 0$

$2p + 3 = 0$ or $p - 5 = 0$

$p = -\frac{3}{2}$ or $p = 5$

If $p = -\frac{3}{2}$, then $x^2 = -\frac{3}{2}$ has no real number solution.

If $p = 5$, then $x^2 = 5$

$$x = \pm\sqrt{5}.$$

The solutions are $\sqrt{5}$ and $-\sqrt{5}$.

187. $3 = \frac{13}{z} + \frac{10}{z^2}$

$3z^2 = 13z + 10$ *Multiply by z2*

$3z^2 - 13z - 10 = 0$

$(3z + 2)(z - 5) = 0$

$3z + 2 = 0$ or $z - 5 = 0$

$z = -\frac{2}{3}$ or $z = 5$

The solutions are -2/3 and 5.

189. $2 + \frac{15}{x-1} + \frac{18}{(x-1)^2} = 0$

Let $p = x - 1$. Then $x = p + 1$.

$2 + \frac{15}{p} + \frac{18}{p^2} = 0$

$2p^2 + 15p + 18 = 0$ *Multiply by p^2*

$(2p + 3)(p + 6) = 0$

$2p + 3 = 0$ or $p + 6 = 0$

$p = -\frac{3}{2}$ or $p = -6$

If $p = -\frac{3}{2}$, then $x = -\frac{3}{2} + 1 = -\frac{1}{2}$.

If $p = -6$, then $x = -6 + 1 = -5$.

The solutions are -1/2 and -5.

191. $p = \frac{E^2R}{(r + R)^2}$ for r

$p(r + R)^2 = E^2R$

$p(r^2 + 2rR + R^2) = E^2R$

$pr^2 + 2rpR + R^2 = E^2R$

$pr^2 + 2rpR + R^2 - E^2R = 0$

Use the quadratic formula to solve for r.

$$r = \frac{-2pR \pm \sqrt{4p^2R^2 - 4p(R^2 - E^2R)}}{2p}$$

$$r = \frac{-2pR \pm \sqrt{4pE^2R}}{2p}$$

$$= \frac{-pR \pm E\sqrt{pR}}{p}$$

193. $K = s(s - a)$ for s

$K = s^2 - as$

Solve by quadratic formula.

$s^2 - as - K = 0$

$$s = \frac{a \pm \sqrt{a^2 - 4(-K)}}{2}$$

$$s = \frac{a \pm \sqrt{a^2 + 4K}}{2}$$

195. $r^2 + r - 6 < 0$

Write corresponding equation and solve to determine values of r where polynomial changes sign.

$r^2 + r - 6 = 0$

$(r + 3)(r - 2) = 0$

$r = -3$ or $r = 2$

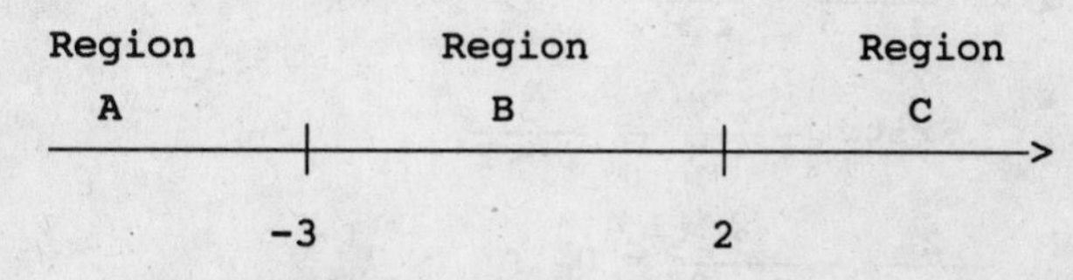

For Region A, test -4:

$(-4)^2 + (-4) - 6 = 6 > 0$

For Region B, test 0:

$0^2 + 0 - 6 = -6 < 0$

For Region C, test 3:

$3^2 + 3 - 6 = 6 > 0$

The solution is $(-3, 2)$.

197. $2z^2 + 7z \geq 15$

Write equation and solve.

$2z^2 + 7z = 15$

$2z^2 + 7z - 15 = 0$

$(2z - 3)(z + 5) = 0$

$z = \frac{3}{2}$ or $z = -5$

These numbers are solutions of the inequality.

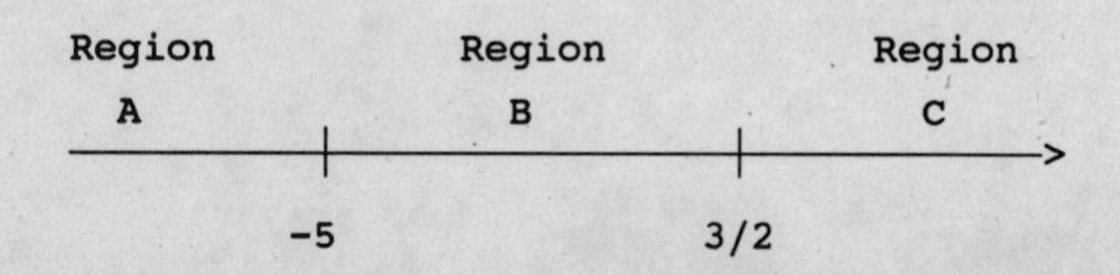

For Region A, test -6:

$2(-6)^2 + 7(-6) = 30 > 15$

For Region B, test 0:

$2 \cdot 0^2 + 7 \cdot 0 = 0 < 15$

For Region C, test 2:

$2 \cdot 2^2 + 7 \cdot 2 = 22 > 15$

The solution is $(-\infty, -5]$ or $[3/2, \infty)$.

199. $8a^2 + 10a > 3$

Solve $8a^2 + 10a = 3$

$8a^2 + 10a - 3 = 0$

$(2a + 3)(4a - 1) = 0$

$a = -\frac{3}{2}$ or $a = \frac{1}{4}$

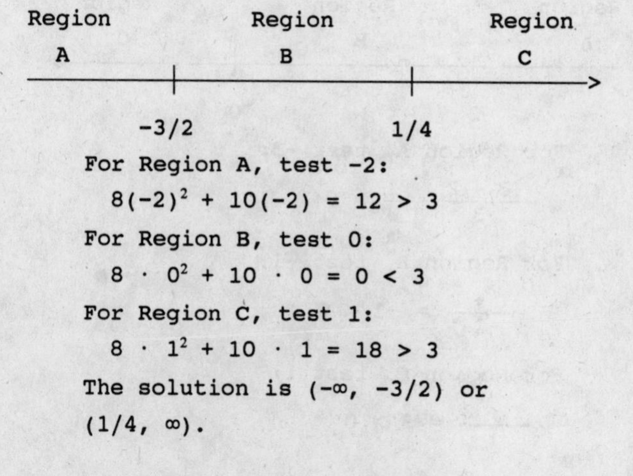

For Region A, test -2:

$8(-2)^2 + 10(-2) = 12 > 3$

For Region B, test 0:

$8 \cdot 0^2 + 10 \cdot 0 = 0 < 3$

For Region C, test 1:

$8 \cdot 1^2 + 10 \cdot 1 = 18 > 3$

The solution is $(-\infty, -3/2)$ or $(1/4, \infty)$.

201. $3r^2 - 5r \leq 0$

Solve $3r^2 - 5r = 0$

$r(3r - 5) = 0$

$r = 0$ or $r = \frac{5}{3}$

These are solutions of the inequality.

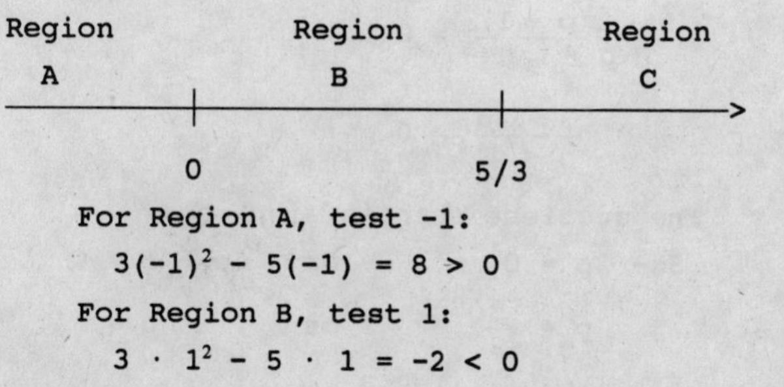

For Region A, test -1:

$3(-1)^2 - 5(-1) = 8 > 0$

For Region B, test 1:

$3 \cdot 1^2 - 5 \cdot 1 = -2 < 0$

For Region C, test 2:

$3 \cdot 2^2 - 5 \cdot 2 = 2 > 0$

The solution is [0, 5/3].

203. $\frac{m + 2}{m} \le 0$

Write $\frac{m + 2}{m} = 0$.

The quotient changes sign when

$m + 2 = 0$ or $m = 0$.

$m = -2$

-2 is a solution of the inequality but the inequality is undefined when $m = 0$.

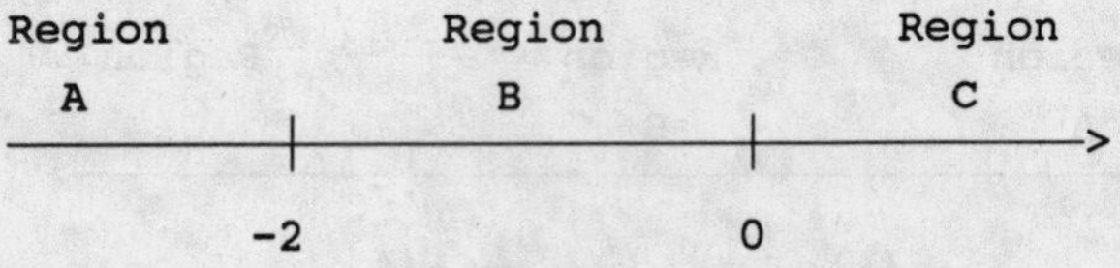

For Region A, test -3:

$\frac{-3 + 2}{-3} = \frac{1}{3} > 0$

For Region B, test -1:

$\frac{-1 + 2}{-1} = -1 < 0$

For Region C, test 1:

$\frac{1 + 2}{1} = 3 > 0$

The solution is [-2, 0).

205. $\frac{5}{p + 1} > 2$

Write $\frac{5}{p + 1} = 2$

$\frac{5}{p + 1} - 2 = 0$

$\frac{5 - 2(p + 1)}{p + 1} = 0$

$\frac{3 - 2p}{p + 1} = 0$

The quotient changes sign when

$3 - 2p = 0$ or $p + 1 = 0$

$p = \frac{3}{2}$ or $p = -1$.

Neither solution is a solution of the inequality.

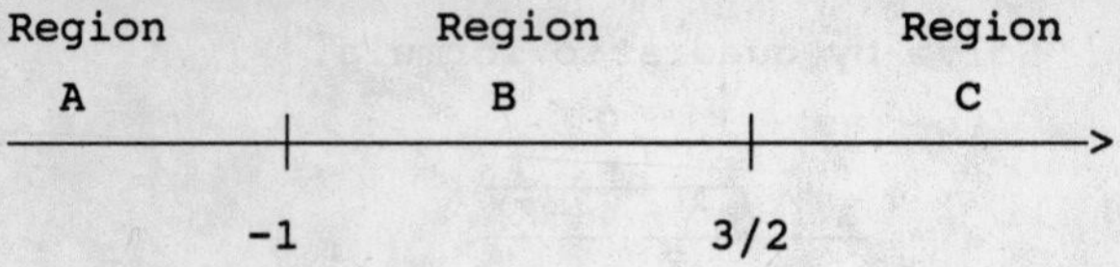

For Region A, test -2:

$\frac{5}{-2 + 1} = -5 < 2$

For Region B, test 0:

$\frac{5}{0 + 1} = 5 > 2$

For Region C, test 2:

$\frac{5}{2 + 1} = \frac{5}{3} < 2$

The solution is (-1, 3/2).

207. $\frac{2}{r + 5} \le \frac{3}{r - 2}$

Write $\frac{2}{r + 5} = \frac{3}{r - 2}$.

$\frac{2}{r + 5} - \frac{3}{r - 2} = 0$

$\frac{2(r - 2) - 3(r + 5)}{(r + 5)(r - 2)} = 0$

$\frac{2r - 4 - 3r - 15}{(r + 5)(r - 2)} = 0$

$\frac{-r - 19}{(r + 5)(r - 2)} = 0$

The quotient changes sign when

$-r - 19 = 0$ or $r + 5 = 0$ or $r - 2 = 0$

$r = -19$ or $r = -5$ or $r = 2$.

-19 is a solution of the inequality but the inequality is undefined when $r = -5$ or $r = 2$.

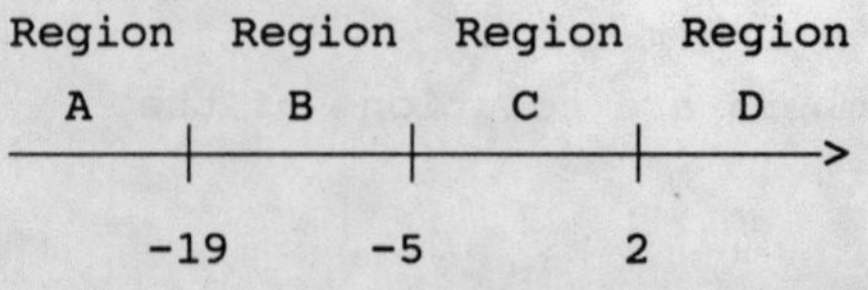

For Region A, test -20:

$$\frac{2}{-20+5} = -\frac{2}{15} \approx -.13 \text{ and}$$

$$\frac{3}{-19-2} = -\frac{1}{7} \approx -.14$$

Since $-.13 > -.14$, -20 is not a solution of the inequality.

For Region B, test -6:

$$\frac{2}{-6+5} = -2 \text{ and } \frac{3}{-6-2} = -\frac{3}{8}$$

Since $-2 < -\frac{3}{8}$, -6 is a solution.

For Region C, test 0:

$$\frac{2}{0+5} = \frac{2}{5} \text{ and } \frac{3}{0-2} = -\frac{3}{2}$$

Since $\frac{2}{5} > -\frac{3}{2}$, 0 is not a solution.

For Region D, test 3:

$$\frac{2}{3+5} = \frac{1}{4} \text{ and } \frac{3}{3-2} = 3$$

Since $\frac{1}{4} < 3$, 3 is a solution.

The solution is $[-19, -5)$ or $(2, \infty)$.

209. Let x = one dimension of the playground. Suppose the other dimension is along the side of the building. Therefore, that dimension has fence on only one side. Then $325 - 2x$ = the other dimension. The area is 11,250 square meters so

$$x(325 - 2x) = 11{,}250$$

$$325x - 2x^2 = 11{,}250$$

$$2x^2 - 325x + 11{,}250 = 0$$

$$x = \frac{325 \pm \sqrt{325^2 - 4 \cdot 2 \cdot 11{,}250}}{2 \cdot 2}$$

$$x = \frac{325 \pm \sqrt{15{,}625}}{4}$$

$$x = \frac{325 \pm 125}{4}$$

$$x = 112.5 \text{ or } x = 50$$

If $x = 112.5$, then $325 - 2x = 100$.
If $x = 50$, then $325 - 2x = 225$.
The width is 100 meters and the length is 112.5 meters or the width is 50 meters and the length is 225 meters.

Case 1 Exercises

1. $$\frac{AX + Q}{1000} - \frac{1}{2} = \frac{(S - C_p)X - C_A}{(S - C_p + C_C)X}$$

   If $S = 10{,}000$, $Q = 200$, $C_A = 100$, $x = .1$, $C_p = 5000$, and $C_C = 3000$, solve for A.

$$\frac{A(.1)+200}{1000} - \frac{1}{2} = \frac{(10{,}000-5000)(.1)-100}{(10{,}000-5000+3000)(.1)}$$

$$\frac{.1A + 200}{1000} - \frac{1}{2} = \frac{(5000)(.1) - 100}{(8000)(.1)}$$

$$\frac{.1A + 200}{1000} - \frac{1}{2} = \frac{400}{800}$$

$$\frac{.1A + 200}{1000} = \frac{1}{2} + \frac{1}{2} = 1$$

$$.1A + 200 = 1000$$

$$.1A = 1000 - 200$$

$$.1A = 800$$

$$A = 8000$$

2. The number of acres to be planted is A, or 8000 acres.

3. Since .1 ton of seeds are produced per acre and there are 8000 acres planted, then $.1(8000) = 800$ tons of seeds will be produced.

4. Since S = selling price per ton of seed, the revenue is R = S (number of tons produced) = $10,000 · 800 = $8,000,000.

CHAPTER 2 FUNCTIONS AND GRAPHS

Section 2.1

1. $y = 8x - 3$

This is a function since a unique valve of y is determined for each valve of x.

3. $y = x^2$

This is a function since a unique valve of y is determined for each valve of x.

5. $x = |y|$

This is not a function. A given valve of x may determine two valves of y. For example, if $x = 2$, $y = 2$ or $y = -2$.

7. $y = \frac{1}{x + 3}$

This is a function since a unique valve of y is determined for each valve of x.

9. $x = \frac{4}{y - 1}$

Solve this equation for y.

$$(y - 1)x = (y - 1)\frac{4}{y - 1}$$

$$xy - x = 4$$

$$xy = 4 + x$$

$$y = \frac{4 + x}{x}$$

For each value of x there will be a unique value of y. This is a function.

11. $x = \sqrt{y + 2}$

Solve this equation for y.

$$x^2 = (\sqrt{y + 2})^2$$

$$x^2 = y + 2$$

$$y = x^2 - 2$$

For each value of x there will be a unique value of y. This is a function.

13. $f(x) = 2x$

Any number may be multiplied by 2, so the domain is $(-\infty, \infty)$.

Since every real number is 2 times some other real number the range is $(-\infty, \infty)$.

15. $f(x) = x^4$

Any number may be raised to the fourth power, so the domain is $(-\infty, \infty)$.

Since $x^4 \geq 0$, the range is $[0, \infty)$.

17. $f(x) = |2 - x|$

We may use any real number for x in $|2 - x|$.

The domain is $(-\infty, \infty)$.

Since absolute value is never negative, the range is $[0, \infty)$.

19. $f(x) - \frac{1}{1 - x}$

Since a denominator can never be zero, $x \neq 1$. The domain is all real numbers except 1.

Since $\frac{1}{1 - x}$ represents the reciprocal of $1 - x$ and the reciprocal of a real number can never be zero, the range is all nonzero real numbers.

21. $f(x) = 3x + 2$

(a) $f(4) = 3(4) + 2 = 12 + 2 = 14$

(b) $f(-3) = 3(-3) + 2 = -9 + 2 = -7$

(c) $f(0) = 3(0) + 2 = 0 + 2 = 2$

(d) $fa) = 3a + 2$

23. $f(x) = -2x - 4$

(a) $f(4) = -2(4) - 4 = -8 - 4 = -12$

(b) $f(-3) = -2(-3) - 4 = 6 - 4 = 2$

(c) $f(0) = -2(0) - 4 = 0 - 4 = -4$

(d) $f(a) = -2a - 4$

25. $f(x) = 6$

(a) $f(4) = 6$

(b) $f(-3) = 6$

(c) $f(0) = 6$

(d) $f(a) = 6$

27. $f(x) = 2x^2 + 4x$

(a) $f(4) = 2(4^2) + 4(4)$
$= 2(16) + 16 = 32 + 16 = 48$

(b) $f(-3) = 2(-3)^2 + 4(-3)$
$= 2(9) + (-12)$
$= 18 - 12 = 6$

(c) $f(0) = 2(0^2) + 4(0)$
$= 2(0) + 0 = 0 + 0 = 0$

(d) $f(a) = 2a^2 + 4a$

29. $f(x) = -x^2 + 5x + 1$

(a) $f(4) = -4^2 + 5(4) + 1$
$= -16 + 20 + 1 = 5$

(b) $f(-3) = -(-3)^2 + 5(-3) + 1$
$= -9 - 15 + 1 = -23$

(c) $f(0) = -0^2 + 5(0) + 1$
$= 0 + 0 + 1 = 1$

(d) $f(a) = -a^2 + 5a + 1$

31. $f(x) = \sqrt{x + 3}$

(a) $f(4) = \sqrt{4 + 3} = \sqrt{7}$

(b) $f(-3) = \sqrt{-3 + 3} = \sqrt{0} = 0$

(c) $f(0) = \sqrt{0 + 3} = \sqrt{3}$

(d) $f(a) = \sqrt{a + 3}$, $a \geq -3$

33. $f(x) = 2x - 3$

(a) $f(a) = 2a - 3$

(b) $f(-r) = 2(-r) - 3$
$= -2r - 3$

(c) $f(m + 3) = 2(m + 3) - 3$
$= 2m + 6 - 3 = 2m + 3$

(d) $f(a + b) = 2(a + b) - 3$
$= 2a + 2b - 3$

35. $f(x) = 2x^2 - x$

(a) $f(a) = 2a^2 - a$

(b) $f(-r) = 2(-r)^2 - (-r)$
$= 2r^2 + r$

(c) $f(m + 3) = 2(m + 3)^2 - (m + 3)$
$= 2(m^2 + 6m + 9) - m - 3$
$= 2m^2 + 11m + 15$

(d) $f(a + b) = 2(a + b)^2 - (a + b)$
$= 2(a^2 + 2ab + b^2) - a - b$
$= 2a^2 + 4ab + 2b^2 - a - b$

37. $f(x) = x^3 + 1$

(a) $f(a) = a^3 + 1$

(b) $f(-r) = (-r)^3 + 1 = -r^3 + 1$

(c) $f(m + 3) = (m + 3)^3 + 1$
$= m^3 + 9m^2 + 27m + 27 + 1$
$= m^3 + 9m^2 + 27m + 28$

(d) $f(a + b) = (a + b)^3 + 1$
$= a^3 + 3a^2b + 3ab^2 + b^3 + 1$

39. $f(x) = \dfrac{3}{x - 1}$

Note: Assume all denominators are not zero.

(a) $f(a) = \dfrac{3}{a - 1}$

(b) $f(-r) = \dfrac{3}{-r - 1}$ or $-\dfrac{3}{r + 1}$

(c) $f(m + 3) = \dfrac{3}{(m + 3) - 1} = \dfrac{3}{m + 2}$

(d) $f(a + b) = \dfrac{3}{(a + b) - 1}$
$= \dfrac{3}{a + b - 1}$

41. $f(x) = \sqrt{2x}$

Note: Assume all radicands are nonnegative.

(a) $f(a) = \sqrt{2a}$

(b) $f(-r) = \sqrt{2(-r)} = \sqrt{-2r}$

(c) $f(m + 3) = \sqrt{2(m + 3)} = \sqrt{2m + 6}$

(d) $f(a + b) = \sqrt{2(a + b)} = \sqrt{2a + 2b}$

43. (a) $C(\frac{3}{4}) = 40(1) + 40$

$C(\frac{3}{4}) = \$80$

(b) $C(\frac{9}{10}) = 40(1) + 40$

$C(\frac{9}{10}) = \$80$

(c) $C(1) = 40(1) + 40$
$C(1) = \$80$

(d) $C(1\frac{5}{8}) = 40(2) + 40$

$C(1\frac{5}{8}) = \$120$

(e) $C(2\frac{1}{9}) = 40(3) + 40$

$C(2\frac{1}{9}) = \$160$

45. $H(t) = 1000 + 50(t + 1)$

(a) For 1991, $t = 0$.
$H(0) = 1000 + 50(0 + 1)$
$H(0) = 1050$ thousand dollars
Revenue: $1,050,000

(b) for 1992, $t = 1$.
$H(1) = 1000 + 50(1 + 1)$
$H(1) = 1100$ thousand dollars
Revenue: $1,100,000

(c) For 1993, $t = 2$.
$H(2) = 1000 + 50(2 + 1)$
$H(2) = 1150$ thousand dollars
Revenue: $1,150,000

(d) For 1994, $t = 3$.
$H(3) = 1000 + 50(3 + 1)$
$H(3) = 1200$ thousand dollars
Revenue: $1,200,000

47.

$$f(x) = \begin{cases} x^2 & \text{if } x < -1 \\ x + 2 & \text{if } -1 \le x \le 4 \\ 10 & \text{if } x > 4 \end{cases}$$

$f(-3) = (-3)^2 = 9$
$f(0) = 0 + 2 = 2$
$f(2) = 2 + 2 = 4$
$f(4) = 4 + 2 = 6$
$f(10) = 10$

Section 2.2

1. No vertical line will intersect the graph in more than one point, so it is the graph of a function.

3. Notice that the y-axis intersects the graph in two points. It is not the graph of a function.

5. No vertical line will intersect the graph in more than one point so it is the graph of a function.

See the graphs for Exercises 7-49 in the answer section at the back of the textbook.

7. $y = |x + 1|$
If $x = 0$, $y = 1$.
The y-intercept is 1.
If $y = 0$, $x = -1$.
The x-intercept is -1.
Find more points by making a table of values.

x	-3	-2	1	2
y	2	1	2	3

Now, plot these points and draw the graph through them.

9. $y = |2 - x|$
If $x = 0$, $y = 2$.
The y-intercept is 2.
If $y = 0$, $x = 2$.
The x-intercept is 2.
Find more points by making a table of values.

x	1	3	4
y	1	1	2

11. $y = |5x + 4|$
If $x = 0$, $y = 4$.
The y-intercept is 4.
If $y = 0$, $x = -\frac{4}{5}$.
The x-intercept is $-\frac{4}{5}$.
Find more points by making a table of values.

x	-3	-2	-1	1	2
y	11	6	1	9	14

13. $y = -|x|$
If $x = 0$, $y = 0$.
The y-intercept is 0.
If $y = 0$, $x = 0$.
The x-intercept is 0.
Find more points by making a table of values.

x	-3	-2	-1	1	2	3
y	-3	-2	-1	-1	-2	-3

15. $y = |x| + 4$
If $x = 0$, $y = 4$.
The y-intercept is 0.
If $y = 0$, there is no value for x since $|x| \neq -4$.
There is no x-intercept.
Find more points by making a table of values.

x	-3	-2	-1	1	2	3
y	7	6	5	5	6	7

17. $y = \begin{cases} x - 1 & \text{if } x \le 3 \\ 2 & \text{if } x > 3 \end{cases}$

For $x \le 3$, use the equation $y = x - 1$ in the following table.

x	-1	0	1	2	3
y	-2	-1	0	1	2

For $x > 3$, use the equation $y = 2$ in the following table.

x	3.01	4	5	6
y	2	2	2	2

19. $y = \begin{cases} 4 - x \text{ if } x < 2 \\ 1 + 2 \text{ if } x \geq 2 \end{cases}$

For $x < 2$, use the equation $y = 4 - x$ in the following table.

x	-2	-1	0	1	1.99
y	6	5	4	3	2.01

For $x \geq 2$, use the equation $y = 1 + 2x$ in the following table.

x	2	3	4	5
y	5	7	9	11

21. $y = \begin{cases} 2x + 1 \text{ if } x \geq 0 \\ x \text{ if } x < 0 \end{cases}$

For $x \geq 0$, use the equation $y = 2x + 1$ in the following table.

x	0	1	2	3	4
y	1	3	5	7	9

For $x < 0$, use the equation $y = x$ in the following table.

x	-3	-2	-1	-.01
y	-3	-2	-1	-.01

23. $y = \begin{cases} 2 + x \text{ if } x < -4 \\ -x \text{ if } -4 \leq x \leq 5 \\ 3x \text{ if } x > 5 \end{cases}$

For $x < -4$, use the equation $y = 2 + x$ in the following table.

x	-7	-6	-5	-4.01
y	-5	-4	-3	-2.01

For $-4 \leq x \leq 5$, use the equation $y = -x$ in the following table.

x	-4	-2	0	2	4	5
y	4	2	0	-2	-4	-5

For $x > 5$, use the equation $y = 3x$ in the following table.

x	5.01	6	7	8	9
y	15.03	18	21	24	27

25. $y = \begin{cases} |x| \text{ if } x > -2 \\ x \text{ if } x \leq -2 \end{cases}$

For $x > -2$, use the equation $y = |x|$ in the following table.

x	-1.99	-1	0	1	2	3
y	1.99	1	0	1	2	3

For $x \leq -2$, use the equation $y = x$ in the following table.

x	-6	-5	-4	-3	-2
y	-6	-5	-4	-3	-2

27. $y = [-x]$

Use the following table to establish the "step" pattern. Once the pattern is established, simply repeat it.

x	0	$\frac{1}{10}$	$\frac{1}{4}$	$\frac{1}{2}$	$\frac{3}{4}$	$\frac{9}{10}$	1	$\frac{11}{10}$
y	0	-1	-1	-1	-1	-1	-1	-2

29. $y = [2x - 1]$

Use the following table to establish the "step" pattern. Once the pattern is established simply repeat it.

x	-.1	0	.1	.2	.3	.4	.5
y	-2	-1	-1	-1	-1	-1	0

31. $y - [3x]$

Use the following table to establish the "step" pattern. Once the pattern is established, simply repeat it.

x	-.1	0	.1	.2	.3	$\frac{1}{3}$
y	-1	0	0	0	0	1

33. $y = [3x] - 1$

Use the following table to establish the "step" pattern. Once the pattern is established, simply repeat it.

x	-.1	0	.1	.2	.3	$\frac{1}{3}$
y	-2	-1	-1	-1	-1	0

35. $y = \sqrt{x}$

The domain is $[0, \infty)$. With the use of a calculator, complete the following table.

x	0	1	2	3	4	5	7	9
y	0	1	1.4	1.7	2	2.2	2.6	3

37. $y = \sqrt{x} + 1$

The domain is $[0, \infty)$. With the use of a calculator, complete the following table.

x	0	1	2	3	4	5	7	9
y	1	2	2.4	3.7	3	3.2	3.6	4

39. $y = x^2$

We use the following table of values.

x	-4	-3	-2	-1	0	1	2	3	4
y	16	9	4	1	0	1	4	9	16

41. **(a)** 25 + 2(2) = \$29

(b) 25 + 2(2) = \$29

(c) 25 + 4(2) = \$33

(d) 25 + 4(2) = \$33

(e) Let x represent hours and y, cost in dollars.
For x in (0, 1], y = 27.
For x in (1, 2], y = 29.
For x in (2, 3], y = 31.
For x in (3, 4], y = 33.

43. Let x represent hours and y, cost.
For x in (0, 1], y = 10.
For x in (1, 2], y = 17.
For x in (2, 3], y = 24.
For x in (3, 4], y = 31.

45. Let x represent miles and y, cost.
For x in (0, 50], y = 37.
For x in (50, 75], y = 47.
For x in (75, 100], y = 57.
For x in (100, 125], y = 67.

47. **(a)** At 7 A.M., t = 1.
$i(1) = 40(1) + 100 = 140$

(b) At 9 A.M., t = 3.
$i(3) = 40(3) + 100 = 220$

(c) At 10 A.M., t = 4.
$i(4) = 220$

(d) At noon, t = 6.
$i(6) = 220$

(e) At 2 P.M., $t = 8$.
$i(8) = 220$

(f) At 5 P.M., $t = 11$.
$i(11) = 60$

(g) At midnight, $t = 18$.
$i(18) = 60$.

(h) For t in $[0, 3]$, use
$y = 40t + 100$.
For t in $(3, 8]$, use $y = 220$.
For t in $(8, 10]$, use $y = -80t + 800$.
For t in $(10, 24]$, use $y = 60$.

(i) From the graph, the blood sugar level is highest for the interval $3 \le t \le 8$ and lowest for the interval $10 \le t \le 24$.

49. (a) For t in $[0, 2]$, use $y = -.10$.
For t in $(2, 5]$, use $y = -.25$.
For t in $(5, 8]$, use $y = -.50$.
For t in $(8, 12]$, use $y = -.75$.
For t in $(12, 16]$, use $y = -1.10$.

(b) The domain is $[0, 16]$ while the range is
$\{-.10, -.25, -.50, -.75, -1.10\}$.

51. (a) No vertical line intersects the graph in more than one point so it is the graph of a function.

(b) The domain represents the years from 1984 to 1989.

(c) An estimate of the range is $[55, 1200]$.

Section 2.3
See the graphs for Exercises 1-23 in the answer section at the back of the textbook.

1. $y = 2x + 1$
If $x = 0$, $y = 1$.
The y-intercept is 1.
If $y = 0$, $x = -1/2$. The x-intercept is $-1/2$.
Using the points $(0, 1)$ and $(-\frac{1}{2}, 0)$, graph the line.

3. $y = 3x + 2$
If $x = 0$, $y = 2$.
The y-interecept is 2.
If $y = 0$, $x = -2/3$. The x-intercept is $-2/3$.
Using the points $(0, 2)$ and $(-\frac{2}{3}, 0)$, graph the line.

5. $3y + 4x = 12$
If $x = 0$, $y = 4$.
The y-intercept is 4.
If $y = 0$, $x = 3$.
The x-intercept is 3.
Using the points $(0, 4)$ and $(3, 0)$, graph the line.

7. $y = -2$
The graph of $y = -2$ is a horizontal line with y-intercept of -2.

9. $6x + y = 12$
If $x = 0$, $y = 12$.
The y-intercept is 12.
If $y = 0$, $x = 2$.
The x-intercept is 2.
Using the points $(0, 12)$ and $(2, 0)$ we graph the line.

11. $x - 5y = 4$
If $x = 0$, $y = -4/5$. The y-intercept is $-4/5$.
If $y = 0$, $x = 4$.
The x-intercept is 4.

Using the points $(0, -\frac{4}{5})$ and (4, 0) we graph the line.

13. $x + 5 = 0$
If $x + 5 = 0$, $x = -5$. The graph of $x = -5$ is a vertical line with an x-intercept of -5.
This is not a linear function.

15. $5y - 3x = 12$
If $x = 0$, $y = 12/5$. The y-intercept is $12/5$.
If $y = 0$, $x = -4$.
The x-intercept is -4.
Using the points $(0, \frac{12}{5})$ and $(-4, 0)$, graph the line.

17. $8x + 3y = 10$
If $x = 0$, $y = 10/3$. The y-intercept is $10/3$.
If $y = 0$, $x = 10/8$ or $5/4$. The x-intercept is $5/4$.
Using the points $(0, \frac{10}{3})$ and $(\frac{5}{4}, 0)$, graph the line.

19. $y = 2x$
If $x = 0$, $y = 0$. Both intercepts are 0, so find another point. If $x = 2$, $y = 4$.
Using the points (0, 0) and (2, 4), graph the line.

21. $y = -4x$
If $x = 0$, $y = 0$. Both intercepts are 0, so we must find another point. If $x = 1$, $y = -4$.
Using the points (0, 0) and (1, -4), graph the line.

23. $x - 3y = 0$
If $x - 3y = 0$, $x = 3y$. If $x = 0$, $y = 0$. Both intercepts are 0, so find another point. If $y = 2$, $x = 6$.
Using the points (0, 0) and (6, 2), graph the line.

25. On the supply curve, when $x = 20$, $p = 140$. The point (20, 140) is on the graph. When 20 items are supplied, the price is \$140.

27. The two curves intersect at the point (10, 120). The equilibrium supply and equilibrium demand are both $x = 10$ or 10 items.

See the graphs for Exercises 29-35 in the answer section at the back of the textbook.

29. $p = 16 - \frac{5}{4}x$

(a) If $x = 0$, $p = \$16$.

(b) If $x = 4$,
$$P = 16 - \frac{5}{4}(4) = 16 - 5 = \$11$$

(c) If $x = 8$,
$$p = 16 - \frac{5}{4}(8) = 16 - 10 = \$6$$

(d) From Part (c), if $p = 6$, $x = 8$ units.

(e) From part (b), if $p = 11$, $x = 4$ units.

(f) From part (a), if $p = 16$, $x = 0$ units.

(g) To graph $p = 16 - \frac{5}{4}x$, use the points (0, 16) and (8, 6) from parts (a) and (c).

(h) $p = \frac{3}{4}x$

If $p = 0$, $0 = \frac{3}{4}x$, and $x = 0$ units.

(i) If $p = 10$,

$$10 = \frac{3}{4}x$$

$$\frac{4}{3}(10) = \frac{4}{3}\left(\frac{3}{4}x\right)$$

$$\frac{40}{3} = x$$

40/3 units is approximately 13 units.

(j) If $p = 20$,

$$20 = \frac{3}{4}x$$

$$\frac{4}{3}(20) = \frac{4}{3}\left(\frac{3}{4}x\right)$$

$$\frac{80}{3} = x.$$

80/3 units is approximately 27 units

(k) To graph $p = (3/4)x$, use the points (0, 0) and $(\frac{80}{3}, 20)$ from parts (h) and (j).

(l) The two graphs intersect at the point (8, 6). The equilibrium supply is 8 units.

(m) The equilibrium price is 6.

31. (a) To graph $p = (2/5)x$: when $x = 0$, $p = 0$, and when $x = 5$, $p = 2$. Use the points (0, 0) and (5, 2).
To graph $p = 100 - (2/5)x$: when $x = 0$, $p = 100$, and when $x = 20$, $p = 92$. Use the points (0, 100) and (20, 92).

(b) The two graphs intersect at the point (125, 50). The equilibrium demand is 125 units.

(c) The equilibrium price is 50.

33. (a) If $x = 0$, $y = \$135$.

(b) If $x = 1000$,

$$y = .07(1000) + 135 = 70 + 135 = \$205.$$

(c) If $x = 2000$,

$$y = .07(2000) + 135 = 140 + 135 = \$275.$$

(d) If $x = 3000$, $y = .07(3000) + 135 = 210 + 135 = \$345$.

(e) Use the points (o, 135) and (3000, 345) from parts (a) and (d) to graph the line.

35. (a) $R(x) = 125x$ and

$$C(x) = 100x + 5000$$
$$R(x) = C(x)$$
$$125x = 100x + 5000$$
$$25x = 5000$$
$$x = 200$$

The break-even point is $x = 200$ or 200,000 policies.

(b) To graph the revenue function, graph $y = 125x$. If $x = 0$, $y = 0$. If $x = 20$, $y = 2500$. Use the points (0, 0) and (20, 2500) to graph the line.
To graph the cost function, graph $y = 100x + 5000$. If $x = 0$, $y = 5000$. If $x = 30$, $y = 8000$. Use the points (0, 5000) and (30, 8000) to graph the line.

(c) From the graph, when $x = 100$, cost is \$15,000 and revenue is \$12,500.

Section 2.4

1. $(-8, 6)$ and $(2, 4)$

$$m = \frac{4 - 6}{2 - (-8)} = \frac{-2}{10} = -\frac{1}{5}$$

3. $(-1), 4)$ and $(2, 6)$

$$m = \frac{6 - 4}{2 - (-1)} = \frac{2}{3}$$

5. $(0, 0)$ and $(-4, 6)$

$$m = \frac{6 - 0}{-4 - 0} = \frac{6}{-4} = -\frac{3}{2}$$

7. $(-2, 9)$ and $(-2, 11)$

$$m = \frac{11 - 9}{-2 - (-2)} = \frac{2}{0}$$

The slope is undefined.

9. $(3, -6)$ and $(-5, -6)$

$$m = \frac{-6 - (-6)}{-5 - 3} = \frac{0}{-8} = 0$$

11. For $4x - 3y = 6$, solve for y and obtain $y = (4/3)x - 2$.
For $3x + 4y = 8$, solve for y and obtain $y = (-3/4)x + 2$.
The two slopes are $4/3$ and $-3/4$.
Since $(\frac{4}{3})(-\frac{3}{4}) = -1$, the lines are perpendicular.

13. For $3x + 2y = 8$, solve for y and obtain $y = (-3/2)x + 4$.
For $6y = 5 - 9x$, solve for y and obtain $y = (-3/2)x + 5/6$.
Since the two slopes are $-3/2$ and $-3/2$, the lines are parallel.

15. For $4x = 2y + 3$, solve for y and obtain $y = 2x - 3/2$.
For $2y = 2x + 3$, solve for y and obtain $y = x + 3/2$.
Since the two slopes are 2 and 1, the lines are neither parallel nor perpendicular.

17. For $2x - y = 9$, solve for y and obtain $y = 2x - 9$.
For $x = 2y$, solve for y and obtain $y = (1/2)x$.
Since the two slopes are 2 and $1/2$, the lines are neither parallel nor perpendicular.

19. $y = 3x + 4$
$M = 3$ and $b = 4$

21. $y + 4x = 8$
$y = -4x + 8$
$m = -4$ and $b = 8$

23. $3x + 4y = 5$
$4y = -3x + 5$
$y = -\frac{3}{4}x + \frac{5}{4}$

$m = -\frac{3}{4}$ and $b = \frac{5}{4}$

25. $3x + y = 0$
$y = -3x + 0$
$m = -3$ and $b = 0$

27. $2x + 5y = 0$
$5y = -2x + 0$
$y = -\frac{2}{5}x + 0$

$m = -\frac{2}{5}$ and $b = 0$

29. $y = 8$
$y = 0x + 8$
$m = 0$ and $b = 8$

31. $y + 2 = 0$

$y = 0x - 2$

$m = 0$ and $b = -2$

33. $x = -8$

The graph at $x = -8$ is a vertical line. It has undefined slope and no y-intercept.

See the graphs for Exercises 35-45 in the answer section at the back of the textbook.

35. $(-4, 2)$; $m = \frac{2}{3}$

$$m = \frac{\Delta y}{\Delta x} = \frac{2}{3}$$

Use $\Delta y = 2$ and $\Delta x = 3$.

37. $(-5, -3)$; $m = -2$

$$m = \frac{\Delta y}{\Delta x} = -2 = \frac{-2}{1}$$

Use $\Delta y = -2$ and $\Delta x = 1$.

39. $(8, 2)$; $m = 0$

Since $m = 0$, draw a horizontal line through $(8, 2)$.

41. $(6, -5)$; undefined slope

Since the slope is undefined, draw a vertical line through $(6, -5)$.

43. $(0, -2)$; $m = \frac{3}{4}$

$$m = \frac{\Delta y}{\Delta x} = \frac{3}{4}$$

Use $\Delta y = 3$ and $\Delta x = 4$.

45. $(5, 0)$; $m = \frac{1}{4}$

$$m = \frac{\Delta y}{\Delta x} = \frac{1}{4}$$

Use $\Delta y = 1$ and $\Delta x = 4$.

47. If $b = 4$ and $m = -3/4$, an equation is $y = -\frac{3}{4}x + 4$.

Multiplying by 4 gives $4y = -3x + 16$.

49. If $b = -2$ and $m = -1/2$ an equation is $y = -\frac{1}{2}x - 2$.

Multiplying by 2 gives $2y = -x - 4$.

51. If $b = 5/4$ and $m = 3/2$, an equation is $y = \frac{3}{2}x + \frac{5}{4}$.

Multiplying by 4 gives $4y = 6x + 5$.

53. Through $(-4, 1)$, $m = 2$

$$y - 1 = 2[x - (-4)]$$
$$y - 1 = 2x + 8$$
$$y = 2x + 9$$

55. Through $(0, 3)$, $m = -3$

The y-intercept is $b = 3$.

$y = -3x + 3$

57. Through $(3, 2))$, $m = \frac{1}{4}$

$$y - 2 = \frac{1}{4}(x - 3)$$

$$y - 2 = \frac{1}{4}x - \frac{3}{4}$$

Multiplying by 4 gives

$4y - 8 = x - 3$

$4y = x + 5$.

59. Through $(-1, 1)$ and $(2, 5)$

$$m = \frac{5 - 1}{2 - (-1)} = \frac{4}{3}$$

$$y - 5 = \frac{4}{3}(x - 2)$$

$$y - 5 = \frac{4}{3}x - \frac{8}{3}$$

Multiplying by 3 gives

$3y - 15 = 4x - 8$

$3y = 4x + 7$.

61. Through (9, -6) and (12, -8)

$$m = \frac{-8 - (-6)}{12 - 9} = \frac{-2}{3} = -\frac{2}{3}$$

$$y - (-8) = -\frac{2}{3}(x - 12)$$

$$y + 8 = -\frac{2}{3}x + 8$$

$$y = -\frac{2}{3}x$$

Multiplying by 3 gives
$3y = -2x$.

63. Through (-8, 4) and (-8, 6)

$$m = \frac{6 - 4}{-8 - (-8)} = \frac{2}{0}$$

Since the slope is undefined, the line is vertical. Its equation is $x = -8$.

65. Through (-1, 3) and (0, 3)

$$m = \frac{3 - 3}{0 - (-1)} = \frac{0}{1} = 0$$

Since the slope is 0, the line is horizontal. Its equation is $y = 3$.

67. We have (20, 13900) and (10, 7500).

$$m = \frac{7500 - 13900}{10 - 20} = \frac{-6400}{-10} = 640$$

The slope is 640.
$y - 7500 = 640(x - 10)$
$y - 7500 = 640x - 6400$
$y = 640x + 1500$

69. Use the ordered pairs (45, 42.5) and (55, 67.5).

$$m = \frac{67.5 - 42.5}{55 - 45} = \frac{25}{10} = 2.5$$

The slope is 2.5.
$y - 42.5 = 2.5(x - 45)$
$y - 42.5 = 2.5x - 112.5$
$y = 2.5x - 70$

71. Use the ordered pairs (3, 37,000) and (12, 28,000).

$$m = \frac{28{,}000 - 37{,}000}{12 - 3} = \frac{-9000}{9} = -1000$$

The slope is -1000.
$y - 37{,}000 = -1000(x - 3)$
$y - 37{,}000 = -1000x + 3000$
$y = -1000x + 40{,}000$

73. **(a)** Use ordered pairs (r, h), or (24, 167) and (26, 174).

$$m = \frac{174 - 167}{26 - 24} = \frac{7}{2} = 3.5$$

$h - 167 = 3.5(r - 24)$
$h - 167 = 3.5r - 84$
$h = 3.5r + 83$

(b) If $r = 23$, $h = 3.5(23) + 83 = 163.5$ cm
If $r = 27$, $h = 3.5(27) + 83 = 177.5$ cm

(c) If $h = 170$,
$170 = 3.5r + 83$
$87 = 3.5R$
$r = 24.86$
≈ 25 cm.

Section 2.5

1. $S(x) = 300x + 2000$

(a) For 1984, $x = 2$.
$S(2) = 300(2) + 2000 = 2600$
2600

(b) For 1985, $x = 3$.
$S(3) = 300(3) + 2000 = 2900$
2900

(c) For 1986, $x = 4$.
$S(4) = 300(4) + 2000 = 3200$
3200

(d) For 1982, $x = 0$.
$S(0) = 2000$
2000

(e) The annual rate of change of sales is the slope which is 300.

3. $N(x) = -5x + 100$

(a) If $x = 0$, $N(0) = -5(0) + 100 = 100$ thousand

(b) If $x = 6$, $N(6) = -5(6) + 100 = 70$ thousand

(c) If $x = 20$, $N(20) = -5(20) + 100 = 0$

(d) The hourly rate of change is the the slope -5 thousand or -5000. The negative sign indicates a decrease.

5. The ordered pairs are (0, 200,000) and (7, 1,000,000).

(a) $m = \dfrac{1{,}000{,}000 - 200{,}000}{7 - 0} = \dfrac{800{,}00}{7}$

Since the y-intercept is 200,000,

$y = \dfrac{800{,}000}{7}x + 200{,}000.$

Multiplying by 7 gives
$7y = 800{,}000x + 1{,}400{,}000.$

(b) For 1986, $x = 2$. If $x = 2$,

$$7y = 800{,}000(2) + 1{,}400{,}000$$
$$7y = 1{,}600{,}000 + 1{,}400{,}000$$
$$7y = 3{,}000{,}000$$
$$y \approx \$429{,}000.$$

(c) For 1993, $x = 9$. If $x = 9$,

$$7y = 800{,}000(9) + 1{,}400{,}000$$
$$7y = 7{,}200{,}000 + 1{,}400{,}000$$
$$7y = 8{,}600{,}000$$
$$y \approx \$1{,}230{,}000.$$

7. The ordered pairs are (0, 10.3) and (3, 11.2).

(a) $m = \dfrac{11.2 - 10.3}{3 - 0} = \dfrac{.9}{3} = .3$

Since the y-intercept is 10.3, $f(x) = .3x + 10.3$.

(b) The average rate of change is .3%, which is the same as the slope.

9. $V(x) = 600 - 20x$

(a) For June 6, use $x = 6$.
$V(6) = 600 - 20(6) = 600 - 120$
$= 480$ bottles

(b) For June 12, we use $x = 12$.
$V(12) = 600 - 20(12) = 600 - 240$
$= 360$ bottles

(c) For June 24, we use $x = 24$.
$V(24) = 600 - 20(24) = 600 - 480$
$= 120$ bottles

(d) When the last bottle is taken, $V(x) = 0$.

$$0 = 600 - 20x$$
$$20x = 600$$
$$x = 30$$

The last bottle is taken on June 29.

(e) The average daily rate of change is the slope, which is -20.

11. The marginal cost is \$1 while the fixed cost is \$12. $C(x) = 1x + 12$ or $C(x) = x + 12$, where x is the number of hours.

13. The marginal cost is 30¢ while the fixed cost is 35¢. $C(x) = 30x + 35$, where x is the number of half-hours.

15. Since the fixed cost is \$100,

$C(x) = mx + 100.$

$C(50) = 1600$

Therefore,

$1600 = m(50) + 100$

$50m = 1500$

$m = 30.$

$C(x) = 30x + 100$

17. Since the fixed cost is \$1000,

$C(x) = mx + 1000.$

$C(40) = 2000.$

Therefore,

$2000 = m(40) + 1000$

$40m = 1000$

$m = 25$

$C(x) = 25x + 1000$

19. Since the marginal cost is \$50,

$C(x) = 50x + b.$

$C(80) = 4500$

$4500 = 50(80) + b$

$500 = b$

$C(x) = 50x + 500$

21. Since the marginal cost is \$90,

$C(x) = 90x + b.$

$C(150) = 16,000$

$16,000 = 90(150) + b$

$16,000 = 13,500 + b$

$b = 2500$

$C(x) = 90x + 2500$

23. $C(x) = .125x + 20$

(a) $C(1000) = .125(1000) + 20$

$= 125 + 20$

$= \$145$

(b) $C(1001) = .125(1001) + 20$

$= 125.125 + 20$

$= \$145.125$

25. (a) $C(x) = 10x + 500$

(b) $R(x) = 35x$

(c) $35x = 10x + 500$

$25x = 500$

$x = 20$

$R(20) = 35(20) = 700$

The break-even point is (20, 700).

27. (a) $C(x) = 18x + 250$

(b) $R(x) = 28x$

(c) $28x = 18x + 250$

$10x = 250$

$x = 25$

$R(25) = 28(25) = 700$

The break-even point is (25, 700).

29. (a) $C(x) = 100x + 2700$

(b) $R(x) = 125x$

(c) $125x = 100x + 2700$

$25x = 2700$

$x = 108$

$R(140) = 125(108) = 13.500$

The break-even point is (108, 13,500).

31. $C(x) = 800 + 20x$

$$\overline{C}(x) = \frac{C(x)}{x} = \frac{800 + 20x}{x} = \frac{800}{x} + 20$$

(a) $\overline{C}(10) = \frac{800}{10} + 20 = 80 + 20$

$= \$100$

(b) $\overline{C}(50) = \frac{800}{50} + 20 = 16 + 20$

$= \$36$

(c) $\overline{C}(200) = \frac{800}{200} + 20 = 4 + 20$

$= \$24$

33. (a) $\frac{f(20) - f(10)}{20 - 10} = \frac{40 - 30}{10} = 1$

(b) $\frac{f(40) - f(10)}{40 - 10} = \frac{50 - 30}{30} = \frac{20}{30} = \frac{2}{3}$

(c) $\frac{f(30) - f(20)}{30 - 20} = \frac{46 - 40}{10} = \frac{6}{10} = \frac{3}{5}$

(d) $\frac{f(40) - f(30)}{40 - 30} = \frac{50 - 46}{10} = \frac{4}{10} = \frac{2}{5}$

35. $C(x) = 50x + 5000$ and $2(x) = 60x$

$60x = 50x + 5000$

$10x = 5000$

$x = 500$

The break-even point is 500 units.

$R(500) = 60(500) = 30{,}000$

The break-even revenue is $30,000.

37. $C(x) = 105x + 6000$ and $R(x) = 250x$

$250x = 105x + 6000$

$145x = 6000$

$x \approx 41$

Produce the item since 41 < 400.

39. $C(x) = 80x + 7000$ and $R(x) = 95x$

$95x = 80x + 7000$

$15x = 7000$

$x \approx 467$

Do not produce the item since 467 > 400.

41. $C(x) = 140x + 3000$ and $R(x) = 125x$

$125x = 140x + 3000$

$-15x = 3000$

$x = -200$

Since the break-even point is negative, the product will never make a profit. Do not produce it.

43. (a) and (b) See the graph in the answer section at the back of the textbook.

(c) (2, 66) and (5, 90)

$m = \frac{90 - 66}{5 - 2} = \frac{24}{3} = 8$

$y - 90 = 8(x - 5)$

$y - 90 = 8x - 40$

$y = 8x + 50$

(d) See the table in the answer section at the back of the textbook.

(e) $y = 8x + 50$

If $x = 7$, $y = 8(7) + 50$

$= 56 + 50 = 106.$

(f) $y = 8x + 50$

If $x = 9$, $y = 8(9) + 50$

$= 72 + 50 = 122$

Chapter 2 Review Exercises

See the graphs for Exercises 1-9 in the answer section at the back of the textbook.

1. $2x - 5y = 10$

The x-intercept is 5 and the y-intercept is -2.

The domain is $(-\infty, \infty)$. The range is $(-\infty, \infty)$

3. $f(x) = (2x + 1)(x - 1)$

$y = (2x + 1)(x - 1)$

The y-intercept is -1. The x-intercepts are -1/2 and 1.

x	-2	-1	$\frac{1}{4}$	2	3
y	9	2	$-\frac{9}{8}$	5	14

From the graph, the domain is $(-\infty, \infty)$ and the range is $[-\frac{9}{8}, \infty)$.

5. $f(x) = -2 + x^2$

$y = x^2 - 2$

The y-intercept is -2 and the x-intercepts are $\pm\sqrt{2}$.

x	-2	-1	0	1	2
y	2	-1	-2	-1	2

From the graph, the domain is $(-\infty, \infty)$ and the range is $[-2, \infty)$.

7. $f(x) = \dfrac{2}{x^2 + 1}$

$y = \dfrac{2}{x^2 + 1}$

The y-intercept is 2. There is no x-intercept.

x	-4	-2	-1	1	2	4
y	$\frac{2}{17}$	$\frac{2}{5}$	1	1	$\frac{2}{5}$	$\frac{2}{17}$

From the graph, the domain is $(-\infty, \infty)$ and the range is $(0, 2]$.

9. $y + 1 = 0$

The graph of $y = -1$ is a horizontal line.

The domain is $(-\infty, \infty)$ and the range is $\{-1\}$.

11. $f(x) = 4x - 1$

(a) $f(6) = 4(6) - 1 = 24 - 1$
$= 23$

(b) $f(-2) = 4(-2) - 1 = -8 - 1$
$= -9$

(c) $f(p) = 4p - 1$

(d) $f(r + 1) = 4(r + 1) - 1$
$= 4r + 4 - 1 = 4r + 3$

13. $f(x) = -x^2 + 2x - 4$

(a) $f(6) = -6^2 + 2(6) - 4$
$= -36 + 12 - 4 = -28$

(b) $f(-2) = -(-2)^2 + 2(-2) - 4$
$= -4 - 4 - 4 = -12$

(c) $f(p) = -p^2 + 2p - 4$

(d) $f(r + 1)$
$= -(r + 1)^2 + 2(r + 1) - 4$
$= -(r^2 + 2r + 1) + 2r + 2 - 4$
$= r^2 - 2r - 1 + 2r + 2 - 4$
$= -r^2 - 3$

15. $f(x) = 5x - 3$ and $g(x) = -x^2 + 4x$

(a) $f(-2) = 5(-2) - 3$
$= -10 - 3 = -13$

(b) $g(3) = -3^2 + 4(3)$
$= -9 + 12 = 3$

(c) $g(-4) = -(-4)^2 + 4(-4)$
$= -16 - 16 = -32$

(d) $f(5) = 5(5) - 3$
$= 25 - 3 = 22$

(e) $g(-k) = -(-k)^2 + 4(-k)$
$= -k^2 - 4k$

(f) $g(3m) = -(3m)^2 + 4(3m)$
$= -9m^2 + 12m$

(g) $g(k - 5) = -(k - 5)^2 + 4(k - 5)$
$= -(k^2 - 10k + 25) + 4k - 20$
$= -k^2 + 10k - 25 + 4k - 20$
$= -k^2 + 14k - 45$

(h) $f(3 - p) = 5(3 - p) - 3$
$= 15 - 5p - 3$
$= 12 - 5p$

See the graphs for Exercises 17-41 in the answer section at the back of the textbook.

17. $f(x) = |x| - 3$

x	-3	-2	-1	0	1	2	3
f(x)	0	-1	-2	-3	-2	-1	0

19. $f(x) = -|x + 1| + 3$

x	-4	-3	-2	-1	0	1	2
f(x)	0	1	2	3	2	1	0

21. $f(x) = [x - 3]$
Use the following table to establish the "step" pattern. Once the pattern is established, simply repeat it.

x	-.1	0	.2	.5	.7	.9	1
f(x)	-4	-3	-3	-3	-3	-3	-2

23. $f(x) = \begin{cases} -4x + 2 & \text{if } x \leq 1 \\ 3x - 5 & \text{if } x > 1 \end{cases}$

For $x \leq 1$, graph the line $y = -4x + 2$ using the two points (0, 2) and (1, -2). For $x > 1$, graph the line $y = 3x - 5$ using the two points (1.1, -1.7) and (4, 7).

25. $f(x) = \begin{cases} |x| & \text{if } x < 3 \\ 6 - x & \text{if } x \geq 3 \end{cases}$

For $x < 3$, graph $y = |x|$ using points from the following table.

x	-3	-2	-1	0	1	2	2.9
y	3	2	1	0	1	2	2.9

For $x \geq 3$, graph the line $y = 6 - x$ using the two points (3, 3) and (6, 0).

27. Let x represent days.
(a) For x in (0, 1], y = 7.
For x in (1, 2], y = 11.
For x in (2, 3], y = 15.
For x in (3, 4], y = 19.

(b) Domain is $(0, \infty)$
Range is {7, 11, 15, 19, . . . }

29. $y = 4x + 3$
The slope is m = 4.
Use the points (0, 3) and (-2, -5) to graph the line.

31. $3x - 5y = 15$
$-5y = -3x + 15$
$y = \frac{3}{5}x - 3$

The slope is $\frac{3}{5}$.

If y = 0, x = 5. The x-intercept is 5.
If x = 0, y = -3. The y-intercept is -3.

33. $x + 2 = 0$
$x = -2$
The graph of x = -2 is a vertical line, which has undefined slope.

35. $y = 2x$
The slope is m = 2.
If x = 0, y = 0. Both intercepts are zero.
If x = 3, y = 6. Use (0, 0) and (3, 6) to graph the line.

37. Through (2, -4), $m = \frac{3}{4}$

$m = \frac{\Delta y}{\Delta x} = \frac{3}{4}$. Use $\Delta y = 3$ and $\Delta x = 4$ to graph this line.

$y - (-4) = \frac{3}{4}(x - 2)$

$y + 4 = \frac{3}{4}x - \frac{3}{2}$

Multiplying by 4 gives

$4y + 16 = 3x - 6$

$4y = 3x - 22$

$3x - 4y = 22.$

39. Through (-4, 1), m = 3

$m = \frac{\Delta y}{\Delta x} = 3 = \frac{3}{1}$.

Use $\Delta y = 3$ and $\Delta x = 1$ to graph this line.

$y - 1 = 3[x - (-4)]$

$y - 1 = 3x + 12$

$y = 3x + 13$

41. Supply: $p = 6x + 3$;
demand: $p = 19 - 2x$

(a) $p = 10$

Supply: $10 = 6x + 3$

$7 = 6x$

$x = \frac{7}{6}$

Demand: $10 = 19 - 2x$

$2x = 9$

$x = \frac{9}{2}$

(b) $p = 15$

Supply: $15 = 6x + 3$

$6x = 12$

$x = 2$

Demand: $15 = 19 - 2x$

$2x = 4$

$x = 2$

(c) $p = 18$

Supply: $18 = 6x + 3$

$6x = 15$

$x = \frac{15}{6} = \frac{5}{2}$

Demand: $18 = 19 - 2x$

$2x = 1$

$x = \frac{1}{2}$

(d) Graph $p = 6x + 3$ using the points (0, 3) and (2, 15).
Graph $p = 19 - 2x$ using the points (0, 19) and (5, 9).

(e) From the graphs, the two graphs intersect at (2, 15). The equilibrium price is 15.

(f) The equilibrium supply and the equilibrium demand are 2.

43. (-2, 5) and (4, 7)

$m = \frac{7 - 5}{4 - (-2)} = \frac{2}{6} = \frac{1}{3}$

45. (0, 0) and (11, -2)

$m = \frac{-2 - 0}{11 - 0} = \frac{-2}{11} = -\frac{2}{11}$

47. $2x + 3y = 15$

$3y = -2x + 15$

$y = -\frac{2}{3}x + 5$

$m = -\frac{2}{3}$

49. $x + 4 = 9$

$x = 5$

The graph is a vertical line. It has undefined slope.

51. $y = x$
$y = 1x + 0$
$m = 1$

53. $x + 5y = 0$
$5y = -x$
$y = -\frac{1}{5}x + 0$
$m = -\frac{1}{5}$

55. Through (5, -1), slope $\frac{2}{3}$
$y - (-1) = \frac{2}{3}(x - 5)$
$y + 1 = \frac{2}{3}x - \frac{10}{3}$
Multiplying by 3 gives
$3y + 3 = 2x - 10$
$3y = 2x - 13.$

57. Through (5, -2) and (1, 3)
$m = \frac{3 - (-2)}{1 - 5} = \frac{5}{-4} = -\frac{5}{4}$
$y - 3 = -\frac{5}{4}(x - 1)$
$y - 3 = -\frac{5}{4}x + \frac{5}{4}$
Multiplying by 4 gives
$4y - 12 = -5x + 5$
$4y = -5x + 17$
$5x + 4y = 17.$

59. Undefined slope, through (-1, 4)
This is a vertical line. Its equation is $x = -1$.

61. x-intercept -3, y-intercept 5
Use the points (-3, 0) and (0, 5).
$m = \frac{5 - 0}{0 - (-3)} = \frac{5}{3}$
$y = \frac{5}{3}x + 5$
Multiplying by 3 gives
$3y = 5x + 15.$

63. $C(x) = mx + b$
$b = 60$ and $C(8) = 300$
$C(x) = mx + 60$
$300 = m(8) + 60$
$8m = 240$
$m = 30$
$C(x) = 30x + 60$

65. $C(x) = mx + b$
Use the two points (12, 445) and (50, 1585).
$m = \frac{1585 - 445}{50 - 12} = \frac{1140}{38} = 30$
$y - 1585 = 30(x - 50)$
$y - 1585 = 30x - 1500$
$y = 30x + 85$
$C(x) = 30x + 85$

67. $C(x) = mx + b$
Use the two points (80, 2735) and (175, 5775).
$m = \frac{5775 - 2735}{175 - 80} = \frac{3040}{95} = 32$
$y - 2735 = 32(x - 80)$
$y - 2735 = 32x - 2560$
$y = 32x + 175$
$C(x) = 32x + 175$

69. $C(x) = 20x + 100$
and $R(x) = 40x$
(a) $C(x) = R(x)$
$20x + 100 = 40x$
$100 = 20x$
$x = 5$ units

(b) $R(5) = 40(5)$
$R(5) = \$200$

Case 2 Exercises

1. $.133x + 10.09 = 10.73$

$.133x = .64$

$x \approx 4.8$ million units

2. $y = .0667x + 10.29$

$= .0667(3.1) + 10.29$

$= 10.50$

$y = .0667(5.7) + 10.29$

$= 10.67$

The graph is the portion of the line through (3.1, 10.50) and (5.7, 10.67).

3. $.0667x + 10.29 = 9.65$

For x in the interval 3.1 to 5.7, the marginal cost exceeds the selling price.

4. **(a)** $y = .133x + 9.46$

$= .133(3.1) + 9.46$

$= 9.87$

$y = .133(5.7) + 9.46$

$= 10.22$

(b) The graph is the line through (3.1, 9.87) and (5.7, 10.22).

(c) $9.57 = .133x + 9.46$

$x = .83$

CHAPTER 3 POLYNOMIAL AND RATIONAL FUNCTIONS

Section 3.1

1.(a) $f(x) = 2x^2$

Write the function in the form

$y = 2x^2$

and make a table of values.

x	-3	-2	-1	0	1	2	3
y	18	8	2	0	2	8	18

Plot these points to obtain the graph of the function.

(b) $f(x) = 3x^2$

$f(x) = y = 3x^2$

x	-3	-2	-1	0	1	2	3
y	27	12	3	0	3	12	27

(c) $f(x) = \frac{1}{2}x^2$

$f(x) = y = \frac{1}{2}x^2$

x	-3	-2	-1	0	1	2	3
y	$\frac{9}{2}$	2	$\frac{1}{2}$	0	$\frac{1}{2}$	2	$\frac{9}{2}$

(d) $f(x) = \frac{1}{3}x^2$

$f(x) = y = \frac{1}{3}x^2$

x	-3	-2	-1	0	1	2	3
y	3	$\frac{4}{3}$	$\frac{1}{3}$	0	$\frac{1}{3}$	$\frac{4}{3}$	3

(e) The coefficient affects the width of the parabola making it narrower or wider than the graph of $y = x^2$. A coefficient whose absolute value is greater than 1 makes it narrower than the graph of $y = x^2$ and a coefficient whose absolute value is less than 1 makes it wider.

3.(a) $f(x) = x^2 + 2$

$f(x) = y = x^2 + 2$

x	-3	-2	-1	0	1	2	3
y	11	6	3	2	3	6	11

(b) $f(x) = x^2 - 1$

$f(x) = y = x^2 - 1$

x	-3	-2	-1	0	1	2	3
y	8	3	0	-1	0	3	8

(c) $f(x) = x^2 + 1$

$f(x) = y = x^2 + 1$

x	-3	-2	-1	0	1	2	3
y	10	5	2	1	2	5	10

(d) $f(x) = x^2 - 2$

$f(x) = y = x^2 - 2$

x	-3	-2	-1	0	1	2	3
y	7	2	-1	-2	-1	2	7

(e) These graph have the same shape as the graph of $f(x) = x^2$ only they are shifted upward or downward.

5. $f(x) = -x^2$

First, locate the vertex of the parabola and find the axis of the parabola

$y = -x^2$

$y = -1(x - 0)^2 + 0.$

The vertex is at the point with coordinates (0, 0). The axis is the line with equation $x = 0$.

Make a table of values to find points on one side of the axis and then use the axis of symmetry to find the corresponding points on the other side.

x	1	2	3
y	-1	-4	-9

Then plot these points.

7. $f(x) = -x^2 + 1$
$y = -x^2 + 1$
$y = -1(x - 0)^2 + 1$
The vertex is at the point with coordinates (0, 1). The axis is the line with equation x = 0.

x	1	2	3
y	0	-3	-8

9. $f(x) = 3x^2 - 2$
$y = 3x^2 - 2$
$y = 3(x - 0)^2 - 2$
The vertex is at the point with coordinates (0, -2). The axis is the line with equation x = 0.

x	1	2	3
y	1	10	25

11. $f(x) = (x + 2)^2$
$y = (x + 2)^2$
$y = 1(x + 2)^2 + 0$
The vertex is at the point with coordinates (-2, 0). The axis is the line with equation x = -2.

x	-1	0	1
y	1	4	9

13. $f(x) = -(x - 4)^2$
$y = -(x - 4)^2$
$y = -1(x - 4)^2 + 0$
The vertex is at the point with coordinates (4, 0). The axis is the line with equation x = 4.

x	5	6	7
y	-1	-4	-9

15. $f(x) = -2(x - 3)^2$
$y = -2(x - 3)^2$
$y = -2(x - 3)^2 + 0$
The vertex is at the point with coordinates (3, 0). The axis is the line with equation x = 3.

x	4	5	6
y	-2	-8	-18

17. $f(x) = (x - 1)^2 - 3$
$y = (x - 1)^2 - 3$
$y = 1(x - 1)^2 - 3$
The vertex is at the point with coordinates (1, -3). The axis is the line with equation x = 1.

x	2	3	4
y	-1	1	6

19. $f(x) = -(x + 4)^2 + 2$
$y = -(x + 4)^2 + 2$
$y = -1(x + 4) + 2$
The vertex is at the point with coordinates (-4, 2). The axis is the line with equation x = -4.

x	-3	-2	-1
y	1	-2	-7

21. $f(x) = x^2 - 4x + 6$
$y = x^2 - 4x + 6$
$y = 1(x^2 - 4x) + 6$
$y = 1(x^2 - 4x + 4) + 6 - 4$
$y = 1(x - 2)^2 + 2$
The vertex is at the point with coordinates (2, 2). The axis is the line with equation x = 2.

x	3	4	5
y	3	6	11

23. $f(x) = x^2 + 12x + 1$

$y = x^2 + 12x + 1$

$y = 1(x^2 + 12x) + 1$

$y = 1(x^2 + 12x + 36) + 1 - 36$

$y = 1(x + 6)^2 - 35$

The vertex is at the point with coordinates (-6, -35). The axis is the line with equation x = -6.

x	-5	-4	-3
y	-34	-31	-26

25. $f(x) = 2x^2 + 4x + 1$

$y = 2x^2 + 4x + 1$

$y = (2x^2 + 4x) + 1$

$y = 2(x^2 + 2x) + 1$

$y = 2(x^2 + 2x + 1) + 1 - 2$

$y = 2(x + 1)^2 - 1$

The vertex is at the point with coordinates (-1, -1). The axis is the line with equation x = -1.

x	0	1	2
y	1	7	17

27. $f(x) = 2x^2 - 4x + 5$

$y = 2x^2 - 4x + 5$

$y = 2(x^2 - 2x) + 5$

$y = 2(x^2 - 2x + 1) + 5 - 2$

$y = 2(x - 1)^2 + 3$

The vertex is at the point with coordinates (1, 3). The axis is the line with equation x = 1.

x	2	3	4
y	5	11	21

29. $f(x) = -x^2 + 6x - 6$

$y = -x^2 + 6x - 6$

$y = -1(x^2 - 6x + 9) - 6 + 9$

$y = -1(x - 3)^2 + 3$

The vertex is at the point with coordinates (3, 3). The axis is the line with equation x = 3.

x	4	5	6
y	2	-1	-6

31. $f(x) = x^2 - 2x - 15$

$y = x^2 - 2x - 15$

(a) To find x-intercepts we let y = 0 and solve for x.

$0 = x^2 - 2x - 15$

$0 = (x - 5)(x + 3)$

$x = 5$ or $x = -3$

(b) If x = 0, y = -15.

(c) The original equation is in the general form $f(x) = ax^2 + bx + c$ with a = 1 and b = -2. Use h = -b/2a to find the value of x at the vertex.

$$\frac{-(-2)}{2(1)} = 1$$

(d) The minimum value of y is the y-coordinate of the vertex, which is f(1).

$f(1) = 1^2 - 2(1) - 15$

$f(1) = -16$

See the graphs for Exercises 33 and 35 in the answer section at the back of the textbook.

33. $f(x) = .14x^2 + .56x - .3$

$y = .14x^2 + .56x - .3$

$y = .14(x^2 + 4x) - .3$

$y = .14(x^2 + 4x + 4) - .3 - .56$

$y = .14(x + 2)^2 - .86$

The vertex is at the point with coordinates (-2, -.86). The axis is the line with equation x = -2.

x	-1	0	1	2
y	-.72	-.30	.40	1.38

35. $f(x) = -.09x^2 - 1.8x + .5$

$y = -.09x^2 - 1.8x + .5$

$y = -.09(x^2 + 20x) + .5$

$y = -.09(x^2 + 20x + 100) + .5 + 9$

$y = -.09(x + 10)^2 + 9.5$

The vertex is at the point with coordinates (-10, 9.5). The axis is the line with equation $x = -10$.

x	-9	-8	-4	-2	0
y	9.41	9.14	6.26	3.74	.5

Section 3.2

1.**(a)** $C(x) = x^2 - 10 + 40$

$= (x^2 - 10 + 25) + 40 - 25$

$C(x) = (x - 5)^2 + 15$

(b) For $y = (x - 5)^2 + 15$ the vertex is at (5, 15) and the axis is $x = 5$. See the graph in the answer section at the back of the textbook.

(c) From the vertex, $x = 5$. Thus, 5 units must be sold to produce minimum cost.

(d) From the vertex, $y = 15$. The minimum cost is \$15.

3. $p = -x^2 + 6x - 1$

$p = -1(x^2 - 6x) - 1$

$p = -1(x^2 - 6x + 9) - 1 + 9$

$p = -1(x - 3)^2 + 8$

The vertex is at (3, 8).

(a) From the vertex, $x = 3$. Thus, the price per bag that gives maximum profit is 3 dimes or 30¢.

(b) From the vertex, $p = 8$. The maximum profit is $p = 8$ or \$800.

5. $h = 32t - 16t^2$

$h = -16(t^2 + 2t)$

$h = -16(t^2 + 2t + 1) + 16$

$h = -16(t - 1)^2 + 16$

The vertex is at (1, 16). From the vertex, the maximum height is $h = 16$ or 16 feet. When the object hits the ground, $h = 0$.

$0 = 32t - 16t^2$

$0 = 16t(2 - t)$

$t = 0$ or $t = 2$

The object hits the ground in 2 seconds.

7. Let x be the width of the field. Since one side of the field is bordered by a building, the length is $320 - 2x$. The area is given by

$A(x) = (320 - 2x)x$

$= -2x^2 + 320x$

$= -2(x^2 - 160x)$

$= -2(x^2 - 160x + 6400) + 12{,}800$

$= -2(x - 80)^2 + 12{,}800$

The maximum area occurs when $x = 80$. If $x = 80$, $320 - 2x$ is 160. The dimensions are 80 feet (width) by 160 feet (length).

9. $p = 640 - 5x^2$

(a) If $x = 0$, $p = 640$.

(b) If $x = 5$, $p = 640 - 5 \cdot 5^2$

$= 640 - 125 = 515$

(c) If $x = 10$, $p = 640 - 5 \cdot 10^2$

$= 640 - 500 = 140$

(d) $p = -5x^2 + 640$

$p = -5(x + 0)^2 + 640$

The vertex is at (0, 640) and the axis is $x = 0$. See the graph in the answer section at the back of the textbook.

(e) $p = 5x^2$

$p = 5(x + 0)^2 + 0$

The vertex is at (0, 0) and the axis is $x = 0$. See the graph in the answer section at the back of the textbook.

(f)
$$640 - 5x^2 = 5x^2$$
$$640 = 10x^2$$
$$x^2 = 64$$
$$x = \pm 8$$

The equilibrium supply is $x = 8$ or 800 units.

(g) The equilibrium price is $p = 5 \cdot 8^2 = 5 \cdot 64 = 320$

The equilibrium price is 320.

11.
$$640 - 5x^2 = 5x^2$$
$$640 = 10x^2$$
$$x^2 = 64$$
$$x = \pm 8$$

The equilibrium demand is $x = 8$ or 800 units.

13.(a) If R is the revenue function,

$$R(x) = xp = x\left(150 - \frac{x}{4}\right) = 150x - \frac{1}{4}x^2.$$

(b)
$$R(x) = -\frac{1}{4}x^2 + 150x$$
$$= -\frac{1}{4}(x^2 - 600x)$$
$$= -\frac{1}{4}(x^2 - 600x + 90{,}000) + 22{,}500$$
$$= -\frac{1}{4}(x - 300)^2 + 22{,}500$$

The vertex is at (300, 22,500).

From the vertex, the number of bicycle sales that leads to maximum revenue is $x = 300$.

(c) From the vertex, the maximum revenue is $22,500.

15.(a) If R is the revenue function,

$$R(x) = (500 - x)x$$
$$= -x^2 + 500x$$
$$= -1(x^2 - 500x)$$
$$= -1(x^2 - 500x + 62{,}500) + 62{,}500$$
$$= -1(x - 250)^2 + 62{,}500$$

(b) The vertex is at (250, 62,500), and the axis is $x = 250$. See the graph in the answer section at the back of the textbook.

(c) The maximum revenue occurs when $x = 250$. Since $p = 500 - x$, if $x = 250$, $p = 250$.

(d) From the vertex, the maximum revenue is $62,500.

17.(a) The total number of seats is 100. If x is the number of unsold seats, the number of sold seats is $100 - x$.

(b) Since each unsold seat increases the $50 cost by $1, the price per seat is $50 + 1 \cdot x$ or, simply, $50 + x$.

(c) If R is the revenue, then R = (number of sold seats)(price per seat).

Thus,

$$R(x) = (100 - x) \cdot (50 + x).$$

(d) $R(x) = 5000 + 50x - x^2$

$= -1(x^2 - 50x) + 5000$

$= -1(x^2 - 50x + 625) + 5000 + 625$

$= -1(x - 25)^2 + 5625$

The vertex is at (25, 5625). From the vertex, x = 25. Thus, 25 unsold seats will maximize revenue.

(e) From the vertex, the maximum revenue is $5625.

19. For June (x = 1) through October (x = 5), use $C(x) = 10x + 50$. For November (x = 6) and December (x = 7), use $C(x) = -20(x - 5)^2 + 100$. Note that both rules give the same value for October (x = 5). Thus,

$$C(x) = \begin{cases} 10x + 50 \text{ if } x = 1, 2, 3, 4, 5 \\ -20(x - 5)^2 + 100 \text{ if } x = 6,7 \end{cases}$$

(a) $C(1) = 10(1) + 50 = 60$
(b) $C(2) = 10(2) + 50 = 70$
(c) $C(4) = 10(4) + 50 = 90$
(d) $C(5) = 10(5) + 50 = 100$
(e) $C(6) = -20(6 - 5)^2 + 100 = 80$
(f) $C(7) = -20(7 - 5)^2 + 100 = 20$

21. Let $y = ax^2$

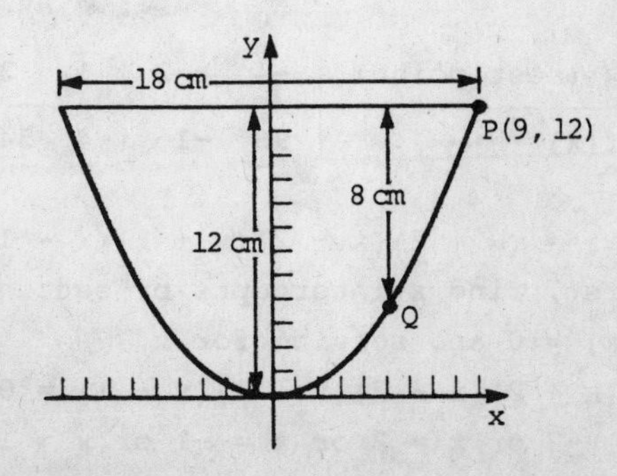

The point P(9, 12) is on the parabola, so

$12 = a \cdot 9^2$

$a = \frac{12}{81} = \frac{4}{27}$.

Thus, $y = \frac{4}{27}x^2$

At 8 centimeters from the top, y = 4. At point Q, y = 4. To find x, use

$4 = \frac{4}{27}x^2$

$x^2 = 27$

$x = \pm 3\sqrt{3}$

The width of the culvert will be twice the x-coordinate of Q.

$2(3\sqrt{3}) = 6\sqrt{3}$.

The width is $6\sqrt{3}$ centimeters.

23. $R(x) = 60x - 2x^2$ and $C(x) = 20x + 80$

(a) $60x - 2x^2 = 20x + 80$

$0 = 2x^2 - 40x + 80$

$0 = x^2 - 20x + 40$

$$x = \frac{20 \pm \sqrt{(-20)^2 - 4(1)(40)}}{2(1)}$$

$$x = \frac{20 \pm \sqrt{240}}{2}$$

$$x = \frac{20 \pm 4\sqrt{15}}{2}$$

$x = 10 \pm 2\sqrt{15}$

$x \approx 17.7$ or $x \approx 2.3$

(b) $R(x) = -2(x^2 - 30x + 225) + 450$

$R(x) = -2(x - 15)^2 + 450$

The vertex is at (15, 450). Revenue is maximum when x = 15.

(c) $P(x) = R(x) - C(x)$

$= (60x - 2x^2) - (20x + 80)$

$= -2x^2 + 40x - 80$

$= -2(x^2 - 20x + 100) - 80 + 200$

$= -2(x - 10)^2 + 120$

The vertex is at (10, 120). Profit is maximum when x = 10.

(d) From the vertex, the maximum profit is 120.

(e) Since the break-even points are 2.3 and 17.7, a loss will occur if $0 < x < 2.3$ and if $x > 17.7$.

(f) A profit will occur for $2.3 < x < 17.7$.

Section 3.3

See the graphs for Exercises 1 - 25 in the answer section at the back of the textbook.

1. $f(x) = x^3 + 2$

Make a table of values and plot the points.

x	-3	-2	-1	0	1	2	3
y	-25	-6	1	2	3	10	29

3. $f(x) = x^4$

x	-3	-2	-1	0	1	2	3
y	81	16	1	0	1	16	81

5. $f(x) = 2x(x - 3)(x + 2)$

First, find x-intercepts by setting $f(x) = 0$ and solving for x.

$2x(x - 3)(x + 2) = 0$

$x = 0$ or $x = 3$ or $x = -2$

These three numbers divide the x-axis into four regions.

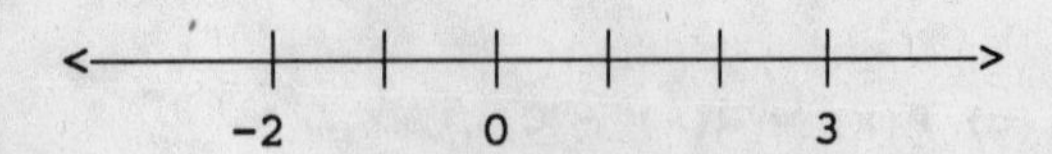

In each region the value of f is either positive or negative. Choose a point in each region as a test point and compute the function value.

x (test point)	-3	-1	1	4
f(x)	-36	8	-12	48

Using these, sketch the graph.

7. $f(x) = (x + 2)(x - 3)(x + 4)$

First, find x-intercepts by setting $f(x) = 0$ and solving for x.

$(x + 2)(x - 3)(x + 4) = 0$

$x = -2$ or $x = 3$ or $x = -4$

These three numbers divide the x-axis into four regions.

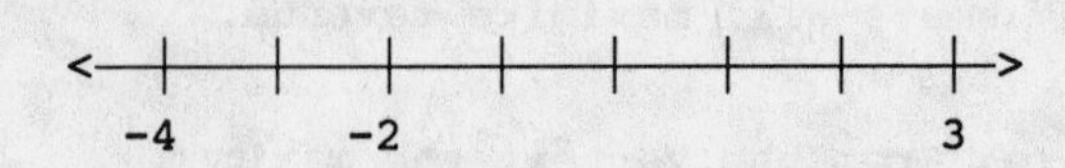

Choose a point in each region as a test point and compute the function value.

x (test point)	-5	-3	0	4
f(x)	-24	6	-24	48

9. $f(x) = x^2(x - 2)(x + 3)$

First, find x-intercepts by setting f(x) and solving for x.

$x^2(x - 2)(x + 3) = 0$

$x = 0$ or $x = 2$ or $x = -3$

These three numbers divide the x-axis into four regions.

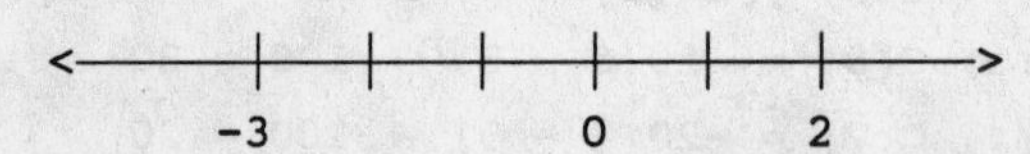

Choose a point in each region as a test point and compute the function value.

x (test point)	-4	-2	1	3
f(x)	96	-16	-4	54

11. $f(x) = (x + 2)(x - 2)(x + 1)(x - 1)$

First, find x-intercepts by setting $f(x) = 0$ and solving for x.

$(x + 2)(x - 2)(x + 1)(x - 1) = 0$

$x = -2$ or $x = 2$ or $x = -1$ or $x = 1$

These four numbers divide the x-axis into five regions.

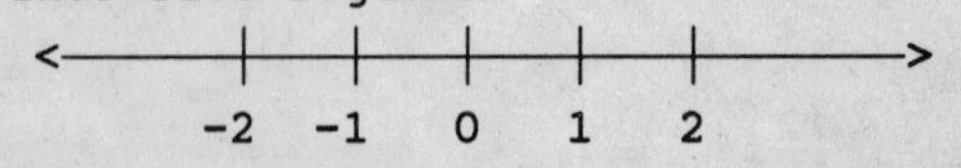

Choose a point in each region as a test point and compute the function value.

x (testpoint)	-3	$-\frac{3}{2}$	0	$\frac{3}{2}$	3
f(x)	40	$-\frac{35}{16}$	4	$-\frac{35}{16}$	40

13. $f(x) = x^3 - 7x^2 + 6x$

$f(x) = x(x^2 - 7x + 6)$

$f(x) = x(x - 1)(x - 6)$

First, find x-intercepts by setting $f(x) = 0$ and solving for x.

$x(x - 1)(x - 6) = 0$

$x = 0$ or $x = 1$ or $x = 6$

These three numbers divide the x-axis into four regions.

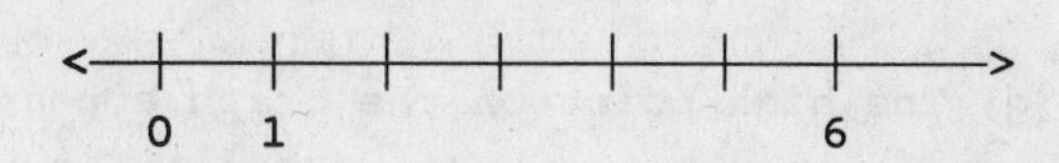

Choose a point in each region as a test point and compute the function value.

x (testpoint)	-2	$\frac{1}{2}$	4	7
f(x)	-48	$\frac{11}{8}$	-24	4

15. $f(x) = x^3 - 2x^2 - 8x$

$f(x) = x(x^2 - 2x - 8)$

$f(x) = x(x - 4)(x + 2)$

First, find x-intercepts by setting $f(x) = 0$ and solving for x.

$x(x - 4)(x + 2) = 0$

$x = 0$ or $x = 4$ or $x = -2$

These three numbers divide the x-axis into four regions.

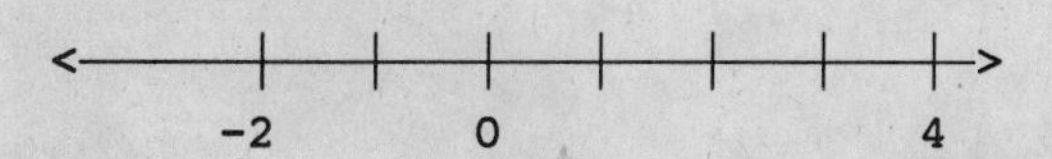

Choose a point in each region as a test point and compute the function value.

x (test point)	-3	-1	2	5
f(x)	-21	5	-16	35

17. $f(x) = x^4 - 5x^2$

$f(x) = x^2(x^2 - 5)$

First, find x-intercepts by setting $f(x) = 0$ and solving for x.

$x^2(x^2 - 5) = 0$

$x^2 = 0$ or $x^2 = 5$

$x = 0$ or $x = \pm\sqrt{5} \approx \pm 2.2$

These three numbers divide the x-axis into four regions.

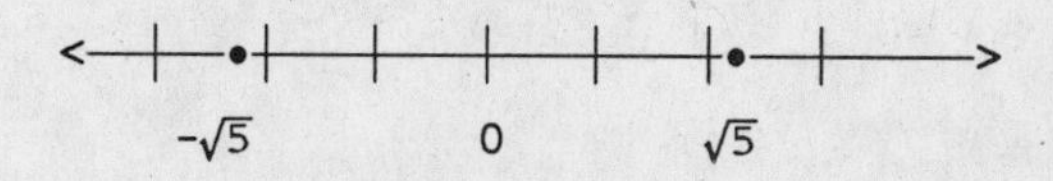

Choose a point in each region as a test point and compute the function value.

x (test point)	-3	-1	1	3
f(x)	36	-4	-4	36

19. $f(x) = x^4 - 7x^2 + 12$

$f(x) = (x^2 - 4)(x^2 - 3)$

$f(x) = (x + 2)(x - 2)(x^2 - 3)$

First, find x-intercepts by setting $f(x) = 0$ and solving for x.

$(x + 2)(x - 2)(x^2 - 3) = 0$

$x = -2$ or $x = 2$ or $x = \pm\sqrt{3} \approx \pm 1.7$

These four numbers divide the x-axis into five regions.

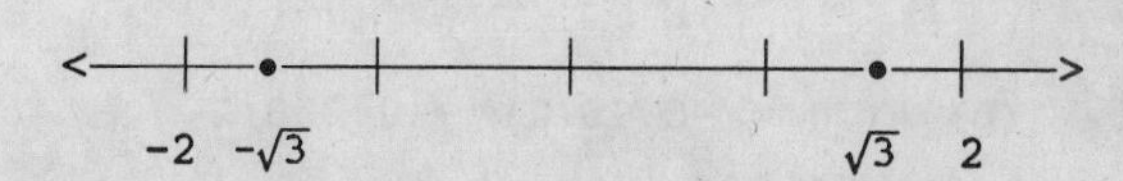

Choose a point in each region as a test point and compute the function value.

x (test point)	-3	-1.9	0	1.9	3
f(x)	30	-.2	12	-.2	30

21. $f(x) = 8x^4 - 2x^3 - x^2$

$f(x) = x^2(8x^2 - 2x - 1)$

$f(x) = x^2(4x + 1)(2x - 1)$

First, find x-intercepts by setting $f(x) = 0$ and solving for x.

$x^2(4x + 1)(2x - 1) = 0$

$x^2 = 0$ or $4x + 1 = 0$ or $2x - 1 = 0$

$x = 0$ or $x = -\frac{1}{4}$ or $x = \frac{1}{2}$

These three numbers divide the x-axis into four regions.

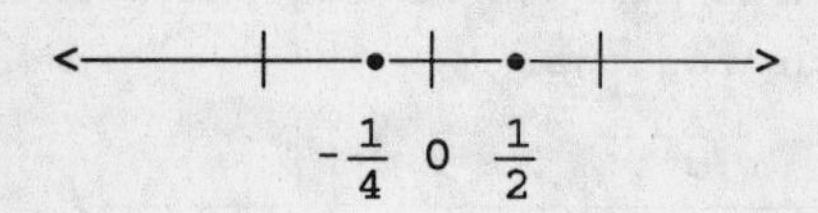

Choose a point in each region as a test point and compute the function value.

x (testpoint)	-1	$-\frac{1}{8}$	$\frac{1}{4}$	1
f(x)	9	$-\frac{5}{512}$	$-\frac{1}{16}$	5

Notice that in both intervals (-1/4, 0) and (0, 1/2), f is negative.

23. $A(x) = -.015x^3 + 1.058x$

(a) $A(1) = -.015(1)^3 + 1.058(1)$

$A(1) = 1.043$

Approximately 1.0 tenths of a percent or .1%

(b) $A(2) = -0.15(2)^3 + 1.058(2)$

$A(2) = 1.996$

Approximately 2.0 tenths of a percent or .2%

(c) $A(4) = -.015(4)^3 + 1.058(4)$

$A(4) = 3.272$

Approximately 3.3 tenths of a percent or .33%

(d) $A(6) = -.015(6)^3 + 1.058(6)$

$A(6) = 3.108$

Approximately 3.1 tenths of a percent or .31%

(e) $A(8) = -.015(8)^3 + 1.058(8)$

$A(8) = 0.784$

Approximately .8 tenths of a percent or .08%

(f) Notice that $A(0) = 0$. Using this value and the values from Parts (a) through (e), plot points and graph the function.

(g) The high point on the graph appears to occur between 4 and 5, closer to 5. The time of maximum alcohol concentration occurs a little under 5 hours.

(h) The points on the graph with y-coordinates at or above 1.5 appear between the x-values of 1.5 and 7.5. The average person is legally drunk between 1.5 hour and 7.5 hours.

25. $P(t) = t^3 - 18t^2 + 81t$

$P(t) = t(t^2 - 18t + 81)$

$P(t) = t(t - 9)^2$

Since t is time in years since the date of the first reading, $t \geq 0$.

(a) $P(0) = 0(0 - 9)^2 = 0$

$P(3) = 3(3 - 9)^2 = 108$

$P(7) = 7(7 - 9)^2 = 28$

$P(10) = 10(10 - 9)^2 = 10$

(b) Notice that $P(9) = 0$. Using this value and the values from Part (a), plot points and graph the function for $t \geq 0$.

(c) From the graph, the pressure is increasing on the interval (0, 3) and from 9 on and decreasing on the interval (3, 4).

27. $f(x) = x^3 + 4x^2 - 8x - 8$
Using a calculator, compute the following function values.
$f(.3) = -10.013$
$f(.4) = -10.496$
$f(.5) = -10.875$
$f(.6) = -11.144$
$f(.7) = -11.297$
$f(.8) = -11.328$
$f(.9) = -11.231$
$f(1.0) = -11$
The maximum value is -10.013 when $x = 3$. The minimum value is -11.328 when $x = .8$.

29. $f(x) = x^4 - 7x^3 + 13x^2 + 6x - 28$
Using a calculator, compute the following function values.
$f(-2) = 84$
$f(-1.9) = 68.5751$
$f(-1.8) = 54.6416$
$f(-1.7) = 42.1131$
$f(-1.6) = 30.9056$
$f(-1.5) = 20.9375$
$f(-1.4) = 12.1296$
$f(-1.3) = 4.4051$
$f(-1.2) = -2.3014$
$f(-1.1) = -8.0889$
$f(-1) = -13$
The maximum value is 84 when $x = -2$. The minimum value is -13 when $x = -1$.

See the graphs for Exercises 31 and 33 in the answer section in the back of the textbook.

31. $f(x) = x^3 + 3x^2 - 2x + 1$
for $-3 \leq x \leq 1.5$

x	-3	-2.5	-2	-1.5	-1	-.5	0	.5
y	7	9.125	9	7.375	5	2.625	1	.875

33. $f(x) = -x^4 + 2x^3 - 3x^2 + 4x - 7$
for $-1 \leq x \leq 2$

x	-1	-.5	0	.5	1	1.5
y	-17	-10.0625	-7	-5.5625	-5	-6.0625

Section 3.4

See the graphs for Exercises 1-25 and 29-35 in the answer section at the back of the textbook.

1. $f(x) = \frac{1}{x + 2}$

First, find the asymptotes. Find the vertical asymptote by setting the denominator equal to zero and solving for x.
$x + 2 = 0$
$x = -2$
The line $x = -2$ is a vertical asymptote. The horizontal asymptote of

$y = \frac{ax + b}{cx + d}$ is $y = \frac{a}{c}$.

So the horizontal asymptote of the function in the form

$y = \frac{0x + 1}{1x + 2}$ is $y = \frac{0}{1}$ or $y = 0$.

If $x = 0$, $y = \frac{1}{2}$.

If $y = 0$, there is no solution for x.

x	-5	-4	-3	-1	1	2
y	$-\frac{1}{3}$	$-\frac{1}{2}$	-1	1	$\frac{1}{3}$	$\frac{1}{4}$

3. $f(x) = \frac{-4}{x - 3}$

Find the vertical asymptote by setting the denominator equal to zero and solving for x.

$x - 3 = 0$

$x = 3$

The line $x = 3$ is a vertical asymptote. The horizontal asymptote of the function in the form

$y = \frac{0x - 4}{1x - 3}$ is

$y = \frac{0}{1}$ or $y = 0$.

if $x = 0$, $y = \frac{4}{3}$.

If $y = 0$, there is no solution for x.

x	-1	1	2	4	5	6
y	1	2	4	-4	-2	$-\frac{4}{3}$

5. $f(x) = \frac{2}{x}$

Find the vertical asymptote by setting the denominator equal to zero and solving for x.

$x = 0$

The line $x = 0$ is a vertical asymptote. The horizontal asymptote of the function in the form

$y = \frac{0x + 2}{1x + 0}$ is

$y = \frac{0}{1}$ or $y = 0$.

If $x = 0$, there is no value for y.

If $y = 0$, there is no solution for x.

x	-3	-2	-1	1	2	3
y	$-\frac{2}{3}$	-1	-2	2	1	$\frac{2}{3}$

7. $f(x) = \frac{2}{3 + 2x}$

Find the vertical asymptote by setting the denominator equal to zero and solving for x.

$3 + 2x = 0$

$x = -\frac{3}{2}$

The line $x = -\frac{3}{2}$ is a vertical asymptote. The horizontal asymptote of the function in the form

$y = \frac{0x + 2}{2x + 3}$ is

$y = \frac{0}{2}$ or $y = 0$.

If $x = 0$, $y = \frac{2}{3}$.

If $y = 0$, there is no solution for x.

x	-4	-3	-2	-1	1	2
y	$-\frac{2}{5}$	$-\frac{2}{3}$	-2	2	$\frac{2}{5}$	$\frac{2}{7}$

9. $f(x) = \frac{3x}{x - 1}$

Find the vertical asymptote by setting the denominator equal to zero and solving for x.

$x - 1 = 0$

$x = 1$

The line $x = 1$ is a vertical asymptote. The horizontal asymptote of the function in the form

$y = \frac{3x + 0}{1x - 1}$ is

$y = \frac{3}{1}$ or $y = 3$.

If $x = 0$, $y = 0$.

If $y = 0$, $x = 0$.

x	-3	-2	-1	2	3	4
y	$\frac{9}{4}$	2	$\frac{3}{2}$	6	$\frac{9}{2}$	4

11. $f(x) = \dfrac{x}{x - 9}$

Find the vertical asymptote by setting the denominator equal to zero and solving for x.

$x - 9 = 0$

$x = 9$

The line $x = 9$ is a vertical asymptote. The horizontal asymptote of the function in the form

$y = \dfrac{1x + 0}{1x - 9}$ is

$y = \dfrac{1}{1}$ or $y = 1$.

If $x = 0$, $y = 0$.
If $y = 0$, $x = 0$.

x	6	7	8	10	11	12
y	-2	$-\frac{7}{2}$	-8	10	$\frac{11}{2}$	4

13. $f(x) = \dfrac{x + 1}{x - 4}$

Find the vertical asymptote by setting the denominator equal to zero and solving for x.

$x - 4 = 0$

$x = 4$

The line $x = 4$ is a vertical asymptote. The horizontal asymptote of the function in the form

$y = \dfrac{1x + 1}{1x - 4}$ is

$y = \dfrac{1}{1}$ or $y = 1$.

If $x = 0$, $y = -\frac{1}{4}$.

If $y = 0$, $x = -1$.

x	1	2	3	5	6	7
y	$-\frac{2}{3}$	$-\frac{3}{2}$	-4	6	$\frac{7}{2}$	$\frac{8}{3}$

15. $f(x) = \dfrac{2x - 1}{4x + 2}$

Find the vertical asymptote by setting the denominator equal to zero and solving for x.

$4x + 2 = 0$

$x = -\frac{1}{2}$

The line $x = -\frac{1}{2}$ is a vertical asymptote. The horizontal asymptote of the function in the form

$y = \dfrac{2x - 1}{4x + 2}$ is

$y = \dfrac{2}{4}$ or $y = \dfrac{1}{2}$.

If $x = 0$, $y = -\frac{1}{2}$.

If $y = 0$, $x = \frac{1}{2}$.

x	-3	-2	-1	1	2	3
y	$\frac{7}{10}$	$\frac{5}{6}$	$\frac{3}{2}$	$\frac{1}{6}$	$\frac{3}{10}$	$\frac{5}{14}$

17. $f(x) = \dfrac{1 - 2x}{5x + 20}$

Find the vertical asymptote by setting the denominator equal to zero and solving for x.

$5x + 20 = 0$

$x = -4$

The line $x = -4$ is a vertical asymptote. The horizontal asymptote of the function in the form

$y = \dfrac{-2x + 1}{5x + 20}$ is

$y = \dfrac{-2}{5}$ or $y = -\dfrac{2}{5}$.

If $x = 0$, $y = \dfrac{1}{20}$.

If $y = 0$, $x = \frac{1}{2}$.

x	-7	-6	-5	-3	-2	-1
y	-1	$-\frac{13}{10}$	$-\frac{11}{5}$	$\frac{7}{5}$	$\frac{1}{2}$	$\frac{1}{5}$

19. $f(x) = \frac{-x - 4}{3x + 6}$

Find the vertical asymptote by setting the denominator equal to zero and solving for x.

$3x + 6 = 0$

$x = -2$

The line $x = -2$ is a vertical asymptote. The horizontal asymptote of the function in the form

$y = \frac{-1x - 4}{3x + 6}$ is

$y = \frac{-1}{3}$ or $y = -\frac{1}{3}$.

If $x = 0$, $y = -\frac{2}{3}$.

If $y = 0$, $x = -4$.

x	-6	-5	-3	-1	1	2
y	$-\frac{1}{6}$	$-\frac{1}{9}$	$-\frac{1}{3}$	-1	$-\frac{5}{9}$	$-\frac{1}{2}$

21. $C(x) = \frac{500}{x + 30}$

(a) $C(10) = \frac{500}{10 + 30}$

$= \frac{50}{4}$ or \$12.50

(b) $C(20) = \frac{500}{20 + 30}$

$= 10$ or \$10.00

(c) $C(50) = \frac{500}{50 + 30}$

$= \frac{50}{8}$ or \$6.25

(d) $C(70) = \frac{500}{70 + 30}$

$= 5$ or \$5.00

(e) $C(100) = \frac{500}{100 + 30}$

$= \frac{50}{13}$ or \$3.85

(f) For $y = \frac{500}{x + 30}$, $x = -30$ is a vertical asymptote and $y = 0$ is a horizontal asymptote. Use the values from (a) through (e) and the asymptotes to graph the function.

23. $f(x) = \frac{6.7x}{100 - x}$

(a) $f(50) = \frac{6.7(50}{100 - 50}$

$= 6.7$ or \$6700

(b) $f(70) = \frac{6.7(70)}{100 - 70}$

$= 15.6\overline{3}$ or \$15,600

(c) $f(80) = \frac{6.7(80)}{100 - 80}$

$= 26.8$ or \$26,800

(d) $f(90) = \frac{6.7(90)}{100 - 90}$

$= 60.3$ or \$60,300

(e) $f(95) = \frac{6.7(95)}{100 - 95}$

$= 127.3$ or \$127,300

(f) $f(98) = \frac{6.7(98)}{100 - 98}$

$= 328.3$ or \$328,300

(g) $f(99) = \frac{6.7(99)}{100 - 99}$

$= 663.3$ or \$663,300

(h) For $y = \frac{6.7x}{100 - x}$ there is no y-value when $x = 100$ so it is not possible to remove all pollutants.

(i) Graph $y = \frac{6.7x}{100 - x}$ for $0 \le x < 100$.

. . .

Note that $x = 100\%$ is a vertical asymptote. Use the values from (a) through (g) and the vertical asymptote, to graph the function.

25. $W = \frac{3(3 - A)}{A}$

(a) If $A = 1$, $W = \frac{3(3 - 1)}{1}$

$= 6$ minutes.

(b) If $A = 2$, $W = \frac{3(3 - 2)}{2}$

$= \frac{3}{2}$ minutes

(c) If $A = 2.5$, $W = \frac{3(3 - 2.5)}{2.5}$

$= \frac{3}{5}$ minute which is 36 seconds.

(d) The vertical asymptote occurs when the denominator is zero or $A = 0$.

(e) Use the vertical asymptote and the values found in (a), (b), and (c) to graph the function for $0 < A \le 3$.

(f) If $A > 3$, the formula does not apply since there will be no waiting if people arrive more than 3 minutes apart.

27. $C(x) = \frac{10x}{49(101 - x)}$

(a) $C(99) = \frac{10(99)}{49(101 - 99)}$

≈ 10.1 or about \$10,100.

(b) $C(100) = \frac{10(100)}{49(101 - 100)}$

≈ 20.4 or about \$20,400.

29. $y = \frac{125{,}000 - 25x}{125 + 2x}$

If $x = 0$, $y = 1000$.
If $y = 0$, $x = 5000$.
$x = -62.5$ is a vertical asymptote and $y = \frac{-25}{2}$ or $y = -\frac{25}{2}$ is a horizontal asymptote. Use the intercepts and asymptotes to graph the quadrant I portion of the graph.
From the intercepts, the maximum value of y is 1000 or 100,000 gallons of oil and the maximum value of x is 5000 or 500,000 gallons of oil.

31. $f(x) = \frac{60x - 6000}{x - 120}$

for $50 \le x \le 100$

(a) $f(50) = \frac{60(50) - 6000}{50 - 120}$

≈ 42.9 or \$42.9 million

(b) $f(60) = \frac{60(60) - 6000}{60 - 120}$

$= 40$ or \$40 million

(c) $f(80) = \frac{60(80) - 6000}{80 - 120}$

$= 30$ or \$30 million

(d) $f(100) = \frac{60(100) - 6000}{100 - 120}$

$= 0$ or \$0 million

(e) For $y = \frac{60x - 6000}{x - 120}$, $x = 120$ is a vertical asymptote and $y = 60$ is a horizontal asymptote. Use the values from (a) through (d) and the asymptotes to graph the function for $50 \le x \le 100$.

33. $f(x) = \frac{-2x^2 + x - 1}{2x + 3}$

There is a vertical asymptote when $2x + 3 = 0$ or when $x = -\frac{3}{2}$. There is no horizontal asymptote.

If $x = 0$, $y = -\frac{1}{3}$.

If $y = 0$, $-2x^2 + x - 1 = 0$.

$2x^2 - x + 1 = 0$.

There are no real solutions so there is no x-intercept.

x	-4	-3	-2	-1	1	2	3	4
y	7.4	7.3	11	-4	-.4	-1	-1.7	-2.63

35. $f(x) = \frac{2x^2 - 5}{x^2 - 1}$

Find the vertical asymptote by setting the denominator equal to zero and solving for x.

$x^2 - 1 = 0$

$x = \pm 1$

The lines $x = 1$ and $x = -1$ are vertical asymptotes. The horizontal asymptote of the function in the form

$y = \frac{2x^2 - 5}{1x^2 - 1}$ is

$y = \frac{2}{1}$ or $y = 2$.

If $x = 0$, $y = 5$.

If $y = 0$, $x = \pm\sqrt{\frac{5}{2}}$

or $\pm\frac{\sqrt{10}}{2}$ (approximately $\pm$ 1.6).

x	-4	-3	-2	-.5	.5	2	3
y	1.8	1.625	1	6	6	1	1.625

Chapter 3 Review Exercises

See the graphs for Exercises 1 - 17 in the answer section at the back of the textbook.

1. $f(x) = x^2 - 4$

First, locate the vertex of the parabola and find the axis of the parabola.

$y = 1(x - 0)^2 - 4$

The vertex is at the point with coordinates (0, -4). The axis is the line with equation $x = 0$.

Make a table of values to find points on one side of the axis and then use the axis of symmetry to find the corresponding points on the other side.

x	1	2	3
y	-3	0	5

Then plot these points.

3. $f(x) = -(x - 1)^2$

$y = -(x - 1)^2$

$y = -1(x - 1)^2 + 0$

The vertex is at the point with coordinates (1, 0). The axis is the line with equation $x = 1$.

x	2	3	4
y	-1	-4	-9

5. $f(x) = 3(x + 1)^2 - 5$

The vertex is at the point with coordinates (-1, -5). The axis is the line with equation x = -1.

x	0	1	2
y	-2	7	22

7. $f(x) = -(x + 3)^2 - 2$

The vertex is at the point with coordinates (-3, -2). The axis is the line with equation x = -3.

x	-2	-1	0
y	-3	-6	-11

9. $f(x) = x^2 - 4x + 2$

$y = x^2 - 4x + 2$

$y = (x^2 - 4x + 4) + 2 - 4$

$y = 1(x - 2)^2 - 2$

The vertex is at the point with coordinates (2, -2). The axis is the line with equation x = 2.

x	3	4	5
y	-1	2	7

11. $f(x) = -x^2 + 6x - 3$

$y = -x^2 + 6x - 3$

$y = -1(x^2 - 6x) - 3$

$y = -1(x^2 - 6x + 9) - 3 + 9$

$y = -1(x - 3)^2 + 6$

The vertex is at the point with coordinates (3, 6). The axis is the line with equation x = 3.

x	4	5	6
y	5	2	-3

13. $f(x) = 4x^2 - 8x + 3$

$y = 4x^2 - 8x + 3$

$y = (4x^2 - 8x) + 3$

$y = 4(x^2 - 2x + 1) + 3 - 4$

$y = 4(x - 1)^2 - 1$

The vertex is at the point with coordinates (1, -1). The axis is the line with equation x = 1.

x	2	3	4
y	3	15	35

15. $f(x) = -3x^2 - 12x - 8$

$y = -3x^2 - 12x - 8$

$y = (-3x^2 - 12x) - 8$

$y = -3(x^2 + 4x + 4) - 8 + 12$

$y = -3(x + 2)^2 + 4$

The vertex is at the point with coordinates (-2, 4). The axis is the line with equation x = -2.

x	-1	0	1
y	1	-8	-23

17. $f(x) = -x^2 + 5x - 2$

$y = -x^2 + 5x - 2$

$y = (-x^2 + 5x) - 2$

$y = -1(x^2 - 5x + \frac{25}{4}) - 2 + \frac{25}{4}$

$y = -1(x - \frac{5}{2})^2 + \frac{17}{4}$

The vertex is at the point with coordinates $(\frac{5}{2}, \frac{17}{4})$. The axis is the line with equation $x = \frac{5}{2}$.

x	3	4	5
y	4	2	-2

19. $f(x) = x^2 - 4x + 1$

$y = x^2 - 4x + 1$

$y = (x^2 - 4x + 4) + 1 - 4$

$y = (x - 2)^2 - 3$

The vertex is at (2, -3) and the parabola opens upward so the minimum value is -3 when x = 2.

21. $f(x) = -3x^2 - 12x - 1$
$y = -3x^2 - 12x - 1$
$y = -3(x^2 + 4x + 4) - 1 + 12$
$y = -3(x + 2)^2 + 11$
The vertex is at (-2, 11) and the parabola opens downward so the maximum value is 11 when x = -2.

23. $f(x) = 4x^2 - 8x + 3$
$y = 4x^2 - 8x + 3$
$y = 4(x^2 - 2x + 1) + 3 - 4$
$y = 4(x - 1)^2 - 1$
The vertex is at (1, -1) and the parabola opens upward so the minimum value is -1 when x = 1.

25. $P = 4t^2 - 29t + 30$ for $t > 0$
If P = 0,
$0 = 4t^2 - 29t + 30$
$0 = (4t - 5)(t - 6)$
$t = \frac{5}{4}$ or $t = 6$

For t > 0, there are three intervals: $(0, \frac{5}{4})$, $(\frac{5}{4}, 6)$ and $(6, \infty)$. Choose a test number in each interval.

Interval	Test Number	P
$(0, \frac{5}{4})$	1	5
$(\frac{5}{4}, 6)$	2	-12
$(6, \infty)$	10	140

$P > 0$ for t in $(0, \frac{5}{4})$ and $(6, \infty)$.

27. $P = 3x^2 - 35x + 50$ for $x \geq 0$.
If P = 0,
$0 = 3x^2 - 35x + 50$
$0 = (3x - 5)(x - 10)$
$x = \frac{5}{3}$ or $x = 10$

For $x \geq 0$ there are three intervals, $[0, \frac{5}{3})$, $(\frac{5}{3}, 10)$, and $(10, \infty)$.

Choose a test number in each interval.

Interval	Test Number	P
$[0, \frac{5}{3})$	1	18
$(\frac{5}{3}, 10)$	5	-50
$(10, +\infty)$	20	550

$P > 0$ for x in $[0, \frac{5}{3})$ and $(10, \infty)$.

29. Let x represent the length of the field. Then the width is $\frac{200 - 2x}{2}$ or $100 - x$,

so the area is given by
$A(x) = x(100 - x)$
$= 100x - x^2$
$= -1(x^2 - 100x + 2500) + 2500$
$= -1(x - 50)^2 + 2500.$
The function has a maximum value when x = 50. If x = 50, 100 - x is also 50. The region of maximum area is 50 m by 50 m.

See the graphs for Exercises 31 - 49 in the answer section at the back of the textbook.

31. $f(x) = x^3 - 2$

x	-3	-2	-1	0	1	2	3
y	-29	-10	-3	-2	-1	6	25

33. $f(x) = -x^4 + 1$

x	-3	-2	-1	0	1	2	3
y	-80	-15	0	1	0	-15	-80

35. $f(x) = -(x - 3)^3$

x	0	1	2	3	4	5	6
y	27	8	1	0	-1	-8	-27

37. $f(x) = (x + 2)^4$

x	-4	-3	-2	-1	0
y	16	1	0	1	16

39. $f(x) = x(2x - 1)(x + 2)$

First, find x-intercepts by setting $f(x) = 0$ and solving for x.

$x(2x - 1)(x + 2) = 0$

$x = 0$ or $x = \frac{1}{2}$ or $x = -2$

These three numbers divide the x-axis into four regions.

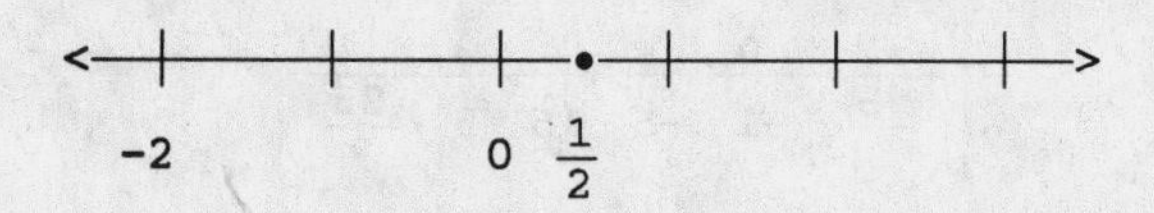

In each region the value of f is either positive or negative. Choose a point in each region as a test point and compute the function value.

x (testpoint)	-3	-1	$\frac{1}{4}$	1
f(x)	-21	3	$-\frac{9}{16}$	3

Use these to sketch the graph.

41. $f(x) = 2x^3 - 3x^2 - 2x$

$f(x) = x(2x^2 - 3x - 2)$

$f(x) = x(2x + 1)(x - 2)$

First, find x-intercepts by setting $f(x) = 0$ and solving for x.

$x(2x + 1)(x - 2) = 0$

$x = 0$ or $x = -\frac{1}{2}$ or $x = 2$

These three numbers divide the x-axis into four regions.

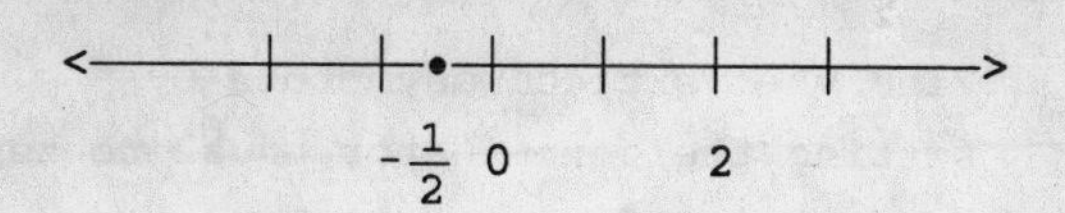

In each region the value of f is either positive or negative. Choose a point in each region as a test point and compute the function value.

x (testpoint)	-1	$-\frac{1}{4}$	1	3
f(x)	-3	$\frac{9}{16}$	-3	21

Use these to sketch the graph.

43. $f(x) = x^4 - 5x^2 - 6$

$f(x) = (x^2 - 6)(x^2 + 1)$

First, find x -intercepts by setting $f(x) = 0$ and solving for x.

$(x^2 - 6)(x^2 + 1) = 0$

$x^2 = 6$

$x = \pm\sqrt{6}$ (approximately $\pm$ 2.4)

These two numbers divide the x-axis into three regions.

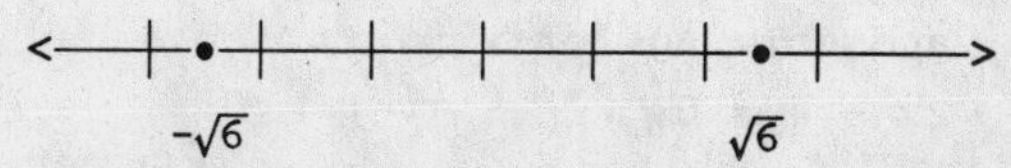

In each region the value of f is either positive or negative. Choose a point in each region as a test point and compute the function value.

x (testpoint)	-3	-1	0	1	3
f(x)	30	-10	-6	-10	30

Use these to sketch the graph.

45. $f(x) = \frac{1}{x - 3}$

Find the vertical asymptote by setting the denominator equal to zero and solving for x.

$x - 3 = 0$

$x = 3$

The line $x = 3$ is a vertical asymptote. The horizontal asymptote of the function in the form

$y = \frac{0x + 1}{1x - 3}$ is

$y = \frac{0}{1}$ or $y = 0$.

If $x = 0$, $y = -\frac{1}{3}$.

If $y = 0$, there is no solution for x.

x	0	1	2	4	5	6
y	$-\frac{1}{3}$	$-\frac{1}{2}$	-1	1	$\frac{1}{2}$	$\frac{1}{3}$

47. $f(x) = \frac{-3}{2x - 4}$

Find the vertical asymptote by setting the denominator equal to zero and solving for x.

$2x - 4 = 0$

$x = 2$

The line $x = 2$ is a vertical asymptote. The horizontal asymptote of the function in the form

$y = \frac{0x - 3}{2x - 4}$ is

$y = \frac{0}{2}$ or $y = 0$.

If $x = 0$, $y = \frac{3}{4}$.

If $y = 0$, there is no solution for x.

x	-1	0	1	3	4	5
y	$\frac{1}{2}$	$\frac{3}{4}$	$\frac{3}{2}$	$-\frac{3}{2}$	$-\frac{3}{4}$	$-\frac{1}{2}$

49. $f(x) = \frac{5x + 5}{3x - 5}$

Find the vertical asymptote by setting the denominator equal to zero and solving for x.

$3x - 5 = 0$

$x = \frac{5}{3}$

The line $x = \frac{5}{3}$ is a vertical asymptote. The horizontal asymptote of the function in the form

$y = \frac{5x + 5}{3x - 5}$ is

$y = \frac{5}{3}$.

If $x = 0$, $y = -1$.

If $y = 0$, $x = -1$.

x	-2	0	1	2	3	4
y	$\frac{5}{11}$	-1	-5	15	5	$\frac{25}{7}$

51. $y = \frac{9.2x}{106 - x}$

(a) If $x = 50$,

$y = \frac{9.2(50)}{106 - 50}$

$y \approx$ 8.2 thousand dollars or about $8200

(b) If $x = 98$,

$y = \frac{9.2(98)}{106 - 98}$

$y =$ 112.7 thousand dollars or about $113,000

(c) If $y = \$22{,}000$, we have

$22 = \frac{9.2x}{106 - x}$

$22(106 - x) = 9.2x$

$2332 - 22x = 9.2x$

$2332 = 31.2x$

$x \approx 74.7$

About 75% of the pollutant can be removed for $22,000.

53. $C(x) = \frac{400}{x}$ and $R(x) = 100x$

We see $P(x) = 100x - \frac{400}{x}$ and that we must use $x > 0$.

(a) $\frac{400}{x} = 100x$

$400 = 100x^2$

$4 = x^2$

$x = \pm 2$

The break-even point is $x = 2$.

(b) $P(1) = 100(1) - \frac{400}{1}$

$P(1) = -300$

Since $P(1)$ is negative, it represents a loss.

(c) $P(4) = 100(4) - \frac{400}{4}$

$P(4) = 300$

Since $P(4)$ is positive, it represents a profit.

CHAPTER 4 EXPONENTIAL AND LOGARITHMIC FUNCTIONS

Section 4.1

1. $f(x) = 5^x - 1$ is an exponential function because the independent variable appears in an exponent.

3. $f(x) = 4x^3 - 1$ is not an exponential function because the independent variable does not appear as an exponent.

5. $f(x) = \left(\frac{2}{3}\right)^x$

$$f(2) = \left(\frac{2}{3}\right)^2 = \frac{2^2}{3^2} = \frac{4}{9}$$

7. $f(x) = \left(\frac{2}{3}\right)^x$

$$f\left(\frac{1}{2}\right) = \left(\frac{2}{3}\right)^{1/2} = \sqrt{\frac{2}{3}} = \frac{\sqrt{2}}{\sqrt{3}} = \frac{\sqrt{6}}{3}$$

See the graphs for Exercises 9-23 in the answer section at the back of the book.

9. $f(x) = 3^x$

First write the function in the form $y = 3^x$.

Then make a table of values.

x	-3	-2	-1	0	1	2	3
y	$\frac{1}{27}$	$\frac{1}{9}$	$\frac{1}{3}$	1	3	9	27

Plot these points to obtain the graph function.

11. $f(x) = 3^{-x}$

$y = 3^{-x}$

x	-3	-2	-1	0	1	2	3
y	27	9	3	1	$\frac{1}{3}$	$\frac{1}{9}$	$\frac{1}{27}$

13. $f(x) = \left(\frac{1}{4}\right)^x$

$y = \left(\frac{1}{4}\right)^x$

x	-3	-2	-1	0	1	2	3
y	64	16	4	1	$\frac{1}{4}$	$\frac{1}{16}$	$\frac{1}{64}$

15. $f(x) = 3^{2x}$

$y = 3^{2x}$

x	-2	-1	0	1	2
y	$\frac{1}{81}$	$\frac{1}{9}$	1	9	81

17. $f(x) = 2^{-x/2}$

$y = 2^{-x/2}$

x	-4	-2	0	2	4
y	4	2	1	$\frac{1}{2}$	$\frac{1}{4}$

19. $f(x) = 3^{x+1}$

$y = 3^{x+1}$

x	-4	-3	-2	-1	0	1	2
y	$\frac{1}{27}$	$\frac{1}{9}$	$\frac{1}{3}$	1	3	9	27

21. $f(x) = 10 - 5 \cdot 2^{-x}$

$y = 10 - 5 \cdot 2^{-x}$

x	-2	-1	0	1	2	4	6
y	-10	0	5	$8\frac{1}{2}$	$8\frac{3}{4}$	$9\frac{11}{16}$	$9\frac{59}{64}$

23. $f(x) = x \cdot 2^x$

$y = x \cdot 2^x$

x	-5	-3	-2	-1	0	1	2
y	$-\frac{5}{32}$	$-\frac{3}{8}$	$-\frac{1}{2}$	$-\frac{1}{2}$	0	2	8

25. $5^x = 25$

$5^x = 5^2$

$x = 2$

27. $2^x = \frac{1}{8}$

$2^x = \frac{1}{2^3}$

$2^x = 2^{-3}$

$x = -3$

29. $a^x = a^2$, $a > 0$

$x = 2$

31. $16^x = 64$

$(4^2)^x = 4^3$

$4^{2x} = 4^3$

$2x = 3$

$x = \frac{3}{2}$

33. $\left(\frac{3}{4}\right)^x = \frac{16}{9}$

$\left(\frac{3}{4}\right)^x = \left(\frac{4}{3}\right)^2$

$\left(\frac{3}{4}\right)^x = \left(\frac{3}{4}\right)^{-2}$

$x = -2$

35. $3^{x-1} = 9$

$3^{x-1} = 3^2$

$x-1 = 2$

$x = 3$

37. $25^{-2x} = 3125$

$(5^2)^{-2x} = 5^5$

$5^{-4x} = 5^5$

$-4x = 5$

$x = -\frac{5}{4}$

39. $81^{-2x} = 3^{x-1}$

$(3^4)^{-2x} = 3^{x-1}$

$3^{-8x} = 3^{x-1}$

$-8x = x-1$

$-9x = -1$

$x = \frac{1}{9}$

41. $2^{|x|} = 16$

$2^{|x|} = 2^4$

$|x| = 4$

$x = 4$ or $x = -4$

43. $2^{x^2-4x} = \frac{1}{16}$

$2^{x^2-4x} = \frac{1}{24}$

$2^{x^2-4x} = 2^{-4}$

$x^2 - 4x = -4$

$x^2 - 4x + 4 = 0$

$(x - 2)^2 = 0$

$x = 2$

45. $8^{x^2} = 2^{5x}$

$(2^3)^{x^2} = 2^{5x}$

$2^{3x^2} = 2^{5x}$

$3x^2 = 5x$

$3x^2 - 5x = 0$

$x(3x - 5) = 0$

$x = 0$ or $3x - 5 = 0$

$x = 0$ or $x = \frac{5}{3}$

47. (a)

t	0	1	2	3	4	5	6	7	8	9	10
y	1	1.05	1.10	1.16	1.22	1.28	1.34	1.41	1.48	1.55	1.63

(b) Plot the points obtained in Part (a) and graph the function. See the graph in the answer section at the back of the textbook.

49. (a) If $y = (.96)^t$, when $t = 10$,
$y = .66$.
If V is the value of the house in 10 years, $.66V = 65{,}000$.
$V = 98{,}480$.
The value is about $98,000.

(b) If $y = (.96)^t$, when $t = 8$,
$y = .72$.
If V is the value of the textbook in 8 years,
$.72V = 20$.
$V = \$27.78$.
The value is almost $28.00.

51. $P(t) = 1{,}000{,}000(2^{.2t})$

(a) $P(0) = 1{,}000{,}000$

(b) $P\left(\frac{5}{2}\right) = 1{,}000{,}000(2^{.5}) \approx 1{,}410{,}000$

(c) $P(5) = 1{,}000{,}000(2^1) = 2{,}000{,}000$

(d) $P(10) = 1{,}000{,}000(2^2) = 4{,}000{,}000$

(e) Using the function values from (a) through (d) above, plot points to obtain the graph. See the graph in the answer section at the back of the textbook.

53. $S = C(1-r)^n$, $C = \$54{,}000$, $n = 8$, $r = .12$

$S = 54{,}000(1 - .12)^8$

$S = 54{,}000(.88)^8$

$S = \$19{,}420.26$

55. $T(x) = \dfrac{2500}{1 + 24 \cdot 2^{-x/4}}$

(a) For 0 catalogs, x = 0.

$T(0) = \dfrac{2500}{1 + 24 \cdot 1} = \dfrac{2500}{25} = 100$

Total sales are $100,000

(b) For 5000 catalogs, x = 5.

$T(5) = \dfrac{2500}{1 + 24 \cdot 2^{-5/4}}$

$= \dfrac{2500}{1 + 10.091} = 225$

Total sales are about $225,000.

(c) For 24,000 catalogs, x = 24.

$T(24) = \dfrac{2500}{1 + 24 \cdot 2^{-6}}$

$= \dfrac{2500}{1 + .375} = 1818$

Total sales are about $1,818,000.

(d) For 49,000 catalogs, x = 49.

$T(49) = \dfrac{2500}{1 + 24 \cdot 2^{-49/4}}$

$= \dfrac{2500}{1 + .00493} = 2488$

Total sales are about $2,488,000.

57. $p(t) = 250 - 120(2.8)^{-.5t}$

(a) $p(2) = 250 - 120(2.8)^{-1}$
$= 250 - 42.86$
$p(2) \approx 207$

(b) $p(4) = 250 - 120(2.8)^{-2}$
$= 250 - 15.31$
$p(4) \approx 235$

(c) $p(10) = 250 - 120(2.8)^{-5}$
$= 250 - .70$
$p(10) \approx 249$

(d) Using $P(0) = 130$ and the function values from (a), (b), **and (c)** above, plot points and obtain the graph. See the graph in the answer section at the back of the textbook.

59. The curve formed is not a smooth curve normally generated by an exponential function.

61. From the graph, the following are estimates on the horizontal axis.

(a) 7 cell types for mushrooms

(b) 9 cell types for kelp

(c) 15 cell types for sequoia

Section 4.2

1. $P(t) = 1{,}000{,}000e^{.02t}$

(a) For P_0, $t = 0$.
$P_0 = 1{,}000{,}000e^0 = 1{,}000{,}000$

(b) $P(2) = 1{,}000{,}000e^{.04} \approx 1{,}040{,}000$

(c) $P(4) = 1{,}000{,}000e^{.08} \approx 1{,}080{,}000$

(d) $P(10) = 1{,}000{,}000e^{.2} \approx 1{,}220{,}000$

3. $Q(t) = 500\ e^{-.05t}$

(a) $t = 0$, $Q(0) = 500e^0 = 500$ grams

(b) $t = 4$, $Q(4) = 500e^{-.2} = 409$ grams

(c) $t = 8$, $Q(8) = 500e^{-.4} = 335$ grams

(d) $t = 20$, $Q(20) = 500e^{-1} = 184$ grams

5. $B(t) = 25{,}000e^{.2t}$

(a) At noon, $t = 0$.
$B(0) = 25{,}000e^0 = 25{,}000$ bacteria

(b) At 1 P.M., $t = 1$.
$B(1) = 25{,}000e^{.2} \approx 30{,}500$ bacteria

(c) At 2 P.M., $t = 2$.
$B(2) = 25{,}000e^{.4} \approx 37{,}300$ bacteria

(d) At 5 P.M., $t = 5$.
$B(5) = 25{,}000e^{1} \approx 68{,}000$ bacteria

7. $A = Pe^{ni}$
Investing $P = \$20{,}000$ at $i = 8\% = .08$ compounded continuously. Use a calculator.

(a) If $n = 1$ year, $A = 20{,}000e^{.08}$
$= \$21{,}665.74$.

(b) If $n = 5$ years, $A = 20{,}000e^{.4}$
$= \$29{,}836.49$.

(c) If $n = 10$ years, $A = 20{,}000e^{.8}$
$= \$44{,}510.82$

9. $W(t) = 60 - 30e^{-.5t}$

(a) $W_0 = 60 - 30e^0 = 30$

(b) $W(1) = 60 - 30e^{-.5} \approx 42$

(c) $W(4) = 60 - 30e^{-2} \approx 56$

(d) $W(6) = 60 - 30e^{-3} \approx 59$

11. $P(x) = 500 - 500e^{-x}$

(a) $P(0) = 500 - 500e^{0} = 0$

(b) $P(1) = 500 - 500e^{-1} \approx 316$

(c) $P(2) = 500 - 500e^{-2} \approx 432$

(d) $P(5) = 500 - 500e^{-5} \approx 497$

(e) $P(10) = 500 - 500e^{-10} \approx 500$

(f) For large values of x, P(x) is very close to 500 so y = 500 is a horizontal asymptote.

(g) Using the function values from (a) through (e) and the horizontal asymptote y = 500, graph y = P(x). See the graph in the answer section at the back of the textbook.

13. In $S(t) = S_0e^{-at}$, if $S_0 = 80{,}000$ and a = .05 we have $S(t) = 80{,}000e^{-.05t}$.

(a) $S(2) = 80{,}000e^{-.1} \approx 72{,}400$

(b) $S(10) = 80{,}000e^{-.5} \approx 48{,}500$

15. $y = 540e^{-1.3x}$

Since y is the number of applications, round all answers to the nearest whole number.

(a) If x = 2.0, $y = 540e^{-2.6} = 40$.

(b) If x = 2.5, $y = 540e^{-3.25} = 21$.

(c) If x = 3.0, $y = 540e^{-3.9} = 11$.

(d) If x = 3.5, $y = 540e^{-4.55} = 6$.

(e) If x = 3.9, $y = 540e^{-5.07} = 3$.

(f) If x = 4.0, $y = 540e^{-5.2} = 3$.

17. $R(t) = 100e^{.18t}$

(a) For 1981, t = 0. $R(0) = 100e^{0}$ = \$100 million

(b) For 1986, t = 5. $R(5) = 100e^{.9}$ ≈ \$246 million

(c) For 1990, t = 9. $R(9) = 100e^{1.62}$ ≈ \$505 million

19. $F(t) = T_0 + Ce^{-kt}$

If $T_0 = 125$, C = .8, and k = .2, then

$F(t) = 125 + .8e^{-.2t}$.

$F(4) = 125 + .8e^{-.8} = 125.36$

The temperature is 125.36°.

21. $F(t) = T_0 + Ce^{-kt}$

We are given the following information:

$F(0) = 300°$, $T_0 = 50°$, and $F(4) = 175°$.

Since $F(0) = 300°$ and $T_0 = 50°$,

$$300 = 50 + Ce^{0}$$
$$250 = C$$

Thus, $F(t) = 50 + 250e^{-kt}$.

Since $F(4) = 175°$,

$$175 = 50 + 250e^{-4k}$$
$$125 = 250e^{-4k}$$
$$.5 = e^{-4k}$$
$$e^{-.6931} = e^{-4k}$$
$$-.6931 = -4k$$

Thus, $k = .1733$

Therefore,

$F(t) = 50 + 250e^{-.1733t}$ so that

$F(12) = 50 + 250e^{-2.0796}$

$F(12) = 81.25$

After 12 minutes its temperature is 81.25° C.

23. $F(t) = T_0 + Ce^{-kt}$

We are given the following information:

$F(0) = 800°$, $T_0 = 20°$, and $F(5) = 410°$

Since $F(0) = 800°$ and $T_0 = 20°$,

$800 = 20 + Ce^0$

$780 = C.$

Thus, $F(t) = 20 + 780e^{-kt}$.

Since $F(5) = 410°$,

$410 = 20 + 780e^{-5k}$

$390 = 780e^{-5k}$

$.5 = e^{-5k}$

$e^{-.6931} = e^{-5k}$

$-.6931 = -5k$

$k = .1386.$

Therefore,

$F(t) = 20 + 780e^{-.1386t}$ so that

$F(15) = 20 + 780e^{-2.079}$

$F(15) = 117.5.$

After 15 hours the temperature is 117.5°C.

25. $p(h) = p_0e^{-kh}$.

Since the pressure at sea level is 15 pounds per square inch, $p_0 = 15$.

Thus, $p(h) = 15e^{-kh}$.

Since the pressure at 12,000 feet is 9 pounds per square inch,

$p(12,000) = 9.$

$9 = 15e^{-12,000k}$

$\frac{9}{15} = e^{-12,000k}$

$.6 = e^{-12,000k}$

$e^{-.51} = e^{-12,000k}$

$-.51 = -12,000k$

Thus, $k = .0000425$

Therefore,

$p(h) = 15e^{-.0000425h}$ and

$p(6000) = 15e^{-.255} = 11.6.$

At 6000 feet the pressure is 11.6 pounds per square inch.

Section 4.3

1. $2^3 = 8$ is equivalent to $\log_2 8 = 3$.

3. $3^4 = 81$ is equivalent to $\log_3 81 = 4$.

5. $\left(\frac{1}{3}\right)^{-2} = 9$ is equivalent to

$\log_{1/3} 9 = -2.$

7. $\log_2 128 = 7$ is equivalent to

$2^7 = 128.$

9. $\log_5 \frac{1}{25} = -2$ is equivalent to

$5^{-2} = \frac{1}{25}.$

11. $\log 10,000 = 4$ is equivalent to

$10^4 = 10,000.$

13. $\log 1000 = x$

$10^x = 1000$

$10^x = 10^3$

$x = 3$

$\log 1000 = 3$

15. $\log .01 = x$

$10^x = .01$

$10^x = 10^{-2}$

$x = -2$

$\log .01 = -2$

17. $\log_5 25 = x$

$5^x = 25$

$5^x = 5^2$

$x = 2$

$\log_5 25 = 2$

19. $\log_4 64 = x$

$4^x = 64$

$4^x = 4^3$

$x = 3$

$\log_4 64 = 3$

21. $\log_2 \frac{1}{4} = x$

$2^x = \frac{1}{4}$

$2^x = \frac{1}{2^2}$

$2^x = 2^{-2}$

$x = -2$

$\log_2 \frac{1}{4} = -2$

23. $\log_e \sqrt{e} = x$

$e^x = \sqrt{e}$

$e^x = e^{1/2}$

$x = \frac{1}{2}$

$\log_e \sqrt{e} = \frac{1}{2}$

See the graphs for Exercises 25-29 in the answer section at back of the textbook.

25. $y = \log_3 x$ is equivalent to $3^y = x$.

x	$\frac{1}{27}$	$\frac{1}{9}$	$\frac{1}{3}$	1	3	9	27
y	-3	-2	-1	0	1	2	3

Plot these points to obtain the graph.

27. $f(x) = \log_4 x$

First write the function in the form

$y = \log_4 x$

and then, equivalently, as

$4^y = x$.

Then make a table of values, choosing values for y and computing x.

x	$\frac{1}{64}$	$\frac{1}{16}$	$\frac{1}{4}$	1	4	16	64
y	-3	-2	-1	0	1	2	3

Plot these points to obtain the graph of the function.

29. $f(x) = \log_3 (x - 1)$

$y = \log_3 (x - 1)$

$3^4 = x - 1$ or

$x = 3^4 + 1$

Then make a table of values, choosing values for y and computing x.

x	$\frac{28}{27}$	$\frac{10}{9}$	$\frac{4}{3}$	2	4	10
y	-3	-2	-1	0	1	2

31. $\log_3 \frac{2}{5} = \log_3 2 - \log_3 5$

33. $\log_9 7m = \log_9 7 + \log_9 m$

35. $\log_3 \frac{3x}{5k} = \log_3 (3x) - \log_3 (5k)$

$= \log_3 3 + \log_3 x - (\log_3 5 + \log_3 k)$

$= 1 + \log_3 x - \log_3 5 - \log_3 k$

37. $\log_k \frac{pq^2}{m} = \log_k (pq^2) - \log_k m$

$= \log_k p + \log_k q^2 - \log_k m$

$= \log_k p + 2 \log_k q - \log_k m$

39. $\log_r (5m + 7n)$ cannot be simplified using the properties of logarithms.

41. $\log_3 \frac{5\sqrt{2}}{\sqrt[4]{7}} = \log_3 (5 \cdot 2^{1/2}) - \log_3 7^{1/4}$

$= \log_3 5 + \log_3 2^{1/2} - \log_3 7^{1/4}$

$= \log_3 5 + \frac{1}{2} \log_3 2 - \frac{1}{4} \log_3 7$

43. Using a calculator gives
$\log 60.4 = 1.7810$.

45. Using a calculator gives
$\log .00156 = -2.8069$.

47. Using a calculator gives
$\ln 58{,}500 = 10.9768$.

49. $\log 56 - \log 8 = \log \frac{56}{8}$

$= \log 7$

$= .8451$. *Using a calculator*

51. $\log 15 - \log 3 = \log \frac{15}{3} = \log 5$

53. $3 \ln 2 + 2 \ln 3$

$= \ln 2^3 + \ln 3^2$

$= \ln(2^3 \cdot 3^2) = \ln 72$

55. $3 \log x - 2 \log y = \log x^3 - \log y^2$

$= \log \left(\frac{x^3}{y^2}\right)$

57. $\ln (3x + 2) + \ln (x + 4)$

$= \ln [(3x + 2)(x + 4)]$

$= \ln (3x^2 + 14x + 8)$

59. $3 \log x - 2 \log (x + 1)^2 + \frac{1}{2} \log (x + 2)$

$= \log x^3 - \log [(x + 1)^2]^2 + \log (x + 2)^{1/2}$

$= \log x^3 - \log (x + 1)^4 + \log \sqrt{x + 2}$

$= \log \left(\frac{x^3}{(x + 1)^4}\right) + \log \sqrt{x + 2}$

$= \log \left(\frac{x^3 \sqrt{x + 2}}{(x + 1)^4}\right)$

61. $\log_b 8 = \log_b 2^3 = 3 \log_b 2 = 3a$

63. $\log_b 54 = \log_b (2 \cdot 27)$

$= \log_b (2 \cdot 3^3)$

$= \log_b 2 + \log_b 3^3$

$= \log_b 2 + 3 \log_b 3$

$= a + 3c$

65. $\log_b (72b) = \log_b (8 \cdot 9 \cdot b)$

$= \log_b (2^3 \cdot 3^2 \cdot b)$

$= \log_b 2^3 + \log_b 3^2 + \log_b b$

$= 3 \log_b 2 + 2 \log_b 3 + \log_b b$

$= 3a + 2c + 1$

67. $\log_x 25 = -2$

$x^{-2} = 25$

$(x^{-2})^{-1/2} = 25^{-1/2}$

$x = \frac{1}{\sqrt{25}}$

$x = \frac{1}{5}$

69. $\log_9 27 = m$

$9^m = 27$

$(3^2)^m = 3^3$

$3^{2m} = 3^3$

$2m = 3$

$m = \frac{3}{2}$

71. $\log_y 8 = \frac{3}{4}$

$y^{3/4} = 8$

$(y^{3/4})^{4/3} = (2^3)^{4/3}$

$y = 2^4$

$y = 16$

73. $\log_3 (5x + 1) = 2$

$5x + 1 = 3^2$

$5x + 1 = 9$

$5x = 8$

$x = \frac{8}{5}$

75. $\log x - \log(x + 3) = -1$

$$\log\left(\frac{x}{x+3}\right) = -1$$

$$\frac{x}{x+3} = 10^{-1}$$

$$\frac{x}{x+3} = \frac{1}{10}$$

$$10x = x + 3$$

$$9x = 3$$

$$x = \frac{1}{3}$$

77. $\log_3(y + 2) = \log_3(y - 7) + \log_3 4$

$$\log_3(y + 2) = \log_3[4(y - 7)]$$

$$y + 2 = 4(y - 7)$$

$$y + 2 = 4y - 28$$

$$30 = 3y$$

$$y = 10$$

79. $\ln k - \ln(k + 1) = \ln 5$

$$\ln\left(\frac{k}{k+1}\right) = \ln 5$$

$$\frac{k}{k+1} = 5$$

$$k = 5k + 5$$

$$-5 = 4k$$

$$k = -\frac{5}{4}$$

There is no solution since $\ln\left(-\frac{5}{4}\right)$ is not defined.

81. $7 + 2\log_5(2x - 1) = 9$

$$2\log_5(2x - 1) = 2$$

$$\log_5(2x - 1) = 1$$

$$2x - 1 = 5^1$$

$$2x = 6$$

$$x = 3$$

83. $S(t) = 125 + 83\log(5t + 1)$

(a) $S(0) = 125 + 83\log 1$
$= \$125$ thousand

(b) $S(2) = 125 + 83\log 11$
$\approx \$211$ thousand

(c) $S(4) = 125 + 83\log 21$
$\approx \$235$ thousand

(d) $S(31) = 125 + 83\log 156$
$\approx \$307$ thousand

(e) Using the values from parts (a) through (d), plot points and graph the function. See the graph in the answer section at the back of the book.

85. For the 4300 model computer a value of 10 on the horizontal axis gives 60 on the vertical axis. The approximate price is \$60,000.
For the system 1370 computer a value of 10 on the horizontal axis gives 450 on the vertical axis. The approximate price is \$450,000.

Section 4.4

1. $3^x = 6$

$$\ln 3^x = \ln 6$$

$$x\ln 3 = \ln 6$$

$$x = \frac{\ln 6}{\ln 3}$$

$$x = \frac{1.7918}{1.0986}$$

$$x = 1.631$$

3. $7^x = 8$

$$\ln 7^x = \ln 8$$

$$x\ln 7 = \ln 8$$

$$x = \frac{\ln 8}{\ln 7}$$

$$x = \frac{2.0794}{1.9459}$$

$$x = 1.069$$

5.
$$3^{a+2} = 5$$
$$\ln 3^{a+2} = \ln 5$$
$$(a + 2) \ln 3 = \ln 5$$
$$a + 2 = \frac{\ln 5}{\ln 3}$$
$$a + 2 = \frac{1.6094}{1.0986}$$
$$a + 2 = 1.4650$$
$$a = .535$$

7.
$$6^{1-2k} = 8$$
$$\ln 6^{1-2k} = \ln 8$$
$$(1 - 2k) \ln 6 = \ln 8$$
$$1 - 2k = \frac{\ln 8}{\ln 6}$$
$$1 - 2k = \frac{2.0794}{1.7918}$$
$$1 - 2k = 1.1606$$
$$-2k = .1606$$
$$k = -.080$$

9.
$$5 \cdot 4^{3m-1} = 5 \cdot 12^{m+2}$$
$$4^{3m-1} = 12^{m+2}$$
$$\ln 4^{3m-1} = \ln 12^{m+2}$$
$$(3m - 1) \ln 4 = (m + 2) \ln 12$$
$$\frac{3m - 1}{m + 2} = \frac{\ln 12}{\ln 4}$$
$$\frac{3m - 1}{m + 2} = \frac{2.4849}{1.3863}$$
$$\frac{3m - 1}{m + 2} = 1.7925$$
$$3m - 1 = 1.7925m + 3.5850$$
$$1.2075m = 4.5850$$
$$m = 3.797$$

11.
$$e^{k-1} = 4$$
$$\ln e^{k-1} = \ln 4$$
$$(k - 1) \ln e = \ln 4$$
$$(k - 1) \cdot 1 = \ln 4$$
$$k - 1 = \ln 4$$
$$k - 1 = 1.3863$$
$$k = 2.386$$

13.
$$2 e^{5a+2} = 8$$
$$e^{5a+2} = 4$$
$$\ln e^{5a+2} = \ln 4$$
$$(5a + 2) \ln e = \ln 4$$
$$(5a + 2) \cdot 1 = \ln 4$$
$$5a + 2 = 1.3863$$
$$5a = -.6137$$
$$a = -.123$$

15. $2^x = -3$

Since $2^x > 0$ for all real numbers x, this equation has no solution.

17.
$$\left(1 + \frac{r}{2}\right)^5 = 9$$
$$\left[\left(1 + \frac{r}{2}\right)^5\right]^{\frac{1}{5}} = 9^{\frac{1}{5}}$$
$$1 + \frac{r}{2} = 9^{.2}$$
$$1 + \frac{r}{2} = 1.5518$$
$$\frac{r}{2} = .5518$$
$$r = 1.104$$

19. $100\ (1 + .02)^{3+n} = 150$

Divide by 100.
$$(1.02)^{3+n} = 1.5$$
$$\ln (1.02)^{3+n} = \ln 1.5$$
$$(3 + n) \ln 1.02 = \ln 1.5$$
$$3 + n = \frac{\ln 1.5}{\ln 1.02}$$
$$3 + n = \frac{.4055}{.0198}$$
$$3 + n = 20.4753$$
$$n = 17.475$$

21.
$$2^{x^2-1} = 12$$
$$\ln 2^{x^2-1} = \ln 12$$
$$(x^2 - 1) \ln 2 = \ln 12$$
$$x^2 - 1 = \frac{\ln 12}{\ln 2}$$
$$x^2 - 1 = \frac{2.4849}{.6931}$$
$$x^2 - 1 = 3.5850$$
$$x^2 = 4.5850$$
$$x = \pm \sqrt{4.5850}$$
$$x = \pm 2.141$$
Both 2.141 and -2.141 are solutions.

23.
$$2(e^x + 1) = 10$$
$$e^x + 1 = 5$$
$$e^x = 4$$
$$\ln e^x = \ln 4$$
$$x \ln e = \ln 4$$
$$x = 1.386$$

25.
$$\log (t - 1) = 1$$
$$t - 1 = 10^1$$
$$t - 1 = 10$$
$$t = 11$$

27.
$$\log (x - 3) = 1 - \log x$$
$$\log (x - 3) + \log x = 1$$
$$\log [(x - 3)x] = 1$$
$$(x - 3)x = 10^1$$
$$x^2 - 3x = 10$$
$$x^2 - 3x - 10 = 0$$
$$(x - 5)(x + 2) = 0$$
$$x = 5 \text{ or } x = -2$$
Since log (-2) is undefined, the only solution is 5.

29.
$$\ln (y + 2) = \ln (y - 7) + \ln 4$$
$$\ln (y + 2) = \ln [(y - 7)4]$$
$$y + 2 = (y - 7)4$$
$$y + 2 = 4y - 28$$
$$30 = 3y$$
$$y = 10$$

31.
$$\ln (3x - 1) - \ln (2 + x) = \ln 2$$
$$\ln \frac{3x - 1}{2 + x} = \ln 2$$
$$\frac{3x - 1}{2 + x} = 2$$
$$3x - 1 = 4 + 2x$$
$$x = 5$$

33.
$$\ln (5 + 4y) - \ln (3 + y) = \ln 3$$
$$\ln \left(\frac{5 + 4y}{3 + y}\right) = \ln 3$$
$$\frac{5 + 4y}{3 + y} = 3$$
$$5 + 4y = 9 + 3y$$
$$y = 4$$

35.
$$\ln x + 1 = \ln (x - 4)$$
$$\ln x + \ln e = \ln (x - 4)$$
$$\ln (ex) = \ln (x - 4)$$
$$ex = x - 4$$
$$ex - x = -4$$
$$(e - 1)x = -4$$
$$x = \frac{-4}{e - 1}$$
$$x = \frac{-4}{1.7183}$$
$$x = -2.328$$
Since ln (-2.328) is undefined, there is no solution.

37.
$$2 \ln (x - 3) = \ln (x + 5) + \ln 4$$
$$\ln (x - 3)^2 = \ln [4(x + 5)]$$
$$(x - 3)^2 = 4(x + 5)$$
$$x^2 - 6x + 9 = 4x + 20$$
$$x^2 - 10x - 11 = 0$$
$$(x - 11)(x + 1) = 0$$
$$x = 11 \text{ or } x = -1$$
Since ln (-4) is undefined, the only solution is x = 11.

39.

$$\log_5 (r + 2) + \log_5 (r - 2) = 1$$
$$\log_5 [(r + 2)(r - 2)] = 1$$
$$(r + 2)(r - 2) = 5^1$$
$$r^2 - 4 = 5$$
$$r^2 = 9$$
$$r = \pm 3$$

Since $\log_5 (-1)$ is undefined, 3 is the only solution.

41. $\log_3 (a - 3) = 1 + \log_3(a + 1)$

$$\log_3 (a - 3) - \log_3 (a + 1) = 1$$
$$\log_3 \left(\frac{a - 3}{a + 1}\right) = 1$$
$$\frac{a - 3}{a + 1} = 3^1$$
$$a - 3 = 3a + 3$$
$$-6 = 2a$$
$$a = -3$$

Since $\log_3 (-6)$ is undefined, there is no solution.

43.

$$\log_2 \sqrt{2y^2} - 1 = \frac{1}{2}$$
$$\log_2 \sqrt{2y^2} = \frac{3}{2}$$
$$\sqrt{2y^2} = 2^{3/2}$$
$$\left(\sqrt{2y^2}\right)^2 = (2^{3/2})^2$$
$$2y^2 = 2^3$$
$$2y^2 = 8$$
$$y^2 = 4$$
$$y = \pm 2$$

Both 2 and -2 are solutions.

45.

$$\log z = \sqrt{\log z}$$
$$(\log z)^2 = \left(\sqrt{\log z}\right)^2$$
$$(\log z)^2 = \log z$$
$$(\log z)^2 - \log z = 0$$
$$(\log z)(\log z - 1) = 0$$
$$\log z = 0 \quad \text{or} \quad \log z - 1 = 0$$
$$z = 1 \quad \text{or} \quad z = 10$$

47.

$$\log_x 5.87 = 2$$
$$x^2 = 5.87$$
$$x = \pm \sqrt{5.87}$$
$$x = \pm 2.423$$

However, since the base of a logarithm must be nonnegative, the only solution is 2.423.

49.

$$1.8^{p+4} = 9.31$$
$$\ln 1.8^{p+4} = \ln 9.31$$
$$(p + 4) \ln 1.8 = \ln 9.31$$
$$p + 4 = \frac{\ln 9.31}{\ln 1.8}$$
$$p + 4 = \frac{2.2311}{.5878}$$
$$p + 4 = 3.7957$$
$$p = -.204$$

51. $A(t) = 5000\, e^{-.02t}$

(a) $A(0) = 5000\, e^0 = 5000$ grams

(b) $A(5) = 5000\, e^{-.1} \approx 4520$ grams

(c) $A(20) = 5000\, e^{-.4} \approx 3350$ grams

(d) Find t so that

$$A(t) = \frac{1}{2}(5000) = 2500.$$
$$2500 = 5000e^{-.02t}$$
$$.5 = e^{-.02t}$$
$$\ln .5 = -.02t$$
$$t = \frac{\ln .5}{-.02}$$
$$t = \frac{-.6931}{-.02}$$
$$t = 34.7 \text{ seconds}$$

53. **(a)**

$$A(t) = 5000e^{-.03t}$$
$$2500 = 5000e^{-.03t}$$
$$.5 = e^{-.03t}$$
$$\ln .5 = -.03t$$
$$t = \frac{\ln .5}{-.03}$$
$$t = \frac{-.6931}{-.03}$$
$$t \approx 23 \text{ days}$$

(b)

$$A(t) = 2350e^{-.08t}$$
$$1175 = 2350e^{-.08t}$$
$$.5 = e^{-.08t}$$
$$\ln .5 = -.08t$$
$$t = \frac{\ln .5}{-.08}$$
$$t = \frac{-.6931}{-.08}$$
$$t \approx 8.7 \text{ days}$$

(c)

$$A(t) = 18{,}000e^{-.0002t}$$
$$9000 = 18{,}000e^{-.0002t}$$
$$.5 = e^{-.0002t}$$
$$\ln .5 = -.0002t$$
$$t = \frac{\ln .5}{-.0002}$$
$$t \approx 3466 \text{ days}$$

55. $y = y_0\, e^{-(\ln 2)(1/5600)t}$ with $y = .8y_0$

$$.8y_0 = y_0\, e^{-(\ln 2)(1/5600)t}$$
$$.8 = 2e^{-(\ln 2)(1/5600)t}$$
$$\ln .8 = -(\ln 2)(1/5600)t$$
$$t = \frac{\ln .8}{-(\ln 2)(1/5600)}$$
$$t = \frac{(\ln .8)(5600)}{-\ln 2}$$
$$t = 1802.8$$

The age is about 1800 years.

57. $y = y_0\, e^{-(\ln 2)(1/5600)t}$ with $y = .1y_0$

$$.1y_0 = y_0\, e^{-(\ln 2)(1/5600)t}$$
$$.1 = e^{-(\ln 2)(1/5600)t}$$
$$\ln .1 = -(\ln 2)(1/5600)t$$
$$t = \frac{\ln .1}{-(\ln 2)(1/5600)}$$
$$t = \frac{(\ln .1)(5600)}{-\ln 2}$$
$$t = 18{,}602.8$$

The age is about 18,600 years.

59. $P(t) = 100\, e^{-.1t}$

(a) $P(4) = 100\, e^{-.4} = 67.03$

About 67% remains after 4 days.

(b) $P(10) = 100\, e^{-1} = 36.79$

About 37% remains after 10 days.

(c)

$$10 = 100\, e^{-.1t}$$
$$.1 = e^{-.1t}$$
$$\ln .1 = -.1t$$
$$t = \frac{\ln .1}{-.1}$$
$$t = 23.03$$

The hay may be used after about 23 days.

(d)

$$1 = 100\, e^{-.1t}$$
$$.01 = e^{-.1t}$$
$$\ln .01 = -.1t$$
$$t = \frac{\ln .01}{-.1}$$
$$t = 46.05$$

The hay may be used after about 46 days.

61. Let $d = 10 \cdot \log \frac{I}{I_0}$ represent the decibel rating.

(a) If $I = 100I_0$,

$$d = 10 \cdot \log \frac{100I_0}{I_0}$$
$$= 10 \log 100$$
$$= 10 \cdot 2$$
$$= 20.$$

(b) If $I = 1000I_0$,

$$d = 10 \cdot \log \frac{1000I_0}{I_0}$$
$$= 10 \log 1000$$
$$= 10 \cdot 3$$
$$= 30$$

(c) If $I = 100{,}000I_0$,

$$d = 10 \cdot \log \frac{100{,}000I_0}{I_0}$$
$$= 10 \cdot \log 100{,}000$$
$$= 10 \cdot 5$$
$$= 50.$$

(d) If $I = 1{,}000{,}000I_0$,

$$d = 10 \cdot \log \frac{1{,}000{,}000I_0}{I_0}$$
$$= 10 \cdot \log 1{,}000{,}000$$
$$= 10 \cdot 6$$
$$= 60.$$

63. Let $r = \log \frac{I}{I_0}$ represent the intensity measured on the Richter scale.

(a) If $I = 1000\ I_0$,

$$r = \log \frac{1000I_0}{I_0}$$
$$= \log 1000$$
$$= 3.$$

(b) If $I = 1{,}000{,}000I_0$,

$$r = \log \frac{1{,}000{,}000I_0}{I_0}$$
$$= \log 1{,}000{,}000$$
$$= 6.$$

(c) If $I = 100{,}000{,}000I_0$,

$$r = \log \frac{100{,}000{,}000I_0}{I_0}$$
$$= \log 100{,}000{,}000$$
$$= 8.$$

65. $M(h) = e^r$ where $r = .65 \ln h - 1.94$

If $h = 9.7$,

$r = .65 \ln 9.7 - 1.94 = -.463.$

$M(9.7) = e^{-.463} = .629$

.629 micrograms per liter is the maximum level permitted.

67. The value .3 kg on the horizontal axis corresponds to 4.3 ml/min on the vertical axis.
The value .7 kg on the horizontal axis corresponds to 7.8 ml/min on the vertical axis.

69. $p = 86.3 \ln h - 680$

(a) If $h = 3000$ feet,

$p = 86.3 \ln 3000 - 680 = 10.94$

or about 11%.

(b) If $h = 4000$ feet,

$p = 86.3 \ln 4000 - 680 = 35.78$

or about 36%.

(c) If $h = 7000$ feet,

$p = 86.3 \ln 7000 - 680 = 84.07$

or about 84%.

Chapter 4 Review Exercises

1. $2^{3x} = \frac{1}{8}$

$2^{3x} = \frac{1}{2^3}$

$2^{3x} = 2^{-3}$

$3x = -3$

$x = -1$

3. $9^{2y-1} = 27^y$

$(3^2)^{2y-1} = (3^3)^y$

$3^{4y-2} = 3^{3y}$

$4y - 2 = 3y$

$y = 2$

See the graphs for Exercises 5 and 7 in the answer section at the back of the textbook.

5. $f(x) = 5^x$

$y = 5^x$

x	-3	-2	-1	0	1	2
y	$\frac{1}{125}$	$\frac{1}{25}$	$\frac{1}{5}$	1	5	25

7. $f(x) = \log_5 x$

$y = \log_5 x$

$x = 5^y$

x	$\frac{1}{125}$	$\frac{1}{25}$	$\frac{1}{5}$	1	5	25
y	-3	-2	-1	0	1	2

9. $P(x) = 100 - 100e^{-.8x}$

(a) $P(0) = 100 - 100e^0 = 0$

(b) $P(1) = 100 - 100e^{-.8} = 55.07$
or about 55

(c) $P(5) = 100 - 100e^{-4} = 98.17$
or about 98

(d) For large values of x, P(x) approximately equals 100. An experienced worker would produce about 100 items per day.

11. $2^6 = 64$ is equivalent to $\log_2 64 = 6$.

13. $e^{.09} = 1.09417$ is equivalent to $\ln 1.09417 = .09$.

15. $\log_2 32 = 5$ is equivalent to $2^5 = 32$.

17. $\ln 82.9 = 4.41763$ is equivalent to $e^{4.41763} = 82.9$.

19. $\log_3 81 = x$

$3^x = 81$

$3^x = 3^4$

$x = 4$

$\log_3 81 = 4$

21. $\log_{32} 16 = x$

$32^x = 16$

$(2^5)^x = 2^4$

$2^{5x} = 2^4$

$5x = 4$

$x = \frac{4}{5}$

$\log_{32} 16 = \frac{4}{5}$

23. $\log_{100} 1000 = x$

$100^x = 1000$

$(10^2)^x = 10^3$

$10^{2x} = 10^3$

$2x = 3$

$x = \frac{3}{2}$

$\log_{100} 1000 = \frac{3}{2}$

25. A calculator gives log 18 = 1.2553.

27. A calculator gives log .83 = -.0809.

29. A calculator gives ln 6.2 = 1.8245.

31. A calculator gives ln 483 = 6.1800.

33. $\log_5 3k + \log_5 7k^3 = \log_5 [(3k)(7k^3)]$
$= \log_5 (21k^4)$

35. $2 \cdot \log_2 x - 3 \cdot \log_2 m$

$= \log_2 x^2 - \log_2 m^3 = \log_2 \left(\frac{x^2}{m^3}\right)$

37.
$$8^p = 19$$
$$\ln 8^p = \ln 19$$
$$p \ln 8 = \ln 19$$
$$p = \frac{\ln 19}{\ln 8}$$
$$p = \frac{2.9444}{2.0794}$$
$$p = 1.416$$

39.
$$5 \cdot 2^{-m} = 35$$
$$2^{-m} = 7$$
$$\ln (2^{-m}) = \ln 7$$
$$-m \ln 2 = \ln 7$$
$$-m = \frac{\ln 7}{\ln 2}$$
$$-m = \frac{1.9459}{.6931}$$
$$m = -2.807$$

41.
$$e^{-5-2x} = 5$$
$$-5 - 2x = \ln 5$$
$$-5 - 2x = 1.6094$$
$$-2x = 6.6094$$
$$x = -3.305$$

43.
$$10^{2x-3} = 17$$
$$2x - 3 = \log 17$$
$$2x - 3 = 1.2304$$
$$2x = 4.2304$$
$$x = 2.115$$

45.
$$6^{2-m} = 2^{3m+1}$$
$$\ln 6^{2-m} = \ln 2^{3m+1}$$
$$(2 - m) \ln 6 = (3m + 1) \ln 2$$
$$\frac{2 - m}{3m + 1} = \frac{\ln 2}{\ln 6}$$
$$\frac{2 - m}{3m + 1} = \frac{.6931}{1.7918}$$
$$\frac{2 - m}{3m + 1} = .3869$$
$$2 - m = 1.1607m + .3869$$
$$1.6131 = 2.1607m$$
$$m = .747$$

47.
$$(1 + .003)^k = 1.089$$
$$1.003^k = 1.089$$
$$\ln 1.003^k = \ln 1.089$$
$$k \ln 1.003 = \ln 1.089$$
$$k = \frac{\ln 1.089}{\ln 1.003}$$
$$k = \frac{.0853}{.0030}$$
$$k = 28.463$$

49.
$$4 \cdot 3^{x^2} = 15$$
$$3^{x^2} = \frac{15}{4}$$
$$\ln 3^{x^2} = \ln 3.75$$
$$x^2 \ln 3 = \ln 3.75$$
$$x^2 = \frac{\ln 3.75}{\ln 3}$$
$$x^2 = \frac{1.3218}{1.0986}$$
$$x^2 = 1.2031$$
$$x = \pm \sqrt{1.2031}$$
$$x = \pm 1.097$$

51. $\log(m + 2) = 1$

$m + 2 = 10^1$

$m + 2 = 10$

$m = 8$

53. $\log_2(3k - 2) = 4$

$3k - 2 = 2^4$

$3k - 2 = 16$

$3k = 18$

$k = 6$

55. $\log x + \log(x + 3) = 1$

$\log[x(x + 3)] = 1$

$x(x + 3) = 10^1$

$x^2 + 3x = 10$

$x^2 + 3x - 10 = 0$

$(x + 5)(x - 2) = 0$

$x = -5$ or $x = 2$

Since log (-5) is undefined, the only solution is 2.

57. $\log(p - 1) = 1 + \log p$

$\log(p - 1) - \log p = 1$

$\log\left(\frac{p-1}{p}\right) = 1$

$\frac{p-1}{p} = 10^1$

$p - 1 = 10p$

$-1 = 9p$

$p = -\frac{1}{9}$

Since $\log(-\frac{1}{9})$ is undefined, there is no solution.

59. $2\ln(y + 1) = \ln(y^2 - 1) + \ln 5$

$\ln(y + 1)^2 = \ln[(y^2 - 1) \cdot 5]$

$(y + 1)^2 = 5(y^2 - 1)$

$y^2 + 2y + 1 = 5y^2 - 5$

$0 = 4y^2 - 2y - 6$

$0 = 2(2y^2 - y - 3)$

$0 = 2(2y - 3)(y + 1)$

$y = \frac{3}{2}$ or $y = -1$

Since ln (0) is undefined, the only solution is $\frac{3}{2}$.

61. $3 - 2\log_4(2x - 1) = 1$

$2\log_4(2x - 1) = -2$

$\log_4(2x - 1) = -1$

$2x - 1 = 4^{-1}$

$2x - 1 = \frac{1}{4}$

$2x = \frac{5}{4}$

$x = \frac{5}{8}$

63. $h = .5 + \log t$

(a) If $t = 2$, $h = .5 + \log 2 = .801$ or about .8 meter.

(b) If $t = 5$, $h = .5 + \log 5 = 1.199$ or about 1.2 meter.

(c) If $t = 10$, $h = .5 + \log 10 = 1.5$ meter.

(d) If $t = 20$, $h = .5 + \log 20 = 1.801$ or about 1.8 meter.

Case 3 Exercises

1. $\lambda = \frac{\ln 2}{\text{half-life}}$, if half-life is 22 years.

$\lambda = \frac{.6931}{22}$

$\lambda = .0315$

2. $\lambda y_0 = \lambda \cdot y(t) \cdot e^{300\lambda} - r(e^{300\lambda} - 1)$

if $\lambda \cdot y(t) = 8.5$, $\lambda = .0315$, and $r = .8$.

$\lambda y_0 = 8.5 \cdot e^{300(.0315)} - .8(e^{300(.0315)} - 1)$

$\lambda y_0 = 97{,}853.67$ or about 98,000

Since 98,000 > 30,000, the painting is a modern forgery.

3. $\lambda \cdot y(t) = 12.6$ and $r = .26$
$\lambda y_0 = 12.6\ e^{300(.0315)} - .26[e^{300(.0315)} - 1]$
$\lambda y_0 = 156{,}819.02$ or about 157,000
Since $157{,}000 > 30{,}000$, the painting is a modern forgery.

4. $\lambda \cdot y(t) = 1.5$ and $r = .4$
$\lambda y_0 = 1.5\ e^{300(.0315)} - .4(e^{300(.0315)} - 1)$
$\lambda y_0 = 13{,}979.38$ or about 14,000
Since $14{,}000 < 30{,}000$, the painting cannot be a modern forgery.

5. $\lambda \cdot yt = 5.2$ and $r = .6$
$\lambda y_0 = 5 \cdot 2\ e^{300(.0315)} - .6[e^{300(.0315)} - 1]$
$\lambda y_0 = 58{,}458.16$ or about 58,000
Since $58{,}000 > 30{,}000$, the painting is a modern forgery.

6. $\lambda \cdot y(t) = 10.3$ and $r = .3$
$\lambda y_0 = 10.3\ e^{300(.0315)} - .3(e^{300(.0315)} - 1)$
$\lambda y_0 = 127{,}081.95$ or about 127,000
Since $127{,}000 > 30{,}000$, the painting is a modern forgery.

CHAPTER 5 MATHEMATICS OF FINANCE

Section 5.1

1. \$1000 at 12% for 1 year
Use I = prt. 12% = .12
$I = (1000)(.12)(1) = \$120$

3. \$25,000 at 11% for 9 months
t must be given in years.
9 months = 9/12 year
$I = (25{,}000)(.11)\left(\frac{9}{12}\right) = \$2062.50$

5. \$1974 at 16.2% for 7 months
7 months = 7/12 year
$I = (1974)(.162)\left(\frac{7}{12}\right) = \$186.54$

7. \$12,000 at 14% for 72 days
72 days = 72/360 year
$I = (12000)(.14)\left(\frac{72}{360}\right) = \$336.$

9. \$5147.18 at 17.3% for 58 days
$I = (5147.18)(.173)\left(\frac{58}{360}\right) = \$143.46$

11. \$7980 at 15%
From May 7 to September 19 is 132 days.
$I = (7980)(.15)\left(\frac{132}{360}\right) = \$438.90$

13. \$7800 at 16%
From July 7 to October 25 is 110 days.
$I = (7800)(.16)\left(\frac{110}{365}\right) = \$376.11$

15. \$2579 at 17.6%
From October 4 to March 15 is 162 days.
$I = (2579)(.176)\left(\frac{162}{365}\right) = \$201.46$

17. \$15,000 in 8 months at 16%
Use $P = \frac{A}{1 + rt}$.
$P = \frac{15{,}000}{1 + (.16)\left(\frac{2}{3}\right)} = \$13{,}554.22$

19. \$5276 in 3 months at 7.4%
$P = \frac{5276}{1 + (.074)\left(\frac{1}{4}\right)} = \$5180.17$

21. \$15,402 in 125 days at 9.3%
$\frac{15{,}402}{1 + (.093)\left(\frac{125}{360}\right)} = \$14{,}920.20$

23. \$7150, discount rate 16%, for 11 months
$\text{Discount} = (7150)(.16)\left(\frac{11}{12}\right)$
$= \$1048.67$
$\text{Proceeds} = 7150 - 1048.67 = \$6101.33$

25. \$358, discount rate 11.6%, for 183 days
$\text{Discount} = (358)(.116)\left(\frac{183}{360}\right) = \$21.11$
$\text{Proceeds} = 358 - 21.11 = \$336.89$

27. \$6000, discount rate 12%, for 6 months
$\text{Discount} = (6000)(.12)\left(\frac{6}{12}\right) = \$360$
$\text{Proceeds} = 6000 - 360 = \$5640$
I = prt
$360 = (5460)(r)\left(\frac{6}{12}\right)$
$r = 12.77\%$

29. \$17,400, discount rate 9.5%, for 1 year

Discount = (17,400)(.095)(1) = \$1653

Proceeds = 17,400 - 1653 = \$15,747

1653 = (15,747)(r)(1)

r = .1050 = 10.50%

31. Interest = $(25{,}900)(.124)\left(\frac{11}{12}\right)$

= \$2943.97

Amount paid back = 25,900 + 2943.97

= \$28,843.97

33. Interest = 21,850 - 20,000 = \$1850

(20,000)(r)(1) = 1850

r = 9.25%

35. Use $P = \frac{A}{1 + rt}$ with A = 1769,

r = .0625, t = 4/12 year.

$$P = \frac{1769}{1 + (.0625)\left(\frac{4}{12}\right)} = \$1732.90$$

37. Interest = $(14{,}500)(.13)\left(\frac{3}{12}\right)$

= \$471.25

Increase in computer price

= 16,000 - 14,500 = \$1500

Amount saved = 1500 - 471.25

= \$1028.75

39. Use P = A(1 - rt) or $A = \frac{P}{1 - rt}$ with

P = 5196, r = .14, and t = 10/12

$$A = \frac{5196}{1 - (.14)\left(\frac{10}{12}\right)}$$

A = \$5882.26 *Amount to be borrowed*

41. Interest = $(4200)(.132)\left(\frac{10}{12}\right)$ = \$462

Proceeds = 4200 - 462 = \$3738

$462 = (3738)(r)\left(\frac{10}{12}\right)$

r = 14.83%

43. Patrick's Interest = $(7000)(.12)\left(\frac{7}{12}\right)$

= 490

Total to be paid back

= 7000 + 490 = 7490

Bank Interest = $(7490)(.137)\left(\frac{2}{12}\right)$

= \$171.02

Proceeds = 7490 - 171.02

= \$7318.98 *Amount store receives*

Section 5.2

1. \$4500 at 8% compounded annually for 20 years. Use

$A = P(1 + r)^t$.

$A = 4500\ (1 + .08)^{20}$ = \$20,974.31

3. \$470 at 12% compounded semiannually for 12 years. Use

$A = P(1 + i)^n$ with

i = .12/2 and n = 12 · 2

$A = 470\left(1 + \frac{.12}{2}\right)^{12 \cdot 2}$ = \$1903.00

5. \$7500 at 10% compounded quarterly for 9 years.

$A = 7500\ \left(1 + \frac{.10}{4}\right)^{9 \cdot 4}$ = \$18,244.01

7. \$6000 at 8% compounded annually for 8 years.

Amount Accrued = $6000\ (1 + .08)^8$

= \$11,105.58

Interest = 11,105.58 - 6000

= \$5105.58

9. \$43,000 at 10% compounded semiannually for 9 years.

Amount Accrued $= 43,000(1 + \frac{.10}{2})^{2 \cdot 9}$

$= \$103,484.63$

Interest = 103,484.63 - 43,000

= \$60,7484.63

11. \$2196.58 at 10.6% compounded quarterly for 4 years

Amount Accrued

$= 2196.58\ (1 + \frac{.108}{4})^{4 \cdot 4}$

= \$3364.147

Interest = 3364.14 - 2196.58

= \$1167.56

13. \$4500 at 8% compounded annually for 9 years

$4500 = P(1 + .08)^9$

P = \$2251.12 *Solve for P*

15. \$15,902.74 at 9.8 compounded annually for 7 years

$15{,}902.74 = P(1 + .098)^7$

P = \$8265.24

17. \$2000 at 8% compounded semiannually for 8 years

$2000 = P(1 + \frac{.08}{2})^{2 \cdot 8}$

P = \$1067.82

19. \$8800 at 12% compounded quarterly for 5 years

$8800 = P\ (1 + \frac{.12}{4})^{4 \cdot 5}$

P = \$4872.35

21. Amount accrued $= 1000(1 + \frac{.08}{4})^{4 \cdot 5}$

= \$1485.95

Since this amount is greater than \$1210, "\$1000 now" is greater.

Use $A = Pe^{rt}$ with P = 20,000 in Exercises 23-27.

23. $A = 20{,}000e^{(.08)(1)} = \$21{,}665.74$

25. $A = 20{,}000e^{(.08)(10)} = \$44{,}510.82$

27. $A = 20{,}000e^{(.08)(3)} = \$25{,}424.98$

29. 4% compounded semiannually

Effective rate

$= (1 + \frac{.04}{2})^2 - 1 = 1.0404 - 1$

= .0404 = 4.04%

31. 8% compounded semiannually

Effective rate

$= (1 + \frac{.08}{2})^2 - 1 = 1.0816 - 1$

= .0816 = 8.16%

33. 12% compounded semiannually

Effective rate

$= (1 + \frac{.12}{2})^2 - 1 = 1.1236 - 1$

= .1236 = 12.36%

35. \$17,200 at 11.4% compounded continuously for 2 years

$17{,}200 = Pe^{(.114)(2)}$

P = \$13,693.34 *Solve for P*

37. \$17,200 at 11.4% compounded continuously for 7 years

$17{,}200 = Pe^{(.114)(7)}$

P = \$7743.93

39. \$150,000 at -2.4% compounded annually for 2 years

$A = 150{,}000\ (\ 1 + (-.024))^2$

= \$142,886.40

41. \$150,000 at -2.4% compounded annually for 2 years

$A = 150{,}000\ (1 + (-.024))^8$

$= \$123{,}506.50$

43. Amount Accrued $= 18{,}000(1 + \frac{.12}{4})^{4 \cdot 7}$

$= \$41{,}182.70$

Interest $= 41{,}182.70 - 18{,}000$

$= \$23{,}182.70$

45. $45(1 + .04)^{10} = \$66.61$

47. $78{,}000(1 + .07)^{12} = \$175{,}670.94$

or about \$176,000

49. Bank 1: $24{,}000(1 + \frac{.08}{2})^{2 \cdot 5}$

$= \$35{,}525.8628$

Bank 2: $24{,}000(1 + \frac{.08}{4})^{4 \cdot 5}$

$= \$35{,}662.7375$

Difference: $35{,}662.7375 - 35{,}525.8628$

$= \$136.87$

51.(a) $18{,}000 = P(1 + .08)^3$

$P = \$14{,}288.98$

(b) Interest $= 18{,}000 - 14{,}288.98$

$= \$3711.02$

(c) $A - 12{,}000(1 + .08)^3$

$= \$15{,}116.54$

Deficit $= 18{,}000 - 15{,}116.54$

$= \$2883.46$

53. $10 = 6(1 + .08)^t$

$\frac{5}{3} = (1.08)^t$

$$t = \frac{\ln \frac{5}{3}}{\ln 1.08}$$

$t = 6.6$ years

55.(a) $2 = (1 + .04)^t$

$$t = \frac{\ln 2}{\ln 1.04}$$

$t = 17.7$ years

(b) $2 = (1 + .06)^t$

$$t = \frac{\ln 2}{\ln 1.06}$$

$t = 11.9$ years

(c) $2 = (1 + .08)^t$

$$t = \frac{\ln 2}{\ln 1.08}$$

$t = 9.0$ years

(d) $2 = (1 + .12)^t$

$$t = \frac{\ln 2}{\ln 1.12}$$

$t = 6.1$ years

57.(a) $2 = (1 + .06)^t$

$$t = \frac{\ln 2}{\ln 1.06}$$

$t \approx 12$ years

(b) $2 = (1 + .02)^t$

$$t = \frac{\ln 2}{\ln 1.02}$$

$t \approx 35$ years

Section 5.3

Use ar^{n-1} with $n = 5$ in Exercises 1-7.

1. $a = 3,\ r = 2$

$3 \cdot 2^4 = 48$

3. $a = -8,\ r = 3$

$(-8) \cdot 3^4 = -648$

5. $a = 1,\ r = -3$

$1 \cdot (-3)^4 = 81$

7. $a = 1024, r = 1/2$

$1024 \cdot \left(\frac{1}{2}\right)^4 = 64$

Use $S_n = \frac{a(r^n - 1)}{r - 1}$ with $n = 4$ in Exercises 9-13.

9. $a = 1, r = 2$

$S_4 = \frac{1(2^4 - 1)}{2 - 1} = 15$

11. $a = 5, r = 1/5$

$$S_4 = \frac{5\left[\left(\frac{1}{5}\right)^4 - 1\right]}{\frac{1}{5} - 1}$$

$$= \frac{5\left(\frac{-624}{625}\right)}{\frac{-4}{5}}$$

$$= \frac{5(-624)}{(-4)(125)}$$

$$= \frac{156}{25}$$

13. $a = 128, r = 3/2$

$$S^4 = \frac{128\left[\left(-\frac{3}{2}\right)^4 - 1\right]}{-\frac{3}{2} - 1}$$

$$= \frac{128\left(\frac{81}{16} - 1\right)}{\frac{-5}{2}}$$

$$= \frac{128\left(\frac{65}{16}\right)}{\frac{-5}{2}}$$

$$= -208$$

Use $s_{\overline{n}|i} = \frac{(1 + .05)^n - 1}{.05}$ in Exercises 15-21

15. $s_{\overline{12}|.05} = \frac{(1 + .05)^{12} - 1}{.05}$

$= 15.91713$

17. $s_{\overline{16}|.04} = \frac{(1 + .04)^{16} - 1}{.04}$

$= 21.82453$

19. $s_{\overline{20}|.01} = \frac{(1 + .01)^{20} - 1}{.01}$

$= 22.01900$

21. $s_{\overline{15}|.04} = \frac{(1 + .04)^{15} - 1}{.04}$

$= 20.02359$

Use $S = Rs_{\overline{n}|i}$ in Exercises 23-35.

23. $R = 100, i = .06, n = 4$

$S = 100\left[\frac{(1 + .06)^4 - 1}{.06}\right]$

$= \$437.46$

25. $R = 10{,}000, i = .05, n = 19$

$S = 10{,}000\left[\frac{(1 + .05)^{19} - 1}{.05}\right]$

$= \$305{,}390.04$

27. $R = 8500, i = .06, n = 30$

$S = 8500\left[\frac{(1 + .06)^{30} - 1}{.06}\right]$

$= \$671{,}994.58$

29. $R = 46{,}000, i = .06, n = 32$

$S = 46{,}000\left[\frac{(1 + .06)^{32} - 1}{.06}\right]$

$= \$4{,}180{,}929.79$

31. R = 9200, at 10% compounded semiannually for 7 years

$$S = 9200\left[\frac{(1 + \frac{.10}{2})^{2 \cdot 7} - 1}{\frac{.10}{2}}\right]$$

= $180,307.41

33. R = 800, at 6% compounded semiannually for 12 years

$$S = 800\left[\frac{(1 + \frac{.06}{2})^{2 \cdot 12} - 1}{\frac{.06}{2}}\right]$$

= $27,541.18

35. R = 15,000, at 12% compounded quarterly for 6 years

$$S = 1500\left[\frac{(1 + \frac{.12}{.4})^{4 \cdot 6} - 1}{\frac{.12}{4}}\right]$$

= $516,397.05

Use $S = Rs_{n+1|i} - r$ for Exercises 37 - 43.

37. R - 600, i = .06, n = 8, compounded annually.

$$S = 600\left[\frac{(1 + .06)^{8+1} - 1}{.06}\right] - 600$$

= $6294.79

39. R = 20,000, i = .08, n = 6, compounded annually

$$S = 20,000\left[\frac{(1 + .08)^{6+1} - 1}{.08}\right] - 20,000$$

= $158,456.07

41. $1000 at the beginning of each year for 9 years at 8% compounded annually.

$$S = 1000\left[\frac{(1 + .08)^{9+1} - 1}{.08}\right] - 1000$$

= $13,486.56

43. $100 at the beginning of each quarter for 9 years at 12% compounded quarterly

$$S = 100\left[\frac{(1 + \frac{.12}{4})^{36+1} - 1}{\frac{.12}{4}}\right] - 100$$

= $6517.42

45. S = $10,000 at 8% compounded annually, at the end of each year for 12 years

$$10,000 = R\left[\frac{(1 + .05)^{12} - 1}{.05}\right]$$

R = $628.25 *Solve for R*

47. S = $147,200 at 12% compounded quarterly, at the end of each quarter for 8 years

$$50,000 = R\left[\frac{(1 + \frac{.12}{4})^{4 \cdot 8} - 1}{\frac{.12}{4}}\right]$$

R = $952.33

49.(a) $S = 12,000\left[\frac{(1 + .08)^{9} - 1}{.08}\right]$

= $149,850.69

(b) $S = 12,000\left[\frac{(1 + .06)^{9} - 1}{.06}\right]$

= $137,895.79

(c) Amount lost

$= 149{,}850.69 - 137{,}895.79$

$= \$11{,}954.90$

51. $$S = 80\left[\frac{(1 + \frac{.075}{12})^{45+1} - 1}{\frac{.075}{12}}\right] - 80$$

$= \$4168.30$

53. For the first 15 years, he deposits \$1000 every quarter.

$$S = 1000\left[\frac{(1 + \frac{.11}{4})^{60} - 1}{\frac{.11}{4}}\right]$$

$= \$148{,}809.14$

In the next 5 years, no more is deposited.

$$S = 148{,}809.14\left[1 + \frac{.11}{4}\right]^{4 \cdot 5}$$

$= \$256{,}015.48$

55. For the first 8 years:

$$S = 2435\left[\frac{(1 + \frac{.06}{2})^{2 \cdot 8+1} - 1}{\frac{.06}{2}}\right] - 2435$$

$= \$50{,}554.47$

In the last 5 years:

$$S = 67{,}940.98\left(1 + \frac{.06}{2}\right)^{2 \cdot 5}$$

$= \$67{,}940.98$

57. (a) $$10{,}000 = P\left[\frac{\left(1 + \frac{.08}{4}\right)^{4 \cdot 8} - 1}{\frac{.08}{4}}\right]$$

$P = \$226.11$

(b) $$10{,}000 = P\left[\frac{\left(1 + \frac{.06}{4}\right)^{4 \cdot 8} - 1}{\frac{.06}{4}}\right]$$

$P = \$245.77$

59. $$18{,}000 = P\left[\frac{\left(1 + \frac{.12}{4}\right)^{4 \cdot 6} - 1}{\frac{.12}{4}}\right]$$

$P = \$522.85$

61. $i = 6\%$

$$S = 1000\left[\frac{\left(1 + \frac{.06}{2}\right)^{2 \cdot 25} - 1}{\frac{.06}{2}}\right]$$

$= \$112{,}796.87$

63. $i = 10\%$

$$S = 1000\left[\frac{\left(1 + \frac{.10}{2}\right)^{2 \cdot 25} - 1}{\frac{.10}{2}}\right]$$

$= \$209{,}348.00$

65. \$11,000, earns 8% compounded annually, 7 annual payments

$$11{,}000 = P\left[\frac{\left(\frac{1 + .06}{2}\right)^{2 \cdot 6} - 1}{\frac{.06}{2}}\right]$$

$P = \$775.08$

67. \$50,000, earns 10% compounded quarterly for 2 1/2 years

$$50{,}000 = P\left[\frac{\left(1 + \frac{.10}{4}\right)^{4(2 \cdot 5)} - 1}{\frac{.10}{4}}\right]$$

$P = \$4462.94$

69. \$6000, earns 8% compounded monthly for 3 years.

$$6000 = P\left[\frac{\left(1 + \frac{.08}{12}\right)^{12 \cdot 3} - 1}{\frac{.08}{12}}\right]$$

$$P = \$148.02$$

71. $$S = 892.17\left[\frac{\left(1 + \frac{.075}{12}\right)^{12 \cdot 2} - 1}{\frac{.075}{12}}\right]$$

$$= \$23,023.98$$

73. (a) Total interest = (60,000)(.08)(7) = 33,600

Each interest payment = $\frac{33,600}{7 \cdot 4}$ = \$1200

(b) $$60,000 = P\left[\frac{\left(1 + \frac{.06}{2}\right)^{2 \cdot 7} - 1}{\frac{.06}{2}}\right]$$

$$P = \$3511.58$$

(c) Use the amount of each deposit as calculated in part (b). The interest earned is calculated as follows:

(Previous total) $\cdot \left(\frac{.06}{2}\right)$

Each total is calculated as follows. (Previous total) + (amount of deposit) + (interest earned)

See the table in the answer section at the back of the textbook. Notice that the last payment in a sinking fund table may differ slightly from the others because of earlier rounding.

Section 5.4

Use $a_{\overline{n}|i} = \frac{1 - (1 + i)^{-n}}{i}$ in Exercises 1-7.

1. $$a_{\overline{15}|.06} = \frac{1 - (1 + .06)^{-15}}{.06} = 9.71225$$

3. $$a_{\overline{18}|.04} = \frac{1 - (1 + .04)^{-18}}{.04} = 12.65930$$

5. $$a_{\overline{16}|.01} = \frac{1 - (1 + .01)^{-16}}{.01} = 14.71787$$

7. $$a_{\overline{6}|.015} = \frac{1 - (1 + .015)^{-6}}{.015} = 5.69719$$

Use $P - R \cdot a_{\overline{n}|i}$ in Exercises 9-15.

9. Payments of \$1000 for 9 years at 8% compounded annually

$$P = 1000\left[\frac{1 - (1 + .08)^{-9}}{.08}\right] = \$6246.89$$

11. Payments of \$890 for 16 years at 8% compounded annually

$$P = 890\left[\frac{1 - (1 + .08)^{-16}}{.08}\right] = \$7877.72$$

13. Payments of \$10,000 for 15 years at 10% compounded semiannually

$$P = 10,000\left[\frac{1 - \left(1 + \frac{.10}{2}\right)^{-30}}{\frac{.10}{2}}\right]$$

$$= \$153,724.51$$

15. Payments of $15,806 for 3 years at 10.8% compounded quarterly

$$P = 15{,}806\left[\frac{1-\left(1+\frac{.108}{4}\right)^{-12}}{\frac{.108}{4}}\right]$$

$= \$160{,}188.18$

17. 4%, compounded annually for 15 years

$$P = 10{,}000\left[\frac{1-(1+.04)^{-15}}{.04}\right]$$

$= \$111{,}183.87$

19. 6%, compounded annually for 15 years

$$P = 10{,}000\left[\frac{1-(1+.06)^{-15}}{.06}\right]$$

$= \$97{,}122.49$

21. 12%, compounded annually for 15 years

$$P = 10{,}000\left[\frac{1-(1+.12)^{-15}}{.12}\right]$$

$= \$68{,}108.64$

23. $P = 2000\left[\frac{1-(1+.08)^{-9}}{.08}\right]$

$= \$12{,}493.78$

25. (a) $6000 = R\left[\frac{1-\left(1+\frac{.12}{12}\right)^{-48}}{.01}\right]$

$R = \$158.00$ *Solve for R*

(b) Total interest
$= (158)(48) - 6000$
$= 7584 - 6000$
$= \$1584.00$

27. $S = 1000\left[\frac{(1+.06)^8 - 1}{.06}\right]$

$= \$9897.47$

$9897.47 = P(1+.05)^8$

$P = \$6699.00$

Use $R = \frac{P}{a_{n|i}}$ in Exercises 29-39.

29. $1000, 9 annual payments at 8%

$$R = \frac{1000}{\left[\frac{1-(1+.08)^{-9}}{.08}\right]}$$

$= \$160.08$

31. $41,000, 10 semiannual payments at 10%

$$R = \frac{41{,}000}{\left[\frac{1-\left(1+\frac{.10}{2}\right)^{-10}}{\frac{.10}{2}}\right]}$$

$= \$5309.69$

33. $140,000, 15 quarterly payments at 12%

$$R = \frac{140{,}000}{\left[\frac{1-\left(1+\frac{.12}{4}\right)^{-15}}{\frac{.12}{4}}\right]}$$

$= \$11{,}727.32$

35. $5500, 24 monthly payments at 12%

$$R = \frac{5500}{\left[\frac{1-\left(1+\frac{.12}{12}\right)^{-24}}{\frac{.12}{12}}\right]}$$

$= \$258.90$

37. \$49,560 at 10.75% for 25 years

$$R = \frac{49{,}560}{\left[\frac{1 - \left(1 + \frac{.1075}{12}\right)^{-300}}{\frac{.1075}{12}}\right]}$$

= \$476.81

39. \$53,762 at 12.45% for 30 years

$$R = \frac{53{,}762}{\left[\frac{1 - \left(1 + \frac{.1245}{12}\right)^{-360}}{\frac{.1245}{12}}\right]}$$

= \$571.69

41. (a) Amount Borrowed

= 285,000 - 60,000

= \$225,000

For 15 years:

$$R = \frac{225{,}000}{\left[\frac{1 - \left(1 + \frac{.095}{12}\right)^{-180}}{\frac{.095}{12}}\right]}$$

= \$2349.51

Interest = (2349.51)(180) - 225,000

= \$197,911.00

(b) For 20 years:

$$R = \frac{225{,}000}{\left[\frac{1 - \left(1 + \frac{.095}{12}\right)^{-240}}{\frac{.095}{12}}\right]}$$

= \$2097.30 *Rounded*

Interest = (2097.295173)(240)

- 225,000

= \$278,350.84

(c) For 25 years:

$$R = \frac{225{,}000}{\left[\frac{1 - \left(1 + \frac{.095}{12}\right)^{-300}}{\frac{.095}{12}}\right]}$$

= \$1965.82 *Rounded*

Interest = (1965.8175)(300) - 225,000

= \$364,745.25

43.

$$R = \frac{4000}{\left[\frac{1 - (1 + .08)^{-4}}{.08}\right]}$$

= \$1207.68

Interest = (Principal at end of previous period) · (.08)

Amount of payment - Interest

= Portion to principal

See the amortization schedule in the answer section at the back of the textbook.

45. Purchase = (1048)(8)

= 8384

Amount financed = 8384 - 1200

= \$7184

$$\text{Payment} = \frac{7184}{\left[\frac{1 - \left(1 + \frac{.18}{12}\right)^{-48}}{\frac{.18}{12}}\right]}$$

= \$211.03

Interest = (Principal at end of previous period) · $\left(\frac{.18}{12}\right)$

Amount of payment - Interest

= Portion to principal

See the amortization schedule in the answer section at the back of the textbook.

47. Amount financed = 81,000 - 20,000 = $61,000

$$\text{Monthly payment} = \frac{61{,}000}{\left[\frac{1-\left(1+\frac{.11}{12}\right)^{-360}}{\frac{.11}{12}}\right]}$$

= $580.92 *Rounded*

Remaining balance

$$= 580.92\left[\frac{1-\left(1+\frac{.11}{12}\right)^{-260}}{\frac{.11}{12}}\right]$$

≈ $57,463

49. Interest in fourth payment is $7.61, read directly from chart.

51. Portion of second payment used to reduce the debt is $79.64, read directly from chart.

53. Interest paid in first four months: 10 + 9.21 + 8.42 + 7.61 = $35.24

55. $$\text{Monthly payment} = \frac{37{,}947.50}{\left[\frac{1-(1+.085)^{-10}}{.085}\right]}$$

= $5783.49

Interest = (Principal at end of previous period) · (.085)

Amount of payment - Interest = Portion to principal

See the amortization schedule in the answer section at the back of the textbook.

Section 5.5

For Exercises 1 and 3, the amount is $82,000 at 12% compounded quarterly.

1. For 3 years:

$$82{,}000 = P\left(1+\frac{.12}{4}\right)^{4\cdot 3}$$

P = $57,513.15

3. For 7 years:

$$82{,}000 = P\left(1+\frac{.12}{4}\right)^{4\cdot 7}$$

P = $35,840.29

5. $72,000 loan, 11 annual payments at 12%

$$R = \frac{72{,}000}{\left[\frac{1-(1+.12)^{-11}}{.12}\right]}$$

= $12,125.91

7. $58,000 loan, 23 quarterly payments at 16%

$$R = \frac{58{,}000}{\left[\frac{1-\left(1+\frac{.16}{4}\right)^{-23}}{\frac{.16}{4}}\right]}$$

= $3903.93

9. $1000 at 10% compunded annually for 17 years

$A = 1000\ (1 + .10)^{17} = \$5054.47$

11. $2500 at 16% compunded quarterly for 3 3/4 years

$$A = 2500\left(1+\frac{.16}{4}\right)^{4(3\ 3/4)}$$

= $4502.36

13. \$2500 at 8.5% for 2 years

$I = (2500)(.085)(2)$

$= \$425.00$

15. \$32,662 at 8.882% for 225 days

$I = (32,662)(.08882)\left(\frac{225}{360}\right)$

$= \$1813.15$

17. \$22,500 at 12% compounded quarterly for 5 1/4 years

Amount accrued

$= 22,500\left(1 + \frac{.12}{4}\right)^{4(5\ 1/4)}$

$= \$41,856.63$

Interest $= 41,856.63 - 22,500$

$= \$19,356.63$

For Exercises 19 and 21, the deposit is \$32,750 at 10% compounded annually.

19. For 2 years:

Compound amount $= 32,750e^{(.10)(2)}$

$= \$40,000.94$

Interest $= 40,000.94 - 32,750$

$= \$7,250.94$

21. For 7 1/2 years:

Compound amount $= 32,750e^{(.10)(7\ 1/2)}$

$= \$69,331.75$

Interest $= 69,331.75 - 32,750$

$= \$36,581.75$

23. \$50,000 for 11 months at 14%

$P = \dfrac{50,000}{1 + (.14)\left(\frac{11}{12}\right)}$

$= \$44,313.15$

25. \$122,300 for 138 days, at 11.75%

$P = \dfrac{122,300}{1 + (.1175)\left(\frac{138}{360}\right)}$

$= \$117,028.83$

27. \$72,113 for 11 months, discount rate 15%

Proceeds $= 72,113\left[1 - .15\left(\frac{11}{12}\right)\right]$

$= \$62,197.46$

29. \$267,561 for 112, discount rate 15.72%

Proceeds $= 267,100\left[1 - .1572\left(\frac{271}{360}\right)\right]$

$= \$235,492.28$

31. \$1000 end of each year for 12 years, at 10% compounded annually

$S = 1000\left[\frac{(1 + .10)^{12} - 1}{.10}\right]$

$= \$21,384.28$

33. \$250 at end of each quarter for 7 1/4 years, at 12% compounded quarterly

$$S = 250\left[\frac{\left(1 + \frac{.12}{4}\right)^{4(7\ 1/4)} - 1}{\frac{.12}{4}}\right]$$

$= \$11,304.71$

35. \$100 at beginning of each year for 5 years, at 8% compounded annually

$S = 100\left[\frac{(1 + .08)^{5+1} - 1}{.08}\right] - 100$

$= \$633.59$

37. \$10,000 loan, 10% compounded annually, 7 annual payments

$10,000 = R\left[\frac{(1 + .10)^{-7} - 1}{.10}\right]$

$R = \$1054.05$

39. \$100,000, 12% compounded semiannually, 9 semiannual payments

$$100{,}000 = R\left[\frac{\left(1 + \frac{.12}{2}\right)^{-9} - 1}{\frac{.12}{2}}\right]$$

R = \$8702.22

41. Annual payments of \$1200 for 7 years, at 10% compounded annually

$$P = 1200\left[\frac{1 - (1 + .10)^{-7}}{.10}\right]$$

= \$5842.10

43. Quarterly payments of \$1500 for 5 1/4 years at 12% compounded quarterly

$$P = 1500\left[\frac{1 - \left(1 + \frac{.12}{4}\right)^{-4(5\ 1/4)}}{\frac{.12}{4}}\right]$$

= \$23,122.54

45. \$8500 loan repaid in semiannual payments for 3 1/2 years at 12%
Monthly payment:

$$8500 = R\left[\frac{1 - \left(1 + \frac{.12}{2}\right)^{-2(3\ 1/2)}}{\frac{.12}{2}}\right]$$

R = \$1522.65
Interest = (Principal at end of previous period) $\cdot \left(\frac{.12}{2}\right)$
Portion to Principal = Amount of payment - Interest

See the amortization schedule in the answer section at the back of the textbook.

47. $5320 = (42{,}000)(r)\left(\frac{10}{12}\right)$

$r = .152 \approx 15\%$

49. Amount due = $7850\left[1 + (.10)\left(\frac{5}{12}\right)\right]$
= \$8177.08
Discount = $(8177.08)(.132)\left(\frac{1}{12}\right)$
= \$89.95
Proceeds = 8177.08 - 89.95
= \$8087.13

51. $$P = \frac{2800}{\left(1 + \frac{.12}{12}\right)^{17}}$$

= \$2364.26

53. Final amount = $$\left[\frac{\left(1 + \frac{.10}{2}\right)^{2(7\ 1/2)} - 1}{\frac{.10}{2}}\right]$$

= \$107,892.82
Amount invested = (5000)(7.5)(2)
= \$75,000
Interest = 107,892.82 - 75,000
= \$32,892.82

55. $$P = \frac{7500}{\left(1 + \frac{.10}{2}\right)^{2\cdot 3}}$$

= \$5596.62

57. $A = 3250(1 + .09)^4 = \$4587.64$

59. Monthly payment:

$$115{,}700 = R\left[\frac{1-\left(1+\frac{.105}{12}\right)^{-300}}{\frac{.105}{12}}\right]$$

$\$1092.42 = R$

Total of all payments

$= (1092.42)(300) = \$327{,}726$

Interest $= 327{,}726 - 115{,}700$

$= \$212{,}026$

Chapter 5 Review Exercises

1. $I = (15{,}903)(.08)\left(\frac{8}{12}\right)$

$= \$848.16$

3. $I = (42{,}368)(.1522)\left(\frac{5}{12}\right)$

$= \$2686.84$

5. $I = (2390)(.187)\left(\frac{86}{365}\right)$

$= \$105.30$

May 3 to July 28 is 86 days

7. $P = \dfrac{25{,}000}{1 + (.07)\left(\frac{10}{12}\right)}$

$= \$23{,}622.05$

9. $P = \dfrac{80{,}612}{1 + (.0677)\left(\frac{128}{360}\right)}$

$= \$78{,}717.19$

11. Proceeds $= 802.34\left[1 - (.186)\left(\frac{11}{12}\right)\right]$

$= \$665.54$

13. Amount due

$= 5800\left[1 + (.10)\left(\frac{10}{12}\right)\right]$

$= \$6283.3\overline{3}$

Proceeds $= \$6283.3\overline{3}\left[1 - (.1445)\left(\frac{3}{12}\right)\right]$

$= \$6056.35$

15. $1000 = A(1 - (.14)(1))$

$1000 = .86A$

$1162.79 = A$

Effective annual rate

$= \dfrac{1162.79-1000}{1000}$

$= .163 = 16.3\%$

17. $A = 1000(1 + .08)^{9}$

$= \$1999.00$

19. $A = 19{,}456.11\left(1 + \frac{.12}{2}\right)^{2\cdot 7}$

$= \$43{,}988.40$

21. $A = 1900\left(1 + \frac{.12}{4}\right)^{4\cdot 9}$

$= \$5506.73$

23. $A = 2500\left(1 + \frac{.12}{12}\right)^{12\cdot 3}$

$= \$3576.92$

25. Amount Accrued $= 3954(1 + .08)^{12}$

$= \$9956.84$

Interest $= 9956.84 - 3954$

$= \$6002.84$

27. Amount Accrued $= 7801.72\left(1 + \frac{.12}{4}\right)^{4\cdot 5}$

$= \$14{,}090.77$

Interest $= 14{,}090.77 - 7801.72$

$= \$6289.05$

29. Amount Accrued

$$= (12{,}903.45)\left(1 + \frac{.1237}{4}\right)^{29}$$

$$= \$31{,}209.79$$

$$\text{Interest} = 31{,}209.79 - 12{,}903.45 = \$18{,}306.34$$

31. $$P = \frac{5000}{(1 + .08)^9} = \$2501.24$$

33. $$P = \frac{42{,}000}{\left(1 + \frac{.18}{12}\right)^{12 \cdot 7}} = \$12{,}025.46$$

35. $$P = \frac{1347.89}{\left(1 + \frac{.1377}{2}\right)^{2(3\ 1/2)}} = \$845.74$$

37. $$P = \frac{5000}{\left(1 + \frac{.12}{2}\right)^{2 \cdot 4}} = \$3137.06$$

39. First five terms:
$2,\ 2 \cdot 3,\ 2 \cdot 3^2,\ 2 \cdot 3^3,\ 2 \cdot 3^4$
or 2, 6, 18, 54, 162

41. Sixth term $= (-3) \cdot 2^5 = (-3)(32) = -96$

43. $$S = \frac{-3(3^4 - 1)}{3 - 1} = \frac{-3(80)}{2} = -120$$

45. $$s_{30|.01} = \frac{(1 + .01)^{30} - 1}{.01} = 34.78489$$

47. $$S = 500\left[\frac{\left(1 + \frac{.06}{2}\right)^{2 \cdot 8} - 1}{\frac{.06}{2}}\right] = \$10{,}078.44$$

49. $$S = 4000\left[\frac{\left(1 + \frac{.08}{4}\right)^{4 \cdot 7} - 1}{\frac{.08}{4}}\right] = \$148{,}204.84$$

51. $$S = 672\left[\frac{\left(1 + \frac{.12}{4}\right)^{28+1} - 1}{\frac{.12}{4}}\right] - 672 = \$29{,}715.07$$

53. $$S = 491\left[\frac{\left(1 + \frac{.094}{4}\right)^{4 \cdot 9} - 1}{\frac{.094}{4}}\right] = \$27{,}320.71$$

55. $$6500 = R\left[\frac{(1 + .08)^6 - 1}{.08}\right]$$

$$R = \$886.05$$

57. $$233{,}188 = R\left[\frac{\left(1 + \frac{.097}{4}\right)^{4(7\ 3/4)} - 1}{\frac{.097}{4}}\right]$$

$$R = \$5132.48$$

59. $$P = 850\left[\frac{1 - (1 + .08)^{-4}}{.08}\right] = \$2815.31$$

61. $P = 4210\left[\dfrac{1 - \left(1 + \dfrac{.086}{2}\right)^{-2 \cdot 8}}{\dfrac{.086}{2}}\right]$

$= \$47,988.11$

63. $20,000 = R\left[\dfrac{1 - (1 + .089)^{-9}}{.089}\right]$

$R = \$3322.43$

65. $80,000 = R\left[\dfrac{1 - (1 + .08)^{-9}}{.08}\right]$

$R = \$12,806.38$

67. $32,000 = R\left[\dfrac{1 - \left(1 + \dfrac{.094}{4}\right)^{-17}}{\dfrac{.094}{4}}\right]$

$R = \$2305.07$

69. $56,890 = R\left[\dfrac{1 - \left(1 + \dfrac{.1474}{12}\right)^{-12 \cdot 25}}{\dfrac{.1474}{12}}\right]$

$R = \$717.21$

71. (a) Monthly payment:

$80,000 = R\left[\dfrac{1 - \left(1 + \dfrac{.12}{12}\right)^{-12 \cdot 30}}{\dfrac{.12}{12}}\right]$

$R = \$822.89$

Total payments
$= (822.89)(12)(30)$
$= \$296,240.40$

Interest $= 296,240.40 - 80,000$
$= \$216,240.40$

(b) Monthly payment:

$80,000 = R\left[\dfrac{1 - \left(1 + \dfrac{.12}{12}\right)^{-12 \cdot 25}}{\dfrac{.12}{12}}\right]$

$R = \$842.58$

Total payments $= (842.58)(12)(25)$
$= \$252,774$

Interest $= 252,774 - 80,000$
$= \$172,774$

(c) Monthly payment:

$80,000 = R\left[\dfrac{1 - \left(1 + \dfrac{.12}{12}\right)^{-12 \cdot 15}}{\dfrac{.12}{12}}\right]$

$R = \$960.13$

Total payments $= (960.13)(12)(15)$
$= \$172,823.40$

Interest $= 172,823.40 - 80,000$
$= \$92,823.40$

Note: Rounding in early steps produces differences in the final answers.

73. Monthly payment:

$$5000 = R\left[\frac{1 - \left(1 + \frac{.10}{2}\right)^{-2 \cdot 3}}{\frac{.10}{2}}\right]$$

$R = 985.09$

Interest = (Principal at end of previous period) $\cdot \left(\frac{.10}{2}\right)$

Portion to Principal = Monthly payment - Interest

Principal at end of period = Period at end of previous period - Portion to principal

See amortization schedule in the answer section at the back of the textbook.

Case 4 Exercises

1. Cash flow
$= -.52(6228 + 2976) + .52(26,251) + .48(10,778) \approx \$14,038$

2. Cash flow
$= -.52(6228 + 2976) + .52(26,251) + .48(1347) = \$9511$

3. Cash flow
$= -.52(2870 + 6386) + .52(26,251) \approx \$8837$

4. Cash flow
$= -.52(6228 + 2976) + .52(10,618) + .48(6736) \approx \$3969$

CHAPTER 6 SYSTEMS OF LINEAR EQUATIONS AND MATRICES

Section 6.1

1. $x + y = 9 \quad (1)$
$2x - 6 = 0 \quad (2)$

Multiply equation (1) by -2 and add the result to equation (2). The new system is

$$\begin{aligned} x + y &= 9 & (1) \\ -3y &= -18. & (3) \end{aligned}$$

Multiply equation (3) by -1/3 to get

$$\begin{aligned} x + y &= 9 & (1) \\ y &= 6. & (4) \end{aligned}$$

Substitute 6 for y in equation (1) to get x = 3. The solution of the system is (3, 6).

3. $5x + 3y = 7 \quad (1)$
$7x - 3y = -19 \quad (2)$

Multiply equation (1) by 1/5. This gives

$$\begin{aligned} x + \frac{3}{5}y &= \frac{7}{5} & (3) \\ 7x - 3y &= -19. & (2) \end{aligned}$$

Multiply equation (3) by -7 and add the result to equation (2). The new system is

$$\begin{aligned} x + \frac{3}{5}y &= \frac{7}{5} & (3) \\ -\frac{36}{5}y &= -\frac{144}{5}. & (4) \end{aligned}$$

Multiply equation (4) by -5/36 to get

$$\begin{aligned} x + \frac{3}{5}y &= \frac{7}{5} & (3) \\ y &= 4. & (5) \end{aligned}$$

Substitute 4 for y in equation (3) to get x = -1. The solution of the system is (-1, 4).

5. $3x + 2y = -6 \quad (1)$
$5x - 2y = -10 \quad (2)$

Multiply equation (1) by 1/3. This gives

$$\begin{aligned} x + \frac{2}{3}y &= -2 & (3) \\ 5x - 2y &= -10. & (2) \end{aligned}$$

Multiply equation (3) by -5 and add the result to equation (2). The new system is

$$\begin{aligned} x + \frac{2}{3}y &= -2 & (3) \\ -\frac{16}{3}y &= 0. & (4) \end{aligned}$$

Multiply equation (4) by -3/16 to get

$$\begin{aligned} x + \frac{2}{3}y &= -2 & (3) \\ y &= 0. & (5) \end{aligned}$$

Substitute 0 for y in equation (3) to get x = -2. The solution of the system is (-2, 0).

7. $2x - 3y = -7 \quad (1)$
$5x + 4y = 17 \quad (2)$

Multiply equation (1) by 1/2. This gives

$$\begin{aligned} x - \frac{3}{2}y &= -\frac{7}{2} & (3) \\ 5x + 4y &= 17. & (2) \end{aligned}$$

Multiply equation (3) by -5 and add the result to equation (2). The new system is

$$\begin{aligned} x - \frac{3}{2}y &= -\frac{7}{2} & (3) \\ \frac{23}{2}y &= \frac{69}{2}. & (4) \end{aligned}$$

Multiply equation (4) by 2/23 to get

$$\begin{aligned} x - \frac{3}{2}y &= -\frac{7}{2} & (3) \\ y &= 3. & (5) \end{aligned}$$

Substitute 3 for y in equation (3) to get x = 1. The solution of the system is (1, 3).

9. $5p + 7q = 6$ *(1)*
$10p - 3q = 46$ *(2)*
Multiply equation (1) by 1/5. The result is

$$p + \frac{7}{5}q = \frac{6}{5} \quad (3)$$
$$10p - 3q = 46. \quad (2)$$

Multiply equation (3) by -10 and add the result to equation (2). The new system is

$$p + \frac{7}{5}q = \frac{6}{5} \quad (3)$$
$$-17q = 34. \quad (4)$$

Multiply equation (4) by -1/17 to get

$$p + \frac{7}{5}q = \frac{6}{5} \quad (3)$$
$$q = -2 \quad (5)$$

Substitute -2 for q in equation (3) to get p = 4. The solution of the system is (4, -2).

11. $6x + 7y = -2$ *(1)*
$7x - 6y = 26$ *(2)*
Multiply equation (1) by 1/6. This gives

$$x + \frac{7}{6}y = -\frac{1}{3} \quad (3)$$
$$7x - 6y = 26. \quad (2)$$

Multiply equation (3) by -7 and add the result to equation (2). The new system is

$$x + \frac{7}{6}y = -\frac{1}{3} \quad (3)$$
$$-\frac{85}{6}y = \frac{85}{3}. \quad (4)$$

Multiply equation (4) by -6/85 to get

$$x + \frac{7}{6}y = -\frac{1}{3} \quad (3)$$
$$y = -2. \quad (5)$$

Substitute -2 for y in equation (3) to get x = 2. The solution for the system is (2, -2).

13. $3x + 2y = 5$ *(1)*
$6x + 4y = 8$ *(2)*
Multiply equation (1) by 1/3. This gives

$$x + \frac{2}{3}y = \frac{5}{3} \quad (3)$$
$$6x + 4y = 8. \quad (2)$$

Multiply equation (3) by -6 and add the result to equation (2). The new system is

$$x + \frac{2}{3}y = \frac{5}{3} \quad (3)$$
$$0y = -5. \quad (4)$$

Since equation (4) has no solution, the system is inconsistent and thus has no solution.

15. $4x - y = 9$ *(1)*
$-8x + 2y = -18$ *(2)*
Multiply equation (1) by 1/4. This gives

$$x - \frac{1}{4}y = \frac{9}{4} \quad (3)$$
$$-8x + 2y = -18. \quad (2)$$

Multiply equation (3) by 8 and add the result to equation (2). The new system is

$$x - \frac{1}{4}y = \frac{9}{4} \quad (3)$$
$$0y = 0. \quad (4)$$

Since equation (4) is true for all values of y, the solution consists of all ordered pairs (x, y) which are solutions of equations (1), (2), or (3), which are dependent equations and represent the same line.

17. $\frac{x}{2} + \frac{y}{3} = 8$ *(1)*

$\frac{2x}{3} + \frac{3y}{2} = 17$ *(2)*

Multiply equation (1) by 2. This gives

$x + \frac{2}{3}y = 16$ *(3)*

$\frac{2x}{3} + \frac{3}{2}y = 17.$ *(2)*

Multiply equation (3) by -2/3 and add the result to equation (2). The new system is

$x + \frac{2}{3}y = 16$ *(3)*

$\frac{19}{18}y = \frac{19}{3}.$ *(4)*

Multiply equation (4) by 18/19 to get

$x + \frac{2}{3}y = 16$ *(3)*

$y = 6.$ *(5)*

Substitute 6 for y in equation (3) to get x = 12. The solution of the system is (12, 6).

19. $\frac{x}{2} + y = \frac{3}{2}$ *(1)*

$\frac{x}{3} + y = \frac{1}{3}$ *(2)*

Multiply equation (1) by 2. This gives

$x + 2y = 3$ *(3)*

$\frac{x}{3} + y = \frac{1}{3}.$ *(2)*

Multiply equation (3) by -1/3 and add the result to equation (2). The new system is

$x + 2y = 3$ *(3)*

$\frac{1}{3}y = -\frac{2}{3}.$ *(4)*

Multiply equation (4) by 3 to get

$x + 2y = 3$ *(3)*

$y = -2.$ *(5)*

Substitute -2 for y in equation (3) to get x = 7. The solution of the system is (7, -2).

21. $x + y + z = 2$ *(1)*

$2x + y - z = 5$ *(2)*

$x - y + z = -2$ *(3)*

Multiply equation (1) by -2 and add the result to equation (2). The new system is

$x + y + z = 2$ *(1)*

$-y - 3z = 1$ *(4)*

$x - y + z = -2.$ *(3)*

Multiply equation (1) by -1 and add the result to equation (3). This gives

$x + y + z = 2$ *(1)*

$-y - 3z = 1$ *(4)*

$-2y = -4.$ *(5)*

Multiply equation (4) by -2 and add the result to equation (5). This gives

$x + y + z = 2$ *(1)*

$-y - 3z = 1$ *(4)*

$6z = -6.$ *(6)*

Multiply equation (6) by 1/6 to get

$x + y + z = 2$ *(1)*

$-y - 3z = 1$ *(4)*

$z = -1.$ *(7)*

Substitute -1 for z in equation (4) to get y = 2. Finally, substitute -1 for z and 2 for y in equation (1) to get x = 1. The solution of the system is (1, 2, -1).

23. $x + 3y + 4z = 14$ *(1)*

$2x - 3y + 2z = 10$ *(2)*

$3x - y + z = 9$ *(3)*

The x-coefficient in the first equation is already 1, so the first equation remains x + 3y + 4z = 14. Multiply this equation by -2, and add the resulting equation to the second equation. Then -9y - 6z = -18 is the result. Next, multiply the first equation by -3, and add it to the third equation. $-10y - 11z = -33$ is the result. We now have the system

$$\begin{aligned} x + 3y + 4z &= 14 \quad (1) \\ -9y - 6z &= -18 \quad (4) \\ -10y - 11z &= -33. \quad (5) \end{aligned}$$

Multiply equation (4) by -1/9 to get

$$\begin{aligned} x + 3y + 4z &= 14 \quad (1) \\ y + \frac{2}{3}z &= 2 \quad (6) \\ -10y - 11z &= -33. \quad (5) \end{aligned}$$

Multiply equation (6) by 10 and add the result to equation (5). This gives

$$\begin{aligned} x + 3y + 4z &= 14 \quad (1) \\ y + \frac{2}{3}z &= 2 \quad (6) \\ -\frac{13}{3}z &= -13. \quad (7) \end{aligned}$$

Multiply equation (7) by -3/13 to get

$$\begin{aligned} x + 3y + 4z &= 14 \quad (1) \\ y + \frac{2}{3}z &= 2 \quad (6) \\ z &= 3. \quad (8) \end{aligned}$$

Substitute 3 for z in equation (6) to get $y = 0$. Now substitute 0 for y and 3 for z in equation (1) to get $x = 2$. The solution is (2, 0, 3).

25.
$$\begin{aligned} x + 2y + 3z &= 8 \quad (1) \\ 3x - y + 2z &= 5 \quad (2) \\ -2x - 4y - 6z &= 5 \quad (3) \end{aligned}$$

Multiply equation (1) by -3 and add the result to equation (2). Also, multiply equation (1) by 2 and add the result to equation (3). The new system is

$$\begin{aligned} x + 2y + 3z &= 8 \quad (1) \\ -7y - 7z &= -19 \quad (4) \\ 0 &= 21. \quad (5) \end{aligned}$$

The last equation indicates that the system is inconsistent and thus has no solution.

27.
$$\begin{aligned} 2x - 4y + z &= -4 \quad (1) \\ x + 2y - z &= 0 \quad (2) \\ -x + y + z &= 6 \quad (3) \end{aligned}$$

First exchange the first and second equations to obtain the new system

$$\begin{aligned} x + 2y - z &= 0 \quad (2) \\ 2x - 4y + z &= -4 \quad (1) \\ -x + y + z &= 6. \quad (3) \end{aligned}$$

Multiply equation (2) by -2 and add the result to equation (1). Also add equation (2) to equation (3). This gives

$$\begin{aligned} x + 2y + z &= 0 \quad (2) \\ -8y + 3z &= -4 \quad (4) \\ 3y &= 6. \quad (5) \end{aligned}$$

Multiply equation (5) by 1/3 to get

$$\begin{aligned} x + 2y - z &= 0 \quad (2) \\ -8y + 3z &= -4 \quad (4) \\ z &= 2. \quad (5) \end{aligned}$$

Substitute 2 for y in equation (4) to get $z = 4$. Substitute 2 for y and 4 for z in equation (2) to get $x = 0$. The solution is (0, 2, 4).

29.
$$\begin{aligned} x + 4y - z &= 6 \quad (1) \\ 2x - y + z &= 3 \quad (2) \\ 3x + 2y + 3z &= 16 \quad (3) \end{aligned}$$

Multiply equation (1) by -2 and add to equation (2), to get

$$\begin{aligned} x + 4y - z &= 6 \quad (1) \\ -9y + 3z &= -9 \quad (4) \\ 3x + 2y + 3z &= 16. \quad (3) \end{aligned}$$

Now multiply (1) by -3, and add to (3).

$$\begin{aligned} x + 4y - z &= 6 \quad (1) \\ -9y + 3z &= -9 \quad (4) \\ -10y + 6z &= -2 \quad (5) \end{aligned}$$

Multiply (4) by -1/9.

$$\begin{aligned} x + 4y - z &= 6 \quad (1) \\ y - \frac{1}{3}z &= 1 \quad (6) \\ -10y + 6z &= -2 \quad (5) \end{aligned}$$

Now multiply (6) by 10, and add to (5).

$$\begin{aligned} x + 4y - z &= 6 \quad (1) \\ y - \frac{1}{3}z &= 1 \quad (6) \\ \frac{8}{3}z &= 8 \quad (7) \end{aligned}$$

Multiply equation (7) by 3/8 to get

$$\begin{aligned} x + 4y - z &= 6 \quad (1) \\ y - \frac{1}{3}z &= 1 \quad (6) \\ z &= 3. \quad (8) \end{aligned}$$

Substitute 3 for z in (6) to get y = 2. Then substitute 2 for y and 3 for z in equation (1) to get x = 1. The solution is (1, 2, 3).

31.

$$\begin{aligned} 5m + n - 3p &= -6 \quad (1) \\ 2m + 3n + p &= 5 \quad (2) \\ -3m - 2n + 4p &= 3 \quad (3) \end{aligned}$$

We can get the coefficient of m in (1) to be 1 by multiplying (1) by 1/5, but the price we pay is immediate introduction of fractions. An alternate approach is to multiply (2) by -2, and add to (1). Then

$$\begin{aligned} m - 5n - 5p &= -16 \quad (4) \\ 2m + 3n + p &= 5 \quad (2) \\ -3m - 2n + 4p &= 3. \quad (3) \end{aligned}$$

Now multiply (4) by -2 and add to (2). At the same time, multiply (4) by 3, and add to (3). The result is

$$\begin{aligned} m - 5n - 5p &= -16 \quad (4) \\ 13n + 11p &= 37 \quad (5) \\ -17n - 11p &= -45. \quad (6) \end{aligned}$$

Normally, we would now multiply (5) by 1/13, but again we can avoid some fractions by first adding (6) to (5):

$$\begin{aligned} m - 5n - 5p &= -16 \quad (4) \\ -4n &= -8 \quad (7) \\ -17n - 11p &= -45. \quad (6) \end{aligned}$$

Now we multiply (7) by -1/4, to obtain

$$\begin{aligned} m - 5n - 5p &= -16 \quad (4) \\ n &= 2 \quad (8) \\ -17n - 11p &= -45. \quad (6) \end{aligned}$$

Then (8), multiplied by 17 and added to (6), becomes

$$\begin{aligned} m - 5n - 5p &= -16 \quad (4) \\ n &= 2 \quad (8) \\ -11p &= -11 \quad (9) \end{aligned}$$

Multiply equation (9) by -1/11 to get

$$\begin{aligned} m - 5n - 5p &= -16 \quad (4) \\ n &= 2 \quad (8) \\ p &= 1. \quad (9) \end{aligned}$$

Substitute n = 2 and p = 1 into equation (4) to get m = -1. The solution is (-1, 2, 1).

33.

$$\begin{aligned} a - 3b - 2c &= -3 \quad (1) \\ 3a + 2b - c &= 12 \quad (2) \\ -a - b + 4c &= 3 \quad (3) \end{aligned}$$

Multiply equation (1) by -3 and add the result to equation (2). Also, add equation (1) to equation (3). The new system is

$$\begin{aligned} a - 3b - 2c &= -3 \quad (1) \\ 11b + 5c &= 21 \quad (4) \\ -4b + 2c &= 0. \quad (5) \end{aligned}$$

Multiply equation (4) by 1/11. This gives

$$\begin{aligned} a - 3b - 2c &= -3 \quad (1) \\ b + \frac{5}{11}c &= \frac{21}{11} \quad (6) \\ -4b + 2c &= 0. \quad (5) \end{aligned}$$

Multiply equation (6) by 4 and add the result to equation (5). This gives

$$\begin{aligned} a - 3b - 2c &= -3 \quad (1) \\ b + \frac{5}{11}c &= \frac{21}{11} \quad (6) \\ \frac{42}{11}c &= \frac{84}{11}. \quad (7) \end{aligned}$$

Multiply equation (7) by 11/42 to get

$$a - 3b - 2c = -3 \quad (1)$$

$$b + \frac{5}{11}c = \frac{21}{11} \quad (6)$$

$$c = 2 \quad (8)$$

Substitute 2 for c in equation (6) to get $b + \frac{10}{11} = \frac{21}{11}$, or $b = \frac{11}{11} = 1$.

Substitute 2 for c and 1 for b in equation (1) to get a = 4. The solution is (4, 1, 2).

35. $5x + 3y + 4z = 19 \quad (1)$

$3x - y + z = -4 \quad (2)$

If we begin by multiplying equation (1) by 1/5, we will immediately introduce fractions. The work will be easier if we first multiply equation (2) by 3 and add the result to the equation (1).

$$14x + 7z = 7 \quad (3)$$

$$3x - y + z = -4 \quad (2)$$

Multiply equation (3) by 1/14.

$$x + \frac{1}{2}z = \frac{1}{2} \quad (4)$$

$$3x - y + z = -4 \quad (2)$$

Since there are only two equations, it is not possible to continue with the echelon method. To complete the solution, solve equation (4) for x in terms of the parameter z.

$$x = \frac{1}{2} - \frac{1}{2}z$$

$$= \frac{1 - z}{2}$$

Now substitute (1 - z)/2 for x in equation (2) and solve for y in terms of the parameter z.

$$3\left(\frac{1 - z}{2}\right) - y + z = -4$$

$$3\left(\frac{1 - z}{2}\right) + z + 4 = y$$

$$\frac{3(1 - z) + 2(z + 4)}{2} = y$$

$$\frac{3 - 3z + 2z + 8}{2} = y$$

$$\frac{11 - z}{2} = y$$

The solution is written ((1 - z)/2, (11 - z)/ 2, z) for any real number z.

37. $x + 2y + 3z = 11 \quad (1)$

$2x - y + z = 2 \quad (2)$

Multiply equation (1) by -2 and add the result to equation (2).

$$x + 2y + 3z = 11 \quad (1)$$

$$-5y - 5z = -20 \quad (2)$$

Multiply equation (2) by -1/5.

$$x + 2y + 3z = 11 \quad (1)$$

$$y + z = 4 \quad (3)$$

Since there are only two equations, it is not possible to continue with the echelon method. To complete the solution, solve equation (3) for y.

$$y = 4 - z$$

Now substitute 4 - z for y in equation (1) and solve for x.

$$x + 2(4 - z) + 3z = 11$$

$$x + 8 - 2z + 3z = 11$$

$$x = 3 - z$$

The solution is written (3 - z, 4 - z, z) for any real number z.

39. $x + y - z = -20 \quad (1)$

$2x - y + z = 11 \quad (2)$

Multiply equation (1) by -2 and add the result to equation (2).

$$x + y - z = -20 \quad (1)$$

$$-3y + 3z = 51 \quad (3)$$

Multiply equation (3) by -1/3.

$$x + y - z = -20 \quad (1)$$

$$y - z = -17 \quad (4)$$

Since there are only two equation, it is not possible to continue the echelon method. Solve equation (4) for y.

$$y = -17 + z$$

Now substitute $-17 + z$ for y in equation (1).

$$x - 17 + z - z = -20$$
$$x = -3$$

The solution is written $(-3, -17 + z, z)$ for any real number z.

41. If $(-2, 1)$ is on the line $ax + by = 5$, then

$-2a + b = 5.$ *(1)*

If $(-1, -2)$ is on the line $ax + by = 5$, then

$-a - 2b = 5.$ *(2)*

Solve the linear system

$-2a + b = 5$ *(1)*

$-a - 2b = 5.$ *(2)*

Use the echelon method. Multiply equation (1) by $-1/2$.

$a - \frac{1}{2}b = -\frac{5}{2}$ *(3)*

$-a - 2b = 5$ *(2)*

Add equation (3) to equation (2).

$a - \frac{1}{2}b = -\frac{5}{2}$ *(3)*

$-\frac{5}{2}b = \frac{5}{2}$ *(4)*

Multiply equation (4) by $-2/5$.

$a - \frac{1}{2}b = -\frac{5}{2}$ *(3)*

$b = -1$ *(5)*

Substitute -1 for b in equation (3) to get

$$a - \frac{1}{2}(-1) = -\frac{5}{2}$$
$$a = -3.$$

Therefore, $a = -3$ and $b = -1$.

43. If $(2, 3)$ is on the graph of $ax^2 + bx + c = y$, then

$4a + 2b + c = 3.$ *(1)*

If $(-1, 0)$ is on the graph, then

$a - b + c = 0.$ *(2)*

If $(-2, 2)$ is on the graph, then

$4a - 2b + c = 2.$ *(3)*

Solve the linear system

$4a + 2b + c = 3$ *(1)*

$a - b + c = 0$ *(2)*

$4a - 2b + c = 2.$ *(3)*

Multiply equation (1) by $1/4$.

$a + \frac{1}{2}b + \frac{1}{4}c = \frac{3}{4}$ *(4)*

$a - b + c = 0$ *(2)*

$4a - 2b + c = 2$ *(3)*

Multiply equation (4) by -1 and add it to equation (2). Multiply equation (1) by -4 and add it to equation (3).

$a + \frac{1}{2}b + \frac{1}{4}c = \frac{3}{4}$ *(1)*

$-\frac{3}{2}b + \frac{3}{4}c = -\frac{3}{4}$ *(5)*

$-4b = -1$ *(6)*

From equation (6) we see $b = 1/4$. Substitute $1/4$ for b in equation (5).

$$-\frac{3}{2}\left(\frac{1}{4}\right) + \frac{3}{4}c = -\frac{3}{4}$$
$$-\frac{3}{8} + \frac{3}{4}c = -\frac{3}{4}$$

Multiply by 8.

$$-3 + 6c = -6$$
$$6c = -3$$
$$c = -\frac{1}{2}$$

Substitute $1/4$ for b and $-1/2$ for c in equation (1).

$$4a + 2\left(\frac{1}{4}\right) + \left(-\frac{1}{2}\right) = 3$$
$$4a + \frac{1}{2} - \frac{1}{2} = 3$$
$$a = \frac{3}{4}$$

Therefore, $a = \frac{3}{4}$, $b = \frac{1}{4}$, and

$c = -\frac{1}{2}$.

45. The linear system is

$$2p = -.2q + 5 \quad (1)$$
$$5p = .3q + 5.3. \quad (2)$$

To eliminate decimals, multiply both equations (1) and (2) by 10 and write the variable terms on one side.

$$20p + 2q = 50 \quad (3)$$
$$50p - 3q = 53 \quad (4)$$

Multiply equation (3) by 1/20.

$$p + \frac{2}{20}q = \frac{50}{20} \quad (5)$$
$$50p - 3q = 53 \quad (4)$$

Multiply equation (5) by -50 and add it to equation (4).

$$p + \frac{2}{20}q = \frac{50}{20} \quad (5)$$
$$-8q = -72 \quad (6)$$

Thus, q = 9. Substitute 9 for q in equation (5).

$$p + \frac{2}{20}(9) = \frac{50}{20}$$
$$p = \frac{32}{20}$$
$$p = 1.6$$

The equilibrium quantity is 9 thousand pounds. The equilibrium price is $1.60 per pound.

47. Let x = number of shares of USAir
y = number of shares of BPAmerica.

Thus, the linear system is

$$30x + 70y = 16,000 \quad (1)$$
$$45x + 210y = 34,500. \quad (2)$$

Multiply equation (1) by 1/30.

$$x + \frac{70}{30}y = \frac{16,000}{30} \quad (3)$$
$$45x + 210y = 34,500 \quad (2)$$

Multiply equation (3) by -45 and add to equation (2).

$$x + \frac{70}{30}y = \frac{16,000}{30} \quad (3)$$
$$105y = 10,500 \quad (4)$$

From equation (4), y = 100.
Substitute 100 for y in equation (1).

$$30x + 70(100) = 16,000$$
$$30x + 7000 = 16,000$$
$$30x = 9000$$
$$x = 300$$

She has 300 shares of USAir and 100 shares of BPAmerica.

49. Let x = original number of skirts.
y = original number of blouses.

Solve the linear system

$$45x + 35y = 51,750 \quad (1)$$
$$45(\frac{1}{2}x) + 35(\frac{2}{3}y) = 30,600. \quad (2)$$

Multiply equation (1) by 1/45 and equation (2) by 6.

$$x + \frac{35}{45}y = \frac{51,750}{45} \quad (3)$$
$$135x + 140y = 183,600 \quad (4)$$

Multiply equation (3) by -135 and add it to equation (4).

$$x + \frac{35}{45}y = \frac{51,750}{45} \quad (3)$$
$$35y = 28,350 \quad (5)$$

From equation (5), y = 810.
Substitute 810 for y in equation (1).

$$45x + 35(810) = 51,750$$
$$45x + 28,350 = 51,750$$
$$45x = 23,400$$
$$x = 520$$

The problem asks for the number of skirts and blouses left in the store. Since (1/2)x = 260 and (1/3)y = 270, there are 260 skirts left and 270 blouses left.

51. First tabulate the given information as follows.

	Model 201	Model 301	Available
Assembly Time (hours)	2	3	34
Cost (dollars)	25	30	365

Let x = the number of model 201 bikes and y = the number of model 301 bikes.

The x model 201 bikes use 2x hours of assembly time, while the y model 301 bikes use 3y hours. Together, they require 2x + 3y hours. Since there are 34 hours available, this leads to the first equation, $2x + 3y = 34$.

The cost of the x model 201 bikes is 25x (dollars) and the cost of the y model 301 bikes is 30y (dollars). The total cost is 25x + 30y. Since $365 is available, this leads to the second equation, $25x + 30y = 365$.

Solve the system

$$2x + 3y = 34 \quad (1)$$
$$25x + 30y = 365. \quad (2)$$

Multiply equation (1) by 1/2 to get

$$x + \frac{3}{2}y = 17 \quad (3)$$
$$25x + 30y = 365. \quad (2)$$

Multiply equation (3) by -25 and add to equation (2) to get

$$x + \frac{3}{2}y = 17 \quad (3)$$
$$-\frac{15}{2}y = -60. \quad (4)$$

Multiply equation (4) by -2/15 to get

$$x + \frac{3}{2}y = 17 \quad (3)$$
$$y = 8. \quad (5)$$

Substitute 8 for y in equation (3) to get x = 5. 5 model 201 bikes and 8 model 301 bikes can be made in one day.

53. Let x = the amount invested at 8%
y = the amount invested at 9%
z = the amount invested at 5%.

Solve the system

$$x + y + z = 10{,}000$$
$$.08x + .09y + .05z = 830$$
$$2x = y.$$

Simplify the second equation to get the system

$$x + y + z = 10{,}000 \quad (1)$$
$$8x + 9y + 5z = 83{,}000 \quad (2)$$
$$2x = y. \quad (3)$$

Multiply equation (1) by -8 and add to equation (2) to get

$$x + y + z = 10{,}000 \quad (1)$$
$$y - 3z = 3000 \quad (4)$$
$$2x = y. \quad (3)$$

Use equation (3) to get x = y/2.

Substitute y/2 for x in equation (1) and simplify

$$\frac{y}{2} + y + z = 10{,}000$$
$$\frac{3y}{2} + z = 10{,}000$$
$$3y + 2z = 20{,}000 \quad (5)$$

Multiply equation (4) by -3 and add to equation (5) to get equation (6).

$$x + y + z = 10{,}000 \quad (1)$$
$$y - 3z = 3000 \quad (4)$$
$$11z = 11{,}000 \quad (6)$$

From equation (6), z = 1000.

Substitute 1000 for z in (4) to get y = 6000. Substitute 6000 and 1000 for y and z in (1) to get x = 3000.

Juanita invested $3000 at 8%, $6000 at 9%, and put $1000 in the bank at 5%.

55. Let A, B, and C = the number of grams of the corresponding food groups. Then

$$A + B + C = 400$$
$$A = \frac{1}{3}B$$
$$A + C = 2B$$

is the symbolic form of the given information, which may be rewritten as

$$A + B + C = 400$$
$$3A - B = 0$$
$$A - 2B + C = 0.$$

This system can now be solved in the usual way to obtain

$$A = \frac{400}{9},$$

$$B = \frac{400}{3}, \text{ and}$$

$$C = \frac{2000}{9}.$$

57. (a) An equation $y = ax^2 + bx + c$ is sought.

When $x = 6$, $y = 2.80$. Therefore,

$$36a + 6b + c = 2.80.$$

When $x = 8$, $y = 2.48$. Therefore,

$$64a + 8b + c = 2.48.$$

When $x = 10$, $y = 2.24$. Therefore,

$$100a + 10b + c = 2.24.$$

Thus, the system of equations is

$$36a + 6b + c = 2.80 \quad (1)$$
$$64a + 8b + c = 2.48 \quad (2)$$
$$100a + 10b + c = 2.24. \quad (3)$$

Solve the system by elimination. Subtract equation (1) from equation (2)

$$28a + 2b = -.32$$

and subtract equation (2) from equation (3)

$$36a + 2b = -.24.$$

Now the system is

$$28a + 2b = -.32 \quad (4)$$
$$36a + 2b = -.24. \quad (5)$$

Subtract equation (5) from equation (4).

$$-8a = -.08$$

Thus, $a = .01$.

Substituting into $28a + 2b = -.32$ gives

$$28(.01) + 2b = -.32$$
$$.28 + 2b = -.32$$
$$2b = -.60$$
$$b = -.30$$

Substituting into equation (1) gives

$$36(.01) + 6(-.30) + c = 2.80$$
$$.36 - 1.80 + c = 2.80$$
$$-1.44 + c = 2.80$$
$$c = 4.24.$$

Thus, the equation is

$$y = .01x^2 - .3x + 4.24.$$

(b) Write $y = .01x^2 - .3x + 4.24$ in the form

$$y = a(x - h)^2 + k.$$
$$y = .01(x^2 - 30x) + 4.24$$
$$y = .01(x^2 - 30x + 225) + 4.24 - 2.25$$
$$y = .01(x - 15)^2 + 1.99$$

The minimum value of y is 1.99 occurring when $x = 15$. Thus, 15 platters should be fired at one time to minimize the fuel cost. The minimum fuel cost is $1.99.

Section 6.2

1.

$$\left[\begin{array}{ccc|c} 2 & 3 & 5 & 4 \\ 6 & 9 & 1 & 10 \\ 1 & 4 & 3 & 8 \end{array}\right]$$

Interchange rows 1 and 3.

$$\left[\begin{array}{ccc|c} 1 & 4 & 3 & 8 \\ 6 & 9 & 1 & 10 \\ 2 & 3 & 5 & 4 \end{array}\right]$$

3.

$$\left[\begin{array}{ccc|c} 3 & 6 & 12 & 18 \\ 0 & 5 & 2 & 9 \\ 4 & 7 & 8 & 15 \end{array}\right]$$

Replace row 1 by row 1 multiplied by 1/3.

$$\left[\begin{array}{ccc|c} 1 & 2 & 4 & 6 \\ 0 & 5 & 2 & 9 \\ 4 & 7 & 8 & 15 \end{array}\right]$$

5.

$$\left[\begin{array}{ccc|c} 1 & 4 & 2 & 9 \\ 0 & 1 & 5 & 14 \\ 0 & 3 & 8 & 16 \end{array}\right]$$

Replace row 1 by row 2 multiplied by -4 and added to row 1.

$$\left[\begin{array}{ccc|c} 1+(-4)(0) & 4+(-4)(1) & 2+(-4)(5) & 9+(-4)(14) \\ 0 & 1 & 5 & 14 \\ 0 & 3 & 8 & 16 \end{array}\right]$$

$$\left[\begin{array}{ccc|c} 1 & 0 & -18 & -47 \\ 0 & 1 & 5 & 14 \\ 0 & 3 & 8 & 16 \end{array}\right]$$

7. $x + 2y = 5$
$2x + y = -2$
corresponds to the augmented matrix

$$\left[\begin{array}{cc|c} 1 & 2 & 5 \\ 2 & 1 & -2 \end{array}\right].$$

Multiplying the first row by -2 and adding the result to the second row, obtain

$$\left[\begin{array}{cc|c} 1 & 2 & 5 \\ 0 & -3 & -12 \end{array}\right] \quad -2R_1 + R_2$$

(Note that $-2R_1 + R_2$ is the abbreviated way to document the row operation that was performed.)

Next, multiply the second row by -1/3, and document that operation in a similar way, to obtain

$$\left[\begin{array}{cc|c} 1 & 2 & 5 \\ 0 & 1 & 4 \end{array}\right] \quad -\frac{1}{3}R_2$$

and continue, to obtain

$$\left[\begin{array}{cc|c} 1 & 0 & -3 \\ 0 & 1 & 4 \end{array}\right]. \quad -2R_2 + R_1$$

This leads to $x = -3$ and $y = 4$, so the solution is $(-3, 4)$.

9. $y = 5 - 4x$
$2x = 3 - y$
Write the system in the appropriate form.
$4x + y = 5$
$2x + y = 3$

$$\left[\begin{array}{cc|c} 4 & 1 & 5 \\ 2 & 1 & 3 \end{array}\right]$$

$$\left[\begin{array}{cc|c} 1 & \frac{1}{4} & \frac{5}{4} \\ 2 & 1 & 3 \end{array}\right] \quad \frac{1}{4}R_1$$

$$\left[\begin{array}{cc|c} 1 & \frac{1}{4} & \frac{5}{4} \\ 2 & \frac{1}{2} & \frac{1}{2} \end{array}\right] \quad -2R_1 + R_2$$

$$\left[\begin{array}{cc|c} 1 & \frac{1}{4} & \frac{5}{4} \\ 0 & 1 & 1 \end{array}\right] \quad 2R_2$$

$$\left[\begin{array}{cc|c} 1 & 0 & 1 \\ 0 & 1 & 1 \end{array}\right] \quad -\frac{1}{4}R_2 + R_1$$

(1, 1) is the solution.

11. $x + 2y = 1$

$2x + 4y = 3$

The matrix is

$$\left[\begin{array}{cc|c} 1 & 2 & 1 \\ 2 & 4 & 3 \end{array}\right].$$

$$\left[\begin{array}{cc|c} 1 & 2 & 1 \\ 0 & 0 & 1 \end{array}\right] R_2 + (-2)R_1$$

Since we cannot get a 1 for the second element in row two, we can go no further. There is no solution.

13. $x - z = -3$

$y + z = 9$

$x + z = 7$

The matrix is

$$\left[\begin{array}{ccc|c} 1 & 0 & -1 & -3 \\ 0 & 1 & 1 & 9 \\ 1 & 0 & 1 & 7 \end{array}\right].$$

$$\left[\begin{array}{ccc|c} 1 & 0 & -1 & -3 \\ 0 & 1 & 1 & 9 \\ 0 & 0 & 2 & 10 \end{array}\right] -R_1 + R_3$$

$$\left[\begin{array}{ccc|c} 1 & 0 & -1 & -3 \\ 0 & 1 & 1 & 9 \\ 0 & 0 & 1 & 5 \end{array}\right] \tfrac{1}{2}R_3$$

$$\left[\begin{array}{ccc|c} 1 & 0 & 0 & 2 \\ 0 & 1 & 0 & 4 \\ 0 & 0 & 1 & 5 \end{array}\right] \begin{array}{l} R_3 + R_1 \\ -R_3 + R_2 \\ \end{array}$$

(2, 4, 5) is the solution.

15. $x = 1 - y$

$2x = z$

$2z = -2 - y$

$$\left[\begin{array}{ccc|c} 1 & 1 & 0 & 1 \\ 2 & 0 & -1 & 0 \\ 0 & 1 & 2 & -2 \end{array}\right]$$

$$\left[\begin{array}{ccc|c} 1 & 1 & 0 & 1 \\ 0 & -2 & -1 & -2 \\ 0 & 1 & 2 & -2 \end{array}\right] -2R_1 + R_2$$

$$\left[\begin{array}{ccc|c} 1 & 1 & 0 & 1 \\ 0 & 1 & \frac{1}{2} & 1 \\ 0 & 1 & 2 & -2 \end{array}\right] -\tfrac{1}{2}R_2$$

$$\left[\begin{array}{ccc|c} 1 & 0 & -\frac{1}{2} & 0 \\ 0 & 1 & \frac{1}{2} & 1 \\ 0 & 0 & \frac{3}{2} & -3 \end{array}\right] \begin{array}{l} -R_2 + R_1 \\ \\ -R_2 + R_3 \end{array}$$

$$\left[\begin{array}{ccc|c} 1 & 0 & -\frac{1}{2} & 0 \\ 0 & 1 & \frac{1}{2} & 1 \\ 0 & 0 & 1 & -2 \end{array}\right] \tfrac{2}{3}R_3$$

$$\left[\begin{array}{ccc|c} 1 & 0 & 0 & -1 \\ 0 & 1 & 0 & 2 \\ 0 & 0 & 1 & -2 \end{array}\right] \begin{array}{l} \frac{1}{2}R_3 + R_1 \\ -\frac{1}{2}R_3 + R_2 \\ \end{array}$$

(-1, 2, -2) is the solution.

17. $x + y = -1$

$y + z = 4$

$x + z = 1$

The matrix is

$$\left[\begin{array}{ccc|c} 1 & 1 & 0 & -1 \\ 0 & 1 & 1 & 4 \\ 1 & 0 & 1 & 1 \end{array}\right].$$

$$\left[\begin{array}{ccc|c} 1 & 1 & 0 & -1 \\ 0 & 1 & 1 & 4 \\ 0 & -1 & 1 & 2 \end{array}\right] R_3 + (-1)R_1$$

$$\left[\begin{array}{ccc|c} 1 & 0 & -1 & -5 \\ 0 & 1 & 1 & 4 \\ 0 & 0 & 2 & 6 \end{array}\right] \begin{array}{l} R_1 + (-1)R_2 \\ \\ R_3 + (-1)R_2 \end{array}$$

$$\left[\begin{array}{ccc|c} 1 & 0 & -1 & -5 \\ 0 & 1 & 1 & 4 \\ 0 & 0 & 1 & 3 \end{array}\right] \frac{1}{2}R_3$$

$$\left[\begin{array}{ccc|c} 1 & 0 & 0 & -2 \\ 0 & 1 & 0 & 1 \\ 0 & 0 & 1 & 3 \end{array}\right] \begin{array}{l} R_1 + R_3 \\ R_2 + (-1)R_3 \\ \\ \end{array}$$

The solution is (-2, 1, 3).

19. First write the system in the appropriate form.

$$\begin{aligned} -x + y \quad\quad &= -1 \\ y - z &= 6 \\ x \quad\quad + z &= -1 \end{aligned}$$

Write the augmented matrix and use row operations.

$$\left[\begin{array}{ccc|c} -1 & 1 & 0 & -1 \\ 0 & 1 & -1 & 6 \\ 1 & 0 & 1 & -1 \end{array}\right]$$

$$\left[\begin{array}{ccc|c} 1 & -1 & 0 & 1 \\ 0 & 1 & -1 & 6 \\ 1 & 0 & 1 & -1 \end{array}\right] \begin{array}{l} -R_1 \\ \\ \\ \end{array}$$

$$\left[\begin{array}{ccc|c} 1 & -1 & 0 & 1 \\ 0 & 1 & -1 & 6 \\ 0 & 1 & 1 & -2 \end{array}\right] -R_1 + R_3$$

$$\left[\begin{array}{ccc|c} 1 & 0 & -1 & 7 \\ 0 & 1 & -1 & 6 \\ 0 & 1 & 1 & -2 \end{array}\right] \begin{array}{l} R_2 + R_1 \\ \\ \\ \end{array}$$

$$\left[\begin{array}{ccc|c} 1 & 0 & -1 & 7 \\ 0 & 1 & -1 & 6 \\ 0 & 0 & 2 & -8 \end{array}\right] -R_2 + R_3$$

$$\left[\begin{array}{ccc|c} 1 & 0 & -1 & 7 \\ 0 & 1 & -1 & 6 \\ 0 & 0 & 1 & -4 \end{array}\right] \frac{1}{2}R_3$$

$$\left[\begin{array}{ccc|c} 1 & 0 & 0 & 3 \\ 0 & 1 & -1 & 6 \\ 0 & 0 & 1 & -4 \end{array}\right] \begin{array}{l} R_3 + R_1 \\ \\ \\ \end{array}$$

$$\left[\begin{array}{ccc|c} 1 & 0 & 0 & 3 \\ 0 & 1 & 0 & 2 \\ 0 & 0 & 1 & -4 \end{array}\right] R_3 + R_2$$

Solution: (3, 2, −4)

21.
$$\begin{aligned} 2x + 3y + z &= 9 \\ 4x + y - 3z &= -7 \\ 6x + 2y - 4z &= -8 \end{aligned}$$

The matrix is

$$\left[\begin{array}{ccc|c} 2 & 3 & 1 & 9 \\ 4 & 1 & -3 & -7 \\ 6 & 2 & -4 & -8 \end{array}\right].$$

$$\left[\begin{array}{ccc|c} 1 & \frac{3}{2} & \frac{1}{2} & \frac{9}{2} \\ 4 & 1 & -3 & -7 \\ 6 & 2 & -4 & -8 \end{array}\right] \begin{array}{l} \frac{1}{2}R_1 \\ \\ \\ \end{array}$$

$$\left[\begin{array}{ccc|c} 1 & \frac{3}{2} & \frac{1}{2} & \frac{9}{2} \\ 0 & -5 & -5 & -25 \\ 0 & -7 & -7 & -35 \end{array}\right] \begin{array}{l} \\ R_2 + (-4)R_1 \\ R_3 + (-6)R_1 \end{array}$$

$$\left[\begin{array}{ccc|c} 1 & \frac{3}{2} & \frac{1}{2} & \frac{9}{2} \\ 0 & 1 & 1 & 5 \\ 0 & 1 & 1 & 5 \end{array}\right] \begin{array}{l} \\ -\frac{1}{5}R_2 \\ -\frac{1}{7}R_3 \end{array}$$

$$\left[\begin{array}{ccc|c} 1 & 0 & -1 & -3 \\ 0 & 1 & 1 & 5 \\ 0 & 0 & 0 & 0 \end{array}\right] \begin{array}{l} R_1 + \left(-\frac{3}{2}\right)R_2 \\ \\ R_3 + (-1)R_2 \end{array}$$

Since the last row is all zeros, the system has no unique solution. Rows 1 and row 2 can be written as the system

$x - z = -3$
$y + z = 5.$

Let z be arbitrary.

$x = z - 3$
$y = -z - 5$

The solution is (z - 3, -z + 5, z).

23. $3x + 2y - z = -16$
$6x - 4y + 3z = 12$
$3x + 3y + z = -11$

The matrix is

$$\left[\begin{array}{ccc|c} 3 & 2 & -1 & -16 \\ 6 & -4 & 3 & 12 \\ 3 & 3 & 1 & -11 \end{array}\right].$$

$$\left[\begin{array}{ccc|c} 1 & \frac{2}{3} & -\frac{1}{3} & -\frac{16}{3} \\ 6 & -4 & 3 & 12 \\ 3 & 3 & 1 & -11 \end{array}\right] \begin{array}{l} \frac{1}{3}R_1 \\ \\ \\ \end{array}$$

$$\left[\begin{array}{ccc|c} 1 & \frac{2}{3} & -\frac{1}{3} & -\frac{16}{3} \\ 0 & -8 & 5 & 44 \\ 0 & 1 & 2 & 5 \end{array}\right] \begin{array}{l} \\ -6R_1 + R_2 \\ -3R_1 + R_3 \end{array}$$

$$\left[\begin{array}{ccc|c} 1 & \frac{2}{3} & -\frac{1}{3} & -\frac{16}{3} \\ 0 & 1 & -\frac{5}{8} & -\frac{11}{2} \\ 0 & 1 & 2 & 5 \end{array}\right] \begin{array}{l} \\ -\frac{1}{8}R_2 \\ \\ \end{array}$$

$$\left[\begin{array}{ccc|c} 1 & 0 & \frac{1}{12} & -\frac{5}{3} \\ 0 & 1 & -\frac{5}{8} & -\frac{11}{2} \\ 0 & 0 & \frac{21}{8} & \frac{21}{2} \end{array}\right] \begin{array}{l} -\frac{2}{3}R_2 + R_1 \\ \\ -R_2 + R_3 \end{array}$$

$$\left[\begin{array}{ccc|c} 1 & 0 & \frac{1}{12} & -\frac{5}{3} \\ 0 & 1 & -\frac{5}{8} & -\frac{11}{2} \\ 0 & 0 & 1 & 4 \end{array}\right] \begin{array}{l} \\ \\ \frac{8}{21}R_3 \end{array}$$

$$\left[\begin{array}{ccc|c} 1 & 0 & 0 & -2 \\ 0 & 1 & 0 & -3 \\ 0 & 0 & 1 & 4 \end{array}\right] \begin{array}{l} -\frac{1}{12}R_3 + R_1 \\ \frac{5}{8}R_3 + R_2 \\ \\ \end{array}$$

(-2, -3, 4) is the solution.

25. Let x_1 = number of cars sent from I to A
x_2 = number of cars sent from II to A
x_3 = number of cars sent from I to B
x_4 = number of cars sent from II to B.

	A	B
I	x_1	x_3
II	x_2	x_4

Plant I has 28 cars, so

$x_1 + x_3 = 28.$ *(1)*

Plant II has 8 cars, so

$x_2 + x_4 = 8.$ *(2)*

Dealer A needs 20 cars, so

$x_1 + x_2 = 20.$ *(3)*

Dealer B needs 16 cars, so

$x_3 + x_4 = 16.$ *(4)*

Total transportation cost is $10,640, so

$220x_1 + 400x_2 + 300x_3 + 180x_4 = 10{,}640.$ *(5)*

The augmented matrix is

$$\left[\begin{array}{cccc|c} 1 & 0 & 1 & 0 & 28 \\ 0 & 1 & 0 & 1 & 8 \\ 1 & 1 & 0 & 0 & 20 \\ 0 & 0 & 1 & 1 & 16 \\ 220 & 400 & 300 & 180 & 10{,}640 \end{array}\right]$$

Clear the first column.

$$\left[\begin{array}{cccc|c} 1 & 0 & 1 & 0 & 28 \\ 0 & 1 & 0 & 1 & 8 \\ 0 & 1 & -1 & 0 & -8 \\ 0 & 0 & 1 & 1 & 16 \\ 0 & 400 & 80 & 180 & 4480 \end{array}\right]$$

Clear the second column.

$$\left[\begin{array}{cccc|c} 1 & 0 & 1 & 0 & 28 \\ 0 & 1 & 0 & 1 & 8 \\ 0 & 0 & -1 & -1 & -16 \\ 0 & 0 & 1 & 1 & 16 \\ 0 & 0 & 80 & -220 & 1280 \end{array}\right]$$

Clear the third column.

$$\left[\begin{array}{cccc|c} 1 & 0 & 0 & -1 & 12 \\ 0 & 1 & 0 & 1 & 8 \\ 0 & 0 & -1 & -1 & -16 \\ 0 & 0 & 0 & 0 & 0 \\ 0 & 0 & 0 & -300 & 0 \end{array}\right]$$

Interchange the fourth and fifth rows.
Multiply row 3 by -1 and new row 4 by -1/300.

$$\left[\begin{array}{cccc|c} 1 & 0 & 0 & -1 & 12 \\ 0 & 1 & 0 & 1 & 8 \\ 0 & 0 & 1 & 1 & 16 \\ 0 & 0 & 0 & 1 & 0 \\ 0 & 0 & 0 & 0 & 0 \end{array}\right]$$

Now clear the fourth column.

$$\left[\begin{array}{cccc|c} 1 & 0 & 0 & 0 & 12 \\ 0 & 1 & 0 & 0 & 8 \\ 0 & 0 & 1 & 0 & 16 \\ 0 & 0 & 0 & 1 & 0 \\ 0 & 0 & 0 & 0 & 0 \end{array}\right]$$

Read the solution from the last column of the final matrix: $x_1 = 12$, $x_2 = 8$, $x_3 = 16$, $x_4 = 0$. 12 cars should be sent from I to A, 8 cars from II to A, 16 cars from I to B, and no cars from II to B.

27. Let x = number of kilograms of the first chemical
y = number of kilograms of the second chemical
z = number of kilograms of the third chemical.

Solve the following system.

$$x = (.108)(750)$$
$$\frac{y}{z} = \frac{4}{3}$$
$$x + y + z = 750.$$

Write the system in the appropriate form.

$$x = 81$$
$$3y - 4z = 0$$
$$x + y + z = 750$$

Substitute 81 for x in the last equation and solve the resulting system consisting of two equations in the variables y and z.

$$3y - 4z = 0$$
$$81 + y + z = 750$$

or the system

$$3y - 4z = 0 \qquad (1)$$
$$y + z = 669. \qquad (2)$$

Multiply equation (2) by 4.

$$3y - 4z = 0 \qquad (1)$$
$$4y + 4z = 2676 \qquad (3)$$

Adding equations (1) and (3) gives

$$7y = 2676.$$

Thus,

$$y = 382.286.$$

From equation (1), $z = \frac{3}{4}y$. Thus,

$$z = \frac{3}{4}(382.286)$$

$$z = 286.714.$$

Thus, 81 kilograms of the first chemical, 382.286 kilograms of the second chemical, and 286.714 kilograms of the third chemical should be mixed.

29. Organize the information into a table.

		FOODS		
		I	II	III
Species of fish	A	1.32	2.9	1.75
	B	2.1	.95	.6
	C	.86	1.52	2.01

Let x = number of species A
y = number of species B
z = number of species C.

Solve the system

$$1.32x + 2.1y + .86z = 490$$
$$2.9x + .95y + 1.52z = 897$$
$$1.75x + .6y + 2.01z = 653.$$

To eliminate the decimals, multiply each equation by 100.

$$132x + 210y + 86z = 49{,}000$$
$$290x + 95y + 152z = 89{,}700$$
$$175x + 60y + 201z = 65{,}300$$

Write the augmented matrix of the system.

$$\left[\begin{array}{ccc|c} 132 & 210 & 86 & 49{,}000 \\ 290 & 95 & 152 & 89{,}700 \\ 175 & 60 & 201 & 65{,}300 \end{array}\right]$$

$$\left[\begin{array}{ccc|c} 1 & 1.591 & .652 & 371.212 \\ 290 & 95 & 152 & 89{,}700 \\ 175 & 60 & 201 & 65{,}300 \end{array}\right] \begin{array}{l} \frac{1}{132}R_1 \\ \\ \\ \end{array}$$

$$\left[\begin{array}{ccc|c} 1 & 1.591 & .652 & 371.212 \\ 0 & -366.39 & -37.08 & -17{,}951.48 \\ 0 & -218.425 & 86.9 & 337.9 \end{array}\right] \begin{array}{l} \\ -290R_1 + R_2 \\ -175R_1 + R_3 \end{array}$$

$$\left[\begin{array}{ccc|c} 1 & 1.591 & .652 & 371.212 \\ 0 & 1 & .101 & 48.996 \\ 0 & -218.425 & 86.9 & 337.9 \end{array}\right] \begin{array}{l} \\ -\frac{1}{366.39}R_2 \\ \\ \end{array}$$

$$\left[\begin{array}{ccc|c} 1 & 1.591 & .652 & 371.212 \\ 0 & 1 & .101 & 48.996 \\ 0 & 0 & 108.961 & 11{,}039.851 \end{array}\right] \begin{array}{l} \\ \\ 218.425R_2 + R_3 \end{array}$$

Thus, the system may be written

$$x + 1.591y + .652z = 371.212 \qquad (1)$$
$$y + .101z = 48.996 \qquad (2)$$
$$108.961z = 11{,}039.851 \qquad (3)$$

From equation (3),
$z = 101.319.$

Substituting into equation (2) gives
$y + .101(101.319) = 48.996$
$y = 38.763.$

Substituting into equation (1) gives
$x + 1.591(38.763) + .652(101.319)$
$= 371.212$
$x = 243.48.$

There should be about 243 of species A, 39 of species B, and 101 of species C.

31. (a) Let x_1 = units from first supplier for Roseville,
x_2 = units from first supplier for Akron,
x_3 = units from second supplier for Roseville,
and x_4 = units from second supplier for Akron.

(b)
$$x_1 + x_2 = 75$$
$$x_3 + x_4 = 40$$
$$x_1 + x_3 = 40$$
$$x_2 + x_4 = 75$$
$$70x_1 + 90x_2 + 80x_3 + 120x_4 = 10{,}750$$

(c) The matrix is

$$\left[\begin{array}{cccc|c} 1 & 1 & 0 & 0 & 75 \\ 0 & 0 & 1 & 1 & 40 \\ 1 & 0 & 1 & 0 & 40 \\ 0 & 1 & 0 & 1 & 75 \\ 70 & 90 & 80 & 120 & 10{,}750 \end{array}\right]$$

$$\left[\begin{array}{cccc|c} 1 & 1 & 0 & 0 & 75 \\ 0 & 0 & 1 & 1 & 40 \\ 0 & -1 & 1 & 0 & -35 \\ 0 & 1 & 0 & 1 & 75 \\ 0 & 20 & 80 & 120 & 5500 \end{array}\right] \begin{array}{l} \\ \\ -R_1 + R_3 \\ \\ -70R_1 + R_5 \end{array}$$

Interchange rows 2 and 4, and continue.

$$\left[\begin{array}{cccc|c} 1 & 0 & 0 & -1 & 0 \\ 0 & 1 & 0 & 1 & 75 \\ 0 & 0 & 1 & 1 & 40 \\ 0 & 0 & 1 & 1 & 40 \\ 0 & 0 & 80 & 100 & 4000 \end{array}\right] \begin{array}{l} -R_2 + R_1 \\ \\ R_2 + R_3 \\ \\ -20R_2 + R_5 \end{array}$$

$$\left[\begin{array}{cccc|c} 1 & 0 & 0 & -1 & 0 \\ 0 & 1 & 0 & 1 & 75 \\ 0 & 0 & 1 & 1 & 40 \\ 0 & 0 & 0 & 0 & 0 \\ 0 & 0 & 0 & 20 & 800 \end{array}\right] \begin{array}{l} \\ \\ \\ -R_3 + R_4 \\ -80R_3 + R_5 \end{array}$$

Interchange rows 4 and 5, and continue.

$$\left[\begin{array}{cccc|c} 1 & 0 & 0 & 0 & 40 \\ 0 & 1 & 0 & 0 & 35 \\ 0 & 0 & 1 & 0 & 0 \\ 0 & 0 & 0 & 1 & 40 \\ 0 & 0 & 0 & 0 & 0 \end{array}\right] \begin{array}{l} R_4 + R_1 \\ -R_4 + R_2 \\ -R_4 + R_3 \\ \frac{1}{20}R_4 \\ \end{array}$$

Each of the original variables has a value, so the last row of all zeros may be ignored. The single solution is (40, 35, 0, 40).

Section 6.3

1. $\begin{bmatrix} 1 & 3 \\ 5 & 7 \end{bmatrix} = \begin{bmatrix} 1 & 5 \\ 3 & 7 \end{bmatrix}$

False.
Not all corresponding elements are equal.

3. $\begin{bmatrix} x \\ y \end{bmatrix} = \begin{bmatrix} 3 \\ 5 \end{bmatrix}$ if $x = 3$ and $y = 5$.

True.
The matrices are of the same order and corresponding elements are equal.

5. $\begin{bmatrix} 1 & 9 & -4 \\ 3 & 7 & 2 \\ -1 & 1 & 0 \end{bmatrix}$

True.
This matrix has 3 rows and 3 columns.

7. $\begin{bmatrix} -4 & 8 \\ 2 & 3 \end{bmatrix}$

is a 2×2 square matrix. Its inverse is

$\begin{bmatrix} 4 & -8 \\ -2 & -3 \end{bmatrix}$.

9. $\begin{bmatrix} -6 & 8 & 0 & 0 \\ 4 & 1 & 9 & 2 \\ 3 & -5 & 7 & 1 \end{bmatrix}$

is a 3×4 matrix. Its inverse is

$\begin{bmatrix} 6 & -8 & 0 & 0 \\ -4 & -1 & -9 & -2 \\ -3 & 5 & -7 & -1 \end{bmatrix}$.

11. $\begin{bmatrix} 2 \\ 4 \end{bmatrix}$

is a 2×1 column matrix. Its inverse is

$\begin{bmatrix} -2 \\ -4 \end{bmatrix}$.

13. $\begin{bmatrix} 2 & 1 \\ 4 & 8 \end{bmatrix} = \begin{bmatrix} x & 1 \\ y & z \end{bmatrix}$

Corresponding elements must be equal for the matrices to be equal.
Therefore, $x = 2$, $y = 4$, and $z = 8$.

15. $\begin{bmatrix} x+6 & y+2 \\ 8 & 3 \end{bmatrix} = \begin{bmatrix} -9 & 7 \\ 8 & k \end{bmatrix}$

$x + 6 = -9 \qquad y + 2 = 7 \quad k = 3$

$x = -15 \qquad y = 5$

17. $\begin{bmatrix} -7+z & 4r & 8s \\ 6p & 2 & 5 \end{bmatrix} + \begin{bmatrix} -9 & 8r & 3 \\ 2 & 5 & 4 \end{bmatrix}$

$= \begin{bmatrix} 2 & 36 & 27 \\ 20 & 7 & 12a \end{bmatrix}$

$\begin{bmatrix} -7+z & 4r & 8s \\ 6p & 2 & 5 \end{bmatrix} + \begin{bmatrix} -9 & 8r & 3 \\ 2 & 5 & 4 \end{bmatrix}$

$= \begin{bmatrix} (-7+z)+(-9) & 4r+8r & 8s+3 \\ 6p+2 & 7 & 9 \end{bmatrix}$

$= \begin{bmatrix} -16+z & 12r & 8s+3 \\ 6p+2 & 7 & 9 \end{bmatrix}$

$-16 + z = 2 \quad 12r = 36 \quad 8s + 3 = 27$

$z = 18 \quad r = 3 \quad s = 3$

$6p + 2 = 30 \quad 9 = 12a$

$p = 3 \quad a = \frac{3}{4}$

19. $\begin{bmatrix} 1 & 2 & 5 & -1 \\ 3 & 0 & 2 & -4 \end{bmatrix} + \begin{bmatrix} 8 & 10 & -5 & 3 \\ -2 & -1 & 0 & 0 \end{bmatrix}$

$= \begin{bmatrix} 1+8 & 2+10 & 5+(-5) & -1+3 \\ 3+(-2) & 0+(-1) & 2+0 & -4+0 \end{bmatrix}$

$= \begin{bmatrix} 9 & 12 & 0 & 2 \\ 1 & -1 & 2 & -4 \end{bmatrix}$

21. $\begin{bmatrix} 1 & 5 & 7 \\ 2 & 2 & 3 \end{bmatrix} + \begin{bmatrix} 4 & 8 & -7 \\ 1 & -1 & 5 \end{bmatrix}$

$= \begin{bmatrix} 1+4 & 5+8 & 7+(-7) \\ 2+1 & 2+(-1) & 3+5 \end{bmatrix}$

$= \begin{bmatrix} 5 & 13 & 0 \\ 3 & 1 & 8 \end{bmatrix}$

23. $\begin{bmatrix} 1 & 3 & -2 \\ 4 & 7 & 1 \end{bmatrix} + \begin{bmatrix} 3 & 0 \\ 6 & 4 \\ -5 & 2 \end{bmatrix}$

These matrices cannot be added since the first matrix is 2×3, while the second is 3×2. Only matrices of the same order can be added.

25. The matrices have the same order, so the subtraction can be done. Letting A and B represent the matrices and using the definition of subtraction, we have

$A - B = A + (-B)$

$= \begin{bmatrix} 2 & 8 & 12 & 0 \\ 7 & 4 & -1 & 5 \\ 1 & 2 & 0 & 10 \end{bmatrix} + \begin{bmatrix} -1 & -3 & -6 & -9 \\ -2 & 3 & 3 & -4 \\ -8 & 0 & 2 & -17 \end{bmatrix}$

$= \begin{bmatrix} 1 & 5 & 6 & -9 \\ 5 & 7 & 2 & 1 \\ -7 & 2 & 2 & -7 \end{bmatrix}.$

27. $\begin{bmatrix} -4x+2y & -3x+y \\ 6x-3y & 2x-5y \end{bmatrix} + \begin{bmatrix} -8x+6y & 2x \\ 3y-5x & 6x+4y \end{bmatrix}$

$= \begin{bmatrix} (-4x+2y)+(-8x+6y) & (-3x+y)+2x \\ (6x-3y)+(3y-5x) & (2x-5y)+(6x+4y) \end{bmatrix}$

$= \begin{bmatrix} -12x+8y & -x+y \\ x & 8x-y \end{bmatrix}$

29. $X + T = \begin{bmatrix} x & y \\ z & w \end{bmatrix} + \begin{bmatrix} r & s \\ t & u \end{bmatrix}$

$= \begin{bmatrix} x+r & y+s \\ z+t & w+u \end{bmatrix},$

which is another 2×2 matrix. Therefore, the addition of two 2×2 matrices is closed.

31. Show that X + (T + P) = (X + T) + P.

On the left-hand side, the sum T + P is obtained first, and then

$$X + (T + P).$$

This gives the matrix

$$\begin{bmatrix} x + (r + m) & y + (s + n) \\ z + (t + p) & w + (u + q) \end{bmatrix}.$$

For the right-hand side, first the sum X + T is obtained, and then

$$(X + T) + P.$$

This gives the matrix

$$\begin{bmatrix} (x + r) + m & (y + s) + n \\ (z + t) + p & (w + u) + q \end{bmatrix}.$$

Comparing corresponding elements shows that they are equal by the associative property of addition of real numbers.

Thus,

$$X + (T + P) = (X + T) + P.$$

33. Show that P + O = P.

$$P + O = \begin{bmatrix} m & n \\ p & q \end{bmatrix} + \begin{bmatrix} 0 & 0 \\ 0 & 0 \end{bmatrix}$$

$$= \begin{bmatrix} m + 0 & n + 0 \\ p + 0 & q + 0 \end{bmatrix}$$

$$= \begin{bmatrix} m & n \\ p & q \end{bmatrix}$$

$$= P$$

35.

	Initial cost	Guaranteed rent
Market	18	2.7
Barber shop	10	1.5
Variety store	8	1.0
Drug store	10	2.0
Bakery	10	1.7

	Market	Barber shop	Variety store	Drug store	Bakery
Initial cost	18	10	8	10	10
Guaranteed rent	2.7	1.5	1.0	2.0	1.7

37. (a) There are four food groups and three meals. To represent the data by a 3 × 4 matrix, use the rows to correspond to the meals: breakfast, lunch, and dinner, and the columns to correspond to the four food groups. The matrix is

$$\begin{bmatrix} 2 & 1 & 2 & 1 \\ 3 & 2 & 2 & 1 \\ 4 & 3 & 2 & 1 \end{bmatrix}.$$

(b) There are four food groups. These will correspond to the four rows. There are three components in each food group: fat, carbohydrates, and protein. These will correspond to the three columns. The matrix is

$$\begin{bmatrix} 5 & 0 & 7 \\ 0 & 10 & 1 \\ 0 & 15 & 2 \\ 10 & 12 & 8 \end{bmatrix}.$$

(c) The matrix is

$$\begin{bmatrix} 8 \\ 4 \\ 5 \end{bmatrix}.$$

39. (a)

	Bread	Milk	Peanut butter	Cold cuts
Store I	88	48	16	112
Store II	105	72	21	147
Store III	60	40	0	50

(b) For store I,

$1.25(88) = 110$

$1.25(48) = 60$

$1.25(16) = 20$

$1.25(112) = 140.$

For store II,

$\frac{4}{3}(105) = 140$

$\frac{4}{3}(72) = 96$

$\frac{4}{3}(21) = 28$

$\frac{4}{3}(147) = 196.$

For store III,

$1.10(60) = 66$

$1.10(40) = 44$

$1.10(0) = 0$

$1.10(50) = 55.$

The new matrix is

$$\begin{bmatrix} 110 & 60 & 20 & 140 \\ 140 & 96 & 28 & 196 \\ 66 & 44 & 0 & 55 \end{bmatrix}.$$

(c) To find the total sales, add the matrices from parts (a) and (b).

$$\begin{bmatrix} 88 & 48 & 16 & 112 \\ 105 & 72 & 21 & 147 \\ 60 & 40 & 0 & 50 \end{bmatrix} + \begin{bmatrix} 110 & 60 & 20 & 140 \\ 140 & 96 & 28 & 196 \\ 66 & 44 & 0 & 55 \end{bmatrix}$$

$$= \begin{bmatrix} 198 & 108 & 36 & 252 \\ 245 & 168 & 49 & 343 \\ 126 & 84 & 0 & 105 \end{bmatrix}$$

Section 6.4

1. Since A is 2×2 and B is 2×2, we have the following diagram:

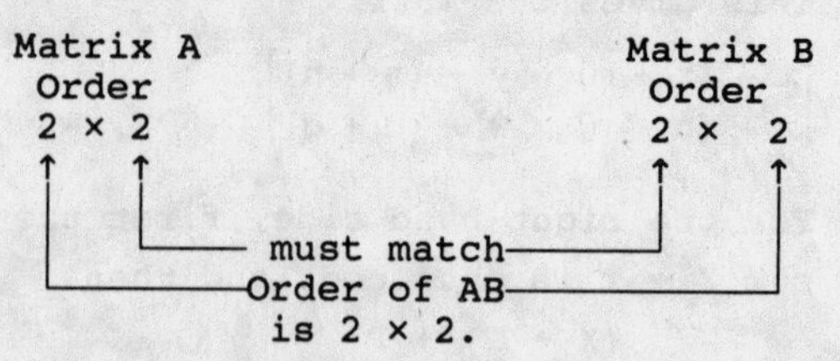

The product AB exists because A has two columns and B has two rows. The order of the product AB is 2×2. A similar diagram with B and A interchanged shows that the product BA exists and its order is also 2×2.

3.

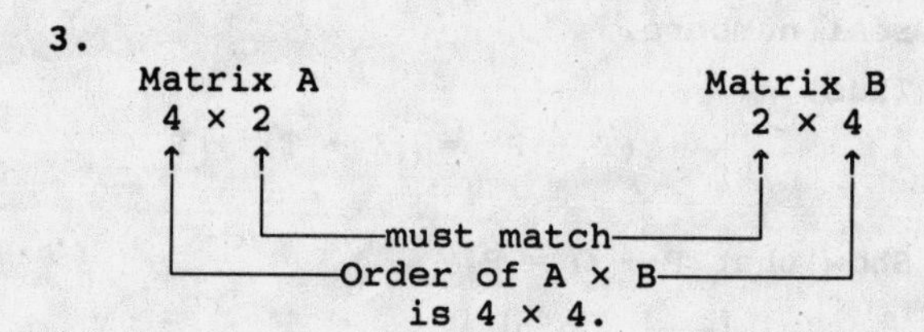

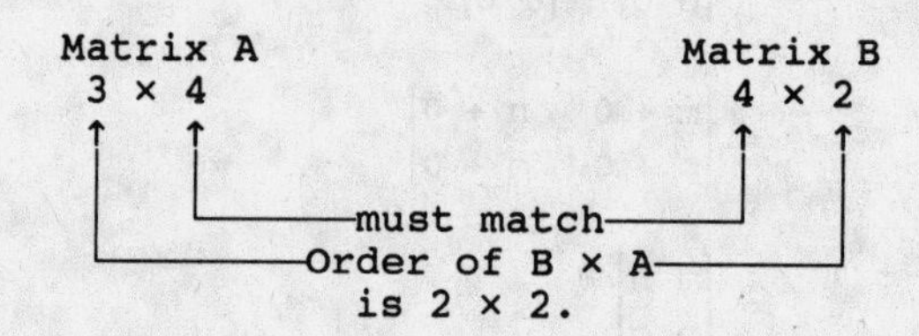

5.

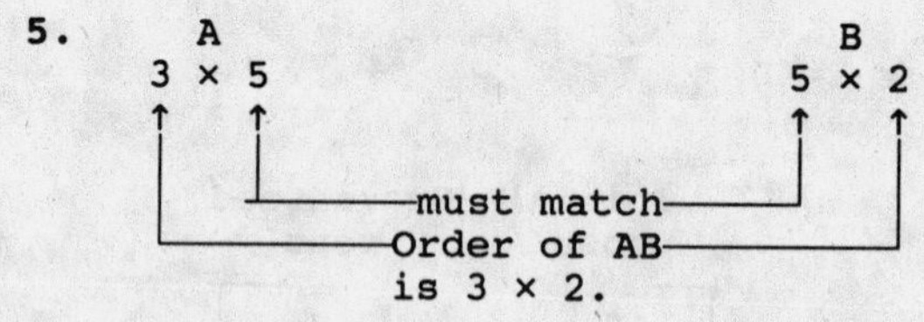

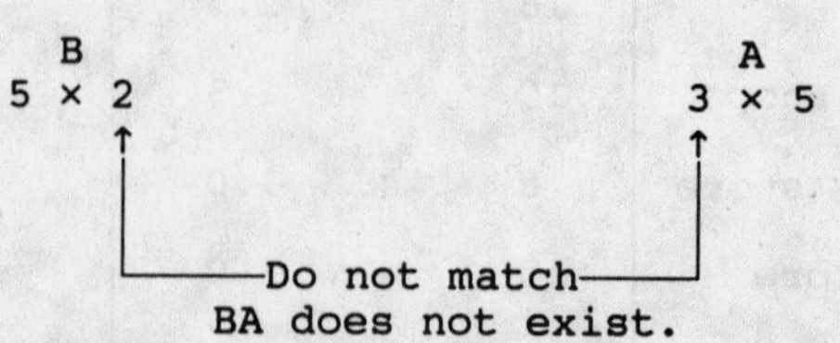

7.

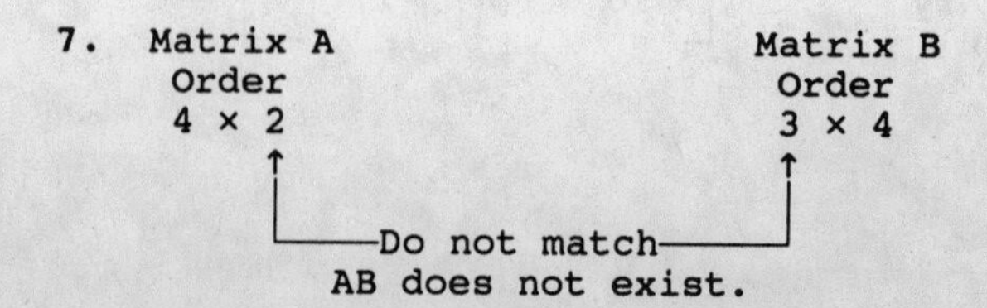

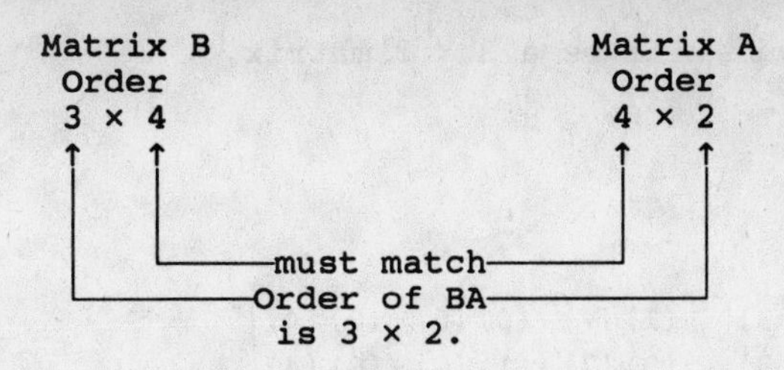

9. $2A = 2\begin{bmatrix} -2 & 4 \\ 0 & 3 \end{bmatrix} = \begin{bmatrix} -4 & 8 \\ 0 & 6 \end{bmatrix}$

11. $-4B = -4\begin{bmatrix} -6 & 2 \\ 4 & 0 \end{bmatrix} = \begin{bmatrix} 24 & -8 \\ -16 & 0 \end{bmatrix}$

13. $-4A + 5B = -4\begin{bmatrix} -2 & 4 \\ 0 & 3 \end{bmatrix} + 5\begin{bmatrix} -6 & 2 \\ 4 & 0 \end{bmatrix}$

$= \begin{bmatrix} 8 & -16 \\ 0 & -12 \end{bmatrix} + \begin{bmatrix} -30 & 10 \\ 20 & 0 \end{bmatrix}$

$= \begin{bmatrix} -22 & -6 \\ 20 & -12 \end{bmatrix}$

15. Call the first matrix A and the second matrix B.

Step 1: Multiply the elements of the first row of A by the corresponding elements of the column of B and add.

$\begin{bmatrix} 1 & 2 \\ 3 & 4 \end{bmatrix}\begin{bmatrix} -1 \\ 7 \end{bmatrix} 1(-1) + 2(7) = 13$

Therefore, 13 is the first row entry of the product matrix AB.

Step 2: Multiply the elements of the second row of A by the corresponding elements of the column of B and add.

$\begin{bmatrix} 1 & 2 \\ 3 & 4 \end{bmatrix}\begin{bmatrix} -1 \\ 7 \end{bmatrix} 3(-1) + 4(7) = 25$

The second row entry of the product is 25.

Step 3: Write the product using the two entries found above.

$AB = \begin{bmatrix} 1 & 2 \\ 3 & 4 \end{bmatrix}\begin{bmatrix} -1 \\ 7 \end{bmatrix} = \begin{bmatrix} 13 \\ 25 \end{bmatrix}$

17. The product of a 2 × 3 matrix and a 3 × 2 matrix will be a 2 × 2 matrix.

$\begin{bmatrix} 2 & 2 & -1 \\ 3 & 0 & 1 \end{bmatrix}\begin{bmatrix} 0 & 2 \\ -1 & 4 \\ 0 & 2 \end{bmatrix}$

$= \begin{bmatrix} (2)(0) + (2)(-1) + (-1)(0) & (2)(2) + (2)(4) + (-1)(2) \\ (3)(0) + (0)(-1) + (1)(0) & (3)(2) + (0)(4) + (1)(2) \end{bmatrix}$

$= \begin{bmatrix} -2 & 10 \\ 0 & 8 \end{bmatrix}$

19. The product of a 2 × 2 matrix and a 2 × 2 matrix will be a 2 × 2 matrix.

$\begin{bmatrix} -4 & 1 \\ 2 & -3 \end{bmatrix}\begin{bmatrix} 1 & 0 \\ 0 & 1 \end{bmatrix}$

$= \begin{bmatrix} (-4)(1) + (1)(0) & (-4)(0) + (1)(1) \\ (2)(1) + (-3)(0) & (2)(0) + (-3)(1) \end{bmatrix}$

$= \begin{bmatrix} -4 & 1 \\ 2 & -3 \end{bmatrix}$

21. The product of a 3 × 3 matrix and a 3 × 3 matrix will be a 3 × 3 matrix.

$$\begin{bmatrix} 1 & 0 & 0 \\ 0 & 1 & 0 \\ 0 & 0 & 1 \end{bmatrix} \begin{bmatrix} 3 & -5 & 7 \\ -2 & 1 & 6 \\ 0 & -3 & 4 \end{bmatrix}$$

$$= \begin{bmatrix} (1)(3)+(0)(-2)+(0)(0) & (1)(-5)+(0)(1)+(0)(-3) & (1)(7)+(0)(6)+(0)(4) \\ (0)(3)+(1)(-2)+(0)(0) & (0)(-5)+(1)(1)+(0)(-3) & (0)(7)+(1)(6)+(0)(4) \\ (0)(3)+(0)(-2)+(1)(0) & (0)(-5)+(0)(1)+(1)(-3) & (0)(7)+(0)(6)+(1)(4) \end{bmatrix}$$

$$= \begin{bmatrix} 3 & -5 & 7 \\ -2 & 1 & 6 \\ 0 & -3 & 4 \end{bmatrix}$$

23.

$$\begin{bmatrix} 1 & 2 \\ 3 & 4 \end{bmatrix} \begin{bmatrix} -1 & 5 \\ 7 & 0 \end{bmatrix}$$

$$= \begin{bmatrix} 1(-1) + 2 \cdot 7 & 1 \cdot 5 + 2 \cdot 0 \\ 3(-1) + 4 \cdot 7 & 3 \cdot 5 + 4 \cdot 0 \end{bmatrix}$$

$$= \begin{bmatrix} 13 & 5 \\ 25 & 15 \end{bmatrix}$$

25.

$$\begin{bmatrix} -2 & -3 & 7 \\ 1 & 5 & 6 \end{bmatrix} \begin{bmatrix} 1 \\ 2 \\ 3 \end{bmatrix}$$

$$= \begin{bmatrix} -2(1) + (-3)(2) + 7 \cdot 3 \\ 1 \cdot 1 + 5 \cdot 2 + 6 \cdot 3 \end{bmatrix}$$

$$= \begin{bmatrix} 13 \\ 29 \end{bmatrix}$$

27. $$\left(\begin{bmatrix} 4 & 3 \\ 1 & 2 \\ 0 & -5 \end{bmatrix} \begin{bmatrix} 2 & -2 \\ 1 & -1 \end{bmatrix}\right) \begin{bmatrix} 10 \\ 0 \end{bmatrix}$$

3 × 2 2 × 2 2 × 1

3 × 2

3 × 1

The parentheses say,

"Do $\begin{bmatrix} 4 & 3 \\ 1 & 2 \\ 0 & -5 \end{bmatrix} \cdot \begin{bmatrix} 2 & -2 \\ 1 & -1 \end{bmatrix}$ first."

$$\begin{bmatrix} 4 & 3 \\ 1 & 2 \\ 0 & -5 \end{bmatrix} \begin{bmatrix} 2 & -2 \\ 1 & -1 \end{bmatrix}$$

$$= \begin{bmatrix} 4 \cdot 2 + 3 \cdot 1 & 4(-2) + 3(-1) \\ 1 \cdot 2 + 2 \cdot 1 & 1(-2) + 2(-1) \\ 0 \cdot 2 - 5 \cdot 1 & 0(-2) + (-5)(-1) \end{bmatrix}$$

$$= \begin{bmatrix} 11 & -11 \\ 4 & -4 \\ -5 & 5 \end{bmatrix},$$

a 3 × 2 matrix, as predicted.

Then

$$\left(\begin{bmatrix} 4 & 3 \\ 1 & 2 \\ 0 & -5 \end{bmatrix} \begin{bmatrix} 2 & -2 \\ 1 & -1 \end{bmatrix}\right) \begin{bmatrix} 10 \\ 0 \end{bmatrix}$$

$$= \begin{bmatrix} 11 & -11 \\ 4 & -4 \\ -5 & 5 \end{bmatrix} \begin{bmatrix} 10 \\ 0 \end{bmatrix}$$

$$= \begin{bmatrix} 11 \cdot 10 + (-11)(0) \\ 4 \cdot 10 + (-4)(0) \\ -5 \cdot 10 + 5 \cdot 0 \end{bmatrix}$$

$$= \begin{bmatrix} 110 \\ 40 \\ -50 \end{bmatrix}, \text{ a } 3 \times 1 \text{ matrix.}$$

29. First, add the second and third matrices.

$$\begin{bmatrix} 2 & -2 \\ 1 & -1 \end{bmatrix}\left(\begin{bmatrix} 4 & 3 \\ 1 & 2 \end{bmatrix} + \begin{bmatrix} 7 & 0 \\ -1 & 5 \end{bmatrix}\right)$$

$$= \begin{bmatrix} 2 & -2 \\ 1 & -1 \end{bmatrix}\begin{bmatrix} 11 & 3 \\ 0 & 7 \end{bmatrix}$$

Now compute the product.

$$\begin{bmatrix} 2 & -2 \\ 1 & -1 \end{bmatrix}\begin{bmatrix} 11 & 3 \\ 0 & 7 \end{bmatrix} = \begin{bmatrix} 22 & -8 \\ 11 & -4 \end{bmatrix}$$

31. $AB = \begin{bmatrix} -3 & -9 \\ 2 & 6 \end{bmatrix}\begin{bmatrix} 4 & 6 \\ 2 & 3 \end{bmatrix} = \begin{bmatrix} -30 & -45 \\ 20 & 30 \end{bmatrix}$

$$BA = \begin{bmatrix} 4 & 6 \\ 2 & 3 \end{bmatrix}\begin{bmatrix} -3 & -9 \\ 2 & 6 \end{bmatrix} = \begin{bmatrix} 0 & 0 \\ 0 & 0 \end{bmatrix}$$

Since $\begin{bmatrix} -30 & -45 \\ 20 & 30 \end{bmatrix} \neq \begin{bmatrix} 0 & 0 \\ 0 & 0 \end{bmatrix}$, $AB \neq BA$.

Therefore, matrix multiplication is not commutative.

33.

$$A + B = \begin{bmatrix} -3 & -9 \\ 2 & 6 \end{bmatrix} + \begin{bmatrix} 4 & 6 \\ 2 & 3 \end{bmatrix} = \begin{bmatrix} 1 & -3 \\ 4 & 9 \end{bmatrix}$$

$$A - B = \begin{bmatrix} -3 & -9 \\ 2 & 6 \end{bmatrix} - \begin{bmatrix} 4 & 6 \\ 2 & 3 \end{bmatrix} = \begin{bmatrix} -7 & -15 \\ 0 & 3 \end{bmatrix}$$

$$(A + B)(A - B) = \begin{bmatrix} 1 & -3 \\ 4 & 9 \end{bmatrix}\begin{bmatrix} -7 & -15 \\ 0 & 3 \end{bmatrix} = \begin{bmatrix} -7 & -24 \\ -28 & -33 \end{bmatrix}$$

$$A^2 = \begin{bmatrix} -3 & -9 \\ 2 & 6 \end{bmatrix}\begin{bmatrix} -3 & -9 \\ 2 & 6 \end{bmatrix} = \begin{bmatrix} -9 & -27 \\ 6 & 18 \end{bmatrix}$$

$$B^2 = \begin{bmatrix} 4 & 6 \\ 2 & 3 \end{bmatrix}\begin{bmatrix} 4 & 6 \\ 2 & 3 \end{bmatrix} = \begin{bmatrix} 28 & 42 \\ 14 & 21 \end{bmatrix}$$

$$A^2 - B^2 =$$

$$\begin{bmatrix} -9 & -27 \\ 6 & 18 \end{bmatrix} - \begin{bmatrix} 28 & 42 \\ 14 & 21 \end{bmatrix} = \begin{bmatrix} -37 & -69 \\ -8 & -3 \end{bmatrix}$$

Since $\begin{bmatrix} -7 & -24 \\ -28 & -33 \end{bmatrix} \neq \begin{bmatrix} -37 & -69 \\ -8 & -3 \end{bmatrix}$,

$(A + B)(A - B) \neq A^2 - B^2$.

35. Verify that $(PX)T = P(XT)$.

$$(PX)T = \left(\begin{bmatrix} m & n \\ p & q \end{bmatrix}\begin{bmatrix} x & y \\ z & w \end{bmatrix}\right)\begin{bmatrix} r & s \\ t & u \end{bmatrix}$$

$$= \begin{bmatrix} mx + nz & my + nw \\ px + qz & py + qw \end{bmatrix}\begin{bmatrix} r & s \\ t & u \end{bmatrix}$$

$$= \begin{bmatrix} (mx + nz)r + (my + nw)t & (mx + nz)s + (my + nw)u \\ (px + qz)r + (py + qw)t & (px + qz)s + (py + qw)u \end{bmatrix}$$

$$= \begin{bmatrix} mxr + nzr + myt + nwt & mxs + nzs + myu + nwu \\ pxr + qzr + pyt + qwt & pxs + qzs + pyu + qwu \end{bmatrix}$$

$P(XT)$ is the same, so $(PX)T = P(XT)$.

37. Verify that PX is a 2×2 matrix.

$PX = \begin{bmatrix} mx + nz & my + nw \\ px + qz & py + qw \end{bmatrix}$, which is a

2×2 matrix.

39. Verify that $(k + h)P = kP + hP$ for any real numbers k and h.

$$(k + h)P = \begin{bmatrix} (k+h)m & (k+h)n \\ (k+h)p & (k+h)q \end{bmatrix}$$

$$= \begin{bmatrix} km + hm & kn + hn \\ kp + hp & kq + hq \end{bmatrix}$$

$$= \begin{bmatrix} km & kn \\ kp & kq \end{bmatrix} + \begin{bmatrix} hm & hn \\ hp & hq \end{bmatrix}$$

$$= kP = hP,$$

so $(k + h)P = kP + hP$.

41. (a) To find the comparative costs, find the matrix product AB.

$$AB = \begin{bmatrix} 1 & 4 & \frac{1}{4} & \frac{1}{4} & 1 \\ 0 & 3 & 0 & \frac{1}{4} & 0 \\ 4 & 3 & 2 & 1 & 1 \\ 0 & 1 & 0 & \frac{1}{3} & 0 \end{bmatrix} \begin{bmatrix} 5 & 5 \\ 8 & 10 \\ 10 & 12 \\ 12 & 15 \\ 5 & 6 \end{bmatrix}$$

$$= \begin{bmatrix} 47.5 & 57.5 \\ 27 & 33.75 \\ 81 & 95 \\ 12 & 15 \end{bmatrix}$$

(b) Let matrix C represent a day's orders.

$C = [20 \quad 200 \quad 50 \quad 60]$

To find the amount of each ingredient to fill a day's order, find the product CA.

$$CA = [20 \quad 200 \quad 50 \quad 60] \begin{bmatrix} 1 & 4 & \frac{1}{4} & \frac{1}{4} & 1 \\ 0 & 3 & 0 & \frac{1}{4} & 0 \\ 4 & 3 & 2 & 1 & 1 \\ 0 & 1 & 0 & \frac{1}{3} & 0 \end{bmatrix}$$

$$= [220 \quad 890 \quad 105 \quad 125 \quad 70]$$

(c) Let D represent the costs under the two purchase options to fill the day's orders.

$D = C(AB)$

$$= [20 \quad 200 \quad 50 \quad 60] \begin{bmatrix} 47.5 & 57.5 \\ 27 & 33.75 \\ 81 & 95 \\ 12 & 15 \end{bmatrix}$$

$$= [11{,}120 \quad 13{,}555]$$

43. (a) The cost of these three assets is given by

$$C = \begin{bmatrix} 90{,}000 \\ 60{,}000 \\ 120{,}000 \end{bmatrix}.$$

(b)

$$PC = \begin{bmatrix} .8 & 0 & 0 \\ 0 & .8 & 0 \\ 0 & 0 & .8 \end{bmatrix} \begin{bmatrix} 90{,}000 \\ 60{,}000 \\ 120{,}000 \end{bmatrix} = \begin{bmatrix} 72{,}000 \\ 48{,}000 \\ 96{,}000 \end{bmatrix}$$

(c) $R = \begin{bmatrix} \frac{1}{2} & \frac{1}{3} & \frac{1}{6} \end{bmatrix}$

(d)

$$(PC)R = \begin{bmatrix} 72{,}000 \\ 48{,}000 \\ 96{,}000 \end{bmatrix} \begin{bmatrix} \frac{1}{2} & \frac{1}{3} & \frac{1}{6} \end{bmatrix}$$

$$= \begin{bmatrix} 36{,}000 & 24{,}000 & 12{,}000 \\ 24{,}000 & 16{,}000 & 8000 \\ 48{,}000 & 32{,}000 & 16{,}000 \end{bmatrix}$$

45. Consider the linear system

$$2x_1 + 4x_2 = 2$$
$$x_1 + 3x_2 = -1.$$

This could be written as two equal matrices.

$$\begin{bmatrix} 2x_1 + 4x_2 \\ x_1 + 3x_2 \end{bmatrix} = \begin{bmatrix} 2 \\ -1 \end{bmatrix}$$

The 2×1 matrix on the left could be written as a product of two matrices.

$$\begin{bmatrix} 2 & 4 \\ 1 & 3 \end{bmatrix} \begin{bmatrix} x_1 \\ x_2 \end{bmatrix} = \begin{bmatrix} 2 \\ -1 \end{bmatrix}$$

Let

$$A = \begin{bmatrix} 2 & 4 \\ 1 & 3 \end{bmatrix}, \; X = \begin{bmatrix} x_1 \\ x_2 \end{bmatrix}, \text{ and } B = \begin{bmatrix} 2 \\ -1 \end{bmatrix},$$

which gives AX = B.

Solve the system, using augmented matrix.

$$\left[\begin{array}{cc|c} 2 & 4 & 2 \\ 1 & 3 & -1 \end{array}\right]$$

$$\left[\begin{array}{cc|c} 1 & 3 & -1 \\ 2 & 4 & 2 \end{array}\right] \text{ Interchange } R_1 \text{ and } R_2$$

$$\left[\begin{array}{cc|c} 1 & 3 & -1 \\ 0 & -2 & 4 \end{array}\right] -2R_1 + R_2$$

$$\left[\begin{array}{cc|c} 1 & 3 & -1 \\ 0 & 1 & -2 \end{array}\right] -\frac{1}{2}R_2$$

$$\left[\begin{array}{cc|c} 1 & 0 & 5 \\ 0 & 1 & -2 \end{array}\right] -3R_2 + R_1$$

The solution is (5, -2).

Thus,

$$AX = \begin{bmatrix} 2 & 4 \\ 1 & 3 \end{bmatrix} \begin{bmatrix} 5 \\ -2 \end{bmatrix} = \begin{bmatrix} 2 \\ -1 \end{bmatrix} = B.$$

47.

$$CD = \begin{bmatrix} -6 & 8 & 2 & 4 & -3 \\ 1 & 9 & 7 & -12 & 5 \\ 15 & 2 & -8 & 10 & 11 \\ 4 & 7 & 9 & 6 & -2 \\ 1 & 3 & 8 & -23 & 4 \end{bmatrix} \begin{bmatrix} 5 & -3 & 7 & 9 & 2 \\ 6 & 8 & -5 & 2 & 1 \\ 3 & 7 & -4 & 2 & 11 \\ 5 & -3 & 9 & 4 & -1 \\ 0 & 3 & 2 & 5 & 1 \end{bmatrix}$$

$$= \begin{bmatrix} 44 & 75 & -60 & -33 & 11 \\ 20 & 169 & -164 & 18 & 105 \\ 113 & -82 & 239 & 218 & -55 \\ 119 & 83 & 7 & 82 & 106 \\ 162 & 20 & 175 & 143 & 74 \end{bmatrix}$$

49. Since C is a 5 × 5 matrix and A is a 4 × 5 matrix, the product CA cannot be found.

51. The product AC cannot be found as A is a 4 × 5 matrix and C is a 5 × 5 matrix. Since the product AC cannot be found, AC ≠ CA.

53.

$$C + D = \begin{bmatrix} -6 & 8 & 2 & 4 & -3 \\ 1 & 9 & 7 & -12 & 5 \\ 15 & 2 & -8 & 10 & 11 \\ 4 & 7 & 9 & 6 & -2 \\ 1 & 3 & 8 & 23 & 4 \end{bmatrix} + \begin{bmatrix} 5 & -3 & 7 & 9 & 2 \\ 6 & 8 & -5 & 2 & 1 \\ 3 & 7 & -4 & 2 & 11 \\ 5 & -3 & 9 & 4 & -1 \\ 0 & 3 & 2 & 5 & 1 \end{bmatrix}$$

$$= \begin{bmatrix} -1 & 5 & 9 & 13 & -1 \\ 7 & 17 & 2 & -10 & 6 \\ 18 & 9 & -12 & 12 & 22 \\ 9 & 4 & 18 & 10 & -3 \\ 1 & 6 & 10 & 28 & 5 \end{bmatrix}$$

$$(C + D)B = \begin{bmatrix} -1 & 5 & 9 & 13 & -1 \\ 7 & 17 & 2 & -10 & 6 \\ 18 & 9 & -12 & 12 & 22 \\ 9 & 4 & 18 & 10 & -3 \\ 1 & 6 & 10 & 48 & 5 \end{bmatrix} \begin{bmatrix} 9 & 3 & 7 & -6 \\ -1 & 0 & 4 & 2 \\ -10 & -7 & 6 & 9 \\ 8 & 4 & 2 & -1 \\ 2 & -5 & 3 & 7 \end{bmatrix}$$

$$= \begin{bmatrix} -2 & -9 & 90 & 77 \\ -42 & -63 & 127 & 62 \\ 413 & 76 & 180 & -56 \\ -29 & -44 & 198 & 85 \\ 137 & 20 & 162 & 103 \end{bmatrix}$$

$$CB = \begin{bmatrix} -6 & 8 & 2 & 4 & -3 \\ 1 & 9 & 7 & -12 & 5 \\ 15 & 2 & -8 & 10 & 11 \\ 4 & 7 & 9 & 6 & -2 \\ 1 & 3 & 8 & 23 & 4 \end{bmatrix} \begin{bmatrix} 9 & 3 & 7 & -6 \\ -1 & 0 & 4 & 2 \\ -10 & -7 & 6 & 9 \\ 8 & 4 & 2 & -1 \\ 2 & -5 & 3 & 7 \end{bmatrix}$$

$$= \begin{bmatrix} -56 & -1 & 1 & 45 \\ -156 & -119 & 76 & 122 \\ 315 & 86 & 118 & -91 \\ -17 & -17 & 116 & 51 \\ 118 & 19 & 125 & 77 \end{bmatrix}$$

$$DB = \begin{bmatrix} 5 & -3 & 7 & 9 & 2 \\ 6 & 8 & -5 & 2 & 1 \\ 3 & 7 & -4 & 2 & 11 \\ 5 & -3 & 9 & 4 & -1 \\ 0 & 3 & 2 & 5 & 1 \end{bmatrix} \begin{bmatrix} 9 & 3 & 7 & -6 \\ -1 & 0 & 4 & 2 \\ -10 & -7 & 6 & 9 \\ 8 & 4 & 2 & -1 \\ 2 & -5 & 3 & 7 \end{bmatrix}$$

$$= \begin{bmatrix} 54 & -8 & 89 & 32 \\ 114 & 56 & 51 & -60 \\ 98 & -10 & 62 & 35 \\ -12 & -27 & 82 & 34 \\ 19 & 1 & 37 & 26 \end{bmatrix}$$

CB + DB

$$= \begin{bmatrix} -56 & -1 & 1 & 45 \\ -156 & -119 & 76 & 122 \\ 315 & 86 & 118 & -91 \\ -17 & -17 & 116 & 51 \\ 118 & 19 & 125 & 77 \end{bmatrix} + \begin{bmatrix} 54 & -8 & 89 & 32 \\ 114 & 56 & 51 & -60 \\ 98 & -10 & 62 & 35 \\ -12 & -27 & 82 & 34 \\ -19 & 1 & 37 & 26 \end{bmatrix}$$

$$= \begin{bmatrix} -2 & -9 & 90 & 77 \\ -42 & -63 & 127 & 62 \\ 413 & 76 & 180 & -56 \\ -29 & -44 & 198 & 85 \\ 137 & 20 & 162 & 103 \end{bmatrix}$$

From the results of (C + D)B and CB + DB above, we see that (C + D)B = CB + DB.

Section 6.5

1. $\begin{bmatrix} 2 & 4 \\ 4 & 7 \end{bmatrix}$

$\begin{bmatrix} 2 & 4 \\ 0 & -1 \end{bmatrix}$ $-2R_1 + R_2$

3. $\begin{bmatrix} 1 & \frac{1}{5} \\ 5 & 2 \end{bmatrix}$

$\begin{bmatrix} 1 & \frac{1}{5} \\ 0 & 1 \end{bmatrix}$ $-5R_1 + R_2$

5. $\begin{bmatrix} 1 & 5 & 6 \\ -2 & 3 & -1 \\ 4 & 7 & 0 \end{bmatrix}$

$\begin{bmatrix} 1 & 5 & 6 \\ 0 & 13 & 11 \\ 4 & 7 & 0 \end{bmatrix}$ $2R_1 + R_2$

7. $\begin{bmatrix} -3 & 1 & -4 \\ 2 & 1 & 3 \\ -7 & 5 & 2 \end{bmatrix}$

$\begin{bmatrix} -3 & 1 & -4 \\ 2 & 1 & 3 \\ -17 & 0 & -13 \end{bmatrix}$ $-5R_2 + R_3$

9. $\begin{bmatrix} 2 & 3 \\ 1 & 1 \end{bmatrix} \begin{bmatrix} -1 & 3 \\ 1 & -2 \end{bmatrix} = \begin{bmatrix} 1 & 0 \\ 0 & 1 \end{bmatrix} = I$

$\begin{bmatrix} -1 & 3 \\ 1 & -2 \end{bmatrix} \begin{bmatrix} 2 & 3 \\ 1 & 1 \end{bmatrix} = \begin{bmatrix} 1 & 0 \\ 0 & 1 \end{bmatrix} = I$

Yes, they are inverses of each other.

11. $\begin{bmatrix} 2 & 1 \\ 3 & 2 \end{bmatrix}\begin{bmatrix} 2 & 1 \\ -3 & 2 \end{bmatrix} = \begin{bmatrix} 1 & 4 \\ 0 & 7 \end{bmatrix} \neq I$

No, they are not inverses of each other.

13. $\begin{bmatrix} 1 & 2 & 0 \\ 0 & 1 & 0 \\ 0 & 1 & 0 \end{bmatrix}\begin{bmatrix} 1 & -2 & 0 \\ 0 & 1 & 0 \\ 0 & -1 & 1 \end{bmatrix}$

$= \begin{bmatrix} 1 & 0 & 0 \\ 0 & 1 & 0 \\ 0 & 1 & 0 \end{bmatrix} \neq I$

No, they are not inverses of each other.

15. $\begin{bmatrix} 1 & 3 & 3 \\ 1 & 4 & 3 \\ 1 & 3 & 4 \end{bmatrix}\begin{bmatrix} 7 & -3 & -3 \\ -1 & 1 & 0 \\ -1 & 0 & 1 \end{bmatrix}$

$= \begin{bmatrix} 1 & 0 & 0 \\ 0 & 1 & 0 \\ 0 & 0 & 1 \end{bmatrix} = I$

$\begin{bmatrix} 7 & -3 & -3 \\ -1 & 1 & 0 \\ -1 & 0 & 1 \end{bmatrix}\begin{bmatrix} 1 & 3 & 3 \\ 1 & 4 & 3 \\ 1 & 3 & 4 \end{bmatrix}$

$= \begin{bmatrix} 1 & 0 & 0 \\ 0 & 1 & 0 \\ 0 & 0 & 1 \end{bmatrix} = I$

Yes, they are inverses of each other.

17. Let $A = \begin{bmatrix} 1 & -1 \\ 2 & 0 \end{bmatrix}$.

Form the augmented matrix $[A|I]$.

$[A|I] = \left[\begin{array}{cc|cc} 1 & -1 & 1 & 0 \\ 2 & 0 & 0 & 1 \end{array}\right]$.

Perform row operations on $[A|I]$ to get a matrix of the form $[I|B]$.

$\left[\begin{array}{cc|cc} 1 & -1 & 1 & 0 \\ 2 & 0 & 0 & 1 \end{array}\right]$

$\left[\begin{array}{cc|cc} 1 & -1 & 1 & 0 \\ 0 & 2 & -2 & 1 \end{array}\right] \; -2R_1 + R_2$

$\left[\begin{array}{cc|cc} 1 & -1 & 1 & 0 \\ 0 & 1 & -1 & \frac{1}{2} \end{array}\right] \; \frac{1}{2}R_1$

$[I|B] = \left[\begin{array}{cc|cc} 1 & 0 & 0 & \frac{1}{2} \\ 0 & 1 & -1 & \frac{1}{2} \end{array}\right] \; R_2 + R_1$

$A^{-1} = \begin{bmatrix} 0 & \frac{1}{2} \\ -1 & \frac{1}{2} \end{bmatrix}$

19. Let $A = \begin{bmatrix} 3 & -1 \\ -5 & 2 \end{bmatrix}$.

$[A|I] = \left[\begin{array}{cc|cc} 3 & -1 & 1 & 0 \\ -5 & 2 & 0 & 1 \end{array}\right]$

$\left[\begin{array}{cc|cc} 1 & -\frac{1}{3} & \frac{1}{3} & 0 \\ -5 & 2 & 0 & 1 \end{array}\right] \; \frac{1}{3}R_1$

$\left[\begin{array}{cc|cc} 1 & -\frac{1}{3} & \frac{1}{3} & 0 \\ 0 & \frac{1}{3} & \frac{5}{3} & 1 \end{array}\right] \; 5R_1 + R_3$

$\left[\begin{array}{cc|cc} 1 & 0 & 2 & 1 \\ 0 & \frac{1}{3} & \frac{5}{3} & 1 \end{array}\right] \; R_2 + R_1$

$[I|B] = \left[\begin{array}{cc|cc} 1 & 0 & 2 & 1 \\ 0 & 1 & 5 & 3 \end{array}\right] \; 3R_2$

$A^{-1} = \begin{bmatrix} 2 & 1 \\ 5 & 3 \end{bmatrix}$

21. Let $A = \begin{bmatrix} -6 & 4 \\ -3 & 2 \end{bmatrix}$.

$[A|I] = \left[\begin{array}{cc|cc} -6 & 4 & 1 & 0 \\ -3 & 2 & 0 & 1 \end{array}\right]$

$\left[\begin{array}{cc|cc} 1 & -\frac{2}{3} & -\frac{1}{6} & 0 \\ -3 & 2 & 0 & 1 \end{array}\right] \; -\frac{1}{6}R_1$

$$\left[\begin{array}{cc|cc} 1 & -\frac{2}{3} & -\frac{1}{6} & 0 \\ 0 & 0 & -\frac{1}{2} & 1 \end{array}\right] \begin{array}{l} \\ 3R_1 + R_2 \end{array}$$

There is no way to complete the desired transformation, so A has no inverse.

23. Let $A = \begin{bmatrix} 1 & 0 & 0 \\ 0 & -1 & 0 \\ 1 & 0 & 1 \end{bmatrix}$

$$[A|I] = \left[\begin{array}{ccc|ccc} 1 & 0 & 0 & 1 & 0 & 0 \\ 0 & -1 & 0 & 0 & 1 & 0 \\ 1 & 0 & 1 & 0 & 0 & 1 \end{array}\right]$$

$$[I|B] = \left[\begin{array}{ccc|ccc} 1 & 0 & 0 & 1 & 0 & 0 \\ 0 & 1 & 0 & 0 & -1 & 0 \\ 0 & 0 & 1 & -1 & 0 & 1 \end{array}\right] \begin{array}{l} \\ -R_2 \\ -R_1 + R_3 \end{array}$$

$$A^{-1} = \begin{bmatrix} 1 & 0 & 0 \\ 0 & -1 & 0 \\ -1 & 0 & 1 \end{bmatrix}$$

25. Let $A = \begin{bmatrix} -1 & -1 & -1 \\ 4 & 5 & 0 \\ 0 & 1 & -3 \end{bmatrix}$.

$$[A|I] = \left[\begin{array}{ccc|ccc} -1 & -1 & -1 & 1 & 0 & 0 \\ 4 & 5 & 0 & 0 & 1 & 0 \\ 0 & 1 & -3 & 0 & 0 & 1 \end{array}\right]$$

$$\left[\begin{array}{ccc|ccc} 1 & 1 & 1 & -1 & 0 & 0 \\ 4 & 5 & 0 & 0 & 1 & 0 \\ 0 & 1 & -3 & 0 & 0 & 1 \end{array}\right] \begin{array}{l} -R_1 \\ \\ \\ \end{array}$$

$$\left[\begin{array}{ccc|ccc} 1 & 1 & 1 & -1 & 0 & 0 \\ 0 & 1 & -4 & 4 & 1 & 0 \\ 0 & 1 & -3 & 0 & 0 & 1 \end{array}\right] \begin{array}{l} \\ -4R_1 + R_2 \\ \\ \end{array}$$

$$\left[\begin{array}{ccc|ccc} 1 & 0 & 5 & -5 & -1 & 0 \\ 0 & 1 & -4 & 4 & 1 & 0 \\ 0 & 0 & 1 & -4 & -1 & 1 \end{array}\right] \begin{array}{l} -R_2 + R_1 \\ \\ -R_2 + R_3 \end{array}$$

$$[I|B] = \left[\begin{array}{ccc|ccc} 1 & 0 & 0 & 15 & 4 & -5 \\ 0 & 1 & 0 & -12 & -3 & 4 \\ 0 & 0 & 1 & -4 & -1 & 1 \end{array}\right] \begin{array}{l} -5R_3 + R_1 \\ 4R_3 + R_2 \\ \\ \end{array}$$

$$A^{-1} = \begin{bmatrix} 15 & 4 & -5 \\ -12 & -3 & 4 \\ -4 & -1 & 1 \end{bmatrix}$$

27. Let $A = \begin{bmatrix} 1 & 2 & 3 \\ -3 & -2 & -1 \\ -1 & 0 & 1 \end{bmatrix}$

$$[A|I] = \left[\begin{array}{ccc|ccc} 1 & 2 & 3 & 1 & 0 & 0 \\ -3 & -2 & -1 & 0 & 1 & 0 \\ -1 & 0 & 1 & 0 & 0 & 1 \end{array}\right]$$

$$\left[\begin{array}{ccc|ccc} 1 & 2 & 3 & 1 & 0 & 0 \\ 0 & 4 & 8 & 3 & 1 & 0 \\ 0 & 2 & 4 & 1 & 0 & 1 \end{array}\right] \begin{array}{l} \\ 3R_1 + R_2 \\ R_1 + R_3 \end{array}$$

$$\left[\begin{array}{ccc|ccc} 1 & 2 & 3 & 1 & 0 & 0 \\ 0 & 1 & 2 & \frac{3}{4} & \frac{1}{4} & 0 \\ 0 & 2 & 4 & 1 & 0 & 1 \end{array}\right] \begin{array}{l} \\ \frac{1}{4}R_2 \\ \\ \end{array}$$

$$\left[\begin{array}{ccc|ccc} 1 & 0 & -1 & -\frac{1}{2} & -\frac{1}{2} & 0 \\ 0 & 1 & 2 & \frac{3}{4} & \frac{1}{4} & 0 \\ 0 & 0 & 0 & -\frac{1}{2} & -\frac{1}{2} & 1 \end{array}\right] \begin{array}{l} -2R_2 + R_1 \\ \\ -2R_2 + R_3 \end{array}$$

Since the third row can never become [0 0 1], A has no inverse.

29. Let $A = \begin{bmatrix} 2 & 4 & 6 \\ -1 & -4 & -3 \\ 0 & 1 & -1 \end{bmatrix}$.

$$[A|I] = \left[\begin{array}{ccc|ccc} 2 & 4 & 6 & 1 & 0 & 0 \\ -1 & -4 & -3 & 0 & 1 & 0 \\ 0 & 1 & -1 & 0 & 0 & 1 \end{array}\right]$$

$$\left[\begin{array}{ccc|ccc} 1 & 0 & 3 & 1 & 1 & 0 \\ -1 & -4 & -3 & 0 & 1 & 0 \\ 0 & 1 & -1 & 0 & 0 & 1 \end{array}\right] \begin{array}{l} R_2 + R_1 \\ \\ \\ \end{array}$$

$$\left[\begin{array}{ccc|ccc} 1 & 0 & 3 & 1 & 1 & 0 \\ 0 & -4 & 0 & 1 & 2 & 0 \\ 0 & 1 & -1 & 0 & 0 & 1 \end{array}\right] \begin{array}{l} \\ R_1 + R_2 \\ \\ \end{array}$$

$$\left[\begin{array}{ccc|ccc} 1 & 0 & 3 & 1 & 1 & 0 \\ 0 & 1 & -1 & 0 & 0 & 1 \\ 0 & -4 & 0 & 1 & 2 & 0 \end{array}\right] \begin{array}{l} \text{Interchange} \\ R_2 \text{ and } R_3 \\ \\ \end{array}$$

$$\left[\begin{array}{ccc|ccc} 1 & 0 & 3 & 1 & 1 & 0 \\ 0 & 1 & -1 & 0 & 0 & 1 \\ 0 & 0 & 1 & -\frac{1}{4} & -\frac{1}{2} & -1 \end{array}\right] \begin{array}{l} \\ \\ -\frac{1}{4}R_3 \end{array}$$

$$\left[\begin{array}{ccc|ccc} 1 & 0 & 3 & 1 & 1 & 0 \\ 0 & 1 & -1 & 0 & 0 & 1 \\ 0 & 0 & -4 & 1 & 2 & 4 \end{array}\right] \begin{array}{l} \\ \\ 4R_2 + R_3 \end{array}$$

$$[I|B] = \left[\begin{array}{ccc|ccc} 1 & 0 & 0 & \frac{7}{4} & \frac{5}{2} & 3 \\ 0 & 1 & 0 & -\frac{1}{4} & -\frac{1}{2} & 0 \\ 0 & 0 & 1 & -\frac{1}{4} & -\frac{1}{2} & -1 \end{array}\right] \begin{array}{l} -3R_3 + R_1 \\ R_3 + R_2 \\ \\ \end{array}$$

$$A^{-1} = \begin{bmatrix} \frac{7}{4} & \frac{5}{2} & 3 \\ -\frac{1}{4} & -\frac{1}{2} & 0 \\ -\frac{1}{4} & -\frac{1}{2} & -1 \end{bmatrix}$$

31. $$[A|I] = \left[\begin{array}{cccc|cccc} 1 & -2 & 3 & 0 & 1 & 0 & 0 & 0 \\ 0 & 1 & -1 & 1 & 0 & 1 & 0 & 0 \\ -2 & 2 & -2 & 4 & 0 & 0 & 1 & 0 \\ 0 & 2 & -3 & 1 & 0 & 0 & 0 & 1 \end{array}\right]$$

$$\left[\begin{array}{cccc|cccc} 1 & -2 & 3 & 0 & 1 & 0 & 0 & 0 \\ 0 & 1 & -1 & 1 & 0 & 1 & 0 & 0 \\ 0 & -2 & 4 & 4 & 0 & 0 & 1 & 0 \\ 0 & 2 & -3 & 1 & 0 & 0 & 0 & 1 \end{array}\right] \begin{array}{l} \\ \\ 2R_1 + R_3 \\ \\ \end{array}$$

$$\left[\begin{array}{cccc|cccc} 1 & 0 & 1 & 2 & 1 & 2 & 0 & 0 \\ 0 & 1 & -1 & 1 & 0 & 1 & 0 & 0 \\ 0 & 0 & 1 & 5 & 2 & 0 & 1 & 1 \\ 0 & 0 & -1 & -1 & 0 & -2 & 0 & 1 \end{array}\right] \begin{array}{l} 2R_2 + R_1 \\ \\ R_4 + R_3 \\ -2R_2 + R_4 \end{array}$$

$$\left[\begin{array}{cccc|cccc} 1 & 0 & 0 & -3 & -1 & 2 & -1 & -1 \\ 0 & 1 & 0 & 6 & 2 & 1 & 1 & 1 \\ 0 & 0 & 1 & 5 & 2 & 0 & 1 & 1 \\ 0 & 0 & 0 & 4 & 2 & -2 & 1 & 2 \end{array}\right] \begin{array}{l} -R_3 + R_4 \\ R_3 + R_2 \\ \\ R_3 + R_4 \end{array}$$

$$\left[\begin{array}{cccc|cccc} 1 & 0 & 0 & -3 & -1 & 2 & -1 & -1 \\ 0 & 1 & 0 & 6 & 2 & 1 & 1 & 1 \\ 0 & 0 & 1 & 5 & 2 & 0 & 1 & 1 \\ 0 & 0 & 0 & 1 & \frac{1}{2} & -\frac{1}{2} & \frac{1}{4} & \frac{1}{2} \end{array}\right] \begin{array}{l} \\ \\ \\ \frac{1}{4}R_4 \end{array}$$

$$[I|B] = \left[\begin{array}{cccc|cccc} 1 & 0 & 0 & 0 & \frac{1}{2} & \frac{1}{2} & -\frac{1}{4} & \frac{1}{2} \\ 0 & 1 & 0 & 0 & -1 & 4 & -\frac{1}{2} & -2 \\ 0 & 0 & 1 & 0 & -\frac{1}{2} & \frac{5}{2} & -\frac{1}{4} & -\frac{3}{2} \\ 0 & 0 & 0 & 1 & \frac{1}{2} & -\frac{1}{2} & \frac{1}{4} & \frac{1}{2} \end{array}\right] \begin{array}{l} 3R_4 + R_1 \\ -6R_4 + R_2 \\ -5R_4 + R_3 \\ \\ \end{array}$$

$$A^{-1} = \begin{bmatrix} \frac{1}{2} & \frac{1}{2} & -\frac{1}{4} & \frac{1}{2} \\ -1 & 4 & -\frac{1}{2} & -2 \\ -\frac{1}{2} & \frac{5}{2} & -\frac{1}{4} & -\frac{3}{2} \\ \frac{1}{2} & -\frac{1}{2} & \frac{1}{4} & \frac{1}{2} \end{bmatrix}$$

33. $$IA = \begin{bmatrix} 1 & 0 \\ 0 & 1 \end{bmatrix}\begin{bmatrix} a & b \\ c & d \end{bmatrix} = \begin{bmatrix} a & b \\ c & d \end{bmatrix} = A$$

35. $$AO = \begin{bmatrix} a & b \\ c & d \end{bmatrix}\begin{bmatrix} 0 & 0 \\ 0 & 0 \end{bmatrix} = \begin{bmatrix} 0 & 0 \\ 0 & 0 \end{bmatrix} = O$$

37. From Exercise 36,

$$A^{-1} = \begin{bmatrix} \frac{d}{ad - bc} & -\frac{b}{ad - bc} \\ -\frac{c}{ad - bc} & \frac{a}{ad - bc} \end{bmatrix}$$

$$A^{-1}A = \begin{bmatrix} \frac{d}{ad - bc} & -\frac{b}{ad - bc} \\ -\frac{c}{ad - bc} & \frac{a}{ad - bc} \end{bmatrix} \begin{bmatrix} a & b \\ c & d \end{bmatrix}$$

$$= \begin{bmatrix} \frac{ad - bc}{ad - bc} & \frac{db - bd}{ad - bc} \\ \frac{-ca + ac}{ad - bc} & \frac{-cb + ad}{ad - bc} \end{bmatrix}$$

$$= \begin{bmatrix} 1 & 0 \\ 0 & 1 \end{bmatrix}$$

$$= I$$

39. If

$$C = \begin{bmatrix} -6 & 8 & 2 & 4 & -3 \\ 1 & 9 & 7 & -12 & 5 \\ 15 & 2 & -8 & 10 & 11 \\ 4 & 7 & 9 & 6 & -2 \\ 1 & 3 & 8 & 23 & 4 \end{bmatrix},$$

use a computer to find that

$$C^{-1} = \begin{bmatrix} -.0447 & -.0230 & .0292 & .0895 & .0402 \\ .0921 & .0150 & .0321 & .0209 & -.0276 \\ -.0678 & .0315 & -.0404 & .0326 & .0373 \\ .0171 & -.0248 & .0069 & -.0003 & .0246 \\ -.0208 & .0740 & .0096 & -.1018 & .0646 \end{bmatrix}.$$

41. If

$$D = \begin{bmatrix} 5 & -3 & 7 & 9 & 2 \\ 6 & 8 & -5 & 2 & 1 \\ 3 & 7 & -4 & 2 & 11 \\ 5 & -3 & 9 & 4 & -1 \\ 0 & 3 & 2 & 5 & 1 \end{bmatrix},$$

use a computer to find that

$$D^{-1} = \begin{bmatrix} .0394 & .0880 & .0033 & .0530 & -.1499 \\ -.1492 & .0289 & .0187 & .1033 & .1668 \\ -.1330 & -.0543 & .0356 & .1768 & .1055 \\ .1407 & .0175 & -.0453 & -.1344 & .0655 \\ .0102 & -.0653 & .0993 & .0085 & -.0388 \end{bmatrix}$$

43. Use a computer to multiply C times C^{-1} which was found in Exercises 39.

$$C^{-1}C = \begin{bmatrix} 1 & 1.596046 \times 10^{-7} & .89407 \times 10^{-7} & 0 & -.149012 \times 10^{-7} \\ .540167 \times 10^{-7} & 1 & .596046 \times 10^{-7} & 0 & 0 \\ -.484288 \times 10^{-7} & .372529 \times 10^{-7} & 1 & .596046 \times 10^{-7} & 0 \\ -.372529 \times 10^{-8} & 0 & -.149012 \times 10^{-7} & 1 & .745058 \times 10^{-8} \\ .149012 \times 10^{-7} & -.447035 \times 10^{-7} & 0 & 0 & 1 \end{bmatrix}$$

Because of round-off error by the computer, this product does not equal I exactly.

Section 6.6

1. Since $AX = B$, if A^{-1} exists, $X = A^{-1}B$.

If $A = \begin{bmatrix} 1 & 3 \\ -2 & 4 \end{bmatrix}$,

then $A^{-1} = \begin{bmatrix} \frac{4}{10} & -\frac{3}{10} \\ \frac{2}{10} & \frac{1}{10} \end{bmatrix}$.

Thus, $X = A^{-1}B$

$$= \begin{bmatrix} \frac{4}{10} & -\frac{3}{10} \\ \frac{2}{10} & \frac{1}{10} \end{bmatrix} \begin{bmatrix} 15 \\ 10 \end{bmatrix}$$

$$= \begin{bmatrix} 3 \\ 4 \end{bmatrix}.$$

3. Since $AX = B$, if A^{-1} exists, $X = A^{-1}B$.

If $A = \begin{bmatrix} -2 & 4 \\ 3 & -1 \end{bmatrix}$,

then $A^{-1} = \begin{bmatrix} \frac{1}{10} & \frac{4}{10} \\ \frac{3}{10} & \frac{2}{10} \end{bmatrix}$.

Thus, $X = A^{-1}B$

$$= \begin{bmatrix} \frac{1}{10} & \frac{4}{10} \\ \frac{3}{10} & \frac{2}{10} \end{bmatrix} \begin{bmatrix} 40 & -20 \\ 80 & 20 \end{bmatrix}$$

$$= \begin{bmatrix} 36 & 6 \\ 28 & -2 \end{bmatrix}.$$

5. Since $AX = B$, if A^{-1} exists, $X = A^{-1}B$.

If $A = \begin{bmatrix} 1 & 0 & 2 \\ -1 & 1 & 0 \\ 3 & 0 & 4 \end{bmatrix}$,

then $A^{-1} = \begin{bmatrix} -2 & 0 & 1 \\ -2 & 1 & 1 \\ \frac{3}{2} & 0 & -\frac{1}{2} \end{bmatrix}$.

Thus, $X = A^{-1}B$

$$= \begin{bmatrix} -2 & 0 & 1 \\ -2 & 1 & 1 \\ \frac{3}{2} & 0 & -\frac{1}{2} \end{bmatrix} \begin{bmatrix} 8 \\ 4 \\ -6 \end{bmatrix}$$

$$= \begin{bmatrix} -22 \\ -18 \\ 15 \end{bmatrix}.$$

7. Since $AX = B$, if A^{-1} exists, $X = A^{-1}B$.

If $A = \begin{bmatrix} -3 & 0 & 6 \\ 1 & 1 & 0 \\ 0 & 2 & 5 \end{bmatrix}$,

then $A^{-1} = \begin{bmatrix} -\frac{5}{3} & -4 & 2 \\ \frac{5}{3} & 5 & -2 \\ -\frac{2}{3} & -2 & 1 \end{bmatrix}$.

Thus, $X = A^{-1}B$

$$= \begin{bmatrix} -\frac{5}{3} & -4 & 2 \\ \frac{5}{3} & 5 & -2 \\ -\frac{2}{3} & -2 & 1 \end{bmatrix} \begin{bmatrix} 12 & 3 \\ -6 & 0 \\ 0 & -3 \end{bmatrix}$$

$$= \begin{bmatrix} 4 & -11 \\ -10 & 11 \\ 4 & -5 \end{bmatrix}.$$

9. Since $N = X - MX$,

$N = IX - MX$

$N = (I - M)X$

$(I - M)^{-1}N = (I - M)^{-1}(I - M)X$

$(I - M)^{-1}N = IX$.

Thus, $X = (I - M)^{-1} N$.

$$I - M = \begin{bmatrix} 1 & 0 \\ 0 & 1 \end{bmatrix} - \begin{bmatrix} 0 & 1 \\ -2 & 1 \end{bmatrix} = \begin{bmatrix} 1 & -1 \\ 2 & 0 \end{bmatrix}$$

If $I - M = \begin{bmatrix} 1 & -1 \\ 2 & 0 \end{bmatrix}$,

$$(I - M)^{-1} = \begin{bmatrix} 0 & \frac{1}{2} \\ -1 & \frac{1}{2} \end{bmatrix}.$$

Since $X = (I - M)^{-1} N$

$$= \begin{bmatrix} 0 & \frac{1}{2} \\ -1 & \frac{1}{2} \end{bmatrix} \begin{bmatrix} 8 \\ -12 \end{bmatrix}$$

$$= \begin{bmatrix} -6 \\ -14 \end{bmatrix}.$$

As a check, compute $X - MX$.

$$X - MX = \begin{bmatrix} -6 \\ -14 \end{bmatrix} - \begin{bmatrix} 0 & 1 \\ -2 & 1 \end{bmatrix} \begin{bmatrix} -6 \\ -14 \end{bmatrix}$$

$$= \begin{bmatrix} -6 \\ -14 \end{bmatrix} - \begin{bmatrix} -14 \\ -2 \end{bmatrix}$$

$$= \begin{bmatrix} 8 \\ -12 \end{bmatrix}$$

Thus, $X - MX = N$.

11. $-x - y - z = 1$

$4x + 5y = -2$

$y - 3z = 3$

Matrix has coefficient

$$\begin{bmatrix} -1 & -1 & -1 \\ 4 & 5 & 0 \\ 0 & 1 & -3 \end{bmatrix} \begin{bmatrix} x \\ y \\ z \end{bmatrix} = \begin{bmatrix} 1 \\ -2 \\ 3 \end{bmatrix}$$

$A \cdot B = B$.

In Exercise 25 of Section 6.5, it was found that $A^{-1} = \begin{bmatrix} 15 & 4 & -5 \\ -12 & -3 & 4 \\ -4 & -1 & 1 \end{bmatrix}$.

Since $X = A^{-1}B$,

$$X = \begin{bmatrix} x \\ y \\ z \end{bmatrix} = \begin{bmatrix} 15 & 4 & -5 \\ -12 & -3 & 4 \\ -4 & -1 & 1 \end{bmatrix} \begin{bmatrix} 1 \\ -2 \\ 3 \end{bmatrix}$$

$$= \begin{bmatrix} -8 \\ 6 \\ 1 \end{bmatrix}.$$

The solution of the system is (-8, 6, 1).

13. $2x + 4y + 6z = 4$

$-x - 4y - 3z = 8$

$y - z = -4$

has coefficient matrix

$$A = \begin{bmatrix} 2 & 4 & 6 \\ -1 & -4 & -3 \\ 0 & 1 & -1 \end{bmatrix}.$$

From Exercise 29 of Section 6.5,

$$A^{-1} = \begin{bmatrix} 2 & 4 & 6 \\ -1 & -4 & -3 \\ 0 & 1 & -1 \end{bmatrix}^{-1} = \begin{bmatrix} \frac{7}{4} & \frac{5}{2} & 3 \\ -\frac{1}{4} & -\frac{1}{2} & 0 \\ -\frac{1}{4} & -\frac{1}{2} & -1 \end{bmatrix}.$$

$$X = A^{-1}B = \begin{bmatrix} \frac{7}{4} & \frac{5}{2} & 3 \\ -\frac{1}{4} & -\frac{1}{2} & 0 \\ -\frac{1}{4} & -\frac{1}{2} & -1 \end{bmatrix} \begin{bmatrix} 4 \\ 8 \\ -4 \end{bmatrix}$$

$$= \begin{bmatrix} 15 \\ -5 \\ -1 \end{bmatrix}$$

The solution is (15, -5, -1).

15. $x + 2y + 3z = 5$
$2x + 3y + 2z = 2$
$-x - 2y - 4z = -1$
has coefficient matrix

$$A = \begin{bmatrix} 1 & 2 & 3 \\ 2 & 3 & 2 \\ -1 & -2 & -4 \end{bmatrix}.$$

Find A^{-1}:

$$[A|I] = \left[\begin{array}{ccc|ccc} 1 & 2 & 3 & 1 & 0 & 0 \\ 2 & 3 & 2 & 0 & 1 & 0 \\ -1 & -2 & -4 & 0 & 0 & 1 \end{array}\right]$$

$$\left[\begin{array}{ccc|ccc} 1 & 2 & 3 & 1 & 0 & 0 \\ 0 & -1 & -4 & -2 & 1 & 0 \\ 0 & 0 & -1 & 1 & 0 & 1 \end{array}\right] \begin{array}{l} \\ -2R_1 + R_1 \\ R_1 + R_3 \end{array}$$

$$\left[\begin{array}{ccc|ccc} 1 & 0 & -5 & -3 & 2 & 0 \\ 0 & 1 & 4 & 2 & -1 & 0 \\ 0 & 0 & 1 & -1 & 0 & -1 \end{array}\right] \begin{array}{l} -2R_2 + R_1 \\ -R_2 \\ -R_3 \end{array}$$

$$\left[\begin{array}{ccc|ccc} 1 & 0 & 0 & -8 & 2 & -5 \\ 0 & 1 & 0 & 6 & -1 & 4 \\ 0 & 0 & 1 & -1 & 0 & -1 \end{array}\right] \begin{array}{l} 5R_3 + R_1 \\ -4R_3 + R_2 \\ \end{array}$$

$$A^{-1} = \begin{bmatrix} -8 & 2 & -5 \\ 6 & -1 & 4 \\ -1 & 0 & -1 \end{bmatrix}$$

$$X = A^{-1}B = \begin{bmatrix} -8 & 2 & -5 \\ 6 & -1 & 4 \\ -1 & 0 & -1 \end{bmatrix} \begin{bmatrix} 5 \\ 2 \\ -1 \end{bmatrix}$$

$$= \begin{bmatrix} -31 \\ 24 \\ -4 \end{bmatrix}$$

The solution is (-31, 24, -4).

17. $x + 2y = -10$
$-x + 4z = 8$
$-y + z = 4$
has the coefficient matrix

$$A = \begin{bmatrix} 1 & 2 & 0 \\ -1 & 0 & 4 \\ 0 & -1 & 1 \end{bmatrix}.$$

$$A^{-1} = \begin{bmatrix} \frac{4}{6} & -\frac{2}{6} & \frac{8}{6} \\ \frac{1}{6} & \frac{1}{6} & -\frac{4}{6} \\ \frac{1}{6} & \frac{1}{6} & \frac{2}{6} \end{bmatrix}$$

Since $\overline{X} = A^{-1}B$,

$$\overline{X} = \begin{bmatrix} \frac{4}{6} & -\frac{2}{6} & \frac{8}{6} \\ \frac{1}{6} & \frac{1}{6} & -\frac{4}{6} \\ \frac{1}{6} & \frac{1}{6} & \frac{2}{6} \end{bmatrix} \begin{bmatrix} -10 \\ 8 \\ 4 \end{bmatrix} = \begin{bmatrix} -4 \\ -3 \\ 1 \end{bmatrix}.$$

The solution is (-4, -3, 1).

19. $2x - 2y = 5$
$4y + 8z = 7$
$x + 2z = 1$
has the coefficient matrix

$$\begin{bmatrix} 2 & -2 & 0 \\ 0 & 4 & 8 \\ 1 & 0 & 2 \end{bmatrix} \begin{bmatrix} x \\ y \\ z \end{bmatrix} = \begin{bmatrix} 5 \\ 7 \\ 1 \end{bmatrix}$$

$A \cdot X = B$.
However, using row operations on $[A|I]$ shows A^{-1} does not exist, so another method must be used. Try the Gauss-Jordon method. The augmented matrix is

$$\left[\begin{array}{ccc|c} 2 & -2 & 0 & 5 \\ 0 & 4 & 8 & 7 \\ 1 & 0 & 2 & 1 \end{array}\right].$$

After several row operations, we obtain

$$\left[\begin{array}{ccc|c} 1 & 0 & 2 & 1 \\ 0 & 1 & 2 & \frac{7}{4} \\ 0 & 0 & 0 & \frac{26}{4} \end{array}\right].$$

The bottom row of the final matrix shows that the system has no solution.

21. The coefficient matrix is

$$A = \begin{bmatrix} 1 & -2 & 3 & 0 \\ 0 & 1 & -1 & 1 \\ -2 & 2 & -2 & 4 \\ 0 & 2 & -3 & 1 \end{bmatrix}.$$

From Exercise 31 in Section 6.5,

$$A^{-1} = \begin{bmatrix} \frac{1}{2} & \frac{1}{2} & -\frac{1}{4} & \frac{1}{2} \\ 1 & 4 & -\frac{1}{2} & -2 \\ -\frac{1}{2} & \frac{5}{2} & -\frac{1}{4} & -\frac{3}{2} \\ \frac{1}{2} & -\frac{1}{2} & \frac{1}{4} & \frac{1}{2} \end{bmatrix}.$$

$$X = \begin{bmatrix} \frac{1}{2} & \frac{1}{2} & -\frac{1}{4} & \frac{1}{2} \\ 1 & 4 & -\frac{1}{2} & -2 \\ -\frac{1}{2} & \frac{5}{2} & -\frac{1}{4} & -\frac{3}{2} \\ \frac{1}{2} & -\frac{1}{2} & \frac{1}{4} & \frac{1}{2} \end{bmatrix} \begin{bmatrix} 4 \\ -8 \\ 12 \\ -4 \end{bmatrix}.$$

$$= \begin{bmatrix} -7 \\ -34 \\ -19 \\ 7 \end{bmatrix}.$$

The solution is (-7, -34, -19, 7).

23.

$$A = \begin{bmatrix} \frac{1}{2} & \frac{2}{5} \\ \frac{1}{4} & \frac{1}{5} \end{bmatrix}, D = \begin{bmatrix} 2 \\ 4 \end{bmatrix}$$

To find the production matrix, first calculate I - A.

$$I - A = \begin{bmatrix} 1 & 0 \\ 0 & 1 \end{bmatrix} - \begin{bmatrix} \frac{1}{2} & \frac{2}{5} \\ \frac{1}{4} & \frac{1}{5} \end{bmatrix}$$

$$= \begin{bmatrix} \frac{1}{2} & -\frac{2}{5} \\ -\frac{1}{4} & \frac{4}{5} \end{bmatrix}$$

Using row operations, find the inverse of I - A.

$$(I - A)^{-1} = \begin{bmatrix} \frac{8}{3} & \frac{4}{3} \\ \frac{5}{6} & \frac{5}{3} \end{bmatrix}$$

Since $X = (I - A)^{-1}D$,

$$X = \begin{bmatrix} \frac{8}{3} & \frac{4}{3} \\ \frac{5}{6} & \frac{5}{3} \end{bmatrix} \begin{bmatrix} 2 \\ 4 \end{bmatrix}$$

$$= \begin{bmatrix} \frac{32}{3} \\ \frac{25}{3} \end{bmatrix}.$$

25. $A = \begin{bmatrix} .1 & .03 \\ .07 & .6 \end{bmatrix}, D = \begin{bmatrix} 5 \\ 10 \end{bmatrix}$

First, calculate I - A.

$$I - A = \begin{bmatrix} .9 & -.03 \\ -.07 & .4 \end{bmatrix}$$

Use row operations to find the inverse of I - A.

$$(I - A)^{-1} = \begin{bmatrix} 1.118 & .084 \\ .195 & 2.515 \end{bmatrix}$$

Since $X = (I - A)^{-1}D$,

$$X = \begin{bmatrix} 1.118 & .084 \\ .195 & 2.515 \end{bmatrix} \begin{bmatrix} 5 \\ 10 \end{bmatrix} = \begin{bmatrix} 6.43 \\ 26.12 \end{bmatrix}.$$

27.

$$I - A = \begin{bmatrix} 1 & 0 & 0 \\ 0 & 1 & 0 \\ 0 & 0 & 1 \end{bmatrix} - \begin{bmatrix} .4 & 0 & .3 \\ 0 & .8 & .1 \\ 0 & .2 & .4 \end{bmatrix}$$

$$= \begin{bmatrix} .6 & 0 & -.3 \\ 0 & .2 & -.1 \\ 0 & -.2 & .6 \end{bmatrix}$$

Now use row operations to find $(I - A)^{-1}$.

$$\left[\begin{array}{ccc|ccc} .6 & 0 & -.3 & 1 & 0 & 0 \\ 0 & .2 & -.1 & 0 & 1 & 0 \\ 0 & -.2 & .6 & 0 & 0 & 1 \end{array}\right]$$

$$\left[\begin{array}{ccc|ccc} 1 & 0 & -.5 & 1.67 & 0 & 0 \\ 0 & .2 & -.1 & 0 & 1 & 0 \\ 0 & -.2 & .6 & 0 & 0 & 1 \end{array}\right]$$

$$\left[\begin{array}{ccc|ccc} 1 & 0 & -.5 & 1.67 & 0 & 0 \\ 0 & 1 & -.5 & 0 & 5 & 0 \\ 0 & 0 & .5 & 0 & 1 & 1 \end{array}\right]$$

$$\left[\begin{array}{ccc|ccc} 1 & 0 & 0 & 1.67 & 1 & 1 \\ 0 & 1 & 0 & 0 & 6 & 1 \\ 0 & 0 & 1 & 0 & 2 & 2 \end{array}\right]$$

$$(I - A)^{-1} = \begin{bmatrix} 1.67 & 1 & 1 \\ 0 & 6 & 1 \\ 0 & 2 & 2 \end{bmatrix}.$$

Since $X = (I - A)^{-1}D$,

$$X = \begin{bmatrix} 1.67 & 1 & 1 \\ 0 & 6 & 1 \\ 0 & 2 & 2 \end{bmatrix} \begin{bmatrix} 1 \\ 3 \\ 2 \end{bmatrix} = \begin{bmatrix} 6.67 \\ 20 \\ 10 \end{bmatrix}$$

29. From Example 5,

$$(I - A)^{-1} = \begin{bmatrix} 1.39 & .13 \\ .51 & 1.17 \end{bmatrix}.$$

If the demand is changed to 690 metric tons of wheat and 920 metric tons of oil,

$$D = \begin{bmatrix} 690 \\ 920 \end{bmatrix}.$$

Since $X = (I - A)^{-1}D$,

$$X = \begin{bmatrix} 1.39 & .13 \\ .51 & 1.17 \end{bmatrix} \begin{bmatrix} 690 \\ 920 \end{bmatrix} = \begin{bmatrix} 1078.7 \\ 1428.3 \end{bmatrix}.$$

Thus, about 1079 metric tons of wheat and about 1428 metric tons of oil should be produced.

31. First, write the input-output matrix A.

$$A = \begin{bmatrix} .4 & .6 \\ .5 & .25 \end{bmatrix}$$

$$= \begin{bmatrix} \frac{2}{5} & \frac{3}{5} \\ \frac{1}{2} & \frac{1}{4} \end{bmatrix}$$

$$I - A = \begin{bmatrix} 1 & 0 \\ 0 & 1 \end{bmatrix} - \begin{bmatrix} \frac{2}{5} & \frac{3}{5} \\ \frac{1}{2} & \frac{1}{4} \end{bmatrix} = \begin{bmatrix} \frac{3}{5} & -\frac{3}{5} \\ -\frac{1}{2} & \frac{3}{4} \end{bmatrix}$$

Form $[A - I|I]$.

$$\left[\begin{array}{cc|cc} \frac{3}{5} & -\frac{3}{5} & 1 & 0 \\ -\frac{1}{2} & \frac{3}{4} & 0 & 1 \end{array}\right]$$

$$\left[\begin{array}{cc|cc} 1 & -1 & \frac{5}{3} & 0 \\ -\frac{1}{2} & \frac{3}{4} & 0 & 1 \end{array}\right] \begin{matrix} \frac{5}{3}R_1 \\ \\ \end{matrix}$$

$$\left[\begin{array}{cc|cc} 1 & -1 & \frac{5}{3} & 0 \\ 0 & \frac{1}{4} & \frac{5}{6} & 1 \end{array}\right] \begin{matrix} \\ \frac{1}{2}R_1 + R_2 \end{matrix}$$

$$\left[\begin{array}{cc|cc} 1 & -1 & \frac{5}{3} & 0 \\ 0 & 1 & \frac{10}{3} & 4 \end{array}\right] \begin{matrix} \\ 4R_2 \end{matrix}$$

$$\left[\begin{array}{cc|cc} 1 & 0 & \frac{15}{3} & 4 \\ 0 & 1 & \frac{10}{3} & 4 \end{array}\right] \begin{matrix} 1R_2 + R_1 \\ \\ \end{matrix}$$

Thus, $(I - A)^{-1} = \begin{bmatrix} \frac{15}{3} & 4 \\ \frac{10}{3} & 4 \end{bmatrix}$.

Since $D = \begin{bmatrix} 12 \text{ million} \\ 15 \text{ million} \end{bmatrix}$ and $X = (I - A^{-1})D$,

$$X = \begin{bmatrix} \frac{15}{3} & 4 \\ \frac{10}{3} & 4 \end{bmatrix} \begin{bmatrix} 15 \text{ million} \\ 12 \text{ million} \end{bmatrix} = \begin{bmatrix} 123 \text{ million} \\ 98 \text{ million} \end{bmatrix}.$$

The output should be $123 million of electricity and $98 million of gas.

33. Referring to Exercise 32, use a computer to find the product $(I - A)^{-1}$ times D where

$$D = \begin{bmatrix} 500 \\ 500 \\ 500 \end{bmatrix}$$

and find that

$$X = (I - A)^{-1}D = \begin{bmatrix} 1538.5 \\ 1282.1 \\ 1589.7 \end{bmatrix}.$$

Therefore, 1538.5 units of agriculture, 1282.1 units of manufacturing, and 1589.7 units of transportation should be produced.

35. Break the message into groups of two letters and form a 2 × 1 matrix for each group of two letters.

ar th ur -i s- he re

$$\begin{bmatrix} 1 \\ 18 \end{bmatrix} \begin{bmatrix} 20 \\ 8 \end{bmatrix} \begin{bmatrix} 21 \\ 18 \end{bmatrix} \begin{bmatrix} 27 \\ 9 \end{bmatrix} \begin{bmatrix} 19 \\ 27 \end{bmatrix} \begin{bmatrix} 8 \\ 5 \end{bmatrix} \begin{bmatrix} 18 \\ 5 \end{bmatrix}$$

We multiply each of these 2 × 1 on the left by M.

$$M\begin{bmatrix} 1 \\ 18 \end{bmatrix} = \begin{bmatrix} -1 & 2 \\ 2 & 5 \end{bmatrix}\begin{bmatrix} 1 \\ 18 \end{bmatrix} = \begin{bmatrix} 35 \\ 92 \end{bmatrix}$$

$$M\begin{bmatrix} 20 \\ 8 \end{bmatrix} = \begin{bmatrix} -1 & 2 \\ 2 & 5 \end{bmatrix}\begin{bmatrix} 20 \\ 8 \end{bmatrix} = \begin{bmatrix} -4 \\ 80 \end{bmatrix}$$

$$M\begin{bmatrix} 21 \\ 18 \end{bmatrix} = \begin{bmatrix} -1 & 2 \\ 2 & 5 \end{bmatrix}\begin{bmatrix} 21 \\ 18 \end{bmatrix} = \begin{bmatrix} 15 \\ 132 \end{bmatrix}$$

$$M\begin{bmatrix} 27 \\ 9 \end{bmatrix} = \begin{bmatrix} -1 & 2 \\ 2 & 5 \end{bmatrix}\begin{bmatrix} 27 \\ 9 \end{bmatrix} = \begin{bmatrix} -9 \\ 99 \end{bmatrix}$$

$$M\begin{bmatrix} 19 \\ 27 \end{bmatrix} = \begin{bmatrix} -1 & 2 \\ 2 & 5 \end{bmatrix}\begin{bmatrix} 19 \\ 27 \end{bmatrix} = \begin{bmatrix} 35 \\ 173 \end{bmatrix}$$

$$M\begin{bmatrix} 8 \\ 5 \end{bmatrix} = \begin{bmatrix} -1 & 2 \\ 2 & 5 \end{bmatrix}\begin{bmatrix} 8 \\ 5 \end{bmatrix} = \begin{bmatrix} 2 \\ 41 \end{bmatrix}$$

$$M\begin{bmatrix} 18 \\ 5 \end{bmatrix} = \begin{bmatrix} -1 & 2 \\ 2 & 5 \end{bmatrix}\begin{bmatrix} 18 \\ 5 \end{bmatrix} = \begin{bmatrix} -8 \\ 61 \end{bmatrix}$$

The coded message is

$$\begin{bmatrix} 35 \\ 92 \end{bmatrix}, \begin{bmatrix} -4 \\ 80 \end{bmatrix}, \begin{bmatrix} 15 \\ 132 \end{bmatrix}, \begin{bmatrix} -9 \\ 99 \end{bmatrix}, \begin{bmatrix} 35 \\ 173 \end{bmatrix}, \begin{bmatrix} 2 \\ 41 \end{bmatrix}, \begin{bmatrix} -8 \\ 61 \end{bmatrix}.$$

37. Multiply the matrix

$$M = \begin{bmatrix} 1 & 3 & 3 \\ 1 & 4 & 3 \\ 1 & 3 & 4 \end{bmatrix}$$

times each of the 3 × 1 column matrices of the coded message

$$\begin{bmatrix} 13 \\ 1 \\ 20 \end{bmatrix}, \begin{bmatrix} 8 \\ 5 \\ 13 \end{bmatrix}, \begin{bmatrix} 1 \\ 20 \\ 9 \end{bmatrix}, \begin{bmatrix} 3 \\ 19 \\ 27 \end{bmatrix}, \begin{bmatrix} 9 \\ 19 \\ 27 \end{bmatrix}, \begin{bmatrix} 6 \\ 15 \\ 18 \end{bmatrix}, \begin{bmatrix} 27 \\ 20 \\ 8 \end{bmatrix}, \begin{bmatrix} 5 \\ 27 \\ 2 \end{bmatrix}, \begin{bmatrix} 9 \\ 18 \\ 4 \end{bmatrix}, \begin{bmatrix} 19 \\ 27 \\ 27 \end{bmatrix}$$

to obtain the column matrices

$$\begin{bmatrix} 76 \\ 77 \\ 96 \end{bmatrix}, \begin{bmatrix} 62 \\ 67 \\ 75 \end{bmatrix}, \begin{bmatrix} 88 \\ 108 \\ 97 \end{bmatrix}, \begin{bmatrix} 141 \\ 160 \\ 168 \end{bmatrix}, \begin{bmatrix} 147 \\ 166 \\ 174 \end{bmatrix}, \begin{bmatrix} 105 \\ 120 \\ 123 \end{bmatrix}, \begin{bmatrix} 111 \\ 131 \\ 119 \end{bmatrix}, \begin{bmatrix} 92 \\ 119 \\ 94 \end{bmatrix}, \begin{bmatrix} 75 \\ 93 \\ 79 \end{bmatrix}, \begin{bmatrix} 181 \\ 208 \\ 208 \end{bmatrix}.$$

39. From the example in the textbook,

$$A = \begin{bmatrix} 0 & 1 & 2 & 2 \\ 1 & 0 & 1 & 0 \\ 2 & 1 & 0 & 1 \\ 2 & 0 & 1 & 0 \end{bmatrix}.$$

Thus,

$$A^2 = AA = \begin{bmatrix} 0 & 1 & 2 & 2 \\ 1 & 0 & 1 & 0 \\ 2 & 1 & 0 & 1 \\ 2 & 0 & 1 & 0 \end{bmatrix}\begin{bmatrix} 0 & 1 & 2 & 2 \\ 1 & 0 & 1 & 0 \\ 2 & 1 & 0 & 1 \\ 2 & 0 & 1 & 0 \end{bmatrix} = \begin{bmatrix} 9 & 2 & 3 & 2 \\ 2 & 2 & 2 & 3 \\ 3 & 2 & 6 & 4 \\ 2 & 3 & 4 & 5 \end{bmatrix}$$

and

$$A^3 = AA^2 = \begin{bmatrix} 0 & 1 & 2 & 2 \\ 1 & 0 & 1 & 0 \\ 2 & 1 & 0 & 1 \\ 2 & 0 & 1 & 0 \end{bmatrix}\begin{bmatrix} 9 & 2 & 3 & 2 \\ 2 & 2 & 2 & 3 \\ 3 & 2 & 6 & 4 \\ 2 & 3 & 4 & 5 \end{bmatrix} = \begin{bmatrix} 12 & 12 & 22 & 21 \\ 12 & 4 & 9 & 3 \\ 22 & 9 & 13 & 12 \\ 21 & 6 & 12 & 8 \end{bmatrix}.$$

(a) The number of ways to travel between cities 1 and 4 by passing through exactly 2 cities is the entry in row 1, column 4 of A^3 which is 21.

(b) The number of ways to travel between cities 1 and 4 by passing through at most two cities is the sum of the entries in row 1, column 4 of A, A^2, and A^3 which is 21 + 2 + 2 or 25.

41. **(a)**

$$A = \begin{matrix} & \begin{matrix} S & J & NO & H \end{matrix} \\ \begin{matrix} S \\ J \\ NO \\ H \end{matrix} & \begin{bmatrix} 0 & 1 & 2 & 1 \\ 1 & 0 & 1 & 0 \\ 2 & 1 & 0 & 1 \\ 1 & 0 & 1 & 0 \end{bmatrix} \end{matrix}$$

(b) The number of one-stop flights is given by A^2.

$$A^2 = \begin{bmatrix} 0 & 1 & 2 & 1 \\ 1 & 0 & 1 & 0 \\ 2 & 1 & 0 & 1 \\ 1 & 0 & 1 & 0 \end{bmatrix}\begin{bmatrix} 0 & 1 & 2 & 1 \\ 1 & 0 & 1 & 0 \\ 2 & 1 & 0 & 1 \\ 1 & 0 & 1 & 0 \end{bmatrix} = \begin{bmatrix} 6 & 2 & 2 & 2 \\ 2 & 2 & 2 & 2 \\ 2 & 2 & 6 & 2 \\ 2 & 2 & 2 & 2 \end{bmatrix}$$

The number of one-stop flights between Houston and Jackson is the entry in row 4, column 2 of A^2 which is 2.

(c) The number of flights between Houston and Shreveport which require at most one stop is the sum of the entries in row 4, column 1 of A and A^2 which is 1 + 2 or 3.

(d) The number of one-stop flights between New Orleans and Houston is the entry in row 3, column 4 of A^2 which is 2.

Chapter 6 Review Exercises

1. $3x - 5y = -18$ *(1)*
   $2x + 7y = 19$ *(2)*

   Multiply equation (1) by $\frac{1}{3}$.

   $x - \frac{5}{3}y = -6$ *(3)*

   $2x + 7y = 19$ *(2)*

   Multiply equation (3) by -2 and add this to equation (2).

   $x - \frac{5}{3}y = -6$ *(3)*

   $\frac{31}{3}y = 31$ *(4)*

   From (4), y = 3. Substituting into (3) gives

   $x - \frac{5}{3}(3) = -6$

   $x - 5 = -6$

   $x = -1.$

   The solution is (-1, 3).

3. $\frac{2}{3}x - \frac{3}{4}y = 13$ *(1)*

   $\frac{1}{2}x + \frac{2}{3}y = -5$ *(2)*

   First eliminate fractions. Multiply equation (1) by 12 and equation (2) by 6.

   $8x - 9y = 156$ *(3)*
   $3x + 4y = -30$ *(4)*

   Multiply equation (3) by $\frac{1}{8}$.

   $x - \frac{9}{8}y = \frac{39}{2}$ *(5)*

   $3x + 4y = -30$ *(4)*

   Multiply equation (5) by -3 and add this to equation (4).

   $x - \frac{9}{8}y = \frac{39}{2}$ *(5)*

   $\frac{59}{8}y = \frac{177}{2}$ *(6)*

   From equation (6), y = -12.

   Substituting into equation (5) gives

   $x - \frac{9}{8}(-12) = \frac{39}{2}$

   $x = \frac{39}{2} - \frac{27}{2}$

   $x = 6.$

   The solution is (6, -12)

5. $4x + 8y = 15$ *(1)*

   $\frac{4}{3}x + \frac{8}{3}y = 5$ *(2)*

   Multiply equation (2) by 3 to obtain
   $4x + 8y = 15$
   which is equation (1). The equations are dependent. Let x be arbitrary.
   $8y = 15 - 4x$

   $y = \frac{15}{8} - \frac{1}{2}x$

   The solution is $(x, -\frac{1}{2}x + \frac{15}{8}x)$.

7. $2x - 3y + z = -5$ *(1)*
   $x + 4y + 2z = 13$ *(2)*
   $5x + 5y + 3z = 14$ *(3)*

   Interchange equations (1) and (2).

   $x + 4y + 2z = 13$ *(2)*
   $2x - 3y + z = -5$ *(1)*
   $5x + 5y + 3z = 14$ *(3)*

   Multiply equation (2) by -2 and add the result to equation (1). Multiply equation (2) by -5 and add the result to equation (3).

   $x + 4y + 2z = 13$ *(2)*
   $-11y - 3z = -31$ *(4)*
   $-15y - 7z = -51$ *(5)*

   Multiply equation (5) by -1 and add the result to equation (4).

   $x + 4y + 2z = 13$ *(2)*
   $4y + 4z = 20$ *(6)*
   $-15y - 7z = -51$ *(5)*

   Multiply equation (6) by 1/4.

   $x + 4y + 2z = 13$ *(2)*
   $y + z = 5$ *(7)*
   $-15y - 7z = -51$ *(5)*

Multiply equation (7) by 15 and add the result to equation (5).

$$x + 4y + 2z = 13 \quad (2)$$
$$y + z = 5 \quad (7)$$
$$8z = 24 \quad (8)$$

Multiply equation (8) by 1/8.

$$x + 4y + 2z = 13 \quad (2)$$
$$y + z = 5 \quad (7)$$
$$z = 3 \quad (9)$$

Substituting z = 3 in equation (7) to get y = 2. Substitute y = 2 and z = 3 in equation (2) to get x = -1. The solution is (-1, 2, 3).

9. Let x = the number of 25¢ candy bars
y = the number of 50¢ candy bars.
The system is

$$x + y = 22 \quad (1)$$
$$.25x + .5y = 8.5. \quad (2)$$

Multiply equation (1) by -.25 and add it to equation (2).

$$x + y = 22 \quad (1)$$
$$.25y = 3 \quad (3)$$

From equation (3), y = 12.
Substituting into equation (1) gives

$$x + 12 = 22$$
$$x = 10.$$

The student bought 10 25¢ candy bars and 12 50¢ candy bars.

11. Let x = amount invested at 6%
y = amount invested at 7%
z = amount invested at 9%.
The system is

$$x + y + z = 50{,}000$$
$$y = 2x$$
$$.06x + .07y + .09z = 3800$$

writing the system in proper form,

$$x + y + z = 50{,}000 \quad (1)$$
$$-2x + y = 0 \quad (2)$$
$$.06x + .07y + .09z = 3800 \quad (3)$$

Multiply equation (1) by 2 and add it to equation (2) and multiply equation (1) by -.06 and add it to equation (3).

$$x + y + z = 50{,}000 \quad (1)$$
$$3y + 2z = 100{,}000 \quad (4)$$
$$.01y + .03z = 800 \quad (5)$$

Multiply equation (4) by $\frac{1}{3}$.

$$x + y + z = 50{,}000 \quad (1)$$
$$y + \frac{2}{3}z = \frac{100{,}000}{3} \quad (6)$$
$$.01y + .03z = 800 \quad (5)$$

Multiply equation (6) by -.01 and add it to equation (5).

$$x + y + z = 50{,}000 \quad (1)$$
$$y + \frac{2}{3}z = \frac{100{,}000}{3} \quad (6)$$
$$\frac{7}{300}z = \frac{1400}{3} \quad (7)$$

From equation (7), z = 20,000.
Substituting into equation (6) gives

$$y + \frac{2}{3}(200{,}000) = \frac{100{,}000}{3}$$
$$y = \frac{60{,}000}{3}$$
$$y = 20{,}000.$$

Substituting into equation (1) gives

$$x + 20{,}000 + 20{,}000 = 50{,}000$$
$$x = 10{,}000$$

Thus, $10,000 was invested at 6%, $20,000 at 7%, and $20,000 at 9%.

13. Organize the information into a table.

	I	II	III
Food	100	200	150
Shelter	250	0	200
Counseling	0	100	100

Let x = the number of clients from source I
y = the number of clients from source II
z = the number of clients from source III.

The system is

$100x + 200y + 150z = 50{,}000$ *(1)*

$250x + 200z = 32{,}500$ *(2)*

$100y + 100z = 25{,}000.$ *(3)*

Multiply equation (1) by $\frac{1}{100}$.

$x + 2y + 1.5z = 500$ *(4)*

$250x + 200z = 32{,}500$ *(2)*

$100y + 100z = 25{,}000$ *(3)*

Multiply equation (4) by -250 and add it to equation (2).

$x + 2y + 1.5z = 500$ *(4)*

$-500y - 175z = -92{,}500$ *(5)*

$100y + 100z = 25{,}000$ *(3)*

Multiply equation (5) by $-\frac{1}{500}$.

$x + 2y + 1.5z = 500$ *(4)*

$y + .35z = 185$ *(6)*

$100y + 100z = 25{,}000$ *(3)*

Multiply equation (6) by -100 and add it to equation (3).

$x + 2y + 1.5z = 500$ *(4)*

$y + .35z = 185$ *(6)*

$65z = 6500$ *(7)*

From equation (7), $z = 100$.

Substituting into equation (6) gives

$y + .35(100) = 185$

$y = 150.$

Substituting into equation (4) gives

$x + 2(150) + 1.5(100) = 500$

$x + 300 + 150 = 500$

$x = 50.$

Thus, the agency can serve 50 clients from source I, 150 from source II, and 100 from source III.

15. $2x + 3y - z = 5$ *(1)*

$-x + 2y + 4z = 8$ *(2)*

Multiply equation (1) by $\frac{1}{2}$.

$x + \frac{3}{2}y - \frac{1}{2}z = \frac{5}{2}$ *(3)*

$-x + 2y + 4z = 8$ *(2)*

Add equation (3) to equation (2).

$x + \frac{3}{2}y - \frac{1}{2}z = \frac{5}{2}$ *(3)*

$\frac{7}{2}y + \frac{7}{2}z = \frac{21}{2}$ *(4)*

Multiply equation (4) by $\frac{2}{7}$.

$x + \frac{3}{2}y - \frac{1}{2}z = \frac{5}{2}$ *(3)*

$y + z = 3$ *(5)*

From equation (5), $y = 3 - z$.

Substituting this into equation (1) gives

$2x + 3(3 - z) - z = 5$

$2x + 9 - 3z - z = 5$

$2x = 4z - 4$

$x = 2z - 2.$

The solution is

$(2z - 2, -z + 3, z)$.

17. $5x + 2y = -10$

$3x - 5y = -6$

The augmented matrix is

$$\left[\begin{array}{cc|c} 5 & 2 & -10 \\ 3 & -5 & -6 \end{array}\right].$$

$$\left[\begin{array}{cc|c} 1 & \frac{2}{5} & -2 \\ 3 & -5 & -6 \end{array}\right] \quad \frac{1}{5}R_1$$

$$\left[\begin{array}{cc|c} 1 & \frac{2}{5} & -2 \\ 0 & -\frac{36}{5} & 0 \end{array}\right] \quad -3R_1 + R_2$$

$$\left[\begin{array}{cc|c} 1 & \frac{2}{5} & -2 \\ 0 & 1 & 0 \end{array}\right] \quad -\frac{5}{36}R_2$$

$$\left[\begin{array}{cc|c} 1 & 0 & -2 \\ 0 & 1 & 0 \end{array}\right] \quad -\frac{2}{5}R_1 + R_2$$

The solution is (-2, 0).

19.
$$\begin{aligned} x \quad\quad - z &= -3 \\ y + z &= 6 \\ 2x \quad\quad - 3z &= -9 \end{aligned}$$

The augmented matrix is

$$\left[\begin{array}{ccc|c} 1 & 0 & -1 & -3 \\ 0 & 1 & 1 & 6 \\ 2 & 0 & -3 & -9 \end{array}\right].$$

$$\left[\begin{array}{ccc|c} 1 & 0 & -1 & -3 \\ 0 & 1 & 1 & 6 \\ 0 & 0 & -1 & -3 \end{array}\right] -2R_1 + R_2$$

$$\left[\begin{array}{ccc|c} 1 & 0 & -1 & -3 \\ 0 & 1 & 1 & 6 \\ 0 & 0 & 1 & 3 \end{array}\right] \quad -1R_3$$

$$\left[\begin{array}{ccc|c} 1 & 0 & 0 & 0 \\ 0 & 1 & 0 & 3 \\ 0 & 0 & 1 & 3 \end{array}\right] \quad \begin{array}{l} 1R_3 + R_1 \\ -1R_3 + R_2 \end{array}$$

The solution is (0, 3, 3).

21.
$$\begin{aligned} 5x - 8y + z &= 1 \\ 3x - 2y + 4z &= 3 \\ 10x - 16y + 2z &= 3 \end{aligned}$$

The augmented matrix is

$$\left[\begin{array}{ccc|c} 5 & -8 & 1 & 1 \\ 3 & -2 & 4 & 3 \\ 10 & -16 & 2 & 3 \end{array}\right].$$

$$\left[\begin{array}{ccc|c} 1 & -\frac{8}{5} & \frac{1}{5} & \frac{1}{5} \\ 3 & -2 & 4 & 3 \\ 10 & -16 & 2 & 3 \end{array}\right] \quad \frac{1}{5}R_1$$

$$\left[\begin{array}{ccc|c} 1 & -\frac{8}{5} & \frac{1}{5} & \frac{1}{5} \\ 0 & \frac{14}{5} & \frac{17}{5} & \frac{12}{5} \\ 0 & 0 & 0 & 1 \end{array}\right] \quad \begin{array}{l} -3R_1 + R_2 \\ -10R_1 + R_3 \end{array}$$

Since row 3 has all zeros except for the last entry, there is no solution.

23. $\begin{bmatrix} 2 & 3 \\ 5 & q \end{bmatrix} = \begin{bmatrix} a & b \\ c & 9 \end{bmatrix}$

The matrices are both 2 × 2.
2 = a, 3 = b, 5 = c, q = 9
Thus, a is 2, b is 3, c is 5, and q is 9. Since the matrices are 2 × 2, they are square.

25. $[m \;\; 4 \;\; z \;\; -1] = [12 \;\; k \;\; -8 \;\; r]$
The matrices are both 1 × 4.
m = 12, 4 = k, z = -8, and -1 = r.
Thus, m is 12, k is 4, z is -8, and r is -1.
The matrices are row matrices.

27. $\begin{bmatrix} 6 & m \\ k & 5 - x \end{bmatrix} + \begin{bmatrix} r & m + 3 \\ k + 1 & 2x \end{bmatrix} = \begin{bmatrix} 8 & 10 \\ 12 & -2 \end{bmatrix}$

The matrices are all 2 × 2.

$$\begin{bmatrix} 6 + r & 2m + 3 \\ 2k + 1 & 5 + x \end{bmatrix} = \begin{bmatrix} 8 & 10 \\ 12 & -2 \end{bmatrix}$$

6 + r = 8 Thus, r = 2.

2m + 3 = 10 Thus, $m = \frac{7}{2}$.

2k + 1 = 12 Thus, $k = \frac{11}{2}$.

5 + x = -2 Thus, x = -7.
The matrices are square.

29.

	Grazing	Moving	Resting
Horses	8	8	8
Cattle	10	5	9
Sheep	7	10	7
Goats	8	9	7

31. If $A = \begin{bmatrix} 4 & 10 \\ -2 & -3 \\ 6 & 9 \end{bmatrix}$,

then

$$-A = \begin{bmatrix} -4 & -10 \\ 2 & 3 \\ -6 & -9 \end{bmatrix}.$$

33. If $A = \begin{bmatrix} 4 & 10 \\ -2 & -3 \\ 6 & 9 \end{bmatrix}$ and $C = \begin{bmatrix} 5 & 0 \\ -1 & 3 \\ 4 & 7 \end{bmatrix}$,

then

$$C + 2A = \begin{bmatrix} 5 & 0 \\ -1 & 3 \\ 4 & 7 \end{bmatrix} + 2\begin{bmatrix} 4 & 10 \\ -2 & -3 \\ 6 & 9 \end{bmatrix}$$

$$= \begin{bmatrix} 5 & 0 \\ -1 & 3 \\ 4 & 7 \end{bmatrix} + \begin{bmatrix} 8 & 20 \\ -4 & -6 \\ 12 & 18 \end{bmatrix}$$

$$= \begin{bmatrix} 13 & 20 \\ -5 & -3 \\ 16 & 25 \end{bmatrix}.$$

35. If $A = \begin{bmatrix} 4 & 10 \\ -2 & -3 \\ 6 & 9 \end{bmatrix}$ and $C = \begin{bmatrix} 5 & 0 \\ -1 & 3 \\ 4 & 7 \end{bmatrix}$,

then

$$2A - 5c = 2\begin{bmatrix} 4 & 10 \\ -2 & -3 \\ 6 & 9 \end{bmatrix} - 5\begin{bmatrix} 5 & 0 \\ -1 & 3 \\ 4 & 7 \end{bmatrix}$$

$$= \begin{bmatrix} 8 & 20 \\ -4 & -6 \\ 12 & 18 \end{bmatrix} - \begin{bmatrix} 25 & 0 \\ -5 & 15 \\ 20 & 35 \end{bmatrix}$$

$$= \begin{bmatrix} -17 & 20 \\ 1 & -21 \\ -8 & -17 \end{bmatrix}.$$

37. The first day matrix is the following.

$$\begin{array}{c} \\ \text{ATT} \\ \text{GE} \\ \text{GO} \\ \text{S} \end{array} \begin{array}{c} \begin{array}{cc} \text{Sales} & \text{Price change} \end{array} \\ \begin{bmatrix} 2532 & -\frac{1}{4} \\ 1464 & \frac{1}{8} \\ 4974 & -\frac{3}{2} \\ 1754 & \frac{1}{2} \end{bmatrix} \end{array}$$

The next day matrix is the following.

$$\begin{bmatrix} 2310 & -\frac{1}{4} \\ 1258 & -\frac{1}{4} \\ 5061 & \frac{1}{2} \\ 1812 & \frac{1}{2} \end{bmatrix}$$

The total sales and price changes for the two days are given by the sum

$$\begin{bmatrix} 2532 & -\frac{1}{4} \\ 1464 & \frac{1}{8} \\ 4974 & -\frac{3}{2} \\ 1754 & \frac{1}{2} \end{bmatrix} + \begin{bmatrix} 2310 & -\frac{1}{4} \\ 1258 & -\frac{1}{4} \\ 5061 & \frac{1}{2} \\ 1812 & \frac{1}{2} \end{bmatrix} = \begin{bmatrix} 4842 & -\frac{1}{2} \\ 2722 & -\frac{1}{8} \\ 10{,}035 & -1 \\ 3566 & 1 \end{bmatrix}.$$

39. $AF = \begin{bmatrix} 4 & 10 \\ -2 & -3 \\ 6 & 9 \end{bmatrix} \begin{bmatrix} -1 & 4 \\ 3 & 7 \end{bmatrix} = \begin{bmatrix} 26 & 86 \\ -7 & -29 \\ 21 & 87 \end{bmatrix}$

41. $ED = [1 \ 3 \ -4] \begin{bmatrix} 6 \\ 1 \\ 0 \end{bmatrix} = [9]$

43. $AGF = A(GF)$

$$= \begin{bmatrix} 4 & 10 \\ -2 & -3 \\ 6 & 9 \end{bmatrix} \left(\begin{bmatrix} -1 & 4 \\ 3 & 7 \end{bmatrix} \begin{bmatrix} 2 & 5 \\ 1 & 6 \end{bmatrix} \right)$$

$$= \begin{bmatrix} 4 & 10 \\ -2 & -3 \\ 6 & 9 \end{bmatrix} \begin{bmatrix} 2 & 19 \\ 13 & 57 \end{bmatrix}$$

$$= \begin{bmatrix} 138 & 646 \\ -43 & -209 \\ 129 & 627 \end{bmatrix}$$

45. **(a)**

$$\begin{array}{lcc} & \text{Standard} & \text{Extra large} \\ \text{Cutting} & \frac{1}{4} & \frac{1}{3} \\ \text{Shaping} & \frac{1}{2} & \frac{1}{3} \end{array}$$

(b) $[12 \quad 6] \begin{bmatrix} \frac{1}{4} & \frac{1}{3} \\ \frac{1}{2} & \frac{1}{3} \end{bmatrix} = [6 \quad 6]$

6 units of standard clips and 6 units of extra large clips could be made.

47. Many examples are possible.

Suppose $A = \begin{bmatrix} 1 & 1 \\ 1 & 0 \end{bmatrix}$ and $B = \begin{bmatrix} 0 & 1 \\ 1 & 1 \end{bmatrix}$.

$$AB = \begin{bmatrix} 1 & 1 \\ 1 & 0 \end{bmatrix} \begin{bmatrix} 0 & 1 \\ 1 & 1 \end{bmatrix} = \begin{bmatrix} 1 & 2 \\ 0 & 1 \end{bmatrix}$$

$$BA = \begin{bmatrix} 0 & 1 \\ 1 & 1 \end{bmatrix} \begin{bmatrix} 1 & 1 \\ 1 & 0 \end{bmatrix} = \begin{bmatrix} 1 & 0 \\ 2 & 1 \end{bmatrix}$$

Thus, $AB \neq BA$.

49. Let $A = \begin{bmatrix} -4 & 2 \\ 0 & 3 \end{bmatrix}$.

$$[A|I] = \left[\begin{array}{cc|cc} -4 & 2 & 1 & 0 \\ 0 & 3 & 0 & 1 \end{array}\right]$$

$$\left[\begin{array}{cc|cc} 1 & -\frac{1}{2} & -\frac{1}{4} & 0 \\ 0 & 1 & 0 & \frac{1}{3} \end{array}\right] \begin{array}{l} -\frac{1}{4}R_1 \\ \frac{1}{3}R_2 \end{array}$$

$$[I|B] = \left[\begin{array}{cc|cc} 1 & 0 & -\frac{1}{4} & \frac{1}{6} \\ 0 & 1 & 0 & \frac{1}{3} \end{array}\right] \begin{array}{l} \frac{1}{2}R_2 + R_1 \\ \end{array}$$

$$A^{-1} = \begin{bmatrix} -\frac{1}{4} & \frac{1}{6} \\ 0 & \frac{1}{3} \end{bmatrix}$$

51.

Let $A = \begin{bmatrix} 6 & 4 \\ 3 & 2 \end{bmatrix}$.

$$[A|I] = \left[\begin{array}{cc|cc} 6 & 4 & 1 & 0 \\ 3 & 2 & 0 & 1 \end{array}\right]$$

$$\left[\begin{array}{cc|cc} 1 & \frac{2}{3} & \frac{1}{6} & 0 \\ 3 & 2 & 0 & 1 \end{array}\right] \begin{array}{l} \frac{1}{6}R_1 \\ \end{array}$$

$$\left[\begin{array}{cc|cc} 1 & \frac{2}{3} & \frac{1}{6} & 0 \\ 0 & 0 & -\frac{1}{2} & 1 \end{array}\right] \begin{array}{l} \\ -3R_1 + R_2 \end{array}$$

The second row can never become [0 1], so A has no inverse.

53.

Let $A = \begin{bmatrix} 2 & 0 & 4 \\ 1 & -1 & 0 \\ 0 & 1 & -2 \end{bmatrix}$.

$$[A|I] = \left[\begin{array}{ccc|ccc} 2 & 0 & 4 & 1 & 0 & 0 \\ 1 & -1 & 2 & 0 & 1 & 0 \\ 0 & 1 & -2 & 0 & 0 & 1 \end{array}\right]$$

$$\left[\begin{array}{ccc|ccc} 1 & -1 & 0 & 0 & 1 & 0 \\ 0 & 1 & -2 & 0 & 0 & 1 \\ 2 & 0 & 4 & 1 & 0 & 0 \end{array}\right] \text{Interchange rows}$$

$$\left[\begin{array}{ccc|ccc} 1 & -1 & 0 & 0 & 1 & 0 \\ 0 & 1 & -2 & 0 & 0 & 1 \\ 0 & 2 & 4 & 1 & -2 & 0 \end{array}\right] \begin{array}{l} \\ \\ -2R_1 + R_3 \end{array}$$

$$\left[\begin{array}{ccc|ccc} 1 & 0 & -2 & 0 & 1 & 1 \\ 0 & 1 & -2 & 0 & 0 & 1 \\ 0 & 0 & 8 & 1 & -2 & -2 \end{array}\right] \begin{array}{l} R_2 + R_1 \\ \\ -2R_2 + R_3 \end{array}$$

$$\left[\begin{array}{ccc|ccc} 1 & 0 & -2 & 0 & 1 & 1 \\ 0 & 1 & -2 & 0 & 0 & 1 \\ 0 & 0 & 1 & \frac{1}{8} & -\frac{1}{4} & -\frac{1}{4} \end{array}\right] \begin{array}{l} \\ \\ \frac{1}{8}R_3 \end{array}$$

$$[I|B] = \left[\begin{array}{ccc|ccc} 1 & 0 & 0 & \frac{1}{4} & \frac{1}{2} & \frac{1}{2} \\ 0 & 1 & 0 & \frac{1}{4} & -\frac{1}{2} & \frac{1}{2} \\ 0 & 0 & 1 & \frac{1}{8} & -\frac{1}{4} & -\frac{1}{4} \end{array}\right] \begin{array}{l} 2R_3 + R_1 \\ 2R_3 + R_2 \\ \\ \end{array}$$

$$A^{-1} = \begin{bmatrix} \frac{1}{4} & \frac{1}{2} & \frac{1}{2} \\ \frac{1}{4} & -\frac{1}{2} & \frac{1}{2} \\ \frac{1}{8} & -\frac{1}{4} & -\frac{1}{4} \end{bmatrix}$$

55.

Let $A = \begin{bmatrix} 2 & 3 & 5 \\ -2 & -3 & -5 \\ 1 & 4 & 2 \end{bmatrix}$.

$$[A|I] = \left[\begin{array}{ccc|ccc} 2 & 3 & 5 & 1 & 0 & 0 \\ -2 & -3 & -5 & 0 & 1 & 0 \\ 1 & 4 & 2 & 0 & 0 & 1 \end{array}\right]$$

$$\left[\begin{array}{ccc|ccc} 1 & 4 & 2 & 0 & 0 & 1 \\ -2 & -3 & -5 & 0 & 1 & 0 \\ 2 & 3 & 5 & 1 & 0 & 0 \end{array}\right] \text{Interchange } R_1 \text{ and } R_3$$

$$\left[\begin{array}{ccc|ccc} 1 & 4 & 2 & 0 & 0 & 1 \\ -2 & -3 & -5 & 0 & 1 & 0 \\ 0 & 0 & 0 & 1 & 1 & 0 \end{array}\right] \begin{array}{l} \\ \\ R_2 + R_3 \end{array}$$

The third row can never become [0 0 1], so A does not have an inverse.

57. $G = \begin{bmatrix} 2 & 5 \\ 1 & 6 \end{bmatrix}$

$$\left[\begin{array}{cc|cc} 2 & 5 & 1 & 0 \\ 1 & 6 & 0 & 1 \end{array}\right]$$

$$\left[\begin{array}{cc|cc} 1 & \frac{5}{2} & \frac{1}{2} & 0 \\ 1 & 6 & 0 & 1 \end{array}\right] \begin{array}{l} \frac{1}{2}R_1 \\ \\ \end{array}$$

$$\left[\begin{array}{cc|cc} 1 & \frac{5}{2} & \frac{1}{2} & 0 \\ 0 & \frac{7}{2} & -\frac{1}{2} & 1 \end{array}\right] \begin{array}{l} \\ -1R_1 + R_2 \end{array}$$

$$\left[\begin{array}{cc|cc} 1 & \frac{5}{2} & \frac{1}{2} & 0 \\ 0 & 1 & -\frac{1}{7} & \frac{2}{7} \end{array}\right] \begin{array}{l} \\ \frac{2}{7}R_2 \end{array}$$

$$\left[\begin{array}{cc|cc} 1 & 0 & \frac{6}{7} & -\frac{5}{7} \\ 0 & 1 & -\frac{1}{7} & \frac{2}{7} \end{array}\right] \begin{array}{l} -\frac{5}{2}R_2 + R_1 \\ \\ \end{array}$$

$$G^{-1} = \begin{bmatrix} \frac{6}{7} & -\frac{5}{7} \\ -\frac{1}{7} & \frac{2}{7} \end{bmatrix}$$

59.

$$A + C = \begin{bmatrix} 4 & 10 \\ -2 & -3 \\ 6 & 9 \end{bmatrix} + \begin{bmatrix} 5 & 0 \\ -1 & 3 \\ 4 & 7 \end{bmatrix} = \begin{bmatrix} 9 & 10 \\ -3 & 0 \\ 10 & 16 \end{bmatrix}$$

Since A + C is not square, it is not possible to find $(A + C)^{-1}$.

61.

$$F = \begin{bmatrix} -1 & 4 \\ 3 & 7 \end{bmatrix}$$

$$\left[\begin{array}{cc|cc} -1 & 4 & 1 & 0 \\ 3 & 7 & 0 & 1 \end{array}\right]$$

$$\left[\begin{array}{cc|cc} 1 & -4 & -1 & 0 \\ 3 & 7 & 0 & 1 \end{array}\right] \; -1R_1$$

$$\left[\begin{array}{cc|cc} 1 & -4 & -1 & 0 \\ 0 & 19 & 3 & 1 \end{array}\right] \; -3R_1 + R_2$$

$$\left[\begin{array}{cc|cc} 1 & -4 & -1 & 0 \\ 0 & 1 & \frac{3}{19} & \frac{1}{19} \end{array}\right] \; \frac{1}{19}R_2$$

$$\left[\begin{array}{cc|cc} 1 & 0 & -\frac{7}{19} & \frac{4}{19} \\ 0 & 1 & \frac{3}{19} & \frac{1}{19} \end{array}\right] \; 4R_2 + R_1$$

Thus, $F^{-1} = \begin{bmatrix} -\frac{7}{19} & \frac{4}{19} \\ \frac{3}{19} & \frac{1}{19} \end{bmatrix}$.

$$G = \begin{bmatrix} 2 & 5 \\ 1 & 6 \end{bmatrix}$$

From Exercise 57,

$$G^{-1} = \begin{bmatrix} \frac{6}{7} & -\frac{5}{7} \\ -\frac{1}{7} & \frac{2}{7} \end{bmatrix}.$$

Therefore,

$$F^{-1} - G^{-1} = \begin{bmatrix} -\frac{7}{19} & \frac{4}{19} \\ \frac{3}{19} & \frac{1}{19} \end{bmatrix} - \begin{bmatrix} \frac{6}{7} & -\frac{5}{7} \\ -\frac{1}{7} & \frac{2}{7} \end{bmatrix}$$

$$= \begin{bmatrix} -\frac{7}{19} - \frac{6}{7} & \frac{4}{19} + \frac{5}{7} \\ \frac{3}{19} + \frac{1}{7} & \frac{1}{19} - \frac{2}{7} \end{bmatrix}$$

$$= \begin{bmatrix} -\frac{163}{133} & \frac{123}{133} \\ \frac{40}{133} & -\frac{31}{133} \end{bmatrix}$$

63. If $AX = B$, then $X = A^{-1}B$.

If $A = \begin{bmatrix} 2 & 4 \\ -1 & -3 \end{bmatrix}$,

then $A^{-1} = \begin{bmatrix} \frac{3}{2} & 2 \\ -\frac{1}{2} & -1 \end{bmatrix}$.

Thus,

$$X = \begin{bmatrix} \frac{3}{2} & 2 \\ -\frac{1}{2} & -1 \end{bmatrix} \begin{bmatrix} 8 \\ 3 \end{bmatrix}$$

$$X = \begin{bmatrix} 18 \\ -7 \end{bmatrix}.$$

65. If $AX = B$,
then $X = A^{-1}B$.

If $A = \begin{bmatrix} 1 & 0 & 2 \\ -1 & 1 & 0 \\ 3 & 0 & 4 \end{bmatrix}$,

then $A^{-1} = \begin{bmatrix} -2 & 0 & 1 \\ -2 & 1 & 1 \\ \frac{3}{2} & 0 & -\frac{1}{2} \end{bmatrix}$.

Thus,

$$X = \begin{bmatrix} -2 & 0 & 1 \\ -2 & 1 & 1 \\ \frac{3}{2} & 0 & -\frac{1}{2} \end{bmatrix} \begin{bmatrix} 8 \\ 4 \\ -6 \end{bmatrix}$$

$$X = \begin{bmatrix} -22 \\ -18 \\ 15 \end{bmatrix}$$

67. $x + y = 4$
$2x + 3y = 10$
The system as a matrix equation is

$$\begin{bmatrix} 1 & 1 \\ 2 & 3 \end{bmatrix} \begin{bmatrix} x \\ y \end{bmatrix} = \begin{bmatrix} 4 \\ 10 \end{bmatrix}.$$

Let $A = \begin{bmatrix} 1 & 1 \\ 2 & 3 \end{bmatrix}$.

$$A^{-1} = \begin{bmatrix} 3 & -1 \\ -2 & 1 \end{bmatrix}.$$

Thus,

$$\begin{bmatrix} x \\ y \end{bmatrix} = \begin{bmatrix} 3 & -1 \\ -2 & 1 \end{bmatrix} \begin{bmatrix} 4 \\ 10 \end{bmatrix}$$

$$\begin{bmatrix} x \\ y \end{bmatrix} = \begin{bmatrix} 2 \\ 2 \end{bmatrix}.$$

The solution is (2, 2).

69. $2x + y = 5$
$3x - 2y = 4$
The system as a matrix equation is

$$\begin{bmatrix} 2 & 1 \\ 3 & -2 \end{bmatrix} \begin{bmatrix} x \\ y \end{bmatrix} = \begin{bmatrix} 5 \\ 4 \end{bmatrix}.$$

Let $A = \begin{bmatrix} 2 & 1 \\ 3 & -2 \end{bmatrix}$.

$$A^{-1} = \begin{bmatrix} \frac{2}{7} & \frac{1}{7} \\ \frac{3}{7} & -\frac{2}{7} \end{bmatrix}.$$

Thus,

$$\begin{bmatrix} x \\ y \end{bmatrix} = \begin{bmatrix} \frac{2}{7} & \frac{1}{7} \\ \frac{3}{7} & -\frac{2}{7} \end{bmatrix} \begin{bmatrix} 5 \\ 4 \end{bmatrix}$$

$$\begin{bmatrix} x \\ y \end{bmatrix} = \begin{bmatrix} 2 \\ 1 \end{bmatrix}.$$

The solution is (2, 1).

71. $x + y + z = 1$
$2x - y = -2$
$3y + z = 2$
The system as a matrix equation is

$$\begin{bmatrix} 1 & 1 & 1 \\ 2 & -1 & 0 \\ 0 & 3 & 1 \end{bmatrix} \begin{bmatrix} x \\ y \\ z \end{bmatrix} = \begin{bmatrix} 1 \\ -2 \\ 2 \end{bmatrix}.$$

Let $A = \begin{bmatrix} 1 & 1 & 1 \\ 2 & -1 & 0 \\ 0 & 3 & 1 \end{bmatrix}$.

$$A^{-1} = \begin{bmatrix} -\frac{1}{3} & \frac{2}{3} & \frac{1}{3} \\ -\frac{2}{3} & \frac{1}{3} & \frac{2}{3} \\ 2 & -1 & -1 \end{bmatrix}.$$

Thus,

$$\begin{bmatrix} x \\ y \\ z \end{bmatrix} = \begin{bmatrix} -\frac{1}{3} & \frac{2}{3} & \frac{1}{3} \\ -\frac{2}{3} & \frac{1}{3} & \frac{2}{3} \\ 2 & -1 & -1 \end{bmatrix} \begin{bmatrix} 1 \\ -2 \\ 2 \end{bmatrix}$$

$$\begin{bmatrix} x \\ y \\ z \end{bmatrix} = \begin{bmatrix} -1 \\ 0 \\ 2 \end{bmatrix}$$

The solution is (-1, 0, 2).

73. $3x - 2y + 4z = 4$
$4x + y - 5z = 2$
$-6x + 4y - 8z = -2$
The system as a matrix equation is

$$\begin{bmatrix} 3 & -2 & 4 \\ 4 & 1 & -5 \\ -6 & 4 & -8 \end{bmatrix}\begin{bmatrix} x \\ y \\ z \end{bmatrix} = \begin{bmatrix} 4 \\ 2 \\ -2 \end{bmatrix}.$$

Let $A = \begin{bmatrix} 3 & -2 & 4 \\ 4 & 1 & -5 \\ -6 & 4 & -8 \end{bmatrix}$.

Since row 3 is -2 times row 1, the matrix will have no inverse, and the system cannot be solved by this method.

75. Let x = number of grams of 12 carat gold
y = number of grams of 22 carat gold.
The system is
$x + y = 25$ *(1)*
$\frac{12}{24}x + \frac{22}{24}y = \frac{15}{24}(25)$ *(2)*

Multiply equation (2) by 24.
$x + y = 25$ *(1)*
$12x + 22y = 375$ *(3)*
From equation (1), $x = 25 - y$.
Substitute this into equation (3).
$12(25 - y) + 22y = 375$
$300 - 12y + 22y = 375$
$10y = 75$
$y = 7.5$
$x = 25 - 7.5$
$x = 17.5$
The merchant should mix 17.5 grams of 12 carat gold and 7.5 grams of 22 carat gold.

77. Let x = number of pounds of tea worth \$4.60 a pound
y = number of pounds of tea worth \$6.50 a pound.
The system is
$x + y = 10$ *(1)*
$4.60x + 6.50y = 5.74(10)$ *(2)*
Solve equation (1) for x.
$x = 10 - y$
and substitute $10 - y$ for x in equation (2).
$4.6(10 - y) + 6.5y = 57.4$
$46 - 4.6y + 6.5y = 57.4$
$1.9y = 11.4$
$y = 6$
Since $x = 10 - y$, $x = 4$.
Thus, 4 pounds of the tea worth \$4.60 a pound should be used.

79. Let x = the speed of the boat
y = the speed of the current.
The system is
$3(x + y) = 57$ *(1)*
$5(x - y) = 55.$ *(2)*
Multiplying equation (1) by $\frac{1}{3}$ and equation (2) by $\frac{1}{5}$ gives

$x + y = 19$ *(3)*
$x - y = 11.$ *(4)*
Adding equation (1) to equation (2) gives
$2x = 30$
$x = 15.$
Substituting into (3) gives
$15 + y = 19$
$y = 4.$
The speed of the boat is 15 kilometers per hour, and the speed of the current is 4 kilometers per hour.

81. Let x = the amount invested at 8%

y = the amount invested at $8\frac{1}{2}$%

z = the amount invested at 11%.

The system is

$$\begin{aligned} x + y + z &= 50{,}000 \\ .08x + .085y + .11z &= 4436.25 \\ .11z &= .08x + 80. \end{aligned}$$

Write the system in the proper form.

$$\begin{aligned} x + y + z &= 50{,}000 \quad (1) \\ .08x + .085y + .11z &= 4436.25 \quad (2) \\ -.08x + .11z &= 80 \quad (3) \end{aligned}$$

Write the augmented matrix of the system.

$$\left[\begin{array}{ccc|c} 1 & 1 & 1 & 50{,}000 \\ .08 & .085 & .11 & 4436.25 \\ -.08 & 0 & .11 & 80 \end{array}\right]$$

$$\left[\begin{array}{ccc|c} 1 & 1 & 1 & 50{,}000 \\ 0 & .005 & .03 & 436.25 \\ 0 & .08 & .19 & 4080 \end{array}\right] \begin{array}{l} \\ -.08R_1 + R_2 \\ .08R_1 + R_3 \end{array}$$

$$\left[\begin{array}{ccc|c} 1 & 1 & 1 & 50{,}000 \\ 0 & 1 & 6 & 87{,}250 \\ 0 & .08 & .19 & 4080 \end{array}\right] \frac{1}{.005}R_2$$

$$\left[\begin{array}{ccc|c} 1 & 0 & -5 & -37{,}250 \\ 0 & 1 & 6 & 87{,}250 \\ 0 & 0 & -.29 & -2900 \end{array}\right] \begin{array}{l} -1R_2 + R_1 \\ \\ -.08R_2 + R_3 \end{array}$$

$$\left[\begin{array}{ccc|c} 1 & 0 & -5 & -37{,}250 \\ 0 & 1 & 6 & 87{,}250 \\ 0 & 0 & 1 & 10{,}000 \end{array}\right] -\frac{1}{.29}R_3$$

$$\left[\begin{array}{ccc|c} 1 & 0 & 0 & 12{,}750 \\ 0 & 1 & 0 & 27{,}250 \\ 0 & 0 & 1 & 10{,}000 \end{array}\right] \begin{array}{l} 5R_3 + R_1 \\ -6R_3 + R_2 \\ \\ \end{array}$$

Thus, x = 12,750, y = 27,250, and z = 10,000. Mrs. Levy invested $12,750 at 8%, $27,250 at 8 1/2%, and $10,000 at 11%.

83. $A = \begin{bmatrix} .01 & .05 \\ .04 & .03 \end{bmatrix}$, $D = \begin{bmatrix} 200 \\ 300 \end{bmatrix}$

$$I - A = \begin{bmatrix} 1 & 0 \\ 0 & 1 \end{bmatrix} - \begin{bmatrix} .01 & .05 \\ .04 & .03 \end{bmatrix}$$

$$= \begin{bmatrix} .99 & -.05 \\ -.04 & .97 \end{bmatrix}$$

Use row operations to find $(I - A)^{-1}$.

$$(I - A)^{-1} = \begin{bmatrix} 1.01221 & .05217 \\ .04174 & 1.03306 \end{bmatrix}$$

Since $X = (I - A)^{-1}D$,

$$X = \begin{bmatrix} 1.01221 & .05217 \\ .04174 & 1.03306 \end{bmatrix} \begin{bmatrix} 200 \\ 300 \end{bmatrix}$$

$$= \begin{bmatrix} 218.09 \\ 318.27 \end{bmatrix}.$$

85. **(a)** $I - A = \begin{bmatrix} 1 & 0 \\ 0 & 1 \end{bmatrix} - \begin{bmatrix} 0 & \frac{1}{4} \\ \frac{1}{2} & 0 \end{bmatrix} = \begin{bmatrix} 1 & -\frac{1}{4} \\ -\frac{1}{2} & 1 \end{bmatrix}$

(b) $\left[\begin{array}{cc|cc} 1 & -\frac{1}{4} & 1 & 0 \\ -\frac{1}{2} & 1 & 0 & 1 \end{array}\right]$

$$\left[\begin{array}{cc|cc} 1 & -\frac{1}{4} & 1 & 0 \\ 0 & \frac{7}{8} & \frac{1}{2} & 1 \end{array}\right] \begin{array}{l} \\ \frac{1}{2}R_1 + R_2 \end{array}$$

$$\left[\begin{array}{cc|cc} 1 & -\frac{1}{4} & 1 & 0 \\ 0 & 1 & \frac{4}{7} & \frac{8}{7} \end{array}\right] \begin{array}{l} \\ \frac{8}{7}R_2 \end{array}$$

$$\left[\begin{array}{cc|cc} 1 & 0 & \frac{8}{7} & \frac{2}{7} \\ 0 & 1 & \frac{4}{7} & \frac{8}{7} \end{array}\right] \begin{array}{l} \frac{1}{4}R_2 + R_1 \\ \\ \end{array}$$

Thus,

$$(I - A)^{-1} = \begin{bmatrix} \frac{8}{7} & \frac{2}{7} \\ \frac{4}{7} & \frac{8}{7} \end{bmatrix}.$$

(c) $X = (I - A)^{-1}D$

$$X = \begin{bmatrix} \frac{8}{7} & \frac{2}{7} \\ \frac{4}{7} & \frac{8}{7} \end{bmatrix} \begin{bmatrix} 2100 \\ 1400 \end{bmatrix}$$

$$X = \begin{bmatrix} 2800 \\ 2800 \end{bmatrix}$$

87. Write the input-output matrix.

$$A = \begin{matrix} & \begin{matrix} A & & M \end{matrix} \\ \begin{matrix} A \\ M \end{matrix} & \begin{bmatrix} .10 & .70 \\ .40 & .20 \end{bmatrix} \end{matrix} = \begin{bmatrix} \frac{1}{10} & \frac{7}{10} \\ \frac{4}{10} & \frac{2}{10} \end{bmatrix}$$

$$I - A = \begin{bmatrix} 1 & 0 \\ 0 & 1 \end{bmatrix} - \begin{bmatrix} \frac{1}{10} & \frac{7}{10} \\ \frac{4}{10} & \frac{2}{10} \end{bmatrix} = \begin{bmatrix} \frac{9}{10} & -\frac{7}{10} \\ -\frac{4}{10} & \frac{8}{10} \end{bmatrix}$$

Find $(I - A)^{-1}$.

$$\left[\begin{array}{cc|cc} \frac{9}{10} & -\frac{7}{10} & 1 & 0 \\ -\frac{4}{10} & \frac{8}{10} & 0 & 1 \end{array}\right]$$

$$\left[\begin{array}{cc|cc} 1 & -\frac{7}{9} & \frac{10}{9} & 0 \\ -\frac{4}{10} & \frac{8}{10} & 0 & 1 \end{array}\right] \frac{10}{9}R_1$$

$$\left[\begin{array}{cc|cc} 1 & -\frac{7}{9} & \frac{10}{9} & 0 \\ 0 & \frac{44}{90} & \frac{4}{9} & 1 \end{array}\right] \frac{4}{10}R_1 + R_2$$

$$\left[\begin{array}{cc|cc} 1 & -\frac{7}{9} & \frac{10}{9} & 0 \\ 0 & 1 & \frac{10}{11} & \frac{90}{44} \end{array}\right] \frac{9}{44}R_2$$

$$\left[\begin{array}{cc|cc} 1 & 0 & \frac{180}{99} & \frac{70}{44} \\ 0 & 1 & \frac{10}{11} & \frac{90}{44} \end{array}\right] \frac{7}{9}R_2 + R_1$$

$$(I - A)^{-1} = \begin{bmatrix} \frac{180}{99} & \frac{70}{44} \\ \frac{10}{11} & \frac{90}{44} \end{bmatrix}$$

$X = (I - A)^{-1}D$

$$X = \begin{bmatrix} \frac{180}{99} & \frac{70}{44} \\ \frac{10}{11} & \frac{90}{44} \end{bmatrix} \begin{bmatrix} 60{,}000 \\ 20{,}000 \end{bmatrix}$$

$$X = \begin{bmatrix} 140{,}909 \\ 95{,}455 \end{bmatrix} \text{ *Rounded*}$$

The agriculture industry should produce \$140,909 while the manufacturing industry should produce \$95,455.

89. (a) Break the message into groups of two letters and form a 2×1 matrix for each group of two letters.

it -i s- to o- la te

$$\begin{bmatrix} 9 \\ 20 \end{bmatrix}, \begin{bmatrix} 27 \\ 9 \end{bmatrix}, \begin{bmatrix} 19 \\ 27 \end{bmatrix}, \begin{bmatrix} 20 \\ 15 \end{bmatrix}, \begin{bmatrix} 15 \\ 27 \end{bmatrix}, \begin{bmatrix} 12 \\ 1 \end{bmatrix}, \begin{bmatrix} 20 \\ 5 \end{bmatrix}$$

Multiply each of these matrices on the left by the matrix M.

$$M\begin{bmatrix} 9 \\ 20 \end{bmatrix} = \begin{bmatrix} 2 & 5 \\ -3 & 1 \end{bmatrix}\begin{bmatrix} 9 \\ 20 \end{bmatrix} = \begin{bmatrix} 118 \\ -7 \end{bmatrix}$$

$$M\begin{bmatrix} 27 \\ 9 \end{bmatrix} = \begin{bmatrix} 2 & 5 \\ -3 & 1 \end{bmatrix}\begin{bmatrix} 27 \\ 9 \end{bmatrix} = \begin{bmatrix} 99 \\ -72 \end{bmatrix}$$

$$M\begin{bmatrix} 19 \\ 27 \end{bmatrix} = \begin{bmatrix} 2 & 5 \\ -3 & 1 \end{bmatrix}\begin{bmatrix} 19 \\ 27 \end{bmatrix} = \begin{bmatrix} 173 \\ -30 \end{bmatrix}$$

$$M\begin{bmatrix} 20 \\ 15 \end{bmatrix} = \begin{bmatrix} 2 & 5 \\ -3 & 1 \end{bmatrix}\begin{bmatrix} 20 \\ 15 \end{bmatrix} = \begin{bmatrix} 115 \\ -45 \end{bmatrix}$$

$$M\begin{bmatrix} 15 \\ 27 \end{bmatrix} = \begin{bmatrix} 2 & 5 \\ -3 & 1 \end{bmatrix}\begin{bmatrix} 15 \\ 27 \end{bmatrix} = \begin{bmatrix} 165 \\ -18 \end{bmatrix}$$

$$M\begin{bmatrix}12\\1\end{bmatrix} = \begin{bmatrix}2 & 5\\-3 & 1\end{bmatrix}\begin{bmatrix}12\\1\end{bmatrix} = \begin{bmatrix}29\\-35\end{bmatrix}$$

$$M\begin{bmatrix}20\\5\end{bmatrix} = \begin{bmatrix}2 & 5\\-3 & 1\end{bmatrix}\begin{bmatrix}20\\5\end{bmatrix} = \begin{bmatrix}65\\-55\end{bmatrix}$$

Thus, the coded message is

$$\begin{bmatrix}118\\-7\end{bmatrix}, \begin{bmatrix}99\\-72\end{bmatrix}, \begin{bmatrix}173\\-30\end{bmatrix}, \begin{bmatrix}115\\-45\end{bmatrix}, \begin{bmatrix}165\\-18\end{bmatrix}, \begin{bmatrix}29\\-35\end{bmatrix}, \begin{bmatrix}65\\-55\end{bmatrix}.$$

(b) To decode the message, multiply each matrix in the coded message on the left by M^{-1}.

$$\left[\begin{array}{cc|cc}2 & 5 & 1 & 0\\-3 & 1 & 0 & 1\end{array}\right]$$

$$\left[\begin{array}{cc|cc}1 & \frac{5}{2} & \frac{1}{2} & 0\\-3 & 1 & 0 & 1\end{array}\right] \quad \frac{1}{2}R_1$$

$$\left[\begin{array}{cc|cc}1 & \frac{5}{2} & \frac{1}{2} & 0\\0 & \frac{17}{2} & \frac{3}{2} & 1\end{array}\right] \quad 3R_1 + R_2$$

$$\left[\begin{array}{cc|cc}1 & \frac{5}{2} & \frac{1}{2} & 0\\0 & 1 & \frac{3}{17} & \frac{2}{17}\end{array}\right] \quad \frac{2}{17}R_3$$

$$\left[\begin{array}{cc|cc}1 & 0 & \frac{1}{17} & \frac{-5}{17}\\0 & 1 & \frac{3}{17} & \frac{2}{17}\end{array}\right] \quad -\frac{5}{2}R_2 + R_1$$

Thus, the matrix to be multiplied by is

$$M^{-1} = \begin{bmatrix}\frac{1}{17} & -\frac{5}{17}\\\frac{3}{17} & \frac{2}{17}\end{bmatrix}.$$

Case 5 Exercises

1.

$$PQ = \begin{bmatrix}1&0&0&1&0\\0&0&1&1&0\\1&1&0&0&0\end{bmatrix}\begin{bmatrix}1&1&0&1&1&1\\0&0&0&0&1&0\\0&0&0&0&0&0\\0&1&0&1&0&0\\1&0&0&0&1&0\end{bmatrix}$$

$$3 \times 5 \qquad 5 \times 6$$

$$= \begin{bmatrix}1&2&0&2&1&1\\0&1&0&1&0&0\\1&1&0&1&2&1\end{bmatrix}$$

$$3 \times 6$$

2. The number of second-order contacts between the second contagious person and the third person in the third group is found in the second row, third column of PQ, or 0.

3. The third column in PQ is all zeros, so the third person in the third group had no contact with anyone in the first.

4. The second and fourth persons in the third group each had four contacts in all.

Case 6 Exercises

1. (a)

$$A = \begin{bmatrix}.245 & .102 & .051\\.099 & .291 & .279\\.433 & .372 & .011\end{bmatrix},$$

$$D = \begin{bmatrix}2.88\\31.45\\30.91\end{bmatrix}, \; X = \begin{bmatrix}x_1\\x_2\\x_3\end{bmatrix}$$

(b) $I - A = \begin{bmatrix} 1 & 0 & 0 \\ 0 & 1 & 0 \\ 0 & 0 & 1 \end{bmatrix} - \begin{bmatrix} .245 & .102 & .051 \\ .099 & .291 & .279 \\ .433 & .372 & .011 \end{bmatrix}$

$= \begin{bmatrix} .755 & -.102 & -.051 \\ -.099 & .709 & -.279 \\ -.433 & -.372 & .989 \end{bmatrix}$

(c) $(I - A)^{-1}(I - A)$

$= \begin{bmatrix} 1.454 & .291 & .157 \\ .533 & 1.763 & .525 \\ .837 & .791 & 1.278 \end{bmatrix} \begin{bmatrix} .755 & -.102 & -.051 \\ -.099 & .709 & -.279 \\ -.433 & -.372 & .989 \end{bmatrix}$

$= \begin{bmatrix} 1.00098 & .0004 & .0811 \\ .00055 & 1.0003 & .00017 \\ .00025 & .00003 & 1.00057 \end{bmatrix} \approx I$

(d) $X = (I - A)^{-1}D$

$= \begin{bmatrix} 1.454 & .291 & .157 \\ .533 & 1.763 & .525 \\ .837 & .791 & 1.278 \end{bmatrix} \begin{bmatrix} 2.88 \\ 31.45 \\ 30.91 \end{bmatrix}$

$= \begin{bmatrix} 18.2 \\ 73.2 \\ 66.8 \end{bmatrix}$

(e) $18.2 billion of agriculture, $73.2 billion of manufacturing, and $66.8 billion of household would be required to support a demand of $2.88 billion, $31.45 billion, and $30.91 billion, respectively.

2\. **(a)** $A = \begin{bmatrix} .293 & 0 & 0 \\ .014 & .207 & .017 \\ .044 & .010 & .217 \end{bmatrix}$

$D = \begin{bmatrix} 138,213 \\ 17,597 \\ 1787 \end{bmatrix}$

(b) $I - A = \begin{bmatrix} 1 & 0 & 0 \\ 0 & 1 & 0 \\ 0 & 0 & 1 \end{bmatrix} - \begin{bmatrix} .293 & 0 & 0 \\ .014 & .207 & .017 \\ .044 & .010 & .217 \end{bmatrix}$

$= \begin{bmatrix} .707 & 0 & 0 \\ -.014 & .793 & -.017 \\ -.044 & -.010 & .784 \end{bmatrix}$

(c) $(I - A)^{-1}(I - A)$

$= \begin{bmatrix} 1.414 & 0 & 0 \\ .027 & 1.261 & .027 \\ .080 & .016 & 1.276 \end{bmatrix} \begin{bmatrix} .707 & 0 & 0 \\ -.014 & .793 & -.017 \\ -.044 & -.010 & .784 \end{bmatrix}$

$= \begin{bmatrix} .99970 & 0 & 0 \\ .00025 & .99970 & -.00027 \\ .00019 & -.00007 & 1.00011 \end{bmatrix} \approx I$

(d) $X = (I - A)^{-1}D$

$= \begin{bmatrix} 1.414 & 0 & 0 \\ .027 & 1.261 & .027 \\ .080 & .016 & 1.276 \end{bmatrix} \begin{bmatrix} 138,213 \\ 17,597 \\ 1787 \end{bmatrix}$

$= \begin{bmatrix} 195,000 \\ 26,000 \\ 13,600 \end{bmatrix}$ Rounded

About 195,000 thousand (or 195 million) pounds worth of agricultural products, 26,000 thousand (or 26 million) pounds worth of manufactured products, and 13,600 thousand (or 13.6 million) pounds worth of energy are required.

CHAPTER 7 LINEAR PROGRAMMING

Section 7.1

See the graphs for this section in the answers at the back of the text book.

1. $x + y \le 2$
   First graph the boundary line $x + y = 2$ using the points (2, 0) and (0, 2). Since the points on this line satisfy $x + y \le 2$, draw a solid line. To find the correct region to shade, choose any point not on the line. If (0, 0) is used as the test point, we have
   $x + y \le 2$
   $0 + 0 \le 2$ *Let x = 0, y = 0*
   $0 \le 2.$ *True*
   Shade the side of the line which includes (0, 0). This is the region below the line.

3. $x \ge 3 + y$
   First graph the boundary line $x = 3 + y$ using the points (0, -3) and (3, 0). This will be a solid line. Choose (0, 0) as a test point.
   $x \ge 3 + y$
   $0 \ge 3 + 0$ *Let x = 0, y = 0*
   $0 \ge 3$ *False*
   Shade the side which does not include (0, 0). This is the region below the line.

5. $4x - y < 6$
   Graph $4x - y = 6$ as a dashed line, since the points on the line are not part of the solution. Use the test point (0, 0) to get $0 - 0 < 6$, a true sentence. Shade the side of the boundary line that includes the origin (0, 0), that is, the side above the line.

7. $3x + y < 6$
   Graph $3x + y = 6$ as a dashed line. Use the test point (0, 0) to get $3(0) + 0 < 6$ or $0 < 6$, a true sentence. Shade the region that contains the origin, that is, the side below the line.

9. $x + 3y \ge -2$
   The graph includes the line $x + 3y = -2$, whose intercepts are the points (0, -2/3) and (-2, 0). Graph $x + 3y = -2$ as a solid line and use the origin as a test point. Since $0 + 3(0) \ge -2$ is true, shade the region which includes the origin, the side above the line.

11. $4x + 3y > -3$
    Graph $4x + 3y = -3$ as a dashed line. The intercepts are (-3/4, 0) and (0, -1). Use the origin as a test point. Since $4(0) + 3(0) > -3$ is true, the origin will be included in the region, so shade the half plane above the line.

13. $2x - 4y < 3$
    Graph $2x - 4y = 3$ as a dashed line. The intercepts are (3/2, 0) and (0, -3/4). Use the origin as a test point. $2(0) - 4(0) < 3$ is true, so the region above the line, which includes the origin, is the correct region to shade.

15. $x \le 5y$
    Graph $x = 5y$ as a solid line. Since this line contains the origin, some point other than (0, 0) must be used as a test point. The point (1, 2) gives $1 \le 5(2)$ or $1 \le 10$, a true sentence. Shade the side of the line containing (1, 2), that is, the side above the line.

17. $-3x < y$
Graph $y = -3x$ as a dashed line. Since this line contains the origin, use some point other than (0, 0) as a test point. (1, 1), used as a test point, gives $-3 < 3$, a true sentence. Shade the region containing (1, 1), which is the region above the line.

19. $y < x$
Graph $y = x$ as a dashed line. Since this line contains the origin, choose a point other than (0, 0) as a test point. (2, 3) gives $3 < 2$, which is false. Shade the region that does not contain the test point, that is, the region below the line.

21. $y \le -2$
Graph $y = -2$ as a solid horizontal line. The origin, used as a test point, gives $0 \le -2$, which is false. Shade the region which does not contain the origin. This is the region below the line.

23. $x - y \ge 0$
$x \le 4$
Graph $x - y \ge 0$ as the region below the solid line $x - y = 0$. Graph $x \le 4$ as the region to the left of the solid vertical line $x = 4$. Shade the overlapping part of these two regions to show the feasible region.

25. $3x + 2y \ge 18$
$x - 2y \ge 8$
Graph $3x + 2y \ge 18$ as the region above the solid line $3x + 2y = 18$. Graph $x - 2y \ge 8$ as the region below the solid line $x - 2y = 8$. Shade the overlapping part of these two regions to show the feasible region.

27. $x + y \le 1$
$x - y \ge 2$
Graph each inequality separately, using solid boundary lines. In both cases, the graph is the region below the line. Shade the overlapping part of these two regions, which is the region below both graphs. The shaded region is the feasible region for this system.

29. $2x - y < 1$
$3x + y < 6$
Graph $2x - y < 1$ as the region above the dashed line $2x - y = 1$. Graph $3x + y < 6$ as the region below the dashed line $3x + y < 6$. Shade the overlapping part of these two regions to show the feasible region.

31. $-x - y < 5$
$2x - y < 4$
Graph $-x - y < 5$ as the region above the dashed line $-x - y = 5$. Graph $2x - y < 4$ as the region above the dashed line $2x - y = 4$. Shade the overlapping part of these two regions to show the feasible region.

33. The graph of $x + y \le 4$ consists of the solid line $x + y = 4$ and the points below it. The graph of $x - y \le 5$ consists of the solid line $x - y = 5$ and all the points above it. The graph of $4x + y \le -4$ consists of the solid line $4x + y = -4$ and all the points below it. The feasible region is the overlapping part of these three graphs.

35. The graph of $-2 < x < 3$ is the region between the vertical line $x = -2$ and $x = 3$, but not including the lines themselves. The graph of $-1 \le y \le 5$ is the region between the horizontal lines $y = -1$ and $y = 5$, including the lines. The graph of $2x + y < 6$ is the region below the line $2x + y = 6$. Shade the region common to all three graphs to show the feasible region.

37. The graph of $2y + x \ge -5$ consists of the boundary line $2y + x = 5$ and the region above it. The graph of $y \le 3 + x$ consists of the boundary line $y = 3 + x$ and the region below it. The inequalities $x \ge 0$ and $y \ge 0$ restrict the feasible region to the first quadrant. Shade the feasible region.

39. $3x + 4y > 12$ is the set of points above the dashed line $3x + 4y = 12$. $2x - 3y < 6$ is the set of points above the dashed line $2x - 3y = 6$. $0 \le y \le 2$ is the rectangular strip of points lying on or between the horizontal lines $y = 0$ and $y = 2$. $x \ge 0$ consists of all the points on and to the right of the y-axis. The feasible region is the triangular region satisfying all the inequalities.

41. Use the completed chart shown in the answers in the textbook to write the system of inequalities. On the wheel, x glazed pots require

$\frac{1}{2} \cdot x = \frac{1}{2}x$ hours and y unglazed

pots require $1 \cdot y = y$ hours. Since the wheel is available for at most 8 hours per day,

$$\frac{1}{2}x + y \le 8.$$

In the kiln, x glazed pots require $1 \cdot x = x$ hours and y unglazed pots require $6 \cdot y = 6y$ hours. Since the kiln is available for at most 20 hours per day,

$$x + 6y \le 20.$$

Since it is not possible to produce a negative number of pots,

$$x \ge 0 \text{ and } y \ge 0.$$

Thus, the system is

$$\begin{aligned} \frac{1}{2}x + y &\le 8 \\ x + 6y &\le 20 \\ x &\ge 0 \\ y &\ge 0. \end{aligned}$$

43. The system is

$$\begin{aligned} x &\ge 3000 \\ y &\ge 5000 \\ x + y &\le 10{,}000. \end{aligned}$$

The first inequality gives the region to the right of the vertical line $x = 3000$, including the points on the line. The second inequality gives the region above the horizontal line $y = 5000$, including the points on the line. The third inequality gives the region below the line $x + y = 10{,}000$, including the points on the line.

45. The system is

$$\begin{aligned} x + y &\le 25 \\ x &\ge 4y \\ .12x + .10y &\ge 2.8 \\ x &\le 0 \\ y &\le 0 \end{aligned}$$

where x and y are in millions. The first inequality gives the region below the solid line $x + y = 25$, including the points on the line. The second inequality gives the region below the solid line $x = 4y$, including the points on the line. The third inequality gives the region below the solid line

.12x + .10y = 2.8, including the points on the line.

Section 7.2

1. Make a table indicating the value of the objective function z = 3x + 5y at each corner point.

Corner point	Value of z = 3x + 5y
(1, 1)	3(1) + 5(1) = 8 *Minimum*
(2, 7)	3(2) + 5(7) = 41
(5, 10)	3(5) + 5(10) = 65 *Maximum*
(6, 3)	3(6) + 5(3) = 33

The maximum value of 65 occurs at (5, 10). The minimum value of 8 occurs as (1, 1).

3.

Corner point	Value of z = .40x + .75y	
(0, 0)	0	*Minimum*
(0, 12)	9	*Maximum*
(4, 8)	7.6	
(7, 3)	5.05	
(9, 0)	3.6	

The maximum is 9 at (0, 12); the minimum is 0 at (0, 0).

5.

Corner point	Value of z = 2x + 3y	
(0, 8)	24	
(3, 4)	18	*Minimum*
(13/2, 2)	19	
(12, 0)	24	

The minimum is 18 at (3, 4). There is no maximum because the region of feasible solutions is unbounded.

7. First draw the region using the methods of Section 7.1.

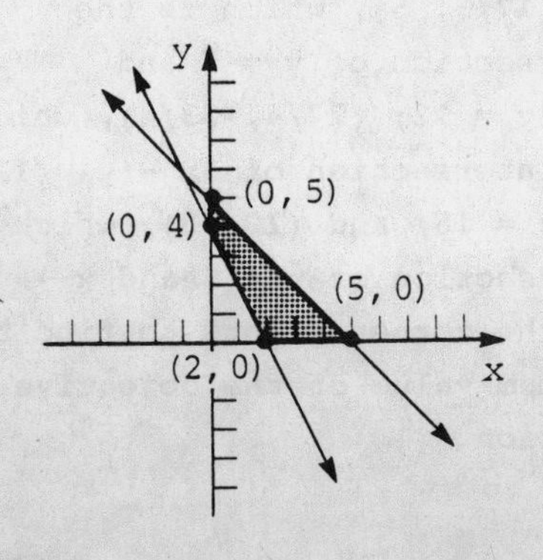

Make a table indicating the value of the objective function z = 4x + 5 at each corner point.

Corner point	Value of z = 4x + 5y
(2, 0)	4(2) + 5(0) = 8 *Minimum*
(5, 0)	4(5) + 5(0) = 20
(0, 4)	4(0) + 5(4) = 20
(0, 5)	4(0) + 5(5) = 25 *Maximum*

The maximum value of 25 occurs at (0, 5). The minimum value of 8 occurs at (2, 0).

9. First draw the region.

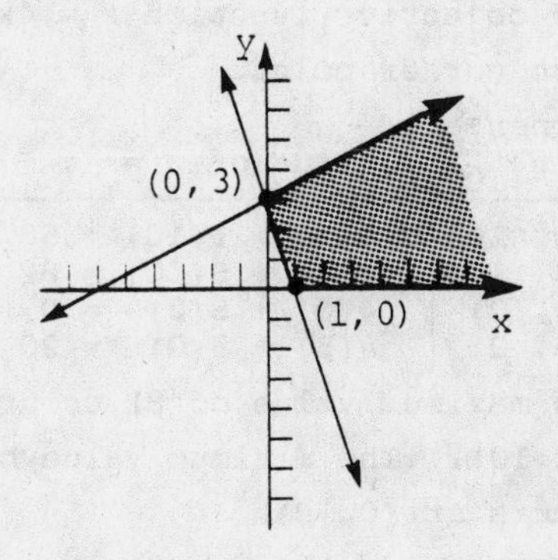

The region is unbounded so there is no maximum value. Make a table indicating the value of the objective function z = 4x + 5y at each corner point.

Corner point	Value of z = 4x + 5y
(1, 0)	4(1) + 5(0) = 4 *Minimum*
(0, 3)	4(0) + 5(3) = 15

The minimum value of 4 occurs at (1, 0).

11. First draw the region.

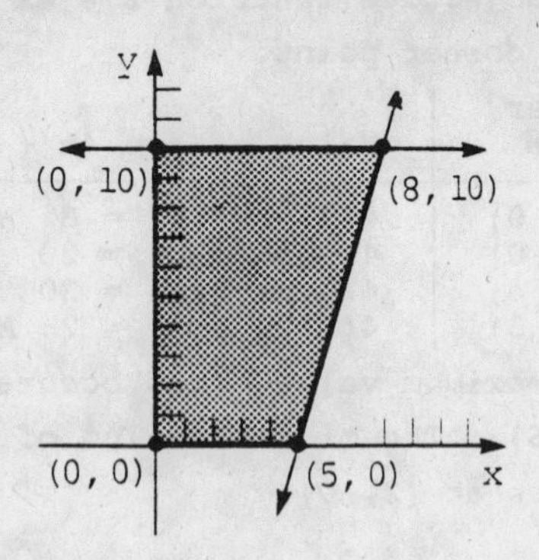

Make a table indicating the value of the objective function $z = 4x + 5y$ at each corner point.

Corner point	Value of $z = 4x + 5y$	
(0, 10)	$4(0) + 5(10) = 50$	
(8, 10)	$4(8) + 5(10) = 82$	*Maximum*
(0, 0)	$4(0) + 5(0) = 0$	*Minimum*
(5, 0)	$4(5) + 5(0) = 20$	

The maximum value of 82 occurs at (8, 10). The minimum value of 0 occurs at (0, 0).

13. Maximize $z = 5x + 2y$

subject to:
$$2x + 3y \le 6$$
$$4x + y \le 6$$
$$x \ge 0$$
$$y \ge 0.$$

Sketch the feasible region.

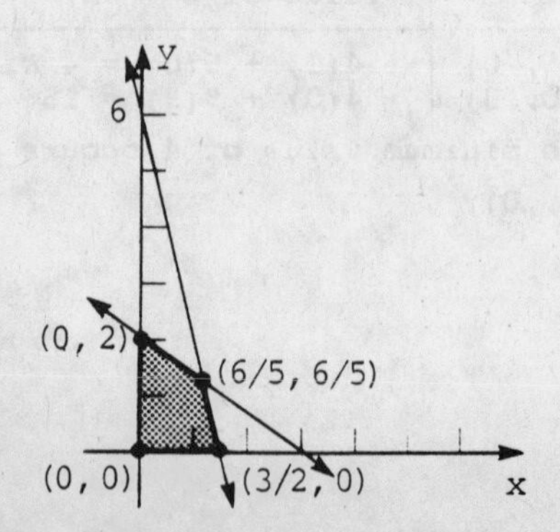

The graph shows that the feasible region is bounded.
The corner points are (0, 0), (0, 2), (3/2, 0), and (6/5, 6/5), which is the intersection of $2x + 3y = 6$ and $4x + y = 6$. Use the corner points to find the maximum value of the objective function.

Corner point	Value of $z = 5x + 2y$	
(0, 0)	0	
(0, 2)	4	
(6/5, 6/5)	42/5	*Maximum*
(3/2, 0)	15/2	

The maximum value of $z = 5x + 2y$ is 42/5 at the corner point (6/5, 6/5).

15. Maximize $z = 2x + y$

subject to:
$$3x - y \le 12$$
$$x + y \le 15$$
$$x \ge 2$$
$$y \ge 5.$$

Sketch the feasible region.

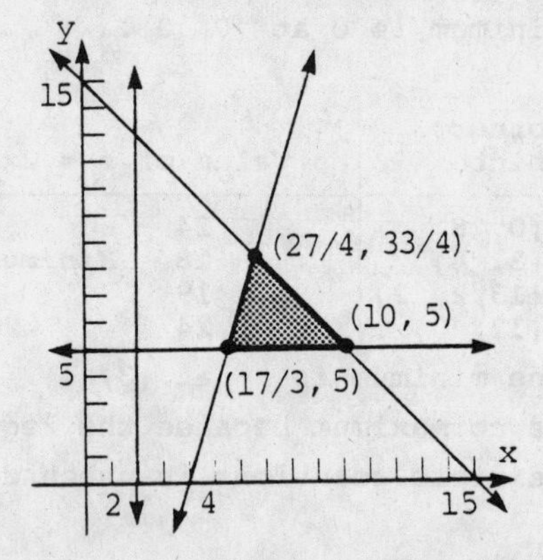

The graph shows that the feasible region is bounded. The corner points are (17/3, 5), which is the intersection of $y = 5$ and $3x - y = 12$; (27/4, 33/4), which is the intersection of $3x - y = 12$ and $x + y = 15$; and (10, 5), which is the intersection of $y = 5$ and $x + y = 15$. Use the corner points to find the maximum value of the objective function.

Corner point	Value of $z = 2x + y$
(17/3, 5)	49/3
(27/4, 33/4)	87/4
(10, 5)	25 *Maximum*

The maximum value of $z = 2x + y$ is 25 at (10, 5).

17. Maximize $z = 4x + 2y$
subject to: $x - y \le 10$
$5x + 3y \le 75$
$x \ge 0$
$y \ge 0.$

Sketch the feasible region.

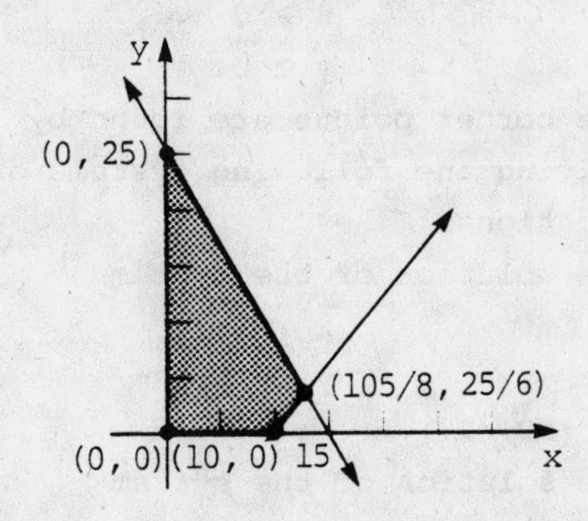

This region is bounded, with corner points (0, 0), (0, 25), (105/8, 25/8), and (10, 0).

Corner point	Values of $z = 4x + 2y$
(0, 0)	0
(0, 25)	50
(105/8, 25/8)	235/4 *Maximum*
(10, 0)	40

The maximum value of $z = 4x + 2y$ is 235/4 at (105/8, 25/8).

19. (a) $x + y \le 20$
$x + 3y \le 24$

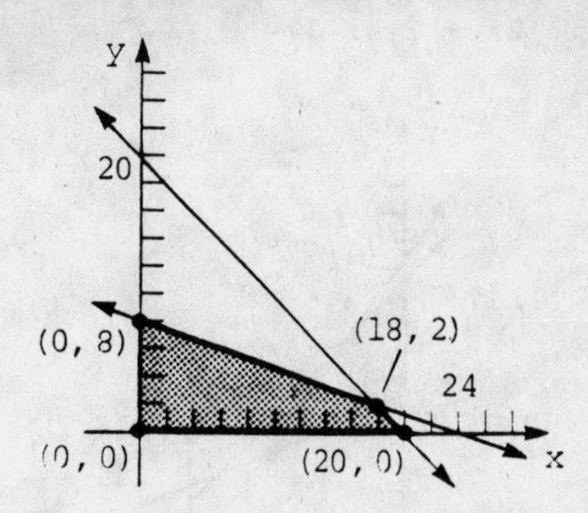

Corner point	Value of $z = 10x + 12y$
(0, 0)	0
(0, 8)	96
(18, 2)	204 *Maximum*
(20, 0)	200

The maximum value of 204 occurs when $x = 18$ and $y = 2$, or at (18, 2).

(b) $3x + y \le 15$
$x + 2y \le 18$

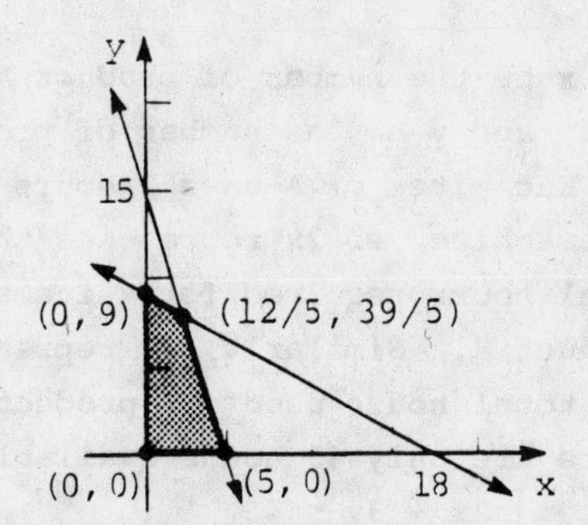

Corner point	Value of $z = 10x + 12y$
(0, 0)	0
(0, 9)	108
(12/5, 39/5)	588/5 *Maximum*
(5, 0)	50

The maximum value of 588/5 (or 117 3/5) occurs when $x = 12/5$ and $y = 39/5$, or at (12/5, 39/5).

(c) $2x + 5y \geq 22$

$4x + 3y \leq 28$

$2x + 2y \leq 17$

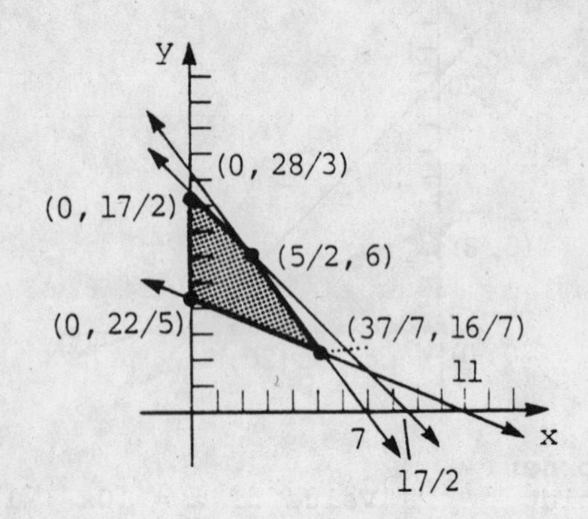

Corner point	Value of $z = 10x + 12y$	
(0, 22/5)	264/5	
(0, 17/2)	102	*Maximum*
(5/2, 6)	97	
(37/7, 16/7)	562/7	

The maximum value of 102 occurs when $x = 0$ and $y = 17/2$, or at (0, 17/2).

Section 7.3

1. Let x be the number of product A made, and y be the number of product B. Each item of A uses 2 hours on the machine, so 2x represents the total hours required for x items of product A. Similarly, 3y represents the total hours used for product B. There are only 45 hours available, so

$$2x + 3y \leq 45.$$

3. Let x be the number of green pills and y be the number of red pills. Then 4x represents the number of vitamin units provided by the green pills, and y represents the vitamin units provided by the red ones. Since at least 25 units are needed per day,

$$4x + y \geq 25.$$

5. Let x be the number of pounds of $6 coffee and y be the number of pounds of $5 coffee.
Since the mixture must weigh at least 50 pounds,

$$x + y \geq 50.$$

7. Let x be the number of engines sent to plant I and y be the number of engines sent to plant II.

Minimize $z = 20x + 35y$

subject to:

$$x \geq 50$$
$$y \geq 27$$
$$x + y \leq 85$$
$$x \geq 0$$
$$y \geq 0.$$

The corner points are found by solving the following systems of equations.

The solution of the system

$$x = 50$$
$$y = 27$$

is (50, 27).

The solution of the system

$$x = 50$$
$$x + y = 85$$

is (50, 35).

The solution of the system

$$y = 27$$
$$x + y = 85$$

is (58, 27).

The feasible region is a triangular region with the corner points (50, 27), (50, 35), and (58, 27).
Use the corner points to find the minimum value of the objective function.

Corner point	Value of $z = 20z + 35y$	
(50, 27)	1945	*Minimum*
(50, 35)	2225	
(58, 27)	2105	

50 engines should be shipped to plant I and 27 engines should be shipped to plant II. The minimum cost is $1945.

(In Exercise 7, it is not surprising that costs are minimized by sending the minimum number required.)

9. Let x be the number of units of policy A to purchase and y be the number of units of policy B to purchase.

Minimize $z = 50x + 40y$
subject to: $10x + 15y \geq 100$
$80x + 120y \geq 1000$
$x \geq 0$
$y \geq 0.$

In the first quadrant, the graph of $80x + 120y = 100$ lies entirely above the graph of $10x + 15y = 100$, so the corner points are the x- and y-intercepts of $80x + 120y = 1000$. These are (0, 25/3) and (25/2, 0). Use the corner points to find the minimum value of the objective function.

Corner point	Value of $z = 50x + 40y$
(0, 25/3)	1000/3 = 333 1/3 *Minimum*
(25/2, 0)	625

In order to minimize the premium costs, 0 units of policy A and 25/3 (or 8 1/3) units of policy B should be purchased for a minimum premium cost of $333.33.

11. Let x be the number of type 1 bolts and y be the number of type 2 bolts.

Maximize $z = .10x + .12y$
subject to: $.1x + .1y \leq 240$
$.1x + .4y \leq 720$
$.1x + .5y \leq 160$
$x \geq 0$
$y \geq 0.$

The first two inequalities do not affect the solution. The feasible region is the triangular region bounded by the x- and y-axes and the line $.1x + .5y = 160$. The corner points are (0, 0), (0, 320), and (1600, 0). Use the corner points to find the maximum value of the objective function.

Corner point	Value of $z = .10x + .12y$
(0, 0)	0
(0, 320)	38.40
(1600, 0)	160.00 *Maximum*

1600 type I bolts and 0 type II bolts should be manufactured for a maximum revenue of $160 per day.

13. Let x be the number of kilograms of mix with 1/2 nuts and 1/2 raisins and y be the number of kilograms of the other mix. The problem requests maximum revenue, so a revenue function is needed:

$z = 6x + 4.8y.$

The constraints are on the available nuts and raisins that make up the mixes.

Maximize $z = 6x + 4.8y$

subject to: $\frac{1}{2}x + \frac{1}{3}y \leq 100$

$\frac{1}{2}x + \frac{2}{3}y \leq 125$

$x \geq 0$
$y \geq 0.$

Three of the corner points are (0, 0), (0, 187.5), (200, 0). The fourth corner point is the point of intersection of $\frac{1}{2}x + \frac{1}{3}y = 100$ and $\frac{1}{2}x + \frac{2}{3}y = 125.$

Find this point by solving the system:

$$\frac{1}{2}x + \frac{1}{3}y = 100 \quad (1)$$

$$\frac{1}{2}x + \frac{2}{3}y = 125 \quad (2)$$

Multiply both equations by 6.

$3x + 2y = 600$ *(3)*

$3x + 4y = 750$ *(4)*

Multiply equation (3) by -1 and add to equation (4).

$-3x - 2y = -600$ *(5)*
$3x + 4y = 750$ *(6)*

$2y = 150$
$y = 75$

Substitute y = 75 into equation (1) to find x.

$$\frac{1}{2}x + \frac{1}{3}(75) = 100$$

$$\frac{1}{2}x = 75$$

$$x = 150$$

Thus, the fourth corner point is (150, 75). The four corner points are (0, 0), (0, 187.5), (150, 75), and (200, 0). Use the corner points to find the maximum value of the objective function.

Corner point	Value of z = 6x + 4.8y	
(0, 0)	0	
(0, 187.5)	900	
(150, 75)	1260	*Maximum*
(200, 0)	1200	

The company should prepare 150 kilograms of the mix with 1/2 nuts and 1/2 raisins and 75 kilograms of the other mix for a maximum revenue of $1260.

15. Let x be the number of gallons of milk from Diary II and y be the number of gallons of milk from Dairy I.

Maximize $z = .032x + .037y$
subject to: $x \le 80$
$y \le 50$
$x + y \le 100$
$x \ge 0$
$y \ge 0.$

The corner points are (0, 0), (80, 0), (80, 20), (50, 50), and (0, 50). Use the corner points to find the maximum value of the objective function.

Corner point	Value of z = .032x + .037y
(0, 0)	0
(80, 0)	2.56
(80, 20)	3.30
(50, 50)	3.45
(0, 50)	1.85

50 gallons from Diary I and 50 gallons from Diary II should be used to get milk with a maximum of 3.45 gallons of butterfat.

17. Let x be the amount invested in bonds and y be the amount invested in mutual funds. (Both are in millions of dollars.)

The amount of annual interest is

$.12x + .08y.$

Maximize $z = .12x + .08y$
subject to: $x \ge 20$
$y \ge 15$
$x + y \le 40.$

The corner points are (20, 15), (20, 20), and (25, 15). Use the corner points to find the maximum value of the objective function.

Corner Point	Value of z = .12x + .08y
(20, 15)	3.6
(20, 20)	4.0
(20, 15)	4.2

He should invest $25 million in bonds and $15 million in mutual funds for maximum annual interest of $4.20 million.

19. Let x be the number of species I prey and y be the number of species II prey.

	Protein	Fat
Species I	5	2
Species II	3	4

Minimize $z = 2x + 3y$

subject to:
$$5x + 3y \geq 10$$
$$2x + 4y \geq 8$$
$$x \geq 0$$
$$y \geq 0.$$

The corner points are (0, 10/3), (4, 0), and the intersection of $5x + 3y = 10$ and $2x + 4y = 8$, which is (8/7, 10/7). Use the corner points to find the minimum value of the objective function.

Corner point	Value of $z = 2x + 3y$	
(0, 10/3)	10	
(4, 0)	8	
(8/7, 10/7)	46/7	*Minimum*

The minimum value of z is $46/7 \approx 6.57$. 8/7 units of species I and 10/7 units of species II will meet the daily food requirements with the least expenditure of energy. However, a predator probably can catch and digest only whole numbers of prey. This problem shows that it is important to consider whether a model produces a realistic answer to a problem.

21. Let x be the number of servings of product A and y be the number of servings of product B. The cost function (in dollars) is $z = .25x + .40y$.

Minimize $z = .25x + .40y$

subject to:
$$3x + 2y \geq 15$$
$$2x + 4y \geq 15$$
$$x \geq 0$$
$$y \geq 0.$$

The corner points are (0, 15/2), (15/4, 15/8), and (15/2, 0). Use the corner points to find the minimum value of the objective function.

Corner point	Value of $z = .25x + .40y$	
(0, 15/2)	3	
(15/4, 15/8)	1.6875	*Minimum*
(15/2, 0)	1.875	

15/4 (or 3 3/4) servings of A and 15/8 (or 1 7/8) servings of B will satisfy the requirements at a minimum cost of \$1.69 (rounded to the nearest cent).

23. 1 Zeta + 2 Beta must not exceed 1000; thus (b) is the correct solution.

25. \$4 Zeta + \$5.25 Beta equals the total contribution margin; (c) is the correct answer.

Section 7.4

1. $x_1 + 2x_2 \leq 6$ becomes
$x_1 + 2x_2 + x_3 = 6.$

3. $2x_1 + 4x_2 + 3x_3 \leq 100$ becomes
$2x_1 + 4x_2 + 3x_3 + x_4 = 100.$

5. **(a)** Since there are three inequalities, 3 slack variables are needed.

(b) The original problem uses x_1 and x_2, so use x_3, x_4 and x_5 for the slack variables.

(c)
$$\begin{aligned} 4x_1 + 2x_2 + x_3 &= 20 \\ 5x_1 + x_2 + x_4 &= 50 \\ 2x_1 + 3x_2 + x_5 &= 25 \end{aligned}$$

7. **(a)** There are two constraints to be turned into equalities, so 2 slack variables are needed.

(b) x_1, x_2, and x_3 are already used in the problem; call the slack variables x_4 and x_5.

(c)
$$\begin{aligned} 7x_1 + 6x_2 + 8x_3 + x_4 &= 118 \\ 4x_1 + 5x_2 + 10x_3 + x_5 &= 220 \end{aligned}$$

9. $x_1 = 0$, $x_2 = 0$, $x_3 = 20$, $x_4 = 0$, $x_5 = 15$

11. $x_1 = 0$, $x_2 = 0$, $x_3 = 8$, $x_4 = 0$, $x_5 = 6$, $x_6 = 7$

13.
$$\begin{array}{c} \begin{array}{ccccc} x_1 & x_2 & x_3 & x_4 & x_5 \end{array} \\ \left[\begin{array}{ccccc|c} 1 & 2 & 4 & 1 & 0 & 56 \\ 2 & \textcircled{2} & 1 & 0 & 1 & 40 \\ \hline -1 & -3 & -2 & 0 & 0 & 0 \end{array}\right] \end{array}$$

$$\begin{array}{c} \begin{array}{ccccc} x_1 & x_2 & x_3 & x_4 & x_5 \end{array} \\ \left[\begin{array}{ccccc|c} -1 & 0 & 3 & 1 & -1 & 16 \\ 2 & \textcircled{2} & 1 & 0 & 1 & 40 \\ \hline -1 & -3 & -2 & 0 & 0 & 0 \end{array}\right] \begin{array}{l} -R_2 + R_1 \\ \\ \\ \end{array} \end{array}$$

$$\begin{array}{c} \begin{array}{ccccc} x_1 & x_2 & x_3 & x_4 & x_5 \end{array} \\ \left[\begin{array}{ccccc|c} -1 & 0 & 3 & 1 & -1 & 16 \\ 1 & 1 & \frac{1}{2} & 0 & \frac{1}{2} & 20 \\ \hline 2 & 0 & \frac{1}{2} & 0 & \frac{3}{2} & 60 \end{array}\right] \begin{array}{l} \\ \frac{1}{2}R_2 \\ 3R_2+R_3 \end{array} \end{array}$$

$x_1 = 0$, $x_2 = 20$, $x_3 = 0$, $x_4 = 16$, $x_5 = 0$

15.
$$\begin{array}{c} \begin{array}{cccccc} x_1 & x_2 & x_3 & x_4 & x_5 & x_6 \end{array} \\ \left[\begin{array}{cccccc|c} 2 & 2 & \textcircled{1} & 1 & 0 & 0 & 12 \\ 1 & 2 & 3 & 0 & 1 & 0 & 45 \\ 3 & 1 & 1 & 0 & 0 & 1 & 20 \\ \hline -2 & -1 & -3 & 0 & 0 & 0 & 36 \end{array}\right] \end{array}$$

$$\begin{array}{c} \begin{array}{cccccc} x_1 & x_2 & x_3 & x_4 & x_5 & x_6 \end{array} \\ \left[\begin{array}{cccccc|c} 2 & 2 & 1 & 1 & 0 & 0 & 12 \\ -5 & -4 & 0 & -3 & 1 & 0 & 9 \\ 1 & -1 & 0 & -1 & 0 & 1 & 8 \\ \hline 4 & 5 & 0 & 3 & 0 & 0 & 36 \end{array}\right] \begin{array}{l} \\ -3R_1+R_2 \\ -R_1+R_3 \\ 3R_1+R_4 \end{array} \end{array}$$

$x_1 = 0$, $x_2 = 0$, $x_3 = 12$, $x_4 = 0$, $x_5 = 9$, $x_6 = 8$

17.
$$\begin{array}{c} \begin{array}{cccccc} x_1 & x_2 & x_3 & x_4 & x_5 & x_6 \end{array} \\ \left[\begin{array}{cccccc|c} 1 & 1 & 1 & 1 & 0 & 0 & 60 \\ 3 & 1 & \textcircled{2} & 0 & 1 & 0 & 100 \\ 1 & 2 & 3 & 0 & 0 & 1 & 200 \\ \hline -1 & -1 & -2 & 0 & 0 & 0 & 0 \end{array}\right] \end{array}$$

$$\begin{array}{c} \begin{array}{cccccc} x_1 & x_2 & x_3 & x_4 & x_5 & x_6 \end{array} \\ \left[\begin{array}{cccccc|c} 1 & 1 & 1 & 1 & 0 & 0 & 60 \\ 3 & 1 & \textcircled{2} & 0 & 1 & 0 & 100 \\ 1 & 2 & 3 & 0 & 0 & 1 & 200 \\ \hline 2 & 0 & 0 & 0 & 1 & 0 & 100 \end{array}\right] \begin{array}{l} \\ \\ \\ R_2+R_4 \end{array} \end{array}$$

$$\begin{array}{c} \begin{array}{cccccc} x_1 & x_2 & x_3 & x_4 & x_5 & x_6 \end{array} \\ \left[\begin{array}{cccccc|c} -\frac{1}{2} & \frac{1}{2} & 0 & 1 & -\frac{1}{2} & 0 & 10 \\ \frac{3}{2} & \frac{1}{2} & 1 & 0 & \frac{1}{2} & 0 & 50 \\ -\frac{7}{2} & \frac{1}{2} & 0 & 0 & -\frac{3}{2} & 1 & 50 \\ \hline 2 & 0 & 0 & 0 & 1 & 0 & 100 \end{array}\right] \begin{array}{l} -R_2+R_1 \\ \\ -3R_2+R_3 \\ \\ \end{array} \end{array}$$

$x_1 = 0$, $x_2 = 0$, $x_3 = 50$, $x_4 = 10$, $x_5 = 0$, $x_6 = 50$

19. Since there are 3 constraints, 3 slack variables we needed: x_3, x_4 and x_5.

The constraints are now

$$\begin{aligned} 2x_1 + 3x_2 + x_3 \quad\quad &= 6 \\ 4x_1 + x_2 + \quad x_4 \quad &= 6 \\ 5x_1 + 2x_2 + \quad\quad x_5 &= 15. \end{aligned}$$

The initial simplex tableau is

$$\begin{array}{ccccc|c} x_1 & x_2 & x_3 & x_4 & x_5 & \\ 2 & 3 & 1 & 0 & 0 & 6 \\ 4 & 1 & 0 & 1 & 0 & 6 \\ 5 & 2 & 0 & 0 & 1 & 15 \\ \hline -5 & -1 & 0 & 0 & 0 & 0 \end{array}.$$

21. Since there are 3 constraints, 3 slack variables are needed: x_4, x_5 and x_6

The constraints are now

$$\begin{aligned} 2x_1 + 4x_2 + x_3 + x_4 \quad\quad &= 18 \\ x_1 + 6x_2 + 2x_3 + \quad x_5 \quad &= 45 \\ 5x_1 + 7x_2 + 3x_3 + \quad\quad x_6 &= 60 \end{aligned}$$

The initial simplex tableau is

$$\begin{array}{cccccc|c} x_1 & x_2 & x_3 & x_4 & x_5 & x_6 & \\ 2 & 4 & 1 & 1 & 0 & 0 & 18 \\ 1 & 6 & 2 & 0 & 1 & 0 & 45 \\ 5 & 7 & 3 & 0 & 0 & 1 & 60 \\ \hline -1 & -5 & 10 & 0 & 0 & 0 & 0 \end{array}.$$

23. Since there are 3 constraints, 3 slack variables are needed: x_4, x_5 and x_6.

The constraints are now

$$\begin{aligned} x_1 + 2x_2 + 3x_3 + x_4 \quad\quad &= 10 \\ 2x_1 + x_2 + x_3 + \quad x_5 \quad &= 8 \\ 3x_1 \quad + 2x_3 + \quad\quad x_6 &= 6. \end{aligned}$$

The initial tableau is

$$\begin{array}{cccccc|c} x_1 & x_2 & x_3 & x_4 & x_5 & x_6 & \\ 1 & 2 & 3 & 1 & 0 & 0 & 10 \\ 2 & 1 & 1 & 0 & 1 & 0 & 8 \\ 3 & 0 & 2 & 0 & 0 & 1 & 6 \\ \hline -6 & -2 & -3 & 0 & 0 & 0 & 0 \end{array}.$$

25. Let x_1 be the number of kilograms of the mix 1/2 nuts and 1/2 raisins and x_2 the number of kilograms of the other mix. The problem is to maximize $z = 6x_1 + 4.8x_2$ subject to

$$\frac{1}{2}x_1 + \frac{1}{3}x_2 \le 100$$

$$\frac{1}{2}x_1 + \frac{2}{3}x_2 \le 125,$$

$x_1 \ge 0$, and $x_2 \ge 0$.

Two constraints need to be changed into equalities, 2 slack variables, x_3 and x_4, are needed.

The problem can now be restated as: Find $x_1 \ge 0$, $x_2 \ge 0$, $x_3 \ge 0$, $x_4 \ge 0$ such that

$$\frac{1}{2}x_1 + \frac{1}{3}x_2 + x_3 \quad = 100$$

$$\frac{1}{2}x_1 + \frac{2}{3}x_2 \quad + x_4 = 125$$

and $z = 6x_1 + 4.8x_2$ is maximized.

The initial simplex tableau is

$$\begin{array}{cccc|c} x_1 & x_2 & x_3 & x_4 & \\ \frac{1}{2} & \frac{1}{3} & 1 & 0 & 100 \\ \frac{1}{2} & \frac{2}{3} & 0 & 1 & 125 \\ \hline -6 & -4.8 & 0 & 0 & 0 \end{array}.$$

27. Let x_1 be the number of prams, x_2 the number of runabouts, and x_3 the number of trimarans. Collect some of the information into a table.

	Pram	Run-about	Trimaran	Available time
Sec A	1	2	3	6240
Sec B	2	5	4	10,800
Profit	75	90	100	

This information, together with the fact that the total number of boats cannot exceed 3000, gives the problem:

Find $x_1 \geq 0$, $x_2 \geq 0$, $x_3 \geq 0$ such that

$$x_1 + 2x_2 + 3x_3 \leq 6240$$
$$2x_1 + 5x_2 + 4x_3 \leq 10{,}800$$
$$x_1 + x_2 + x_3 \leq 3000$$

and $z = 75x_1 + 90x_2 + 100x_3$ is maximized. Introduce slack variables x_4, x_5, and x_6. The problem can now be restated as:

Find $x_1 \geq 0$, $x_2 \geq 0$, $x_3 \geq 0$, $x_4 \geq 0$, $x_5 \geq 0$, and $x_6 \geq 0$ such that

$$x_1 + 2x_2 + 3x_3 + x_4 = 6240$$
$$2x_1 + 5x_2 + 4x_3 + x_5 = 10{,}800$$
$$x_1 + x_2 + x_3 + x_6 = 3000$$

and $z = 75x_1 + 90x_2 + 100x_3$ is maximized.

The initial simplex is

x_1	x_2	x_3	x_4	x_5	x_6	
1	2	3	1	0	0	6240
2	5	4	0	1	0	10,800
1	1	1	0	0	1	3000
-75	-90	-100	0	0	0	0

29. Let x_1 = number of Siamese cats and x_2 = number of Persian cats. The problem is to maximize $z = 12x_1 + 10x_2$ subject to $2x_1 + x_2 \leq 90$, $x_1 + 2x_2 \leq 80$, $x_1 + x_2 \leq 50$, $x_1 \geq 0$, and $x_2 \geq 0$. There are three constraints to be changed into equalities, so introduce three slack variables, x_3, x_4, and x_5. The problem can now be restated as:

Find $x_1 \geq 0$, $x_2 \geq 0$, $x_3 \geq 0$, $x_4 \geq 0$, $x_5 \geq 0$, such that

$$2x_1 + x_2 + x_3 = 90$$
$$x_1 + 2x_2 + x_4 = 80$$
$$x_1 + x_2 + x_3 + x_5 = 50$$

and $z = 12x_1 + 10x_2$ is maximized. Then the initial simplex tableau is

x_1	x_2	x_3	x_4	x_5	
2	1	1	0	0	90
1	2	0	1	0	80
1	1	0	0	1	50
-12	-10	0	0	0	0

Section 7.5

1. The initial simplex tableau is as follows.

x_1	x_2	x_3	x_4	x_5	
1	2	4	1	0	8
2	2	1	0	1	10
-2	-5	-1	0	0	0

The most negative indicator is -5, in the second column. 8/2 = 4, and 10/2 = 5. Since 4 is the smallest quotient, 2 in the first row is the pivot.

	x_1	x_2	x_3	x_4	x_5	
8/2 = 4	1	(2)	4	1	0	8
10/2 = 5	2	2	1	0	1	10
	-2	-5	-1	0	0	0

Performing row transformations gives the second tableau.

x_1	x_2	x_3	x_4	x_5	
$\frac{1}{2}$	1	2	$\frac{1}{2}$	0	4
1	0	-3	-1	1	2
$\frac{1}{2}$	0	9	$\frac{5}{2}$	0	20

All numbers in the last row are nonnegative, the problem is completed. The maximum value is 20 and occurs when $x_1 = 0$, $x_2 = 0$, $x_3 = 0$, $x_4 = 0$ and $x_5 = 2$.

3.

x_1	x_2	x_3	x_4	x_5	
1	3	1	0	0	12
2	1	0	1	0	10
1	1	0	0	1	4
-2	-1	0	0	0	0

The most negative indicator is -2, in the first column. 12/1 = 12, 10/2 = 5, and 4/1 = 4. Since 4 is the smallest quotient, 1 in the third row, first column, is the pivot.

$$\begin{array}{c} \begin{array}{ccccc} x_1 & x_2 & x_3 & x_4 & x_5 \end{array} \\ \left[\begin{array}{ccccc|c} 1 & 3 & 1 & 0 & 0 & 12 \\ 2 & 1 & 0 & 1 & 0 & 10 \\ \textcircled{1} & 1 & 0 & 0 & 1 & 4 \\ \hline -2 & -1 & 0 & 0 & 0 & 0 \end{array}\right] \end{array}$$

$$\begin{array}{c} \begin{array}{ccccc} x_1 & x_2 & x_3 & x_4 & x_5 \end{array} \\ \left[\begin{array}{ccccc|c} 0 & 2 & 1 & 0 & -1 & 8 \\ 0 & -1 & 0 & 1 & -2 & 2 \\ 1 & 1 & 0 & 0 & 1 & 4 \\ \hline 0 & 1 & 0 & 0 & 2 & 8 \end{array}\right] \end{array} \begin{array}{l} -R_3+R_1 \\ -2R_3+R_2 \\ \\ 2R_3+R_4 \end{array}$$

This is a final tableau. The maximum value is 8 when $x_1 = 4$, $x_2 = 0$ $x_3 = 8$, $x_4 = 2$, and $x_5 = 0$.

5.

$$\begin{array}{c} \begin{array}{cccccc} x_1 & x_2 & x_3 & x_4 & x_5 & x_6 \end{array} \\ \left[\begin{array}{cccccc|c} 2 & 2 & 8 & 1 & 0 & 0 & 40 \\ 4 & -5 & 6 & 0 & 1 & 0 & 60 \\ 2 & -2 & 6 & 0 & 0 & 1 & 24 \\ \hline -14 & 10 & -12 & 0 & 0 & 0 & 0 \end{array}\right] \end{array}$$

The most negative indicator is -14, in the first column. Find the quotients 40/2 = 20, 60/4 = 15, and 24/2 = 12.
Since 12 is the smallest quotient, 2 in the third row is the pivot.

$$\begin{array}{l} \\ 40/2 = 20 \\ 60/4 = 15 \\ 24/2 = 2 \\ \\ \end{array} \begin{array}{c} \begin{array}{cccccc} x_1 & x_2 & x_3 & x_4 & x_5 & x_6 \end{array} \\ \left[\begin{array}{cccccc|c} 2 & 2 & 8 & 1 & 0 & 0 & 40 \\ 4 & -5 & 6 & 0 & 1 & 0 & 60 \\ \textcircled{2} & -2 & 6 & 0 & 0 & 1 & 24 \\ \hline -14 & -10 & -12 & 0 & 0 & 0 & 0 \end{array}\right] \end{array}$$

Performing row transformations gives the second tableau.

$$\begin{array}{c} \begin{array}{cccccc} x_1 & x_2 & x_3 & x_4 & x_5 & x_6 \end{array} \\ \left[\begin{array}{cccccc|c} 0 & 4 & 2 & 1 & 1 & -1 & 16 \\ 0 & -1 & -6 & 0 & 1 & -2 & 12 \\ 1 & -1 & 3 & 0 & 0 & \frac{1}{2} & 12 \\ \hline 0 & -24 & 30 & 0 & 0 & 7 & 168 \end{array}\right] \end{array} \begin{array}{l} -R_1 - R_3 \\ R_2 - 2R_3 \\ \frac{1}{2}R_3 \\ 7R_3 - R_4 \end{array}$$

Since there is still a negative indicator, repeat the process. The second pivot is the 4 in the second column. Performing row transformations again gives the tableau.

$$\begin{array}{c} \begin{array}{cccccc} x_1 & x_2 & x_3 & x_4 & x_5 & x_6 \end{array} \\ \left[\begin{array}{cccccc|c} 0 & 1 & \frac{1}{2} & \frac{1}{4} & 0 & -\frac{1}{4} & 4 \\ 0 & 0 & -\frac{11}{2} & \frac{1}{4} & 1 & -\frac{9}{4} & 16 \\ 1 & 0 & \frac{7}{2} & \frac{1}{4} & 0 & \frac{1}{4} & 16 \\ \hline 0 & 0 & 42 & 6 & 0 & 1 & 264 \end{array}\right] \end{array} \begin{array}{l} \frac{1}{4}R_1 \\ \frac{1}{4}R_1+R_2 \\ \frac{1}{4}R_1+R_3 \\ 6R_1+R_4 \end{array}$$

This time all entries in the last row are positive. Thus, the maximum value is z = 264 and occurs when $x_1 = 16$, $x_2 = 4$, $x_3 = 0$, $x_4 = 0$, $x_5 = 16$, and $x_6 = 0$.

7. Let x_3 and x_4 be the two slack variables. The constraints are now

$$x_1 + 2x_2 + x_3 = 8$$
$$4x_1 + x_2 + x_4 = 16.$$

The initial simplex tableau is

$$\begin{array}{c} \begin{array}{cccc} x_1 & x_2 & x_3 & x_4 \end{array} \\ \left[\begin{array}{cccc|c} 1 & 2 & 1 & 0 & 8 \\ 4 & 1 & 0 & 1 & 16 \\ \hline -2 & -5 & 0 & 0 & 0 \end{array}\right] \end{array}.$$

Pivot on the 2.

$$\begin{array}{c} \begin{array}{cccc} x_1 & x_2 & x_3 & x_4 \end{array} \\ \left[\begin{array}{cccc|c} \frac{1}{2} & 1 & \frac{1}{2} & 0 & 4 \\ \frac{7}{2} & 0 & -\frac{1}{2} & 1 & 12 \\ \hline \frac{1}{2} & 0 & \frac{5}{2} & 0 & 20 \end{array}\right] \end{array} \begin{array}{l} \frac{1}{2}R_1 \\ R_2 - \frac{1}{2}R_1 \\ \frac{5}{2}R_1 + R_3 \end{array}$$

The maximum value is 20 when $x_1 = 0$, $x_2 = 4$, $x_3 = 0$, $x_4 = 12$.

9. Introducing slack variables gives the problem:
Find $x_1 \ge 0$, $x_2 \ge 0$, $x_3 \ge 0$, $x_4 \ge 0$, $x_5 \ge 0$, so that

$$x_1 + 6x_2 + 8x_3 + x_4 = 118$$
$$x_1 + 5x_2 + 10x_3 + x_5 = 220$$

and $z = 8x_1 + 3x_2 + x_3$ is maximized.
The first tableau follows.

x_1	x_2	x_3	x_4	x_5	
(1)	6	8	1	0	118
1	5	10	0	1	220
-8	-3	-1	0	0	0

The most negative indicator is -8, in the first column. The quotients are 118/1 and 220/1. Since 118 is the smallest quotient, 1 in the first row is the pivot. Performing row transformations gives the second tableau.

x_1	x_2	x_3	x_4	x_5		
1	6	8	1	0	118	
0	-1	2	-1	1	102	$R_2 - R_1$
0	45	63	8	0	944	$8R_1 + R_3$

All numbers in the last row are now nonnegative. Thus, the maximum value is $z = 944$ and occurs when $x_1 = 188$, $x_2 = 0$, $x_3 = 0$, $x_4 = 0$ and $x_5 = 102$.

11. Introduce 2 slack variables x_4 and x_6 so that the constraints become

$$-2x_1 + 3x_2 + 3x_3 + x_4 = 100$$
$$3x_1 + 5x_2 + 10x_3 + x_5 = 150.$$

The initial simplex tableau is

x_1	x_2	x_3	x_4	x_5	
-2	3	3	1	0	100
3	5	10	0	1	150
4	-2	-1	0	0	0

Pivot on the 5.

x_1	x_2	x_3	x_4	x_5		
$-\frac{19}{5}$	0	-3	1	$-\frac{3}{5}$	10	$-\frac{3}{5}R_2 + R_1$
$\frac{3}{5}$	1	2	0	$\frac{1}{5}$	30	$\frac{1}{5}R_2$
$\frac{26}{5}$	0	3	0	$\frac{2}{5}$	60	$\frac{2}{5}R_2 + R_3$

The maximum value is 60 when $x_1 = 0$, $x_2 = 30$, $x_3 = 0$, $x_4 = 10$, and $x_5 = 0$.

13. Introduce 3 slack variables x_4, x_5, and x_6 so that the constraints become

$$x_1 + x_2 + x_3 + x_4 = 100$$
$$2x_1 + 3x_2 + 4x_3 + x_5 = 320$$
$$2x_1 + x_2 + x_3 + x_6 = 160.$$

The initial simplex tableau is

x_1	x_2	x_3	x_4	x_5	x_6	
1	1	1	1	0	0	100
2	3	4	0	1	0	320
2	1	1	0	0	1	160
-300	-200	-100	0	0	0	0

Pivot on the 2 in row 1, column 3.

x_1	x_2	x_3	x_4	x_5	x_6		
0	$\frac{1}{2}$	$\frac{1}{2}$	1	0	$-\frac{1}{2}$	20	$R_1 - \frac{1}{2}R_3$
0	2	3	0	1	-1	160	$R_2 - R_3$
1	$\frac{1}{2}$	$\frac{1}{2}$	0	0	$\frac{1}{2}$	80	$\frac{1}{2}R_3$
0	-50	50	0	0	150	24000	$150R_3 + R_4$

Since there is a negative value in the bottom row, pivot on the $\frac{1}{2}$ in row 1, column 2.

x_1	x_2	x_3	x_4	x_5	x_6		
0	1	1	2	0	-1	40	$2R_1$
0	0	1	-4	1	1	80	$-4R_1 + R_2$
1	0	0	-1	0	1	60	$R_3 - R_1$
0	0	100	100	0	100	26000	$100R_1 + R$

The maximum value is 26,000 when $x_1 = 60$, $x_2 = 40$, $x_3 = x_4 = 0$, $x_5 = 80$, and $x_6 = 0$.

15. Introducing x_5 and x_6 as slack variables gives the first simplex tableau

$$\begin{array}{cccccc|c} x_1 & x_2 & x_3 & x_4 & x_5 & x_6 & \\ 1 & 2 & 1 & 1 & 1 & 0 & 50 \\ 3 & 1 & 2 & 1 & 0 & 1 & 100 \\ \hline -1 & -2 & -1 & -5 & 0 & 0 & 0 \end{array}$$

The pivot element is the 1 in the first row fourth column. The next simplex tableau is

$$\begin{array}{cccccc|cl} x_1 & x_2 & x_3 & x_4 & x_5 & x_6 & & \\ 1 & 2 & 1 & 1 & 1 & 0 & 50 & \\ 2 & -1 & 1 & 0 & -1 & 1 & 50 & -R_1+R_2 \\ \hline 2 & 8 & 4 & 0 & 5 & 0 & 250 & 5R_1+R_3 \end{array}$$

The maximum is 250 when $x_4 = 50$, $x_6 = 50$, $x_1 = x_2 = x_3 = x_5 = 0$.

17. Put the information into a table.

	Church group	Labor union	Maximum time available
Letter writing	2	2	16
Follow-up	1	3	12
Money raised	\$100	\$200	

Let x_1 and x_2 be the number of church groups and labor unions contacted respectively. 2 slack variables are needed. Maximize $z = 100x_1 + 200x_2$ subject to:

$$2x_1 + 2x_2 + x_3 = 16$$
$$x_1 + 3x_2 + x_4 = 12$$
$$x_1 \ge 0,\ x_2 \ge 0,\ x_3 \ge 0,\ x_4 \ge 0$$

The initial tableau is

$$\begin{array}{cccc|c} x_1 & x_2 & x_3 & x_4 & \\ 2 & 2 & 1 & 0 & 16 \\ 1 & \textcircled{3} & 0 & 1 & 12 \\ \hline -100 & -200 & 0 & 0 & 0 \end{array}.$$

Use x_2 and pivot on 3.

$$\begin{array}{cccc|c} x_1 & x_2 & x_3 & x_4 & \\ \frac{4}{3} & 0 & 1 & -\frac{2}{3} & 8 \\ \frac{1}{3} & 1 & 0 & \frac{1}{3} & 4 \\ \hline -\frac{100}{3} & 0 & 0 & \frac{200}{3} & 800 \end{array}$$

Next, use x_1 and pivot on 4/3.

$$\begin{array}{cccc|c} x_1 & x_2 & x_3 & x_4 & \\ 1 & 0 & \frac{3}{4} & -\frac{1}{2} & 6 \\ 0 & 1 & -\frac{1}{4} & \frac{1}{2} & 2 \\ \hline 0 & 0 & 25 & 50 & 1000 \end{array}$$

There are no negative indicators, so this is a final tableau. She should contact 6 churches and 2 labor unions to raise a maximum of \$1000 per month.

19. First, put the information into a table.

	Recording	Mixing	Editing	Income
Jazz	6	12	2	\$6
Blues	6	6	4	\$8
Reggae	3	6	1	\$6
Amount available	69	78	40	

Let x_1 = the number of jazz discs
x_2 = the number of blues discs
x_3 = the number of reggae discs.

Maximize $6x_1 + 8x_2 + 6x_3$
subject to:

$$6x_1 + 6x_2 + 3x_3 \le 69$$
$$12x_1 + 6x_2 + 6x_3 \le 78$$
$$2x_1 + 4x_2 + x_3 \le 40$$
$$x_1 \ge 0,\ x_2 \ge 0,\ x_3 \ge 0.$$

The initial simplex tableau is

$$\begin{array}{cccccc|c} x_1 & x_2 & x_3 & x_4 & x_5 & x_6 & \\ 6 & 6 & 3 & 1 & 0 & 0 & 69 \\ 12 & 6 & 6 & 0 & 1 & 0 & 78 \\ 2 & 4 & 1 & 0 & 0 & 1 & 40 \\ \hline -6 & -8 & -6 & 0 & 0 & 0 & 0 \end{array}.$$

Pivot on the 4 in row 3, column 2.

$$\begin{array}{cccccc|c|l} x_1 & x_2 & x_3 & x_4 & x_5 & x_6 & & \\ -3 & 0 & -\frac{3}{2} & -1 & 0 & \frac{3}{2} & -9 & \frac{3}{2}R_3-R_1 \\ -9 & 0 & -\frac{9}{2} & 0 & -1 & \frac{3}{2} & -18 & \frac{3}{2}R_3-R_2 \\ \frac{1}{2} & 1 & \frac{1}{4} & 0 & 0 & \frac{1}{4} & 10 & R_3 \\ \hline -2 & 0 & -4 & 0 & 0 & 2 & 80 & 2R_3+R_4 \end{array}$$

Since there are still negative values in the bottom row, pivot on the -9/2 in row 2, column 3.

$$\begin{array}{cccccc|c|l} x_1 & x_2 & x_3 & x_4 & x_5 & x_6 & & \\ 0 & 0 & 0 & 3 & -1 & -3 & 9 & -3R_1+R_2 \\ 2 & 0 & 1 & 0 & \frac{2}{9} & -\frac{1}{3} & 4 & -\frac{2}{9}R_2 \\ 0 & 1 & 0 & 0 & -\frac{1}{18} & \frac{1}{3} & 9 & -\frac{1}{18}R_2+R_3 \\ \hline 6 & 0 & 0 & 0 & \frac{8}{9} & \frac{2}{3} & 96 & -\frac{8}{9}R_2+R_4 \end{array}$$

The maximum is $96.00 when $x_2 = 9$, $x_3 = 4$, $x_4 = 3$, and $x_1 = x_5 = x_6 = 0$. (That is, no jazz ablums, 9 blues albums, and 4 reggae albums).

21. First put the information into a table.

	Plastic	Metal	Paint	Profit
Weights	8	3	2	$3
Plaques	4	1	1	$4
Ornaments	2	2	1	$3
Amount Available	36	24	30	

Let x_1 = the number of weights
x_2 = the number of plaques
x_3 = the number of ornaments.

The problem is to maximize
$3x_1 + 4x_2 + 3x_3$ subject to:

$$8x_1 + 4x_2 + 2x_3 \le 36$$
$$3x_1 + x_2 + 2x_3 \le 24$$
$$2x_1 + x_2 + x_3 \le 30,$$
$$x_1 \ge 0,\ x_2 \ge 0,\ x_3 \ge 0.$$

The initial simplex tableau is

$$\begin{array}{cccccc|c} x_1 & x_2 & x_3 & x_4 & x_5 & x_6 & \\ 8 & 4 & 2 & 1 & 0 & 0 & 36 \\ 3 & 1 & 2 & 0 & 1 & 0 & 24 \\ 2 & 1 & 1 & 0 & 0 & 1 & 30 \\ \hline -3 & -4 & -3 & 0 & 0 & 0 & 0 \end{array}$$

Now pivot on the 4 in row 1, column 2.

$$\begin{array}{cccccc|c|l} x_1 & x_2 & x_3 & x_4 & x_5 & x_6 & & \\ 2 & 1 & \frac{1}{2} & \frac{1}{4} & 0 & 0 & 9 & \frac{1}{4}R_1 \\ -1 & 0 & -\frac{3}{2} & \frac{1}{4} & -1 & 0 & -15 & \frac{1}{4}R_1 - R_2 \\ 1 & 0 & 1 & 0 & 1 & -1 & -6 & R_2 - R_3 \\ \hline 5 & 0 & -1 & 1 & 0 & 0 & 36 & R_1 + R_4 \end{array}$$

Since there is a negative indicator in the bottom row, pivot on the -3/2 in row 2, column 3.

$$\begin{array}{cccccc|c|l} x_1 & x_2 & x_3 & x_4 & x_5 & x_6 & & \\ \frac{5}{3} & 1 & 0 & \frac{1}{3} & -\frac{1}{3} & 0 & 4 & \frac{1}{3}R_2 + R_1 \\ \frac{2}{3} & 0 & 1 & -\frac{1}{6} & \frac{2}{3} & 0 & 10 & -\frac{2}{3}R_2 \\ -\frac{1}{3} & 0 & 0 & -\frac{1}{6} & -\frac{1}{3} & 1 & 16 & -\frac{2}{3}R_2 - R_3 \\ \hline \frac{17}{3} & 0 & 0 & \frac{5}{6} & \frac{2}{3} & 0 & 46 & -\frac{2}{3}R_2 + R_4 \end{array}$$

There is a maximum of \$46.00 with $x_2 = 4$, $x_3 = 10$, $x_6 = 16$, and $x_1 = x_4 = x_5 = 0$ (that is, no paper weights, 4 plaques, and 10 ornanments).

23. The information is contained in the table.

	Aluminum	Steel	Profit
1 speed	12	20	\$8
3 speed	21	30	\$12
10 speed	16	40	\$24
Amount Available	42,000	91,800	

Let x_1 = number of 1-speed bikes,

x_2 = number of 3-speed bikes, and

x_3 = number of 10-speed bikes.

The problem is to maximize $8x_1 + 12x_2 + 24x_3$ subject to:

$12x_1 + 21x_2 + 16x_3 \le 42000$

$20x_1 + 30x_2 + 40x_3 \le 91800$

$x_1 \ge 0,\ x_2 \ge 0,\ x_3 \ge 0.$

The initial simplex is

$$\begin{array}{ccccc|c} x_1 & x_2 & x_3 & x_4 & x_5 & \\ 12 & 21 & 16 & 1 & 0 & 42000 \\ 20 & 30 & 40 & 0 & 1 & 91800 \\ \hline -8 & -12 & -24 & 0 & 0 & 0 \end{array}$$

Pivot on the 40 in row 2, column 3.

$$\begin{array}{ccccc|c|l} x_1 & x_2 & x_3 & x_4 & x_5 & & \\ 4 & 9 & 0 & 1 & -\frac{2}{5} & 5280 & -\frac{2}{5}R_2 + R_1 \\ \frac{1}{2} & \frac{3}{4} & 1 & 0 & \frac{1}{40} & 2295 & \frac{1}{40}R_2 \\ \hline 4 & 6 & 0 & 0 & \frac{3}{5} & 55080 & \frac{3}{5}R_2 + R_3 \end{array}$$

Maximum profit of \$55,080 with $x_3 = 2295$, $x_4 = 5280$, and $x_1 = x_2 = x_5 = 0$ (no 1- or 3-speed bikes, 2295 10-speed bikes)

25. Using the data given, the problem should be phrased

maximize $5x_1 + 4x_2 + 3x_3$

subject to:

$2x_1 + 3x_2 + x_3 \le 400$

$4x_1 + 2x_2 + 3x_3 \le 600$

$x_1 \ge 0,\ x_2 \ge 0,\ x_3 \ge 0,$

where x_1 = number of type A lamps,

x_2 = number of type B lamps, and

x_3 = number of type C lamps.

(a) (3) 5, 4, 3

(b) (4) 400, 600

(c) (3) $2X_1 + 3X_2 + 1X_3 \le 400$

27. Let x_1 = the number of toy trucks, and

x_2 = the number of toy fire trucks.

The constraints are

$x_1 \ge \frac{3}{2}x_2$ or $-2x_1 + 3x_2 \le 0$

$x_1 \le 6700$

$x_2 \le 5500$

$x_1 + x_2 \le 12000.$

The objective function is to maximize $8.5x_1 + 12.10x_2$.

The initial simplex tableau is

$$\begin{array}{c} \begin{array}{cccccc} x_1 & x_2 & x_3 & x_4 & x_5 & x_6 \end{array} \\ \left[\begin{array}{cccccc|c} -2 & 3 & 1 & 0 & 0 & 0 & 0 \\ 1 & 0 & 0 & 1 & 0 & 0 & 6700 \\ 0 & 1 & 0 & 0 & 1 & 0 & 5500 \\ 1 & 1 & 0 & 0 & 0 & 1 & 12{,}000 \\ \hline -8.5 & -12.10 & 0 & 0 & 0 & 0 & 0 \end{array}\right]. \end{array}$$

First, pivot on the 1 in column 1, row 2.

$$\begin{array}{c} \begin{array}{cccccc} x_1 & x_2 & x_3 & x_4 & x_5 & x_6 \end{array} \\ \left[\begin{array}{cccccc|c} 0 & 3 & 1 & 2 & 0 & 0 & 13{,}400 \\ 1 & 0 & 0 & 1 & 0 & 0 & 6700 \\ 0 & 1 & 0 & 0 & 1 & 0 & 5500 \\ 0 & 1 & 0 & -1 & 0 & 1 & 5500 \\ \hline 0 & -12 & 0 & 8.5 & 0 & 0 & 56{,}950 \end{array}\right] \begin{array}{l} 2R_2 + R_1 \\ \\ \\ R_4 - R_2 \\ 8.5R_2 + R_5 \end{array} \end{array}$$

Now, pivot on the 3 in row 1, column 2.

$$\begin{array}{c} \begin{array}{cccccc} x_1 & x_2 & x_3 & x_4 & x_5 & x_6 \end{array} \\ \left[\begin{array}{cccccc|c} 0 & 1 & \frac{1}{3} & \frac{2}{3} & 0 & 0 & 4466.7 \\ 1 & 0 & 0 & 1 & 0 & 0 & 6700 \\ 0 & 1 & 0 & 0 & 1 & 0 & 5500 \\ 0 & 1 & 0 & -1 & 0 & 1 & 5500 \\ \hline 0 & -12.1 & 0 & 8.5 & 0 & 0 & 56{,}950 \end{array}\right] \begin{array}{l} \frac{1}{3}R_1 \\ \\ \\ \\ \\ \end{array} \end{array}$$

$$\begin{array}{c} \begin{array}{cccccc} x_1 & x_2 & x_3 & x_4 & x_5 & x_6 \end{array} \\ \left[\begin{array}{cccccc|c} 0 & 1 & \frac{1}{3} & \frac{2}{3} & 0 & 0 & 4466.7 \\ 1 & 0 & 0 & 1 & 0 & 0 & 6700 \\ 0 & 0 & -\frac{1}{3} & -\frac{2}{3} & 1 & 0 & 1033.3 \\ 0 & 0 & -\frac{1}{3} & -\frac{5}{3} & 0 & 1 & 1033.3 \\ \hline 0 & 0 & 4.03 & 16.56 & 0 & 0 & 110{,}997.07 \end{array}\right] \begin{array}{l} \\ \\ R_3 - R_1 \\ R_4 - R_1 \\ \\ \end{array} \end{array}$$

Maximum of \$110,997 with 6700 toy trucks and 4467 toy fire trucks

Section 7.6

1. $2x_1 + 3x_2 \le 8$
   $x_1 + 4x_2 \ge 7$

   $2x_1 + 3x_2 + x_3 = 8$
   $x_1 + 4x_2 - x_4 = 7$

3. $x_1 + x_2 + x_3 \le 100$
   $x_1 + x_2 + x_3 \ge 75$
   $x_1 + x_2 \ge 27$

   $x_1 + x_2 + x_3 + x_4 = 100$
   $x_1 + x_2 + x_3 - x_5 = 75$
   $x_1 + x_2 - x_6 = 27$

5. Minimize $w = 4x_1 + 3x_2 + 2x_3$
   subject to: $x_1 + x_2 + x_3 \ge 5$
   $x_1 + x_2 \ge 4$
   $2x_1 + x_2 + 3x_3 \ge 15$
   with $x_1 \ge 0$, $x_2 \ge 0$, $x_3 \ge 0$.
   Change this to a maximization problem be letting $z = -w$. The problem can now be stated as follows:
   Maximize $z = -4x_1 - 3x_2 - 2x_3$
   subject to: $x_1 + x_2 + x_3 \ge 5$
   $x_1 + x_2 \ge 4$
   $2x_1 + x_2 + 3x_2 \ge 15$
   with $x_1 \ge 0$, $x_2 \ge 0$, $x_3 \ge 0$.

7. Minimize $w = x_1 + 2x_2 + x_3 + 5x_4$
   subject to: $x_1 + x_2 + x_3 + x_4 \ge 50$
   $3x_1 + x_2 + 2x_3 + x_4 \ge 100$
   with $x_1 \ge 0$, $x_2 \ge 0$, $x_3 \ge 0$, $x_4 \ge 0$.
   Change this to a maximization problem by letting $z = -w$. The problem can now be stated as follows
   Maximize $z = -x_1 - 2x_2 - x_3 - 5x_4$
   subject to: $x_1 + x_2 + x_3 + x_4 \ge 50$
   $3x_1 + x_2 + 2x_3 + x_4 \ge 100$
   with $x_1 \ge 0$, $x_2 \ge 0$, $x_3 \ge 0$, $x_4 \ge 0$.

9. Subtracting the surplus variable x_3 and adding the slack variable x_4 leads to the equations

$$x_1 + 2x_2 - x_3 = 24$$
$$x_1 + x_2 + x_4 = 40.$$

The initial tableau follows.

$$\begin{array}{c} \begin{matrix} x_1 & x_2 & x_3 & x_4 \end{matrix} \\ \left[\begin{array}{cccc|c} \textcircled{1} & 2 & -1 & 0 & 24 \\ 1 & 1 & 0 & 1 & 40 \\ \hline -12 & -10 & 0 & 0 & 0 \end{array}\right] \end{array}$$

Since the basic solution (0, 0, -24, 40) is not feasible, row transformations must be used. Pivot on the 1 in the first row, first column. After row transformations, we obtain the second tableau.

$$\begin{array}{c} \begin{matrix} x_1 & x_2 & x_3 & x_4 \end{matrix} \\ \left[\begin{array}{cccc|c} 1 & 2 & -1 & 0 & 24 \\ 0 & -1 & \textcircled{1} & 1 & 16 \\ \hline 0 & 14 & -12 & 0 & 288 \end{array}\right] \end{array}$$

The basic solution is now (24, 0, 0, 16), which is feasible. Continue now in the usual way. The 1 in the third column is the next pivot. After row transformations, get the following tableau.

$$\begin{array}{c} \begin{matrix} x_1 & x_2 & x_3 & x_4 \end{matrix} \\ \left[\begin{array}{cccc|c} 1 & 1 & 0 & 1 & 40 \\ 0 & -1 & 1 & 1 & 16 \\ \hline 0 & 2 & 0 & 12 & 480 \end{array}\right] \end{array}$$

The entries in the last row are all positive. Thus, the maximum value is 480 and occurs when $x_1 = 40$, $x_2 = 0$, $x_3 = 16$, and $x_4 = 0$.

11. The initial tableau for this maximizing problem is

$$\begin{array}{c} \begin{matrix} x_1 & x_2 & x_3 & x_4 & x_5 \end{matrix} \\ \left[\begin{array}{ccccc|c} 1 & 1 & 1 & 1 & 0 & 150 \\ 1 & \textcircled{1} & 1 & 0 & -1 & 100 \\ \hline -2 & -5 & -3 & 0 & 0 & 0 \end{array}\right]. \end{array}$$

x_4 is a slack variable, while x_5 is a surplus variable; $x_5 = -100$ is not feasible.

Phase I: Use the 1 in the second row of x_2 column to clear column two.

$$\begin{array}{c} \begin{matrix} x_1 & x_2 & x_3 & x_4 & x_5 \end{matrix} \\ \left[\begin{array}{ccccc|c} 0 & 0 & 0 & 1 & \textcircled{1} & 50 \\ 1 & 1 & 1 & 0 & -1 & 100 \\ \hline 3 & 0 & 2 & 0 & -5 & 500 \end{array}\right] \begin{array}{l} -R_2+R_1 \\ \\ 5R_2+R_3 \end{array} \end{array}$$

Since this is a feasible solution, we move to Phase II: the only possible pivot in the fifth column is the 1.

$$\begin{array}{c} \begin{matrix} x_1 & x_2 & x_3 & x_4 & x_5 \end{matrix} \\ \left[\begin{array}{ccccc|c} 0 & 0 & 0 & 1 & 1 & 50 \\ 1 & 1 & 1 & 1 & 0 & 150 \\ \hline 3 & 0 & 2 & 5 & 0 & 750 \end{array}\right] \end{array}$$

This is a final tableau. The maximum value is 750 when $x_1 = 0$, $x_2 = 150$, $x_3 = 0$, $x_4 = 0$, and $x_5 = 50$.

13. The initial tableau is

$$\begin{array}{c} \begin{matrix} x_1 & x_2 & x_3 & x_4 & x_5 \end{matrix} \\ \left[\begin{array}{ccccc|c} 1 & 1 & 1 & 0 & 0 & 100 \\ 1 & \textcircled{1} & 0 & -1 & 0 & 50 \\ 2 & 1 & 0 & 0 & 1 & 110 \\ \hline 2 & -3 & 0 & 0 & 0 & 0 \end{array}\right]. \end{array}$$

This give $x_3 = 100$, $x_4 = -50$, $x_5 = 110$, which is not feasible. Pivot on the circled 1 to obtain

$$\begin{array}{c} \begin{matrix} x_1 & x_2 & x_3 & x_4 & x_5 \end{matrix} \\ \left[\begin{array}{ccccc|c} 0 & 0 & 1 & \textcircled{1} & 0 & 50 \\ 1 & 1 & 0 & -1 & 0 & 50 \\ 1 & 0 & 0 & 1 & 1 & 60 \\ \hline 5 & 0 & 0 & -3 & 0 & 150 \end{array}\right]. \end{array}$$

Since this is a feasible solution, move to Phase II: pivot on 1 circled in the preceeding tableau.

$$\begin{array}{ccccc|c} x_1 & x_2 & x_3 & x_4 & x_5 & \\ 0 & 0 & 1 & 1 & 0 & 50 \\ 1 & 1 & 1 & 0 & 0 & 100 \\ 1 & 0 & -1 & 0 & 1 & 10 \\ \hline 5 & 0 & 3 & 0 & 0 & 300 \end{array}$$

The maximum is 300 when $x_1 = 0$, $x_2 = 100$, $x_3 = 50$, $x_4 = 50$, and $x_5 = 10$.

15. The initial tableau follows.

$$\begin{array}{cccc|c} y_1 & y_2 & y_3 & y_4 & \\ 10 & 5 & -1 & 0 & 100 \\ 20 & \textcircled{10} & 0 & -1 & 150 \\ \hline 4 & 5 & 0 & 0 & 0 \end{array}$$

First pivot on the circled 10 and, after row tranformations, obtain the next tableau.

$$\begin{array}{cccc|c} y_1 & y_2 & y_3 & y_4 & \\ 0 & 0 & -1 & \textcircled{.5} & 25 \\ 2 & 1 & 0 & -.1 & 15 \\ \hline -6 & 0 & 0 & .5 & -75 \end{array}$$

The basic solution (0, 15, -25, 0) is still not feasible. So, pivot on the circled .5 to get the next tableau.

$$\begin{array}{cccc|c} y_1 & y_2 & y_3 & y_4 & \\ 0 & 0 & -2 & 1 & 50 \\ \textcircled{2} & 1 & -.2 & 0 & 20 \\ \hline -6 & 0 & 1 & 0 & 100 \end{array}$$

Now, the basic solution (0, 20, 0, 50) is feasible, so continue in the usual way and pivot on the circled 2 to get the following tableau.

$$\begin{array}{cccc|c} y_1 & y_2 & y_3 & y_4 & \\ 0 & 0 & -2 & 1 & 50 \\ 1 & .5 & -.1 & 0 & 10 \\ \hline 0 & 3 & .4 & 0 & -40 \end{array}$$

The entries in the last row are all positive. Thus, the maximum value of $-w = z$ is -40. Hence, the minimum value of w is 40 when $y_1 = 10$, $y_2 = 0$, $y_3 = 0$, and $y_4 = 50$.

17. This is a standard minimizing problem. Change it to the following.
Maximize $z = 2y_1 - y_2 - 3y_3$
subject to:

$$\begin{aligned} y_1 + y_2 + y_3 &\ge 100 \\ 2y_1 + y_2 &\ge 50 \end{aligned}$$

$y_1 \ge 0$, $y_2 \ge 0$, $y_3 \ge 0$.

The initial tableau is

$$\begin{array}{ccccc|c} y_1 & y_2 & y_3 & y_4 & y_5 & \\ \textcircled{1} & 1 & 1 & -1 & 0 & 100 \\ 2 & 1 & 0 & 0 & -1 & 50 \\ \hline 2 & 1 & 3 & 0 & 0 & 0 \end{array}$$

with y_4 and y_5 as surplus variables. It does not have a feasible solution. Using the circled pivot, the next tableau is

$$\begin{array}{ccccc|c} y_1 & y_2 & y_3 & y_4 & y_5 & \\ 1 & \textcircled{1} & 1 & -1 & 0 & 100 \\ 0 & -1 & -2 & 2 & -1 & -150 \\ \hline 0 & -1 & 1 & 2 & 0 & -200 \end{array}.$$

This tableau has a feasible solution, $y_1 = 100$, $y_2 = y_3 = y_4 = 0$, and $y_6 = 150$. Move to Phase II.

$$\begin{array}{ccccc|c} y_1 & y_2 & y_3 & y_4 & y_5 & \\ 1 & 1 & 1 & -1 & 0 & 100 \\ 1 & 0 & -1 & 1 & -1 & -50 \\ \hline 1 & 0 & 2 & 1 & 0 & -100 \end{array}$$

This is a final tableau, with the maximum occurring when $y_2 = 100$, $y_5 = 50$, and $z = -w = -100$, so $w = 100$. Thus, the minimum value is 100 when $y_1 = 0$, $y_2 = 100$, $y_3 = 0$, $y_4 = 0$, $y_5 = 50$.

19. Let x_1 = number of barrels from S_1 to D_1

x_2 = number of barrels from S_2 to D_1

x_3 = number of barrels from S_1 to D_2

x_4 = number of barrels from S_2 to D_2.

$x_1 + x_2 \geq 3000$

$x_3 + x_4 \geq 5000$

$x_1 + x_3 \leq 5000$

$x_2 + x_4 \leq 5000$

$w = 30x_1 + 25x_2 + 20x_3 + 22x_4$.

The initial tableau is

$$\begin{array}{cccccccc|c} x_1 & x_2 & x_3 & x_4 & x_5 & x_6 & x_7 & x_8 & \\ \textcircled{1} & 1 & 0 & 0 & -1 & 0 & 0 & 0 & 3000 \\ 0 & 0 & 1 & 1 & 0 & -1 & 0 & 0 & 5000 \\ 1 & 0 & 1 & 0 & 0 & 0 & 1 & 0 & 5000 \\ 0 & 1 & 0 & 1 & 0 & 0 & 0 & 1 & 5000 \\ \hline 30 & 25 & 20 & 22 & 0 & 0 & 0 & 0 & 0 \end{array}.$$

This gives a nonfeasible solution. Pivot on the circled 1 to obtain

$$\begin{array}{cccccccc|c} x_1 & x_2 & x_3 & x_4 & x_5 & x_6 & x_7 & x_8 & \\ 1 & 1 & 0 & 0 & -1 & 0 & 0 & 0 & 3000 \\ 0 & 0 & 1 & \textcircled{1} & 0 & -1 & 0 & 0 & 5000 \\ 0 & -1 & 1 & 0 & 1 & 0 & 1 & 0 & 2000 \\ 0 & 1 & 0 & 1 & 0 & 0 & 0 & 1 & 5000 \\ \hline 0 & -5 & 20 & 22 & 30 & 0 & 0 & 1 & -90,000 \end{array}.$$

There is still no feasible solution. Pivot on the circled 1.

$$\begin{array}{cccccccc|c} x_1 & x_2 & x_3 & x_4 & x_5 & x_6 & x_7 & x_8 & \\ 1 & 1 & 0 & 0 & -1 & 0 & 0 & 0 & 3000 \\ 0 & 0 & \textcircled{1} & 1 & 0 & -1 & 0 & 0 & 5000 \\ 0 & -1 & 1 & 0 & 1 & 0 & 1 & 0 & 2000 \\ 0 & 1 & -1 & 0 & 0 & 1 & 0 & 1 & 0 \\ \hline 0 & -5 & -2 & 0 & 30 & 22 & 0 & 0 & -200,000 \end{array}$$

Pivot once more on the circled 1.

$$\begin{array}{cccccccc|c} x_1 & x_2 & x_3 & x_4 & x_5 & x_6 & x_7 & x_8 & \\ 1 & 1 & 0 & 0 & -1 & 0 & 0 & 0 & 3000 \\ 0 & 0 & 1 & 1 & 0 & -1 & 0 & 0 & 5000 \\ 0 & \textcircled{-1} & 0 & -1 & 1 & 1 & 1 & 0 & -3000 \\ 0 & 1 & 0 & 1 & 0 & 0 & 0 & 1 & 5000 \\ \hline 0 & -5 & 0 & 2 & 30 & 20 & 0 & 0 & -190,000 \end{array}$$

Pivot on the circled -1.

$$\begin{array}{cccccccc|c} x_1 & x_2 & x_3 & x_4 & x_5 & x_6 & x_7 & x_8 & \\ 1 & 0 & 0 & -1 & 0 & 1 & \textcircled{1} & 0 & 0 \\ 0 & 0 & 1 & 1 & 0 & -1 & 0 & 0 & 5000 \\ 0 & 1 & 0 & 1 & -1 & -1 & -1 & 0 & 3000 \\ 0 & 0 & 0 & 0 & 1 & 1 & 1 & 1 & 2000 \\ \hline 0 & 0 & 0 & 7 & 25 & 15 & -5 & 0 & -175,000 \end{array}$$

Pivot on the circled 1.

$$\begin{array}{cccccccc|c} x_1 & x_2 & x_3 & x_4 & x_5 & x_6 & x_7 & x_8 & \\ 1 & 0 & 0 & -1 & 0 & 1 & 1 & 0 & 0 \\ 0 & 0 & 1 & 1 & 0 & -1 & 0 & 0 & 5000 \\ 1 & 1 & 0 & 0 & -1 & 0 & 0 & 0 & 3000 \\ 1 & 0 & 0 & -1 & -1 & 0 & 0 & -1 & -2000 \\ \hline 5 & 0 & 0 & 2 & 25 & 20 & 0 & 0 & -175,000 \end{array}$$

This gives $x_2 = 3000$, $x_3 = 5000$, minimum cost $175,000. Ship 5000 barrels of oil from supplier 1 to distributor 2; ship 3000 barrels of oil from supplier 2 to distributor 1. The minimum cost is $175,000.

21. Let y_1 be the amount of commercial loans and y_2 be the amount of home loans.

$y_1 + y_2 \leq 25$

$y_2 \geq 4y_1$ or $4y_1 - y_2 \leq 0$

$y_1 + y_2 \geq 10$

Maximize $10y_1 + 12y_2$.

$$\begin{array}{ccccc|c} y_1 & y_2 & y_3 & y_4 & y_5 & \\ 1 & 1 & 1 & 0 & 0 & 25 \\ 4 & -1 & 0 & 1 & 0 & 0 \\ \textcircled{1} & 1 & 0 & 0 & -1 & 10 \\ \hline -10 & -12 & 0 & 0 & 0 & 0 \end{array}$$

This is not feasible; pivot on the circled 1.

$$\begin{array}{ccccc|c} y_1 & y_2 & y_3 & y_4 & y_5 & \\ 0 & 0 & 1 & 0 & 1 & 15 \\ 0 & \textcircled{-5} & 0 & 1 & 4 & -40 \\ 1 & 1 & 0 & 0 & -1 & 10 \\ \hline 0 & -2 & 0 & 0 & -10 & 100 \end{array} \quad \begin{array}{l} R_1 - R_3 \\ -4R_3 + R_2 \\ \\ 10R_3 + R_4 \end{array}$$

Pivot on the -5.

$$\begin{array}{ccccc|c} y_1 & y_2 & y_3 & y_4 & y_5 & \\ 0 & 0 & 1 & 0 & 1 & 15 \\ 0 & \textcircled{1} & 0 & -\frac{1}{5} & -\frac{4}{5} & 8 \\ 1 & 1 & 0 & 0 & -1 & 10 \\ \hline 0 & -2 & 0 & 0 & -10 & 100 \end{array} \quad \begin{array}{l} \\ -\frac{1}{5}R_2 \\ \\ \\ \end{array}$$

$$\begin{array}{ccccc|c} y_1 & y_2 & y_3 & y_4 & y_5 & \\ 0 & 0 & 1 & 0 & 1 & 15 \\ 0 & 1 & 0 & -\frac{1}{5} & -\frac{4}{5} & 8 \\ 1 & 0 & 0 & \textcircled{\frac{1}{5}} & -\frac{1}{5} & 2 \\ \hline 0 & 0 & 0 & -\frac{2}{5} & -\frac{58}{5} & 116 \end{array} \quad \begin{array}{l} \\ \\ R_3 - R_2 \\ 2R_2 + R_4 \end{array}$$

Pivot on the 1/5.

$$\begin{array}{ccccc|c} y_1 & y_2 & y_3 & y_4 & y_5 & \\ 0 & 0 & 1 & 0 & 1 & 15 \\ 0 & 1 & 0 & -\frac{1}{5} & -\frac{4}{5} & 8 \\ 5 & 0 & 0 & \textcircled{1} & -1 & 10 \\ \hline 0 & 0 & 0 & -\frac{2}{5} & -\frac{58}{5} & 116 \end{array} \quad \begin{array}{l} \\ \\ 5R_3 \\ \\ \end{array}$$

$$\begin{array}{ccccc|c} y_1 & y_2 & y_3 & y_4 & y_5 & \\ 0 & 0 & 1 & 0 & \textcircled{1} & 15 \\ 1 & 1 & 0 & 0 & -1 & 10 \\ 5 & 0 & 0 & 1 & -1 & 10 \\ \hline 2 & 0 & 0 & 0 & -12 & 120 \end{array} \quad \begin{array}{l} \\ \frac{1}{5}R_3 + R_2 \\ \\ \frac{2}{5}R_3 + R_4 \end{array}$$

Pivot on the 1.

$$\begin{array}{ccccc|c} y_1 & y_2 & y_3 & y_4 & y_5 & \\ 0 & 0 & 1 & 0 & 1 & 15 \\ 1 & 1 & 1 & 0 & 0 & 25 \\ 5 & 0 & 1 & 1 & 0 & 25 \\ \hline 2 & 0 & 12 & 0 & 0 & 300 \end{array} \quad \begin{array}{l} \\ R_1 + R_2 \\ R_1 + R_3 \\ 12R_1 + R_4 \end{array}$$

This is optimal. The maximum return is .01(300,000,000) = \$3,000,000 when $y_1 = 0$ and $y_2 = \$25,000,000$; that is, when 25 million dollars in home loans and no commercial loans are made.

23. Let y_1 = number of kilograms of sauce and y_2 = number of kilograms of whole tomatoes. The original problem follows.

Minimize $w = 3.25y_1 + 4y_2$

subject to:

$$\begin{aligned} y_1 + y_2 &\le 3{,}000{,}000 \\ y_1 \quad &\ge 80{,}000 \\ y_2 &\ge 800{,}000 \end{aligned}$$

which is changed to the following maximizing problem.

Maximize $z = -w = -3.25y_1 - 4y_2$

subject to:

$$\begin{aligned} y_1 + y_2 &\le 3{,}000{,}000 \\ y_1 \quad &\ge 80{,}000 \\ y_2 &\ge 800{,}000 \end{aligned}$$

Now introduce 1 slack and 2 surplus variables, and set up the initial tableau.

$$\begin{array}{ccccc|c} y_1 & y_2 & y_3 & y_4 & y_5 & \\ 1 & 1 & 1 & 0 & 0 & 3{,}000{,}000 \\ \textcircled{1} & 0 & 0 & -1 & 0 & 80{,}000 \\ 0 & 1 & 0 & 0 & -1 & 800{,}000 \\ \hline 3.25 & 4 & 0 & 0 & 0 & 0 \end{array}$$

Since y_4 and y_5 are both negative, this does not have a feasible solution. Pivot on the 1 in the second row

$$\begin{array}{ccccc|c} y_1 & y_2 & y_3 & y_4 & y_5 & \\ 0 & 1 & 1 & 1 & 0 & 2{,}920{,}000 \\ 1 & 0 & 0 & -1 & 0 & 80{,}000 \\ 0 & \textcircled{1} & 0 & 0 & -1 & 800{,}000 \\ \hline 0 & 4 & 0 & 3.25 & 0 & -260{,}000 \end{array}$$

This still does not have a feasible solution, since $y_5 = -800{,}000$. Thus, pivot on the 1 in the third row.

$$\begin{array}{ccccc|c} y_1 & y_2 & y_3 & y_4 & y_5 & \\ \hline 0 & 0 & 1 & 1 & 1 & 2{,}120{,}000 \\ 1 & 0 & 0 & -1 & 0 & 80{,}000 \\ 0 & 1 & 0 & 0 & -1 & 800{,}000 \\ \hline 0 & 0 & 0 & 3.25 & 4 & -3{,}460{,}000 \end{array}$$

This has a feasible solution. Brand x should use 800,000 kilograms of tomatoes for whole tomatoes and 80,000 kilograms of tomatoes for sauce for a minimum cost of $3,460,000.

25. Let y_1 = number of small tubes and y_2 = number of large ones. The problem is to minimize $.15y_1 + .12y_2$ subject to:

$y_1 \geq 800$

$y_2 \geq 500$

$y_1 + y_2 \geq 1500$

$y_1 \geq 2y_2$ (or $y_1 - 2y_2 \geq 0$).

This becomes the following.

Maximize $z = -w = -.15y_1 + .12y_2$ subject to the same constraints.

Adding surplus variables y_3 through y_6, the initial tableau is

$$\begin{array}{cccccc|c} y_1 & y_2 & y_3 & y_4 & y_5 & y_6 & \\ \hline \textcircled{1} & 0 & -1 & 0 & 0 & 0 & 800 \\ 0 & 1 & 0 & -1 & 0 & 0 & 500 \\ 1 & 1 & 0 & 0 & -1 & 0 & 1500 \\ 1 & -2 & 0 & 0 & 0 & -1 & 0 \\ \hline .15 & .12 & 0 & 0 & 0 & 0 & 0 \end{array}.$$

Since y_3 through y_6 are negative, this does not have a feasible solution. For Phase I, use the 1 in the first row, first column for a pivot:

$$\begin{array}{cccccc|cl} y_1 & y_2 & y_3 & y_4 & y_5 & y_6 & & \\ \hline 1 & 0 & -1 & 0 & 0 & 0 & 800 & \\ 0 & \textcircled{1} & 0 & -1 & 0 & 0 & 500 & \\ 0 & 1 & 1 & 0 & -1 & 0 & 700 & -R_1 + R_3 \\ 0 & -2 & 1 & 0 & 0 & -1 & -800 & -R_1 + R_4 \\ \hline 0 & .12 & .15 & 0 & 0 & 0 & -120 & -.15R_1 + R_5 \end{array}$$

y_4 and y_5 are still negative, so use the 1 in row 2 to pivot again.

$$\begin{array}{cccccc|cl} y_1 & y_2 & y_3 & y_4 & y_5 & y_6 & & \\ \hline 1 & 0 & -1 & 0 & 0 & 0 & 800 & \\ 0 & 1 & 0 & -1 & 0 & 0 & 500 & \\ 0 & 0 & 1 & 1 & -1 & 0 & 200 & -R_2 + R_3 \\ 0 & 0 & \textcircled{1} & -2 & 0 & -1 & 200 & 2R_2 + R_4 \\ \hline 0 & 0 & .15 & .12 & 0 & 0 & -180 & -.12R_2 + R_5 \end{array}$$

Pivot on the 1 in the fourth row.

$$\begin{array}{cccccc|c} y_1 & y_2 & y_3 & y_4 & y_5 & y_6 & \\ \hline 1 & 0 & 0 & -2 & 0 & -1 & 1000 \\ 0 & 1 & 0 & -1 & 0 & 0 & 500 \\ 0 & 0 & 0 & 3 & -1 & 1 & 0 \\ 0 & 0 & 1 & -2 & 0 & -1 & 200 \\ \hline 0 & 0 & 0 & .42 & 0 & .15 & -210 \end{array}$$

1000 small and 500 large test tubes should be ordered for a minimum cost of $210.

27. Let x_1 = amount invested in securities

x_2 = amount in bonds

x_3 = amount in mutual funds.

Maximize $.07x_1 + .06x_2 + .1x_3$ subject to:

$x_1 + x_2 + x_3 \geq 100{,}000$

$x_1 \geq 40{,}000$

$x_2 + x_3 \geq 50{,}000$.

$$\begin{array}{cccccc|c} x_1 & x_2 & x_3 & x_4 & x_5 & x_6 & \\ \hline 1 & 1 & 1 & 1 & 0 & 0 & 100{,}000 \\ \textcircled{1} & 0 & 0 & 0 & -1 & 0 & 40{,}000 \\ 0 & 1 & 1 & 0 & 0 & -1 & 50{,}000 \\ \hline -.07 & -.06 & -.1 & 0 & 0 & 0 & 0 \end{array}$$

Pivot on the 1 in second row.

$$\begin{array}{cccccc|c} x_1 & x_2 & x_3 & x_4 & x_5 & x_6 & \\ 0 & 1 & 1 & 1 & 1 & 0 & 60{,}000 \\ 1 & 0 & 0 & 0 & -1 & 0 & 40{,}000 \\ 0 & \textcircled{1} & 1 & 0 & 0 & -1 & 50{,}000 \\ \hline 0 & -.06 & -.1 & 0 & -.07 & 0 & 2800 \end{array}$$

Pivot on 1 in the third row.

$$\begin{array}{cccccc|c} x_1 & x_2 & x_3 & x_4 & x_5 & x_6 & \\ 0 & 0 & 0 & 1 & \textcircled{1} & 1 & 10{,}000 \\ 1 & 0 & 0 & 0 & -1 & 0 & 40{,}000 \\ 0 & 1 & 1 & 0 & 0 & -1 & 50{,}000 \\ \hline 0 & 0 & -.04 & 0 & -.07 & -.06 & 5800 \end{array}$$

Pivot on the 1 in the first row.

$$\begin{array}{cccccc|c} x_1 & x_2 & x_3 & x_4 & x_5 & x_6 & \\ 0 & 0 & 0 & 1 & 1 & 1 & 10{,}000 \\ 1 & 0 & 0 & 1 & 0 & 1 & 50{,}000 \\ 0 & 1 & \textcircled{1} & 0 & 0 & -1 & 50{,}000 \\ \hline 0 & 0 & -.04 & -.07 & 0 & .01 & 6500 \end{array}$$

Pivot on 1 in third row.

$$\begin{array}{cccccc|c} x_1 & x_2 & x_3 & x_4 & x_5 & x_6 & \\ 0 & 0 & 0 & 1 & 1 & \textcircled{1} & 10{,}000 \\ 1 & 0 & 0 & 1 & 0 & 1 & 50{,}000 \\ 0 & 1 & 1 & 0 & 0 & -1 & 50{,}000 \\ \hline 0 & .04 & 0 & .07 & 0 & -.03 & 8500 \end{array}$$

Pivot on 1 in first row.

$$\begin{array}{cccccc|c} x_1 & x_2 & x_3 & x_4 & x_5 & x_6 & \\ 0 & 0 & 0 & 1 & 1 & 1 & 10{,}000 \\ 1 & 0 & 0 & 0 & -1 & 0 & 40{,}000 \\ 0 & 1 & 1 & 1 & 1 & 0 & 60{,}000 \\ \hline 0 & .04 & 0 & .1 & .03 & 0 & 8800 \end{array}$$

\$40,000 should be invested in government securities and \$60,000 in mutual funds for a maximum annual interest of \$8800.

29. Let y_1 be the number of computers from W_1 to D_1,
y_2 be the number from W_1 to D_2,
y_3 be the number from W_2 to D_1, and
y_4 be the number from W_2 to D_2.

Minimize $w = 14y_1 + 22y_2 + 12y_3 + 10y_4$
subject to:

$$\begin{aligned} y_1 + y_3 &\ge 32 \\ y_2 + y_4 &\ge 20 \\ y_1 + y_2 &\le 25 \\ y_3 + y_4 &\le 30. \end{aligned}$$

This problem becomes the following.
Maximize $z = -14y_1 - 22y_2 - 12y_3 - 10y_4$
subject to:

$$\begin{aligned} y_1 + y_3 - y_5 &= 32 \\ y_2 + y_4 - y_6 &= 20 \\ y_1 + y_2 + y_7 &= 25 \\ y_3 + y_4 + y_8 &= 30. \end{aligned}$$

$$\begin{array}{cccccccc|c} y_1 & y_2 & y_3 & y_4 & y_5 & y_6 & y_7 & y_8 & \\ 1 & 0 & 1 & 0 & -1 & 0 & 0 & 0 & 32 \\ 0 & 1 & 0 & 1 & 0 & -1 & 0 & 0 & 20 \\ \textcircled{1} & 1 & 0 & 0 & 0 & 0 & 1 & 0 & 25 \\ 0 & 0 & 1 & 1 & 0 & 0 & 0 & 1 & 30 \\ \hline 14 & 22 & 12 & 10 & 0 & 0 & 0 & 0 & 0 \end{array}$$

$y_5 = -32$ and $y_6 - -20$ are not feasible.
Pivot on the 1 in row 3, column 1.

$$\begin{array}{cccccccc|cl} y_1 & y_2 & y_3 & y_4 & y_5 & y_6 & y_7 & y_8 & & \\ 0 & -1 & \textcircled{1} & 0 & -1 & 0 & -1 & 0 & 7 & -R_3 + R_1 \\ 0 & 1 & 0 & 1 & 0 & -1 & 0 & 0 & 20 & \\ 1 & 1 & 0 & 0 & 0 & 0 & 1 & 0 & 25 & \\ 0 & 0 & 1 & 1 & 0 & 0 & 0 & 1 & 30 & \\ \hline 0 & 8 & 12 & 10 & 0 & 0 & -14 & 0 & -350 & -14R_3 + R_5 \end{array}$$

$y_5 = -7$ and $y_6 = -20$ are not feasible.
Pivot on the 1 in row 1, in column 3.

$$\begin{array}{cccccccc|cl} y_1 & y_2 & y_3 & y_4 & y_5 & y_6 & y_7 & y_8 & & \\ 0 & -1 & 1 & 0 & -1 & 0 & -1 & 0 & 7 & \\ 0 & \textcircled{1} & 0 & 1 & 0 & -1 & 0 & 0 & 20 & \\ 1 & 1 & 0 & 0 & 0 & 0 & 1 & 0 & 25 & \\ 0 & 1 & 0 & 1 & 1 & 0 & 1 & 1 & 23 & -R_1 + R_4 \\ \hline 0 & 20 & 0 & 10 & 12 & 0 & -2 & 0 & -434 & -12R_1 + R_5 \end{array}$$

$y_6 = -20$ is not feasible. Pivot on the 1 in row 2, column 2.

$$\begin{array}{cccccccc} y_1 & y_2 & y_3 & y_4 & y_5 & y_6 & y_7 & y_8 \end{array}$$

$$\left[\begin{array}{cccccccc|c} 0 & 0 & 1 & 1 & -1 & -1 & -1 & 0 & 27 \\ 0 & 1 & 0 & \textcircled{1} & 0 & -1 & 0 & 0 & 20 \\ 1 & 0 & 0 & -1 & 0 & 1 & 1 & 0 & 5 \\ 0 & 0 & 0 & 0 & 1 & 1 & 1 & 1 & 4 \\ \hline 0 & 0 & 0 & -10 & 12 & 20 & -2 & 0 & -834 \end{array}\right] \begin{array}{l} R_2 + R_1 \\ \\ -R_2 + R_3 \\ -R_2 + R_4 \\ -20R_2 + R_5 \end{array}$$

↑ Most negative indicator

$$\begin{array}{cccccccc} y_1 & y_2 & y_3 & y_4 & y_5 & y_6 & y_7 & y_8 \end{array}$$

$$\left[\begin{array}{cccccccc|c} 0 & -1 & 1 & 0 & -1 & 0 & -1 & 0 & 7 \\ 0 & 1 & 0 & 1 & 0 & -1 & 0 & 0 & 20 \\ 1 & 1 & 0 & 0 & 0 & 0 & 1 & 0 & 25 \\ 0 & 0 & 0 & 0 & 1 & 1 & \textcircled{1} & 1 & 3 \\ \hline 0 & 10 & 0 & 0 & 12 & 10 & -2 & 0 & -634 \end{array}\right] \begin{array}{l} -R_2 + R_1 \\ \\ R_2 + R_3 \\ \\ 10R_2 + R_5 \end{array}$$

↑

Most negative indicator

$$\begin{array}{cccccccc} y_1 & y_2 & y_3 & y_4 & y_5 & y_6 & y_7 & y_8 \end{array}$$

$$\left[\begin{array}{cccccccc|c} 0 & -1 & 1 & 0 & 0 & 1 & 0 & 1 & 10 \\ 0 & 1 & 0 & 1 & 0 & -1 & 0 & 0 & 20 \\ 1 & 1 & 0 & 0 & -1 & -1 & 0 & -1 & 22 \\ 0 & 0 & 0 & 0 & 1 & 1 & 1 & 1 & 3 \\ \hline 0 & 10 & 0 & 0 & 14 & 12 & 0 & 2 & -628 \end{array}\right] \begin{array}{l} R_4 + R_1 \\ \\ -R_4 + R_3 \\ \\ 2R_4 + R_5 \end{array}$$

The minimum is 628 when $y_1 = 22$, $y_2 = 0$, $y_3 = 10$, $y_4 = 20$, $y_5 = 0$, $y_6 = 0$, $y_7 = 3$, and $y_8 = 0$. So, ship 22 computers from W_1 to D_1, 0 from W_1 to D_2, 10 from W_2 to D_1, and 20 from W_2 to D_2 at a minimum cost of \$628.

31. List the constraints where

y_1 = amount of ingredient I

y_2 = amount of ingredient II

y_3 = amount of ingredient III.

$y_1 + y_2 + y_3 \geq 10$

$y_1 + y_2 + y_3 \leq 15$

$y_1 \geq \frac{1}{4}y_2$ or $y_1 - \frac{1}{4}y_2 \geq 0$

$y_3 \geq y_1$ or $-y_1 + y_3 \geq 0$

$y_1 \geq 0$, $y_2 \geq 0$, $y_3 \geq 0$

Minimize $.30y_1 + .09y_2 + .27y_3$ is equivalent to maximizing $z = -w = -.30y_1 - .09y_2 - .27y_3$.

The initial simplex tableau follows.

$$\begin{array}{ccccccc} y_1 & y_2 & y_3 & y_4 & y_5 & y_6 & y_7 \end{array}$$

$$\left[\begin{array}{ccccccc|c} 1 & 1 & 1 & -1 & 0 & 0 & 0 & 10 \\ 1 & 1 & 1 & 0 & 1 & 0 & 0 & 15 \\ 1 & -\frac{1}{4} & 0 & 0 & 0 & -1 & 0 & 0 \\ -1 & 0 & 1 & 0 & 0 & 0 & -1 & 0 \\ \hline .30 & .09 & .27 & 0 & 0 & 0 & 0 & 0 \end{array}\right]$$

Since y_4 has a negative value, pivot on the 1 in row 1, column 1.

$$\begin{array}{ccccccc} y_1 & y_2 & y_3 & y_4 & y_5 & y_6 & y_7 \end{array}$$

$$\left[\begin{array}{ccccccc|c} 1 & 1 & 1 & -1 & 0 & 0 & 0 & 10 \\ 0 & 0 & 0 & 1 & 1 & 0 & 0 & 5 \\ 0 & -\frac{5}{4} & -1 & 1 & 0 & -1 & 0 & -10 \\ 0 & 1 & 2 & -1 & 0 & 0 & -1 & 10 \\ \hline 0 & -.21 & -.03 & .3 & 0 & 0 & 0 & -3 \end{array}\right] \begin{array}{l} \\ R_2 - R_1 \\ R_3 - R_1 \\ R_1 + R_4 \\ -.3R_1 + R_5 \end{array}$$

Since y_7 has a negative value, pivot on the -1 in row 4, column 4.

$$\begin{array}{ccccccc} y_1 & y_2 & y_3 & y_4 & y_5 & y_6 & y_7 \end{array}$$

$$\left[\begin{array}{ccccccc|c} 1 & 0 & -1 & 0 & 0 & 0 & 1 & 0 \\ 0 & 1 & 2 & 0 & 1 & 0 & -1 & 15 \\ 0 & -\frac{1}{4} & 1 & 0 & 0 & -1 & -1 & 0 \\ 0 & -1 & -2 & 1 & 0 & 0 & 1 & -10 \\ \hline 0 & .09 & .57 & 0 & 0 & 0 & -.3 & 0 \end{array}\right] \begin{array}{l} R_1 - R_4 \\ R_2 + R_4 \\ R_3 + R_4 \\ -R_4 \\ .3R_4 + R_5 \end{array}$$

Since y_4 has a negative value, pivot on -2 in column 3, row 4.

$$\begin{array}{ccccccc} y_1 & y_2 & y_3 & y_4 & y_5 & y_6 & y_7 \end{array}$$

$$\left[\begin{array}{ccccccc|c} 1 & \frac{1}{2} & 0 & -\frac{1}{2} & 0 & 0 & \frac{1}{2} & 5 \\ 0 & 0 & 0 & 1 & 1 & 0 & 0 & 5 \\ 0 & -\frac{3}{4} & 0 & \frac{1}{2} & 0 & -1 & -\frac{1}{2} & -5 \\ 0 & \frac{1}{2} & 1 & -\frac{1}{2} & 0 & 0 & -\frac{1}{2} & 5 \\ \hline 0 & -.195 & 0 & .285 & 0 & 0 & -.015 & -2.85 \end{array}\right] \begin{array}{l} -\frac{1}{2}R_4 + R_1 \\ R_2 + R_4 \\ \frac{1}{2}R_4 + R_3 \\ -\frac{1}{2}R_4 \\ \frac{.57}{2}R_4 + R_5 \end{array}$$

Now proceed as usual, pivoting on $-\frac{3}{4}$ in row 3, column 2.

$$\begin{array}{ccccccc|c|l}
y_1 & y_2 & y_3 & y_4 & y_5 & y_6 & y_7 & & \\
1 & 0 & 0 & -\frac{1}{6} & 0 & -\frac{2}{3} & -\frac{1}{6} & \frac{5}{3} & \frac{2}{3}R_3 + R_1 \\
0 & 0 & 0 & 1 & 1 & 0 & 0 & 5 & \\
0 & 1 & 0 & -\frac{2}{3} & 0 & -\frac{4}{3} & -\frac{2}{3} & \frac{20}{3} & -\frac{4}{3}R_3 \\
0 & 0 & 1 & -\frac{1}{6} & 0 & -\frac{2}{3} & -\frac{5}{6} & \frac{5}{3} & \frac{2}{3}R_3 + R_4 \\
\hline
0 & 0 & 0 & .155 & 0 & .26 & -.115 & -1.55 & \frac{4}{3}(.195)R_3 + R_5
\end{array}$$

There is a minimum cost of $1.55 per gallon using 5/3 ounces of ingredient I, 20/3 ounces of ingredient II, and 5/3 ounces of ingredient III.

Section 7.7

In Exercises 1 and 3, interchange the columns and rows.

1. The transpose of $\begin{bmatrix} 1 & 2 & 3 \\ 3 & 2 & 1 \\ 1 & 10 & 0 \end{bmatrix}$ is $\begin{bmatrix} 1 & 3 & 1 \\ 2 & 2 & 10 \\ 3 & 1 & 0 \end{bmatrix}$.

3. The transpose of $\begin{bmatrix} -1 & 4 & 6 & 12 \\ 13 & 25 & 0 & 4 \\ -2 & -1 & 11 & 3 \end{bmatrix}$ is $\begin{bmatrix} -1 & 13 & -2 \\ 4 & 25 & -1 \\ 6 & 0 & 11 \\ 12 & 4 & 3 \end{bmatrix}$.

5. Minimize $w = 2y_1 + 4y_2$

subject to: $3y_1 + y_2 \ge 4$

$y_1 + 2y_2 \ge 6$

$y_1 \ge 0,\ y_2 \ge 0$

Begin by writing the augmented matrix for the given problem.

$$\left[\begin{array}{cc|c} 3 & 1 & 4 \\ 1 & 2 & 6 \\ \hline 2 & 4 & 0 \end{array}\right]$$

Find the transpose of this matrix.

$$\left[\begin{array}{cc|c} 3 & 1 & 2 \\ 1 & 2 & 4 \\ \hline 4 & 6 & 0 \end{array}\right]$$

The dual problem is stated from this second matrix as follows (using x instead of y):

Maximize $z = 4x_1 + 6x_2$

subject to: $3x_1 + x_2 \le 2$

$x_1 + 2x_2 \le 4$

$x_1 \ge 0,\ x_2 \ge 0.$

7. Minimize $w = y_1 + 3y_2 + 4y_3$

subject to: $3y_1 + 4y_2 + 6y_3 \ge 8$

$y_1 + 5y_2 + 2y_3 \ge 12$

$y_1 \ge 0,\ y_2 \ge 0.$

Begin by writing the augmented matrix for the given problem.

$$\left[\begin{array}{ccc|c} 3 & 4 & 6 & 8 \\ 1 & 5 & 2 & 12 \\ \hline 1 & 3 & 4 & 0 \end{array}\right]$$

Find the transpose of this matrix

$$\left[\begin{array}{cc|c} 3 & 1 & 1 \\ 4 & 5 & 3 \\ 6 & 2 & 4 \\ \hline 8 & 12 & 0 \end{array}\right]$$

The dual problem is stated from this second matrix as follows (using x instead of y):

Maximize $z = 8x_1 + 12x_2$

subject to: $3x_1 + x_2 \le 1$

$4x_1 + 5x_2 \le 3$

$6x_1 + 2x_2 \le 4$

$x_1 \ge 0,\ x_2 \ge 0.$

9. Minimize $w = 3y_1 + 4y_2$

subject to: $y_1 + 7y_2 \geq 18$

$4y_1 + y_2 \geq 15$

$5y_1 + 3y_2 \geq 20$

$y_1 \geq 0,\ y_2 \geq 0.$

Begin by writing the augmented matrix for the given problem.

$$\left[\begin{array}{cc|c} 1 & 7 & 18 \\ 4 & 1 & 15 \\ 5 & 3 & 20 \\ \hline 3 & 4 & 0 \end{array}\right]$$

Find the transpose of this matrix

$$\left[\begin{array}{ccc|c} 1 & 4 & 5 & 3 \\ 7 & 1 & 3 & 4 \\ \hline 18 & 15 & 20 & 0 \end{array}\right]$$

The dual problem is stated from this second matrix as follows (using x instead of y):

Maximize $z = 18x_1 + 15x_2 + 20x_3$

subject to: $x_1 + 4x_2 + 5x_3 \leq 3$

$7x_1 + x_2 + 3x_3 \leq 4$

$x_1 \geq 0,\ x_2 \geq 0,\ x_3 \geq 0.$

11. Minimize $w = y_1 + y_2 + 4y_3$

subject to: $y_1 + 2y_2 + 3y_3 \geq 115$

$2y_1 + y_2 + 8y_3 \geq 200$

$y_1 + y_3 \geq 50$

$y_1 \geq 0,\ y_2 \geq 0,\ y_3 \geq 0.$

Write the augmented matrix for the problem.

$$\left[\begin{array}{ccc|c} 1 & 2 & 3 & 115 \\ 2 & 1 & 8 & 200 \\ 1 & 0 & 1 & 50 \\ \hline 1 & 1 & 4 & 0 \end{array}\right]$$

Find the transpose of this matrix

$$\left[\begin{array}{ccc|c} 1 & 2 & 1 & 1 \\ 2 & 1 & 0 & 1 \\ 3 & 8 & 1 & 4 \\ \hline 115 & 200 & 50 & 0 \end{array}\right]$$

The dual problem follows:

Maximize $z = 115x_1 + 200x_2 + 50x_3$

subject to: $x_1 + 2x_2 + x_3 \leq 1$

$2x_1 + x_2 \leq 1$

$3x_1 + 8x_2 + x_3 \leq 4$

$x_1 \geq 0,\ x_2 \geq 0,\ x_3 \geq 0.$

13. The problem in matrix form is

$$\left[\begin{array}{cc|c} 10 & 5 & 100 \\ 20 & 10 & 150 \\ \hline 4 & 5 & 0 \end{array}\right].$$

The transpose matrix is

$$\left[\begin{array}{cc|c} 10 & 20 & 4 \\ 5 & 10 & 5 \\ \hline 100 & 150 & 0 \end{array}\right].$$

The dual is to

Maximize $z = 100x_1 + 150x_2$

subject to: $10x_1 + 20x_2 \leq 4$

$5x_1 + 10x_2 \leq 5$

$x_1 \geq 0,\ x_2 \geq 0.$

The constraints with slack variables will be

$10x_1 + 20x_2 + x_3 = 4$

$5x_1 + 10x_2 + x_4 = 5.$

The initial simplex tableau is

$$\begin{array}{c} \begin{array}{cccc} x_1 & x_2 & x_3 & x_4 \end{array} \\ \left[\begin{array}{cccc|c} 10 & 20 & 1 & 0 & 4 \\ 5 & 10 & 0 & 1 & 5 \\ \hline -100 & -150 & 0 & 0 & 0 \end{array}\right]. \end{array}$$

Pivot on the 20 in row 1, column 2.

$$\begin{array}{c} \begin{array}{cccc} x_1 & x_2 & x_3 & x_4 \end{array} \\ \left[\begin{array}{cccc|c} \frac{1}{2} & 1 & \frac{1}{20} & 0 & \frac{1}{5} \\ 0 & 0 & -\frac{1}{2} & 1 & 3 \\ \hline -25 & 0 & \frac{15}{2} & 0 & 30 \end{array}\right] \begin{array}{l} \frac{1}{20}R_1 \\ -\frac{1}{2}R_1+R_2 \\ \frac{15}{2}R_1+R_3 \end{array} \end{array}$$

Now pivot on the $\frac{1}{2}$ in row 1, column 1.

$$\begin{array}{c} \begin{array}{cccc} x_1 & x_2 & x_3 & x_4 \end{array} \\ \left[\begin{array}{cccc|c} 1 & 2 & \frac{1}{10} & 0 & \frac{2}{5} \\ 0 & 0 & -\frac{1}{2} & 1 & 3 \\ \hline 0 & 50 & 10 & 0 & 40 \end{array}\right] \begin{array}{l} 2R_1 \\ \\ 50R_1 + R_3 \end{array} \end{array}$$

Minimum of 40 when $y_1 = 10$ and $y_2 = 0$

15. The problem in matrix form is

$$\left[\begin{array}{ccc|c} 1 & 1 & 1 & 100 \\ 2 & 1 & 0 & 50 \\ \hline 2 & 1 & 3 & 0 \end{array}\right].$$

The transpose matrix is

$$\left[\begin{array}{cc|c} 1 & 2 & 2 \\ 1 & 1 & 1 \\ 1 & 0 & 3 \\ \hline 100 & 50 & 0 \end{array}\right].$$

The dual problem follows:

Maximize $\quad 100x_1 + 50x_2$

subject to: $x_1 + 2x_2 \le 2$

$x_1 + x_2 \le 1$

$x_1 \le 3$

State the problem with slack variables.

Maximize $100x_1 + 50x_2$

subject to: $x_1 + 2x_2 + x_3 = 2$

$x_1 + x_2 + x_4 = 1$

$x_1 + x_5 = 3$

The first tableau is

$$\begin{array}{ccccc} x_1 & x_2 & x_3 & x_4 & x_5 \end{array}$$
$$\left[\begin{array}{ccccc|c} 1 & 2 & 1 & 0 & 0 & 2 \\ 1 & 1 & 0 & 1 & 0 & 1 \\ 1 & 0 & 0 & 0 & 1 & 3 \\ \hline -100 & -50 & 0 & 0 & 0 & 0 \end{array}\right].$$

Pivot element is 1.

$$\begin{array}{ccccc} x_1 & x_2 & x_3 & x_4 & x_5 \end{array}$$
$$\left[\begin{array}{ccccc|c} 0 & 1 & 1 & -1 & 0 & 1 \\ 1 & 1 & 0 & 1 & 0 & 1 \\ 0 & -1 & 0 & -1 & 1 & 2 \\ \hline 0 & 50 & 0 & 100 & 0 & 100 \end{array}\right]$$

The minimum value is 100 when $y_1 = y_3 = 0$ and $y_2 = 100$.

17. The matrix for the existing problem is $\left[\begin{array}{ccc|c} 2 & 1 & 1 & 8 \\ 1 & 2 & 4 & 12 \\ \hline 1 & 1 & 3 & 0 \end{array}\right]$.

The transpose of this matrix is

$$\left[\begin{array}{cc|c} 2 & 1 & 1 \\ 1 & 2 & 1 \\ 1 & 4 & 3 \\ \hline 8 & 12 & 0 \end{array}\right].$$

The dual problem follows

Maximize $\quad z = 8x_1 + 12x_2$

subject to: $\quad 2x_1 + x_2 \le 1$

$x_1 + 2x_2 \le 1$

$x_1 + 4x_2 \le 3$

$x_1 \ge 0,\ x_2 \ge 0.$

The initial simplex tableau will be

$$\begin{array}{ccccc} x_1 & x_2 & x_3 & x_4 & x_5 \end{array}$$
$$\left[\begin{array}{ccccc|c} 2 & 1 & 1 & 0 & 0 & 1 \\ 1 & 2 & 0 & 1 & 0 & 1 \\ 1 & 4 & 0 & 0 & 1 & 3 \\ \hline -8 & -12 & 0 & 0 & 0 & 0 \end{array}\right].$$

Pivot on the 2 in row 2, column 2.

$$\begin{array}{ccccc} x_1 & x_2 & x_3 & x_4 & x_5 \end{array}$$
$$\left[\begin{array}{ccccc|c} \frac{3}{2} & 0 & 1 & -\frac{1}{2} & 0 & \frac{1}{2} \\ \frac{1}{2} & 1 & 0 & \frac{1}{2} & 0 & \frac{1}{2} \\ -1 & 0 & 0 & -2 & 1 & 1 \\ \hline -2 & 0 & 0 & 6 & 0 & 6 \end{array}\right] \begin{array}{l} -\frac{1}{2}R_2 + R_1 \\ \frac{1}{2}R_2 \\ -2R_2 + R_3 \\ 6R_2 + R_4 \end{array}$$

Pivot on the 3/2 in row 1, column 1.

$$\begin{array}{ccccc} x_1 & x_2 & x_3 & x_4 & x_5 \end{array}$$
$$\left[\begin{array}{ccccc|c} 1 & 0 & \frac{2}{3} & -\frac{1}{3} & 0 & \frac{1}{3} \\ 0 & 1 & -\frac{1}{3} & \frac{2}{3} & 0 & \frac{1}{3} \\ 0 & 0 & \frac{2}{3} & -\frac{7}{3} & 1 & \frac{4}{3} \\ \hline 0 & 0 & \frac{4}{3} & \frac{16}{3} & 0 & \frac{20}{3} \end{array}\right] \begin{array}{l} \frac{2}{3}R_1 \\ -\frac{1}{3}R_1 + R_2 \\ \frac{2}{3}R_1 + R_3 \\ \frac{4}{3}R_1 + R_4 \end{array}$$

There is a minimum of 20/3 when $y_1 = 4/3$ and $y_2 = 16/3$.

19. The matrix for the existing problem is

$$\left[\begin{array}{ccc|c} 2 & 1 & 1 & 7 \\ 1 & 2 & 1 & 4 \\ \hline 12 & 10 & 7 & 0 \end{array}\right].$$

The transpose of this matrix is

$$\begin{bmatrix} 2 & 1 & | & 12 \\ 1 & 2 & | & 10 \\ 1 & 1 & | & 7 \\ \hline 7 & 4 & | & 0 \end{bmatrix}.$$

Maximize $z = 7x_1 + 4x_2$

subject to: $2x_1 + x_2 \le 12$

$x_1 + 2x_2 \le 10$

$x_1 + x_2 \le 7$

$x_1 \ge 0,\ x_2 \ge 0.$

The initial simplex tableau will be

$$\begin{array}{ccccc|c} x_1 & x_2 & x_3 & x_4 & x_5 & \\ 2 & 1 & 1 & 0 & 0 & 12 \\ 1 & 2 & 0 & 1 & 0 & 10 \\ 1 & 1 & 0 & 0 & 1 & 7 \\ \hline -7 & -4 & 0 & 0 & 0 & 0 \end{array}.$$

First pivot on the 2 in row 1, column 1.

$$\begin{array}{ccccc|c|l} x_1 & x_2 & x_3 & x_4 & x_5 & & \\ 1 & \frac{1}{2} & \frac{1}{2} & 0 & 0 & 6 & \frac{1}{2}R_1 \\ 0 & \frac{3}{2} & -\frac{1}{2} & 1 & 0 & 4 & -\frac{1}{2}R_1 + R_2 \\ 0 & \frac{1}{2} & -\frac{1}{2} & 0 & 1 & 1 & -\frac{1}{2}R_1 + R_3 \\ \hline 0 & -\frac{1}{2} & \frac{7}{2} & 0 & 0 & 42 & \frac{7}{2}R_1 + R_4 \end{array}$$

Since the final row still has a negative value, pivot on the $\frac{1}{2}$ in column 2, row 3

$$\begin{array}{ccccc|c|l} x_1 & x_2 & x_3 & x_4 & x_5 & & \\ 1 & 0 & 1 & 0 & -1 & 5 & R_1 - R_3 \\ 0 & 0 & 1 & 1 & -3 & 1 & -3R_3 + R_2 \\ 0 & 1 & -1 & 0 & 2 & 2 & 2R_3 \\ \hline 0 & 0 & 3 & 0 & 1 & 43 & R_3 + R_4 \end{array}$$

The minimum is 43 with $y_1 = 3$, $y_2 = 0$, and $y_3 = 1$.

21. The matrix for the existing problem is

$$\begin{bmatrix} 2 & 1 & 1 & | & 6 \\ 1 & 2 & 1 & | & 8 \\ 2 & 1 & 2 & | & 12 \\ \hline 3 & 1 & 4 & | & 0 \end{bmatrix}.$$

The transpose of the matrix is

$$\begin{bmatrix} 2 & 1 & 2 & | & 3 \\ 1 & 2 & 1 & | & 1 \\ 1 & 1 & 2 & | & 4 \\ \hline 6 & 8 & 12 & | & 0 \end{bmatrix}.$$

Maximize $z = 6x_1 + 8x_2 + 12x_3$

subject to: $2x_1 + x_2 + 2x_3 \le 3$

$x_1 + 2x_2 + x_3 \le 1$

$x_1 + x_2 + 2x_3 \le 4$

$x_1 \ge 0,\ x_2 \ge 0,\ x_3 \ge 0.$

The initial simplex tableau will be

$$\begin{array}{cccccc|c} x_1 & x_2 & x_3 & x_4 & x_5 & x_6 & \\ 2 & 1 & 2 & 1 & 0 & 0 & 3 \\ 1 & 2 & 1 & 0 & 1 & 0 & 1 \\ 1 & 1 & 2 & 0 & 0 & 1 & 4 \\ \hline -6 & -8 & -12 & 0 & 0 & 0 & 0 \end{array}.$$

Pivot on the 1 in column 3, row 2.

$$\begin{array}{cccccc|c|l} x_1 & x_2 & x_3 & x_4 & x_5 & x_6 & & \\ 0 & -3 & 0 & 1 & -2 & 0 & 1 & -2R_2 + R_1 \\ 1 & 2 & 1 & 0 & 1 & 0 & 1 & \\ -1 & -3 & 0 & 0 & -2 & 1 & 2 & -2R_2 + R_3 \\ \hline 6 & 16 & 0 & 0 & 12 & 0 & 12 & 12R_2 + R_4 \end{array}$$

Minimum is 12 with $y_1 = 0$, $y_2 = 12$, and $y_3 = 0$.

23. Let y_1 = number of tables, and

y_2 = number of chairs.

Minimize $w = 152y_1 + 40y_2$

subject to: $y_1 + y_2 \ge 60$

$3y_1 \ge y_2$ or $3y_1 - y_2 \ge 0.$

The matrix for the existing problem is

$$\left[\begin{array}{cc|c} 1 & 1 & 60 \\ 3 & -1 & 0 \\ \hline 152 & 40 & 0 \end{array}\right].$$

The transpose of this matrix is

$$\left[\begin{array}{cc|c} 1 & 3 & 152 \\ 1 & -1 & 40 \\ \hline 60 & 0 & 0 \end{array}\right]$$

Maximize $z = 60x_1 + 0x_2$

subject to: $x_1 + 3x_2 \le 152$

$x_1 - x_2 \le 40$

$x_1 \ge 0,\ x_2 \ge 0.$

The initial simplex tableau is

$$\begin{array}{c} \begin{array}{cccc} x_1 & x_2 & x_3 & x_4 \end{array} \\ \left[\begin{array}{cccc|c} 1 & 3 & 1 & 0 & 152 \\ 1 & -1 & 0 & 1 & 40 \\ \hline -60 & 0 & 0 & 0 & 0 \end{array}\right]. \end{array}$$

Pivot on the 1 in row 2, column 1.

$$\begin{array}{c} \begin{array}{cccc} x_1 & x_2 & x_3 & x_4 \end{array} \\ \left[\begin{array}{cccc|c} 0 & 4 & 1 & -1 & 112 \\ 1 & -1 & 0 & 1 & 40 \\ \hline 0 & -60 & 0 & 60 & 2400 \end{array}\right] \begin{array}{l} R_1 - R_2 \\ \\ 60R_2 + R_3 \end{array} \end{array}$$

Since the bottom row still has a negative value, pivot on the 4 in row 1, column 2.

$$\begin{array}{c} \begin{array}{cccc} x_1 & x_2 & x_3 & x_4 \end{array} \\ \left[\begin{array}{cccc|c} 0 & 1 & \frac{1}{4} & -\frac{1}{4} & 28 \\ 1 & 0 & \frac{1}{4} & \frac{3}{4} & 68 \\ \hline 0 & 0 & 15 & 45 & 4080 \end{array}\right] \begin{array}{l} \frac{1}{4}R_1 \\ \frac{1}{4}R_1 + R_2 \\ 15R_1 + R_3 \end{array} \end{array}$$

The minimum cost is $4080 when 15 tables and 45 chairs are made.

25\. Put the information into a table.

	Vitamins	Calories	Cost
Soymeal	2	5	4¢
Meat by-product	6.5	3	5¢
Grain	5	10	5¢
Minimum	50	60	

Minimize $w = 4y_1 + 5y_2 + 5y_3$

subject to: $2y_1 + 6.5y_2 + 5y_3 \ge 50$

$5y_1 + 3y_2 + 10y_3 \ge 60$

$y_1 \ge 0,\ y_2 \ge 0,\ y_3 \ge 0$ where

y_1 = number of grams of soymeal

y_2 = number of grams of meat byproducts, and

y_3 = number of grams of grain.

The matrix of the problem is

$$\left[\begin{array}{ccc|c} 2 & 6.5 & 5 & 50 \\ 5 & 3 & 10 & 60 \\ \hline 4 & 5 & 5 & 0 \end{array}\right].$$

The transpose is

$$\left[\begin{array}{cc|c} 2 & 5 & 4 \\ 6.5 & 3 & 5 \\ 5 & 10 & 5 \\ \hline 50 & 60 & 0 \end{array}\right].$$

The dual can be stated.

Maximize $z = 50x_1 + 60x_2$

subject to: $2x_1 + 5x_2 \le 4$

$6.5x_1 + 3x_2 \le 5$

$5x_1 + 10x_2 \le 3$

$x_1 \ge 0,\ x_2 \ge 0.$

The initial simplex tableau is

$$\begin{array}{c} \text{Quotients} \quad \begin{array}{ccccc} x_1 & x_2 & x_3 & x_4 & x_5 \end{array} \\ \begin{array}{l} 4/5 \\ 5/3 \\ 5/10 \\ \\ \end{array} \left[\begin{array}{ccccc|c} 2 & 5 & 1 & 0 & 0 & 4 \\ 1 & 3 & 0 & 1 & 0 & 5 \\ 1 & 10 & 0 & 0 & 1 & 5 \\ \hline -50 & -60 & 0 & 0 & 0 & 0 \end{array}\right]. \end{array}$$

-60 is more negative than -50, so the first pivot is in column 2. 5/10 is the least positive quotient so pivot on the 10 in column 2.

Quotients x_1 x_2 x_3 x_4 x_5

$$\begin{array}{c} \\ \frac{7}{2} \div 5 = \frac{7}{10} \\ \frac{1}{2} \div \frac{1}{2} = 1 \\ \hline \end{array} \left[\begin{array}{ccccc|c} -\frac{1}{2} & 0 & 1 & 0 & -\frac{1}{2} & \frac{3}{2} \\ 5 & 0 & 0 & 1 & -\frac{3}{10} & \frac{7}{2} \\ \frac{1}{2} & 1 & 0 & 0 & \frac{1}{10} & \frac{1}{2} \\ \hline -20 & 0 & 0 & 0 & 6 & 30 \end{array}\right]$$

The -20 in the last row indicates that a pivot must be performed in column 1. The least positive quotient occurs in row 2, so pivot on the 5.

$$\begin{array}{ccccc|c} x_1 & x_2 & x_3 & x_4 & x_5 & \\ 0 & 0 & 1 & \frac{1}{10} & -\frac{53}{100} & \frac{37}{20} \\ 1 & 0 & 0 & \frac{1}{5} & -\frac{3}{50} & \frac{7}{10} \\ 0 & 1 & 0 & -\frac{1}{10} & \frac{13}{100} & \frac{3}{20} \\ \hline 0 & 0 & 0 & 4 & \frac{24}{5} & 44 \end{array}$$

The last row of the final tableau shows that the solution of the standard minimization problem is as follows:

the minimum value of $w = 4y_1 + 5y_2 + 5y_3$, subject to the given constraints, is 44 and occurs when $y_1 = 0$, $y_2 = 4$, and $y_3 = 24/5 = 4.8$.

Use no soymeal, 4 grams of meat byproducts, and 4.8 grams of gram for a minimum cost of 44¢.

27. **(a)** From the examples in the test solve this initial tableau.

$$\begin{array}{cccc|c} x_1 & x_2 & x_3 & x_4 & \\ 1 & 2 & 1 & 0 & 3 \\ 3 & 1 & 0 & 1 & 2 \\ \hline -7 & -4 & 0 & 0 & 0 \end{array}$$

Pivot on the 3, row 2, column 1.

$$\begin{array}{cccc|c|l} x_1 & x_2 & x_3 & x_4 & & \\ 0 & \frac{5}{3} & 1 & -\frac{1}{3} & \frac{7}{3} & -\frac{1}{3}R_2 + R_1 \\ 1 & \frac{1}{3} & 0 & \frac{1}{3} & \frac{2}{3} & \frac{1}{3}R_2 \\ \hline 0 & -\frac{5}{3} & 0 & \frac{7}{3} & \frac{14}{3} & \frac{7}{3}R_2 + R_3 \end{array}$$

Now pivot on 5/3 in row 1, column 2.

$$\begin{array}{cccc|c|l} x_1 & x_2 & x_3 & x_4 & & \\ 0 & 1 & \frac{3}{5} & -\frac{1}{5} & \frac{7}{5} & \frac{3}{5}R_1 \\ 1 & 0 & -\frac{1}{5} & \frac{2}{5} & \frac{1}{5} & -\frac{1}{5}R_1 + R_2 \\ \hline 0 & 0 & 1 & 2 & 7 & R_1 + R_3 \end{array}$$

Use 1 bag of Feed 1, 2 bags of Feed 2.

(b) Use the shadow values of $x_1 = 1/5$ and $x_2 = 3/5$.

7 units of A and 4 of B = \$7.00 *Part A*
Less 2 units of A (same amount of B) = -.40
6.60

For a daily cost of \$6.60, use 1.4 bags of Feed 1 and 1.2 bags of Feed 2.

29. **(a)** The initial matrix is

$$\left[\begin{array}{ccc|c} 1 & 1 & 1 & 100 \\ 400 & 160 & 280 & 20000 \\ \hline 120 & 40 & 60 & 0 \end{array}\right].$$

The transpose is

$$\left[\begin{array}{cc|c} 1 & 400 & 120 \\ 1 & 160 & 40 \\ 1 & 280 & 60 \\ \hline 100 & 20000 & 0 \end{array}\right].$$

Minimize $w = 100y_1 + 20{,}000y_2$
subject to: $y_1 + 400y_2 \ge 120$
$y_1 + 160y_2 \ge 40$
$y_1 + 280y_2 \ge 60$
$y_1 \ge 0$, $y_2 \ge 0$.

(b) The land is decreased by 10 acres from 100 to 90. The capital increased by 1000. Solve the dual problem to obtain the shadow values. Write the initial tableau.

$$\begin{array}{c} \\ \\ \\ \\ \\ \end{array}\begin{array}{ccccc|c} y_1 & y_2 & y_3 & y_4 & y_5 & \\ \hline 1 & 400 & -1 & 0 & 0 & 120 \\ 1 & 160 & 0 & -1 & 0 & 40 \\ 1 & 280 & 0 & 0 & -1 & 60 \\ \hline 100 & 20{,}000 & 100 & 0 & 0 & 0 \end{array}$$

$$\begin{array}{ccccc|c|l} y_1 & y_2 & y_3 & y_4 & y_5 & & \\ \hline \textcircled{1} & 400 & -1 & 0 & 0 & 120 & \\ 0 & 240 & -1 & 1 & 0 & 80 & R_1 - R_2 \\ 0 & 120 & -1 & 0 & 1 & 60 & R_1 - R_2 \\ \hline 0 & -20{,}000 & 100 & 0 & 0 & -12{,}000 & -100R_1 + R_4 \end{array}$$

$$\begin{array}{ccccc|c|l} y_1 & y_2 & y_3 & y_4 & y_5 & & \\ \hline \frac{1}{400} & 1 & -\frac{1}{400} & 0 & 0 & \frac{3}{10} & \frac{1}{400}R_1 \\ 0 & 240 & -1 & 1 & 0 & 80 & \\ 0 & 120 & -1 & 0 & 1 & 60 & \\ \hline 0 & -20{,}000 & 100 & 0 & 0 & -12{,}000 & \end{array}$$

$$\begin{array}{ccccc|c|l} y_1 & y_2 & y_3 & y_4 & y_5 & & \\ \hline \frac{1}{400} & \textcircled{1} & -\frac{1}{400} & 0 & 0 & \frac{3}{10} & \\ -\frac{3}{5} & 0 & -\frac{2}{5} & 1 & 0 & 8 & -240R_1 + R_2 \\ -\frac{3}{10} & 0 & -\frac{7}{10} & 0 & 1 & 24 & -120R_1 + R_3 \\ \hline 50 & 0 & 50 & 0 & 0 & -6000 & 20{,}000R_1 + R_4 \end{array}$$

From the dual solutions, the shadow cost of acreage is 0 and of capital is 3/10.

$$\text{Profit} = 6000 + 0(-10) + \frac{3}{10}(1000) = \$6300$$

Now calculate the number of acres of each. Let x_1 = number for potato, x_2 = number for corn, x_3 = number for cabbage.

$$\text{Profit} = 120x_1 + 40x_2 + 60x_3$$
$$6300 = 120x_1 + 40(0) + 60(0)$$

$x_1 = 52.5$ acres of potatoes and no corn and no cabbage.

(c) $\text{Profit} = 6000 + 0(10) + \frac{3}{10}(-1000)$

$= \$5700$

$$\text{Profit} = 120x_1 + 40x_2 + 60x_3$$
$$5700 = 120x_1 + 40(0) + 60(0)$$

$x_1 = 47.5$ acres of potatoes and no corn nor cabbage.

31. Consult Exercise 31, Section 7.6.

The matrix is

$$\left[\begin{array}{ccc|c} 1 & 1 & 1 & 10 \\ -1 & -1 & -1 & -15 \\ 1 & -\frac{1}{4} & 0 & 0 \\ -1 & 0 & 1 & 0 \\ \hline 30 & 9 & 27 & 0 \end{array}\right].$$

The transpose is

$$\left[\begin{array}{cccc|c} 1 & -1 & 1 & -1 & 30 \\ 1 & -1 & -\frac{1}{4} & 0 & 9 \\ 1 & -1 & 0 & 1 & 27 \\ \hline 10 & -15 & 0 & 0 & 0 \end{array}\right].$$

Maximize $z = 10x_1 - 15x_2$

subject to:

$$x_1 - x_2 + x_3 - x_4 \le 30$$
$$x_1 - x_2 - \frac{1}{4}x_3 \le 9$$
$$x_1 - x_2 + x_4 \le 27$$
$$x_1 \ge 0,\ x_2 \ge 0,\ x_3 \ge 0,\ x_4 \ge 0.$$

The initial simplex tableau is

$$\begin{array}{ccccccc|c} x_1 & x_2 & x_3 & x_4 & x_5 & x_6 & x_7 & \\ \hline 1 & -1 & 1 & -1 & 1 & 0 & 0 & 30 \\ 1 & -1 & -\frac{1}{4} & 0 & 0 & 1 & 0 & 9 \\ 1 & -1 & 0 & 1 & 0 & 0 & 1 & 27 \\ \hline -10 & 15 & 0 & 0 & 0 & 0 & 0 & 0 \end{array}.$$

Pivot on row 2, column 1.

$$\begin{array}{ccccccc|c} x_1 & x_2 & x_3 & x_4 & x_5 & x_6 & x_7 & \\ \hline 0 & 0 & \frac{5}{4} & -1 & 1 & -1 & 0 & 21 \\ 1 & -1 & -\frac{1}{4} & 0 & 0 & 1 & 0 & 9 \\ 0 & 0 & \frac{1}{4} & 1 & 0 & -1 & 1 & 18 \\ \hline 0 & 5 & -\frac{5}{2} & 0 & 0 & 10 & 0 & 90 \end{array} \quad \begin{array}{l} \\ R_1 - R_2 \\ \\ R_3 - R_2 \\ 10R_1 + R_4 \end{array}$$

Now pivot on the $\frac{5}{4}$ in row 1, column 3.

$$\begin{array}{ccccccc|c} x_1 & x_2 & x_3 & x_4 & x_5 & x_6 & x_7 & \\ \hline 0 & 0 & 1 & -\frac{4}{5} & \frac{4}{5} & -\frac{4}{5} & 0 & \frac{84}{5} \\ 1 & -1 & 0 & -\frac{1}{5} & \frac{1}{5} & \frac{4}{5} & 0 & \frac{66}{5} \\ 0 & 0 & 0 & \frac{6}{5} & -\frac{1}{5} & -\frac{4}{5} & 1 & \frac{69}{5} \\ \hline 0 & 5 & 0 & -2 & 2 & 8 & 0 & 132 \end{array} \quad \begin{array}{l} \\ \frac{4}{5}R_1 \\ \frac{1}{5}R_1 + R_2 \\ -\frac{1}{5}R_1 + R_3 \\ 2R_1 + R_4 \end{array}$$

Now pivot on the $\frac{6}{5}$ in row 3, column 4.

$$\begin{array}{ccccccc|c} x_1 & x_2 & x_3 & x_4 & x_5 & x_6 & x_7 & \\ \hline 0 & 0 & 1 & 0 & \frac{2}{3} & -\frac{4}{3} & \frac{2}{3} & 26 \\ 1 & -1 & 0 & 0 & \frac{1}{6} & \frac{2}{3} & \frac{1}{6} & \frac{31}{2} \\ 0 & 0 & 0 & 1 & -\frac{1}{6} & -\frac{2}{3} & \frac{5}{6} & \frac{23}{2} \\ \hline 0 & 5 & 0 & 0 & \frac{5}{3} & \frac{20}{3} & \frac{5}{3} & 155 \end{array} \quad \begin{array}{l} \\ \frac{2}{3}R_3 + R_1 \\ \frac{1}{6}R_3 + R_2 \\ \frac{5}{6}R_3 \\ \frac{5}{3}R_3 + R_4 \end{array}$$

Use 5/3 ounces of ingredient I, 20/3 ounces of ingredient II, and 5/3 ounces of ingredient III for a minimum cost of $1.55 per gallon. 10 ounces of the additive should be used per gallon of gasoline.

Chapter 7 Review Exercises

See the graphs for Exercise 1-13 in the answer section at the back of the textbook.

1. $y \geq 2x + 3$
   Graph $y = 2x + 3$ as a solid line. Use the origin as a test point to get $0 \geq 2(0) + 3$ or $0 \geq 3$, which is false. Shade the region that does not contain the origin, that is, the region above the line.

3. $3x + 4y \leq 12$
   Graph $3x + 4y = 12$ as a solid line, using the intercepts (0, 3) and (4, 0). Using the origin as a test point gives $0 \leq 12$, which is true. Shade the region that contains the origin, that is, the region below the line.

5. $y \geq x$
   Graph $y = x$ as a solid line. Since this line contains the origin, choose a point other than (0, 0) as a test point. Use (1, 4) to get $4 \geq 1$, which is true. Shade the region that contains the test point, that is, the region above the line.

7. $x + y \leq 6$ is the region on or below the line $x + y = 6$; $2x - y \geq 3$ is the region on or below the line $2x - y = 3$. The system of inequalities must meet both conditions. The only corner point is the intersection of the two boundary lines, the point (3, 3).

9. $-4 \leq x \leq 2$ is the rectangular region lying on or between the two vertical lines, $x = -4$ and $x = 2$; $-1 \leq y \leq 3$ is the rectangular region lying on or between the two horizontal lines, $y = -1$ and $y = 3$; $x + y \leq 4$ is the region lying on or below the line $x + y = 4$. The corner points are (-4, -1), (-4, 3), (1, 3), (2, 3), and (2, -1).

11. $x + 3y \ge 6$ is the region on or above the line $x + 3y = 6$; $4x - 3y \le 12$ is the region on or above the line $4x - 3y = 12$; $x \ge 0$ and $y \ge 0$ together restrict the graph to the first quadrant. The corner points are (0, 2) and 18/5, 4/5).

13. Let x be the number of batches of cakes and y the number of batches of cookies. Then we have the following inequalities.

$2x + \frac{3}{2}y \le 15$ *Oven time*

$3x + \frac{2}{3}y \le 13$ *Decorating*

$x \ge 0$

$y \ge 0.$

15.

Corner point	Value of z = 2x + 4
(1, 6)	26
(6, 7)	40 *Maximum*
(7, 3)	26
$(1, 2\frac{1}{2})$	12
(2, 1)	8 *Minumum*

The maximum value of 40 occurs at (6, 7), and the minimum value of 8 occurs at (2, 1).

17. Maximize $z = 2x + 4y$

subject to: $3x + 2y \le 12$

$5x + y \ge 5$

$x \ge 0$

$y \ge 0.$

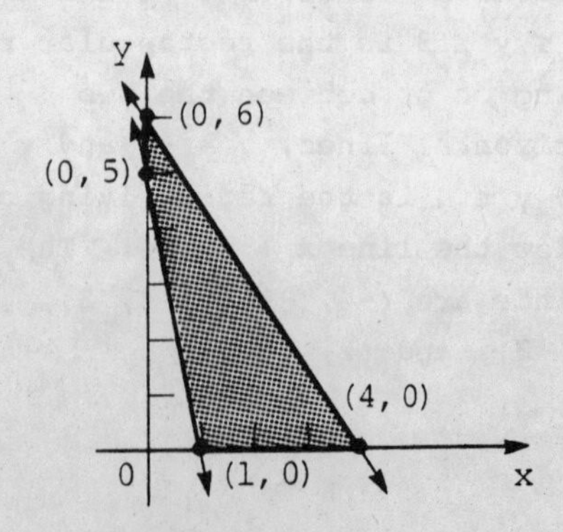

The corner points are (0, 5), (0, 6), (4, 0), and (1, 0).

Corner point	Value of z = 2x + 4y
(0, 5)	20
(0, 6)	24 *Maximum*
(4, 0)	8
(1, 0)	2

The maximum value of $z = 2x + 4y$ is 24 at (0, 6).

19. Maximize $z = 4x + 2y$

subject to: $x + y \le 50$

$2x + y \ge 20$

$x + 2y \ge 30$

$x \ge 0$

$y \ge 0.$

Sketch the feasible region.

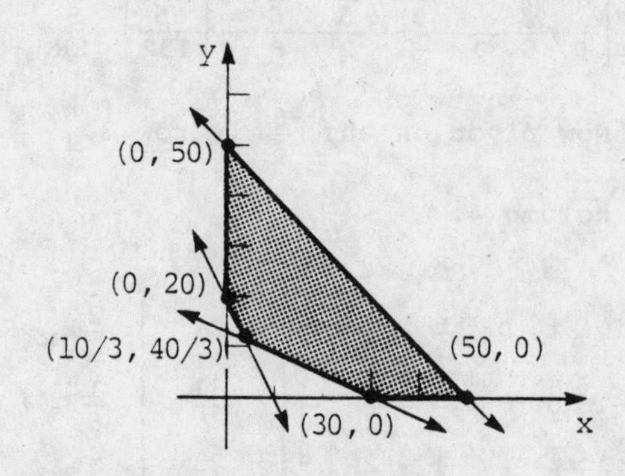

The corner points are (0, 20), (10/3, 40/3), (30, 0), (50, 0), and (0, 50).

Corner point	Value of z = 4x + 2y
(0, 20)	40
(10/3, 40/3)	40
(30, 0)	120
(50, 0)	200
(0, 50)	100

Thus, the minimum value of $4x + 2y$ is 40 and occurs at every point on the line segment joining (0, 20) and (10/3, 40/3).

21. From the graph for Exercise 13, the corner points are (0, 10), (3, 6), (13/3, 0), and (0, 0). Since x was the number of batches of cakes and y the number of batches of cookies, the revenue function is $z = 30x + 20y$. Evaluate this objective function at each corner point.

Corner point	Value of $z = 30x + 20y$
(0, 0)	200
(3, 6)	210 *Maximum*
(13/3, 0)	130
(0, 0)	0

3 batches of cakes and 6 batches of cookies should be made to produce a maximum profit of $210.

23. The information is contained in the following table.

	A	B	C	Available
Buy	5	3	6	1200
Sell	1	2	2	800
Deliver	2	1	5	500
Profit	4	3	3	

(a) Let x_1 = number of item A, x_2 = number of item B, x_3 = number of item C.

(b) The objective function is
$z = 4x_1 + 3x_2 + 3x_3$.

(c)
$$5x_1 + 3x_2 + 6x_3 \le 1200$$
$$x_1 + 2x_2 + 2x_3 \le 800$$
$$2x_1 + x_2 + 5x_3 \le 500$$
$$x_1 \ge 0,\ x_2 \ge 0,\ x_3 \ge 0.$$

25. (a) Let x_1 = number of gallons of Fruity wine, x_2 = number of gallons of Crystal wine to be made.

(b) The profit function is
$z = 12x_1 + 15x_2$.

(c) The ingredients available are the limitations. The constraints are
$$2x_1 + x_2 \le 110$$
$$2x_1 + 3x_2 \le 125$$
$$2x_1 + x_2 \le 90$$
$$x_1 \ge 0,\ x_2 \ge 0.$$

27. (a) Adding the slack variables x_3, x_4, x_5, x_6 gives the equations
$$2x_1 + 5x_2 + x_3 = 50$$
$$x_1 + 3x_2 + x_4 = 25$$
$$4x_1 + x_2 + x_5 = 18$$
$$x_1 + x_2 + x_6 = 12.$$

(b) The initial tableau is

$$\begin{array}{cccccc|c} x_1 & x_2 & x_3 & x_4 & x_5 & x_6 & \\ 2 & 5 & 1 & 0 & 0 & 0 & 50 \\ 1 & 3 & 0 & 1 & 0 & 0 & 25 \\ 4 & 1 & 0 & 0 & 1 & 0 & 18 \\ 1 & 1 & 0 & 0 & 0 & 1 & 12 \\ \hline -5 & -3 & 0 & 0 & 0 & 0 & 0 \end{array}.$$

29. (a) Adding the slack variables x_4, x_5, and x_6. gives the equations
$$x_1 + x_2 + x_3 + x_4 = 90$$
$$2x_1 + 5x_2 + x_3 + x_5 = 120$$
$$x_1 + 3x_2 + x_6 = 80.$$

(b) The initial tableau is

$$\begin{array}{cccccc|c} x_1 & x_2 & x_3 & x_4 & x_5 & x_6 & \\ 1 & 1 & 1 & 1 & 0 & 0 & 90 \\ 2 & 5 & 1 & 0 & 1 & 0 & 120 \\ 1 & 3 & 0 & 0 & 0 & 1 & 80 \\ \hline -5 & -8 & -6 & 0 & 0 & 0 & 0 \end{array}.$$

31.
$$\begin{array}{cccccc|c} x_1 & x_2 & x_3 & x_4 & x_5 & z & \\ 1 & 2 & 3 & 1 & 0 & 0 & 28 \\ \textcircled{2} & 4 & 1 & 0 & 1 & 0 & 32 \\ \hline -5 & -2 & -3 & 0 & 0 & 1 & 0 \end{array}$$

The most negative entry in the last row is -5 and the smaller of the two quotients is 32/2 = 16. Hence, the 2 in the first column is the first pivot. Performing row transformations leads to the second tableau.

$$\begin{array}{c} \begin{array}{ccccc} x_1 & x_2 & x_3 & x_4 & x_5 \end{array} \\ \left[\begin{array}{ccccc|c} 0 & 0 & \textcircled{$\frac{5}{2}$} & 1 & -\frac{1}{2} & 12 \\ 1 & 2 & \frac{1}{2} & 0 & \frac{1}{2} & 16 \\ \hline 0 & 8 & -\frac{1}{2} & 0 & \frac{5}{2} & 80 \end{array}\right] \end{array} \begin{array}{l} R_1 - \frac{1}{2}R_2 \\ \frac{1}{2}R_2 \\ \frac{5}{2}R_2 + R_3 \end{array}$$

The second pivot is 5/2. After row transformations, the tableau is as follows.

$$\begin{array}{c} \begin{array}{ccccc} x_1 & x_2 & x_3 & x_4 & x_5 \end{array} \\ \left[\begin{array}{ccccc|c} 0 & 0 & 1 & \frac{2}{5} & -\frac{1}{5} & \frac{24}{5} \\ 1 & 2 & 0 & -\frac{1}{5} & \frac{3}{5} & \frac{68}{5} \\ \hline 0 & 8 & 0 & \frac{1}{5} & \frac{12}{5} & \frac{412}{5} \end{array}\right] \end{array} \begin{array}{l} \frac{2}{5}R_1 \\ -\frac{1}{5}R_1 + R_2 \\ \frac{1}{5}R_1 + R_3 \end{array}$$

Thus, the maximum value is 412/5 or 82.4 when x_1 = 68/5 or 13.6, x_2 = 0, x_3 = 24/5 or 4.8, x_4 = 0, and x_5 = 0.

33.

$$\begin{array}{c} \begin{array}{cccccc} x_1 & x_2 & x_3 & x_4 & x_5 & x_6 \end{array} \\ \left[\begin{array}{cccccc|c} 1 & 2 & 2 & 1 & 0 & 0 & 50 \\ 3 & 1 & 0 & 0 & 1 & 0 & 20 \\ \textcircled{1} & 0 & 2 & 0 & 0 & -1 & 15 \\ \hline -5 & -3 & -2 & 0 & 0 & 0 & 0 \end{array}\right] \end{array}$$

does not have a feasible solution; x_6 = -15, so pivot on the 1 in the third row to get

$$\begin{array}{c} \begin{array}{cccccc} x_1 & x_2 & x_3 & x_4 & x_5 & x_6 \end{array} \\ \left[\begin{array}{cccccc|c} 0 & 2 & 0 & 1 & 0 & 1 & 35 \\ 0 & 1 & -6 & 0 & 1 & 3 & -25 \\ 1 & 0 & 2 & 0 & 0 & -1 & 15 \\ \hline 0 & -3 & 8 & 0 & 0 & -5 & 75 \end{array}\right]. \end{array} \begin{array}{l} R_1 - R_3 \\ R_2 - R_3 \\ \\ 5R_3 + R_4 \end{array}$$

This does not have a feasible solution either; x_5 = -25, so multiply row 2 by (-1) to get a nonnegative constant, and select another pivot in row 2.

$$\begin{array}{c} \begin{array}{cccccc} x_1 & x_2 & x_3 & x_4 & x_5 & x_6 \end{array} \\ \left[\begin{array}{cccccc|c} 0 & 2 & 0 & 1 & 0 & 1 & 35 \\ 0 & -1 & \textcircled{6} & 0 & -1 & -3 & 25 \\ 1 & 0 & 2 & 0 & 0 & -1 & 15 \\ \hline 0 & -3 & 8 & 0 & 0 & -5 & 75 \end{array}\right] \end{array} \begin{array}{l} \\ -R_2 \\ \\ \\ \end{array}$$

$$\begin{array}{c} \begin{array}{cccccc} x_1 & x_2 & x_3 & x_4 & x_5 & x_6 \end{array} \\ \left[\begin{array}{cccccc|c} 0 & \textcircled{2} & 0 & 1 & 0 & 1 & 35 \\ 0 & -\frac{1}{6} & 1 & 0 & -\frac{1}{6} & -\frac{1}{2} & \frac{25}{6} \\ 1 & \frac{1}{3} & 0 & 0 & \frac{1}{3} & 0 & \frac{20}{3} \\ \hline 0 & -\frac{5}{3} & 0 & 0 & \frac{4}{3} & -1 & \frac{125}{3} \end{array}\right] \end{array} \begin{array}{l} \\ \frac{1}{6}R_2 \\ -\frac{1}{3}R_2 + R_3 \\ -\frac{4}{3}R_2 + R_4 \end{array}$$

This has a feasible solution, so move to Phase II.

$$\begin{array}{c} \begin{array}{cccccc} x_1 & x_2 & x_3 & x_4 & x_5 & x_6 \end{array} \\ \left[\begin{array}{cccccc|c} 0 & 1 & 0 & \frac{1}{2} & 0 & \textcircled{$\frac{1}{2}$} & \frac{35}{2} \\ 0 & 0 & 1 & \frac{1}{12} & -\frac{1}{6} & -\frac{5}{12} & \frac{85}{12} \\ 1 & 0 & 0 & -\frac{1}{6} & \frac{1}{3} & -\frac{1}{6} & \frac{5}{6} \\ \hline 0 & 0 & 0 & \frac{5}{6} & \frac{4}{3} & -\frac{1}{6} & \frac{425}{6} \end{array}\right] \end{array} \begin{array}{l} \frac{1}{2}R_1 \\ \frac{1}{6}R_1 + R_2 \\ -\frac{1}{3}R_1 + R_3 \\ \frac{5}{3}R_1 + R_4 \end{array}$$

$$\begin{array}{c} \begin{array}{cccccc} x_1 & x_2 & x_3 & x_4 & x_5 & x_6 \end{array} \\ \left[\begin{array}{cccccc|c} 0 & 2 & 0 & 1 & 0 & 1 & 35 \\ 0 & \frac{5}{6} & 1 & \frac{1}{2} & -\frac{1}{6} & 0 & \frac{65}{3} \\ 1 & \frac{1}{3} & 0 & 0 & \frac{1}{3} & 0 & \frac{20}{3} \\ \hline 0 & \frac{1}{3} & 0 & 1 & \frac{4}{3} & 0 & \frac{230}{3} \end{array}\right] \end{array} \begin{array}{l} 2R_1 \\ \frac{5}{12}R_1 + R_2 \\ \frac{1}{6}R_1 + R_3 \\ \frac{1}{6}R_1 + R_4 \end{array}$$

The maximum value is 230/3 (or 76 2/3) when x_1 = 20/3 (or 6 2/3), x_2 = 0, x_3 = 65/3 (or 21 2/3), x_4 = 35, x_5 = 0, and x_6 = 0. If the fractions are changed to decimals and rounded to the nearest hundredth, the answer becomes: The maximum value is 76.67 when x_1 = 6.67, x_2 = 0, x_3 = 21.67, x_4 = 35, x_5 = 0, and x_6 = 0.

35. Change the objective function to maximize $z = -w = -10y_1 - 15y_2$, subject to:

$$\begin{aligned} y_1 + y_2 &\le 17 \\ 5y_1 + 8y_2 &\le 42 \\ y_1 \ge 0,\ y_2 &\ge 0. \end{aligned}$$

37. Change the objective function to maximize $z = -w = -7y_1 - 2y_2 - 3y_3$, subject to:

$$\begin{aligned} y_1 + y_2 + 2y_3 &\le 48 \\ y_1 + y_2 &\le 12 \\ y_3 &\le 10 \\ 3y_1 + y_3 &\le 50 \\ y_1 \ge 0,\ y_2 \ge 0,\ y_3 &\ge 0. \end{aligned}$$

39. Maximize $z = -4y_1 - 2y_2$

subject to:

$$\begin{aligned} y_1 + y_2 + y_3 &= 50 \\ 2y_1 + y_2 - y_4 &= 20 \\ y_1 + 2y_2 - y_5 &= 30 \end{aligned}$$

Here y_3 is a slack variable and y_4 and y_5 are surplus variables.

$$\begin{array}{c} \begin{array}{ccccc} y_1 & y_2 & y_3 & y_4 & y_5 \end{array} \\ \left[\begin{array}{ccccc|c} 1 & 1 & 1 & 0 & 0 & 50 \\ 2 & 1 & 0 & -1 & 0 & 20 \\ 1 & 2 & 0 & 0 & -1 & 30 \\ \hline 4 & 2 & 0 & 0 & 0 & 0 \end{array}\right] \end{array}$$

The solution $y_1 = y_2 = 0$, $y_3 = 50$, $y_4 = -20$, $y_5 = -30$ is not a feasible solution. Convert column 2.

$$\begin{array}{c} \begin{array}{ccccc} y_1 & y_2 & y_3 & y_4 & y_5 \end{array} \\ \left[\begin{array}{ccccc|c} -1 & 0 & 1 & 1 & 0 & 30 \\ 2 & 1 & 0 & -1 & 0 & 20 \\ -3 & 0 & 0 & 2 & -1 & -10 \\ \hline 0 & 0 & 0 & 2 & 0 & -40 \end{array}\right] \end{array} \begin{array}{l} R_1 - R_2 \\ \\ R_3 - 2R_2 \\ R_4 - 2R_2 \end{array}$$

The solution is $y_1 = y_4 = 0$, $y_2 = 20$, $y_3 = 30$, and $y_5 = 10$ and the maximum value is -40, that is, the minimum value is 40 at any point on the line segment connecting (0, 20) and (10/3, 40/3).

41. The solution is at (47, 68, 0, 92, 35, 0, 0) and the minimum value is 1957.

43. The solution is at (9, 5, 8, 0, 0, 0) and the minimum value is 62.

45. The first simplex tableau (after introducing slack variables) is

$$\begin{array}{c} \begin{array}{cccccc} x_1 & x_2 & x_3 & x_4 & x_5 & x_6 \end{array} \\ \left[\begin{array}{cccccc|c} 5 & 3 & 6 & 1 & 0 & 0 & 1200 \\ 1 & 2 & 2 & 0 & 1 & 0 & 800 \\ 2 & 1 & 5 & 0 & 0 & 1 & 500 \\ \hline -4 & -3 & -3 & 0 & 0 & 0 & 0 \end{array}\right] \end{array}$$

$$\begin{array}{c} \begin{array}{cccccc} x_1 & x_2 & x_3 & x_4 & x_5 & x_6 \end{array} \\ \left[\begin{array}{cccccc|c} 1 & \frac{3}{5} & \frac{6}{5} & \frac{1}{5} & 0 & 1 & 240 \\ 0 & \frac{7}{5} & \frac{4}{5} & -\frac{1}{5} & 1 & 0 & 560 \\ 0 & -\frac{1}{5} & \frac{13}{5} & -\frac{2}{5} & 0 & 1 & 20 \\ \hline 0 & -\frac{3}{5} & \frac{9}{5} & \frac{4}{5} & 0 & 0 & 960 \end{array}\right] \end{array} \begin{array}{l} \frac{1}{5}R_1 \\ R_2 - \frac{1}{5}R_1 \\ -\frac{2}{5}R_1 + R_3 \\ \frac{4}{5}R_1 + R_4 \end{array}$$

$$\begin{array}{c} \begin{array}{cccccc} x_1 & x_2 & x_3 & x_4 & x_5 & x_6 \end{array} \\ \left[\begin{array}{cccccc|c} \frac{5}{3} & 1 & 2 & \frac{1}{3} & 0 & 0 & 400 \\ -\frac{7}{3} & 0 & -2 & -\frac{2}{3} & 1 & 0 & 0 \\ \frac{1}{3} & 0 & 3 & -\frac{1}{3} & 0 & 1 & 68 \\ \hline 1 & 0 & 3 & 1 & 0 & 0 & 1200 \end{array}\right] \end{array} \begin{array}{l} \frac{5}{3}R_1 \\ -\frac{7}{3}R_1 + R_2 \\ \frac{1}{3}R_1 + R_3 \\ R_1 + R_4 \end{array}$$

Maximum profit is $1200. She should get none of A, 400 items of B, and none of C.

47. First, observe that since $2x + y \le 90$, the other constraint $2x + y \le 110$ is redundant. So, the first simplex tableau after the introduction of slack variables is

$$\begin{array}{c} \begin{array}{cccc} x & y & x_3 & x_4 \end{array} \\ \left[\begin{array}{cccc|c} 2 & 3 & 1 & 0 & 125 \\ 2 & 1 & 0 & 1 & 90 \\ \hline -12 & -15 & 0 & 0 & 0 \end{array}\right]. \end{array}$$

$$\begin{array}{c} \quad x \quad y \quad x_3 \quad x_4 \\ \left[\begin{array}{cccc|c} \frac{2}{3} & 1 & \frac{1}{3} & 0 & \frac{125}{3} \\ \frac{4}{3} & 0 & -\frac{1}{3} & 1 & \frac{145}{3} \\ \hline -2 & 0 & 5 & 0 & 625 \end{array}\right] \begin{array}{l} \frac{1}{3}R_1 \\ R_2 - \frac{1}{3}R_1 \\ 5R_1 + R_3 \end{array} \end{array}$$

$$\begin{array}{c} \quad x \quad y \quad x_3 \quad x_4 \\ \left[\begin{array}{cccc|c} 0 & 1 & \frac{1}{2} & -\frac{1}{2} & \frac{105}{6} \\ 1 & 0 & -\frac{1}{4} & \frac{3}{4} & \frac{145}{4} \\ \hline 0 & 0 & \frac{9}{2} & \frac{3}{2} & 697.5 \end{array}\right] \begin{array}{l} -\frac{1}{2}R_2 + R_1 \\ \frac{3}{4}R_2 \\ \frac{3}{2}R_2 + R_3 \end{array} \end{array}$$

Maximum profit is $697.50 at a production of 36.25 gallons of Fruity and 17.5 gallons of Crystal wine.

49. First put the data into a table.

	Lumber	Con- crete	Ad- vertising	Total spent
Atlantic	1000	3000	2000	$3000
Pacific	2000	3000	3000	$4000
Minimum use	8000	18000	15000	

The problem is to minimize

$w = 3000y_1 + 4000y_2$

subject to: $1000y_1 + 2000y_2 \geq 8000$

$3000y_1 + 3000y_2 \geq 18000$

$2000y_1 + 3000y_2 \geq 15000$

with y_1 = number of Atlantic boats and

y_2 = number of Pacific boats.

$y_1 \geq 0,\ y_2 \geq 0.$

The matrix for this problem is

$$\left[\begin{array}{cc|c} 1000 & 2000 & 8000 \\ 3000 & 3000 & 18,000 \\ 2000 & 3000 & 15,000 \\ \hline 3000 & 4000 & 0 \end{array}\right].$$

The transpose of this matrix is

$$\left[\begin{array}{ccc|c} 1000 & 3000 & 2000 & 3000 \\ 2000 & 3000 & 3000 & 4000 \\ \hline 8000 & 18,000 & 15,000 & 0 \end{array}\right].$$

The dual problem is as follows.

Maximize $z = 8000x_1 + 18000x_2 + 15000x_3$

subject to:

$1000x_1 + 3000x_2 + 2000x_3 \leq 3000$

$2000x_1 + 3000x_2 + 3000x_3 \leq 4000$

$x_1 \geq 0,\ x_2 \geq 0\ x_3 \geq 0.$

The initial simplex tableau is

$$\begin{array}{c} x_1 \quad x_2 \quad x_3 \quad x_4 \quad x_5 \\ \left[\begin{array}{ccccc|c} 1000 & 3000 & 2000 & 1 & 0 & 3000 \\ 2000 & 3000 & 3000 & 0 & 1 & 4000 \\ \hline -8000 & -18,000 & -15,000 & 0 & 0 & 0 \end{array}\right]. \end{array}$$

Pivot on the 3000 in row 1, column 2.

$$\begin{array}{c} x_1 \quad x_2 \quad x_3 \quad x_4 \quad x_5 \\ \left[\begin{array}{ccccc|c} \frac{1}{3} & 1 & \frac{2}{3} & \frac{1}{1000} & 0 & 1 \\ 1000 & 0 & 1000 & -1 & 1 & 1000 \\ \hline -2000 & 0 & -3000 & 6 & 0 & 18,0000 \end{array}\right] \begin{array}{l} \frac{1}{3}R_1 \\ R_2 - R_1 \\ 6R_1 + R_3 \end{array} \end{array}$$

Pivot on the 1000 in row 2, column 3.

$$\begin{array}{c} x_1 \quad x_2 \quad x_3 \quad x_4 \quad x_5 \\ \left[\begin{array}{ccccc|c} -\frac{1}{3} & 1 & 0 & \frac{1}{1000} & -\frac{1}{1500} & \frac{1}{3} \\ 1 & 0 & 1 & -\frac{1}{1000} & \frac{1}{1000} & 1 \\ \hline 1000 & 0 & 0 & 3 & 3 & 21,000 \end{array}\right] \begin{array}{l} -\frac{1}{1500}R_2 + R_1 \\ \frac{1}{1000}R_2 \\ 3R_2 + R_3 \end{array} \end{array}$$

The minimum cost is $21,000, when 3 Atlantic and 3 Pacific models are built.

Case 7 Exercises

1. $w_1 = \frac{570}{600} = .95$

   $w_2 = \frac{500}{600} = .83$

   $w_3 = \frac{450}{600} = .75$

   $w_4 = \frac{600}{600} = 1.00$

   $w_5 = \frac{520}{600} = .87$

   $w_6 = \frac{565}{600} = .94$

2. $x_1 = 100$
   $x_2 = 0$
   $x_3 = 0$
   $x_4 = 90$
   $x_5 = 0$
   $x_6 = 210$

3. Answers may vary. A student might note that a mathematical approach to determining the relative worth of an individual is limited by the difficulty of measuring objectively many qualities such as ability to interact with co-workers, motivation, and so on. On the other hand, if a mathematical approach is used, subjective evaluations can be minimized.

Case 8 Exercises

1. (a)

(i) $.4x_1 + .23x_2 + .805x_3 + .998x_5 + .04x_6 + .5x_{10} + .625x_{11} \geq 14$

(ii) $.054x_1 + .069x_2 + .025x_4 + .078x_6 + .28x_7 + .97x_8 \geq 6$

(iii) $.707x_9 + .1x_{10} \geq 16$

(iv) $.35x_{10} + .315x_{11} \geq .35$

(v) $x_1 + x_2 + x_3 + x_4 + x_5 + x_6 + x_7 + x_8 + x_9 + x_{10} + x_{11} \leq 99.6$

Solving on a computer gives a minimum cost of \$12.55 when $x_2 = 58.6957$, $x_8 = 2.01031$, $x_9 = 22.4895$, $x_{10} = 1$, $x_{12} = 1$, and $x_{13} = 1$.

(b) Replace 14 with 17 in inequality (i) and 16 with 16.5 in inequality (iii).
The minimum cost is \$15.05 when $x_2 = 71.7391$, $x_8 = 3.14433$, $x_9 = 23.1966$, $x_{10} = 1$, $x_{12} = 1$, and $x_{13} = 1$.

CHAPTER 8 SETS AND PROBABILITY

Section 8.1

1. 3 ____ {2, 5, 7, 9, 10}
Since 3 is not an element of the given set, insert ∉.

3. 1 ____ {3, 4, 5, 1, 11}
Since 1 is an element of the given set, insert ∈.

5. 9 ____ {2, 1, 5, 8}
Since 9 is not an element of the given set, insert ∉.

7. {2, 5, 8, 9} ____ {2, 5, 9, 8}
Since both sets contain exactly the same elements, insert =.

9. {5, 8, 9} ____ {5, 8, 9, 0}
Since the first set does not contain 0 but the second set does, insert ≠.

11. {all counting numbers less than 6} ____ {1, 2, 3, 4, 5, 6}
Since 6 is not a counting number less than 6, the first set does not contain 6. But the second set contains 6, so insert ≠.

13. {all whole numbers not greater than 4} ____ {0, 1, 2, 3}
Since 4 is a whole number not greater than 4, it is contained in the first set. But 4 is not contained in the second set, so insert ≠.

15. {x|x is a whole number, x ≤ 5} ____ {0, 1, 2, 3, 4, 5}
Since both sets contain exactly the same elements, insert =.

17. {x|x is an odd integer, 6 ≤ x ≤ 18} ____ {7, 9, 11, 15, 17}
Since 13 is an odd integer between 6 and 18 and is not contained in the second set, insert ≠.

19. {5, 7, 9, 19} ∩ {7, 9, 11, 15} ____ {7, 9}
Since the intersection of the two sets on the left is the set that contains the elements belonging to both sets, or {7, 9}, insert = .

21. {2, 1, 7} ∪ {1, 5, 9} ____ {1}
Since the union of the two sets on the left is the set that contains the elements of the first, the elements of the second, or both, or {2, 1, 7, 5, 9}, insert ≠.

23. {3, 2, 5, 9} ∩ {2, 7, 8, 10} ____ {2}
{3, 2, 5, 9} ∩ {2, 7, 8, 10} is the set that contains the common element, 2, or {2}. Insert = .

25. {3, 5, 9, 10} ∩ ∅ ____ {3, 5, 9, 10}
{3, 5, 9, 10} ∩ ∅ is the set that contains no element since none is common to both sets, or ∅, insert ≠.

27. {1, 2, 4} ∪ {1, 2, 4} ____ {1, 2, 4}
{1, 2, 4} ∪ {1, 2, 4} is the set that contains the common elements 1, 2, 4, or {1, 2, 4}, insert =.

29. ∅ ∪ ∅ ____ ∅
∅ ∪ ∅ is the set that contains no elements since none is common to both sets, or ∅. Insert =.

31. A ____ U
Since every element of A is an element of U, insert ⊆.

33. D ____ B
Since every element of D is an element of B, insert $\subseteq$.

35. A ____ B
Since 6 and 12 are contained in A but not in B, insert $\not\subseteq$.

37. $\emptyset$ ____ A
Since the null set is a subset of every set, insert $\subseteq$.

39. {4, 8, 10} ____ B
Since 4, 8, and 10 are contained in B, insert $\subseteq$.

41. B ____ D
Since 4 and 8 are in B but not in D, insert $\not\subseteq$.

43. There are exactly ____ subsets of A.
Since A has 6 elements, it has $2^6 = 64$ subsets. Insert 64.

45. There are exactly ____ subsets of C.
Since C has 3 elements, it has $2^3 = 8$ subsets. Insert 8.

47. {4, 5, 6} has 3 elements, so it has $2^3 = 8$ subsets.

49. {5, 9, 10, 15, 17} has 5 elements, so it has $2^5 = 32$ subsets.

51. Since $\emptyset$ has no elements, there are $2^0 = 1$ subset.

53. $\{x|x$ is a counting number between (not including) 6 and 12$\}$
= {7, 8, 9, 10, 11}.
It has 5 elements. There are $2^5 = 32$ subsets.

55. $X \cap Y = \{3, 5\}$ since only these elements are contained in both.

57. $X \cup U = \{2, 3, 4, 5, 7, 9\} = U$ since these are the elements in X or U.

59. $X' = \{7, 9\}$ since these are the elements that are in U but not in X.

61. $X' \cap Y' = \{7, 9\} \cap \{2, 4\}$ *No common*
$= \emptyset$ *elements*

63. **(a)** $C \cup D = \{d, e, f\} \cup \{c, f, 3, 5\}$
$= \{c, d, e, f, 3, 5\}$

(b) $B \cup (C \cup D)$
$= \{a, b, c, 1, 2, 3\}$
$\cup (\{d, e, f\} \cup \{c, f, 3, 5\})$
$= \{a, b, c, 1, 2, 3\}$
$\cup \{c, d, e, f, 3, 5\}$
$= \{a, b, c, d, e, f, 1, 2, 3, 5\}$

(c) $B \cup C$
$= \{a, b, c, 1, 2, 3\} \cup \{d, e, f\}$
$= \{a, b, c, d, e, f, 1, 2, 3\}$

(d) $(B \cup C) \cup D$
See Part (c).
$= \{a, b, c, d, e, f, 1, 2, 3\}$
$\cup \{c, f, 3, 5\}$
$= \{a, b, c, d, e, f, 1, 2, 3, 5\}$

65. **(a)** $B \cap (C \cup D)$
See Exercise 63(a).
$= \{a, b, c, 1, 2, 3\}$
$\cap \{c, d, e, f, 3, 5\}$
$= \{c, 3\}$

(b) $(B \cap C) \cup (B \cap D)$
$= (\{a, b, c, 1, 2, 3\}$
$\cap \{d, e, f\})$
$\cup (\{a, b, c, 1, 2, 3\}$
$\cap \{c, f, 3, 5\})$
$= \emptyset \cup \{c, 3\}$
$= \{c, 3\}$

67. (a) $(A \cup B)'$
Since $A \cup B$
$= \{a, c, e, 2, 4, 6\}$
$\cup \{a, b, c, 1, 2, 3\}$
$= \{a, b, c, e, 1, 2, 3, 6\}$,
$(A \cup B)' = \{d, f, 5\}$.

(b) $A' \cup B'$
Since $A = \{a, c, e, 2, 4, 6\}$,
$A' = \{b, d, f, 1, 3, 5\}$.
Since $B = \{a, b, c, 1, 2, 3\}$,
$B' = \{d, e, f, 4, 5, 6\}$.
$A' \cup B' = \{b, d, e, f, 1, 3, 4, 5, 6\}$

(c) $A' \cap B'$ See Part (b).
$= \{b, d, f, 1, 3, 5\}$
$\cap \{d, e, f, 4, 5, 6\}$
$= \{d, f, 5\}$

69. M' is the set of all students in this school not taking this course.

71. $N \cap P$ is the set of all students in this school taking both accounting and zoology.

73. $M \cup P$ is the set of all students in this school taking this course or zoology.

75. (a) $U = \{s, d, c, g, i, m, h\}$
This set includes all the different symptoms.

(b) $O' = \{s, d, c\}$
This set includes all possible symptoms not associated with overactive thyroid.

(c) $N' = \{i, m, h\}$
This set includes all possible symptoms not associated with underactive thyroid.

(d) $N \cap O = \{g\}$
This set includes those symptoms associated with both.

(e) $N \cup O = \{s, d, c, g, i, m, h\}$
This set includes those symptoms associated with either underactive or overactive thyroid.

(f) $N \cap O' = \{s, d, c\}$
This set includes those symptoms associated with an underactive thyroid and not with an overactive thyroid.

77. $A'' = A$. A' are those things in the universal set not in A. Thus A'' are those things in the universal set not in A', which is A.

Section 8.2

See the Venn diagrams for Exercises 1-15 in the answers at the back of the textbook.

1. $B \cap A'$
Shade the area inside B that is outside A.

3. $A' \cup B$
Shade the area outside A and all of B.

5. $B' \cup (A' \cap B')$
First shade the common area that is outside A and outside B.
Then shade the rest of the area outside B.

7. $U' = \emptyset$
since everything outside the universal set is nothing.

9. $(A \cap B) \cap C$
Inside the area that is common to both A and B, shade only the area that is also inside C.

11. $A \cap (B \cup C')$
Inside the area that includes all of B and the area outside C, shade only the area that is also inside A.

13. $(A' \cap B') \cap C$
Inside the common area that is outside A and outside B, shade only the area that is also inside C.

15. $(A \cap B') \cap C$
In the area that is, in common, inside A and outside B, shade only the area that is also inside C.

17. Start with the innermost region: 1 person uses all three. Since 4 people use both gas and electric ranges, $4 - 1 = 3$ people use both gas and electric but not microwaves. Continue to work outward to make the Venn diagram shown as follows.

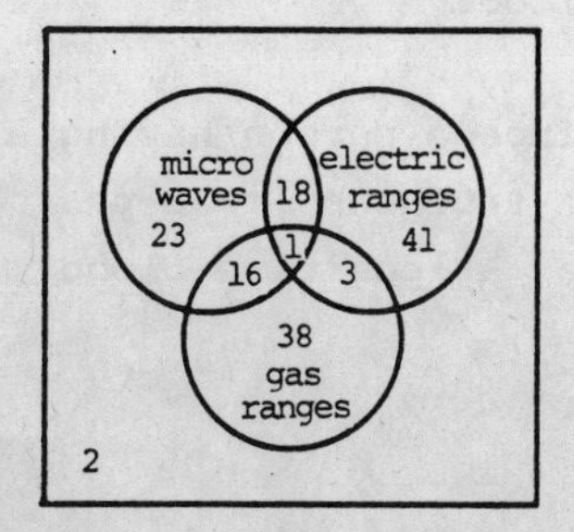

The number of people interviewed is the total of all the numbers in the diagram.

$$2 + 1 + 3 + 18 + 16 + 38 + 41 + 23 = 142$$

Yes, he should be reassigned since his data add up to 142 people not 140.

19. Start with the innermost region: 1% are unemployed black youths. Then work outward.

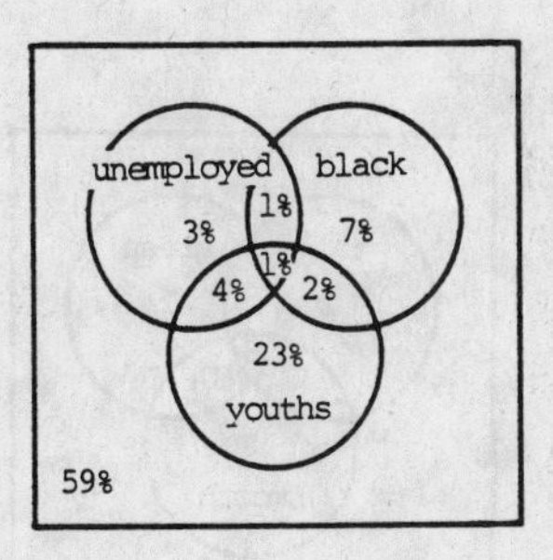

(a) Total the numbers outside the circle for black people.
3% + 4% + 23% + 59% = 89%
89% are not black.

(b) Total the numbers outside the circles for unemployed and youths.
7% + 59% = 66%
66% are employed adults.

(c) Total the numbers inside the unemployed circle but outside the youths circle.
3% + 1% = 4%
4% are unemployed adults.

(d) The number of unemployed black adults is represented by the area inside both the black and unemployed circles, but outside the youth circle, or 1%.

(e) The number of unemployed non-black youths is represented by the area inside both the unemployed and youth circles, but outside the black circle, or 4%.

21. Start with the innermost region: 6 with all three characteristics. Then work outward.

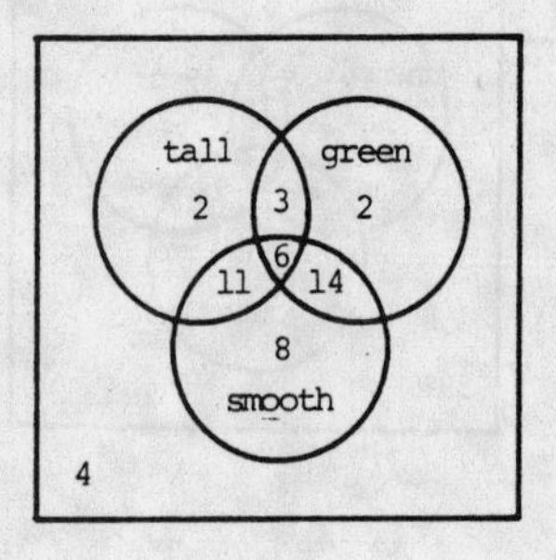

(a) 4 + 6 + 3 + 14 + 11 + 2 + 8 + 2 = 50 plants

(b) The number is inside the tall circle outside the green and smooth circles, or 2 plants.

(c) The number is inside both the green and smooth circles but outside the tall circle, or 14 plants.

23. In the diagram, the areas are labeled (a) to (g). First, the number for each area is found by starting with innermost area (d). 15 had all three; then proceed to work outward as follows:

Area (a) Since 25 had A, 25 - (2 + 15 + 1) = 7.

Area (b) Since 17 had A and B, 17 - 15 = 2.

Area (c) Since 27 had B, 27 - (2 + 15 + 7) = 3.

Area (d) 15 had all three.

Area (e) Since 22 had B and Rh, 22 - 15 = 7.

Area (f) Since 30 had Rh, 30 - (1 + 15 + 7) = 7.

Area (g) 12 had none.

Area (no label)
Since 16 had A and Rh, 16 - 15 = 1.

(a) 7 + 2 + 3 + 15 + 7 + 7 + 12 + 1 = 54
patients were represented.

(b) 7 + 3 + 7 = 17
patients had exactly one antigen.

(c) 2 + 1 + 7 = 10
patients had exactly two antigens.

(d) Since a person having only the Rh antigen has type O-positive blood, 7 had type O-positive blood.

(e) Since a person having A, B, and Rh antigens is AB-positive, 15 had AB-positive blood.

(f) Since a person having only the B antigen is B-negative, 3 had B-negative blood.

(g) Since a person having neither A, B, nor Rh antigens is O-negative, 12 has O-negative blood.

(h) Since a person having A and Rh antigens is A-positive, 1 had A-positive blood.

25. (a) $Y \cap V$ represents 20-25 year olds who drink vodka, or 40 people.

(b) $M \cap B$ represents 26-35 year olds who drink bourbon, or 30 people.

(c) $M \cup (B \cap Y)$ represents all 26-35 year olds plus 20-25 year olds who drink bourbon, or $80 + 15 = 95$ people.

(d) $Y' \cap (B \cup G)$ represents the people over 25 who drink bourbon and gin, or $30 + 20 + 50 + 10 = 110$ people.

(e) $O' \cup G$ represents people under 50 and people 50 or over who drink gin, or $(220 - 70) + 10 = 150 + 10 = 160$ people.

(f) $M' \cap (V' \cup G')$ represents the people who are 20-25 or over 35 who drink bourbon, or $15 + 50 = 65$ people.

27. $n(A) = 5$, $n(B) = 8$, and $n(A \cap B) = 4$

$$n(A \cup B) = n(A) + n(B) - n(A \cap B) = 5 + 8 - 4 = 9$$

29. $n(B) = 7$, $n(A \cap B) = 3$, and $n(A \cup B) = 20$

$$n(A \cup B) = n(A) + n(B) - n(A \cap B)$$
$$20 = n(A) + 7 - 3$$
$$20 = n(A) + 4$$
$$16 = n(A)$$

See the Venn diagrams in Exercises 31-37 in the answer section of the textbook.

31. $n(U) = 38$, $n(A) = 16$, $n(A \cap B) = 12$, $n(B') = 20$

$$n(B) = n(U) - n(B') = 38 - 20 = 18$$
$$n(A \cup B) = n(A) + n(B) - n(A \cap B) = 16 + 18 - 12 = 22$$
$$n(U) - n(A \cup B) = 38 - 22 = 16$$

Use these results to help draw the Venn diagram.

33. $n(A \cup B) = 17$, $n(A \cap B) = 3$, $n(A) = 8$, $n(A' \cup B') = 21$

$$n(A \cup B) = n(A) + n(B) - n(A \cap B)$$
$$n(B) = n(A \cup B) + n(A \cap B) - n(A) = 17 + 3 - 8 = 12$$
$$n(U) = n(A' \cup B') + n(A \cap B) = 21 + 3 = 24$$
$$n(U) - n(A \cup B) = 24 - 17 = 7$$

Use these results to help draw the Venn diagram.

35. $n(A) = 28$, $n(B) = 34$, $n(C) = 25$, $n(A \cap B) = 14$, $n(B \cap C) = 15$, $n(A \cap C) = 11$, $n(A \cap B \cap C) = 9$, $n(U) = 59$

Start with the innermost area, $n(A \cap B \cap C) = 9$, to draw the Venn diagram. Continue outward.
$n(A \cap B) - n(A \cap B \cap C) = 14 - 9 = 5$ and so on.

37. $n(A \cap B) = 6$, $n(A \cap B \cap C) = 4$,
$n(A \cap C) = 7$, $n(B \cap C) = 4$,
$n(A \cap C') = 11$, $n(B \cap C') = 8$,
$n(C) = 15$, $n(A' \cap B' \cap C') = 5$
Start with the innermost area, as before.
Note that

$$n(A) = n(A \cap C') + n(A \cap C) = 11 + 7 = 18.$$

39. Prove $(A \cup B)' = A' \cap B'$.
Since $A \cup B$ is the set containing all of A and B, $(A \cup B)'$ is the set containing that part of the universe not in A or B.

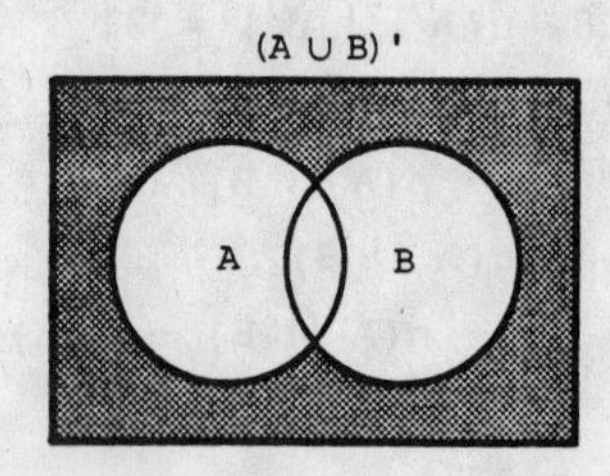

Now shade A'.

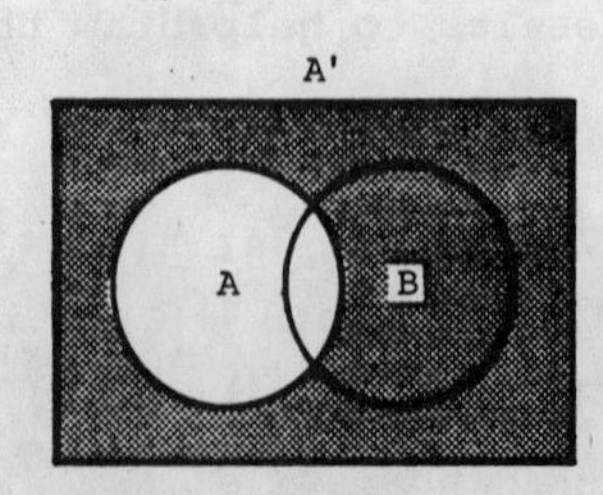

Then shade B'.

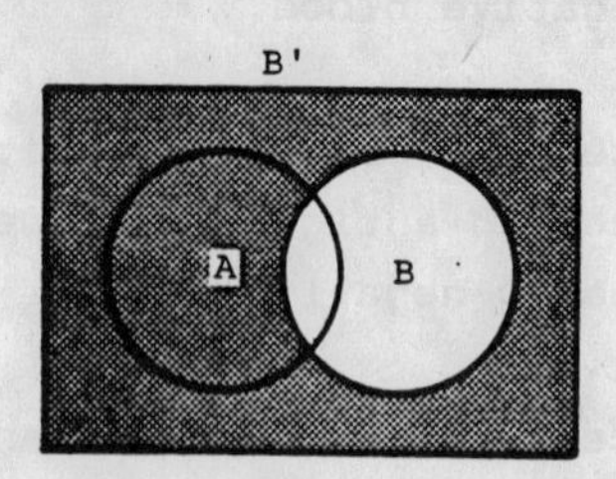

The shaded area that is common to both is the intersection, $A' \cap B'$.

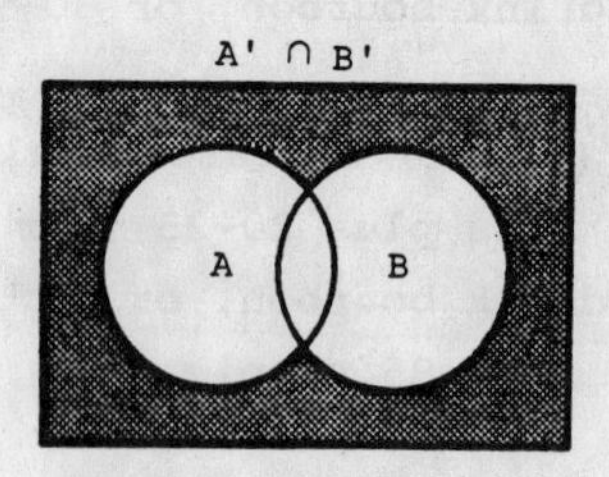

Therefore, $(A \cup B)' = A' \cap B'$.

41. Prove $A \cap (B \cup C) = (A \cap B) \cup (A \cap C)$.
Shade $B \cup C$.

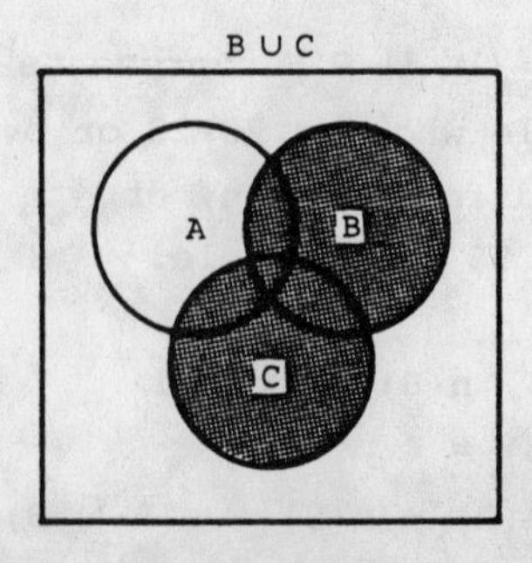

Now shade $A \cap (B \cup C)$.

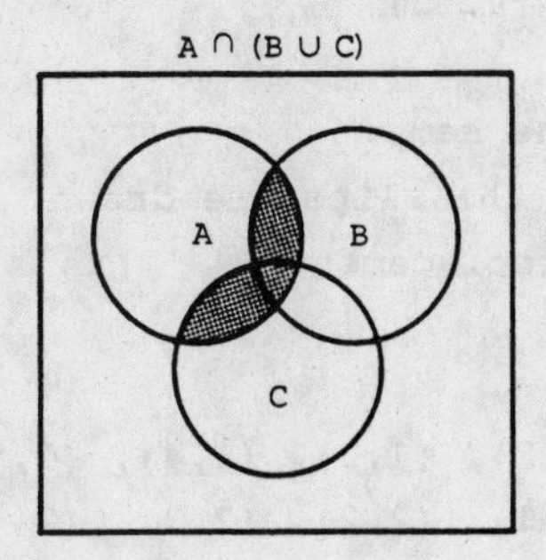

Shade $A \cap B$.

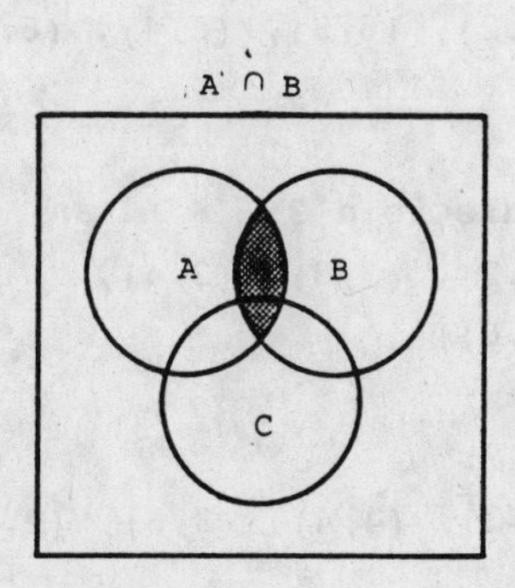

Then shade $A \cap C$.

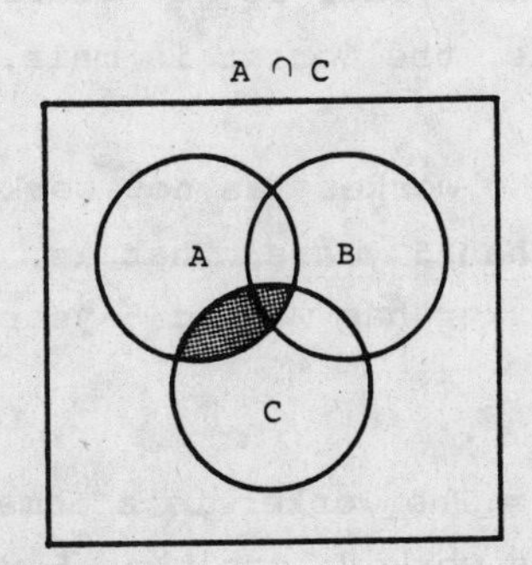

Now shade $(A \cap B) \cup (A \cap C)$.

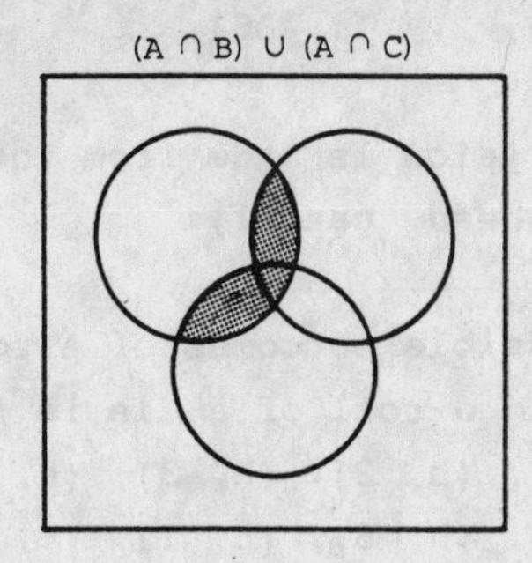

Therefore, $A \cap (B \cup C)$
$= (A \cap B) \cup (A \cap C)$.

43. Show that $n(A') = n(U) - n(A)$.
Let $x = n(A)$, $y = n(A')$, $z = n(U)$.
The figure shows A and A', which compose the universal set, U.

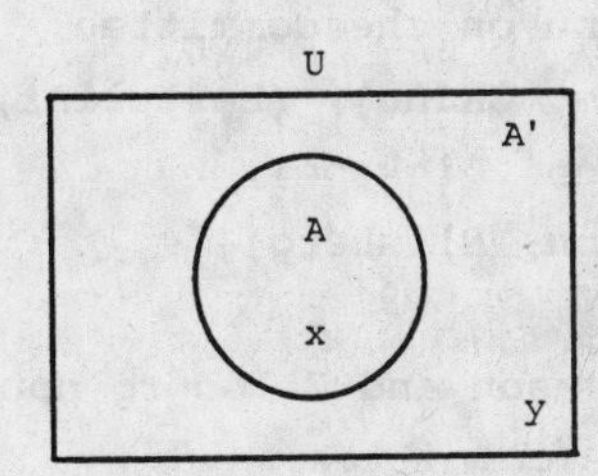

Since $n(U) = z$,
$x + y = z$.
By algebra, subtracting x from each side gives
$y = z - x$.
Therefore, $n(A') = n(U) - n(A)$.

Section 8.3

1. A month of the year is chosen from the sample set
{January, February, March, ..., December}.

3. The set of points earned in an 80-point test is
{0, 1, 2, ..., 80}.

5. The decision is made from the set
{go ahead, cancel}.

7. The possible outcomes of a toss of a coin and a roll of a die is
{(h, 1), (h, 2), (h, 3), (h, 4), (h, 5), (h, 6), (t, 1), (t, 2), (t, 3), (t, 4), (t, 5), (t, 6)}.

9. The sample set is
{(Alam, Bartolini), (Alam, Chinn), (Alam, Dickson), (Alam, Ellsberg), (Bartolini, Chinn), (Bartolini, Dickson), (Bartolini, Ellsberg), (Chinn, Dickson), (Chinn, Ellsberg), (Dickson, Ellsberg)}.

(a) Chinn on the committee:
{(Alam, Chinn), (Bartolini, Chinn), (Chinn, Dickson), (Chinn, Ellsberg)}

(b) Dickson and Ellsberg not both on the committee:
{All of those in the sample set except for (Dickson, Ellsberg)}

(c) Both Alam and Chinn on the committee:
{(Alam, Chinn)}

11. The sample set is
{(1, 2), (1, 3), (1, 4), (1, 5), (2, 3), (2, 4), (2, 5), (3,4), (3, 5), (4, 5)}.

(a) Both even:
{(2, 4)}

(b) One odd, one even:
{(1, 2), (1, 4), (2, 3), (2, 5), (3, 4), (4, 5)}

(c) Both the same:
∅, since the slips are drawn without replacement.

13. Sample set:
{(1,1), (1,2), (1,3), (1,4), (1,5), (1,6), (2,1), (2,2), (2,3), (2,4), (2,5), (2,6), (3,1), (3,2), (3,3), (3,4), (3,5), (3,6), (4,1), (4,2), (4,3), (4,4), (4,5), (4,6), (5,1), (5,2), (5,3), (5,4), (5,5), (5,6), (6,1), (6,2), (6,3), (6,4), (6,5), (6,6)}

(a) First die is a 3:
{(3,1), (3,2), (3,3), (3,4), (3,5), (3,6)}

(b) Sum is 8:
{(6,2), (5,3), (4,4), (3,5), (2,6)}

(c) Sum is 13:
∅, since the largest possible sum on any two die is 12.

15. **(a)** E' = The worker is not female; that is, the worker is male.

(b) F' = The worker has not worked less than 5 years; that is, the worker has worked 5 years or more.

(c) $E \cap F$ = The worker is a female who has worked less than 5 years.

(d) $F \cup G$ = The worker has worked less than 5 years or contributes to a voluntary retirement plan.

(e) $E \cup G'$ = The worker is female or does not contribute to a voluntary retirement plan.

(f) $F' \cap G'$ = The worker has not worked less than 5 years and does not contribute to a voluntary retirement plan; that is the worker has worked 5 or more years and does not contribute to a voluntary retirement plan.

17. P(2)

$$= \frac{\text{number of ways to get a 2}}{\text{total number of possibilities}}$$

$$= \frac{1}{6}$$

19. P(less than 5)

$$= \frac{\text{number of ways to get a number less than 5}}{\text{total number of possibilities}}$$

$$= \frac{4}{6} = \frac{2}{3}$$

21. P(3 or 4)

$$= \frac{\text{number of ways to get a 3 or a 4}}{\text{total number of possibilities}}$$

$$= \frac{2}{6} = \frac{1}{3}$$

23. P(9)

$$= \frac{\text{number of ways to get a 9}}{\text{total number of cards}}$$

$$= \frac{4}{52} = \frac{1}{13}$$

25. P(black 9)

$$= \frac{\text{number of ways to get a black 9}}{\text{total number of cards}}$$

$$= \frac{2}{52} = \frac{1}{26}$$

27. P(9 of hearts)

$$= \frac{\text{number of ways to get a 9 of hearts}}{\text{total number of cards}}$$

$$= \frac{1}{52}$$

29. P(2 or queen)

$$= \frac{\text{number of 2's + number of queens}}{\text{total number of cards}}$$

$$= \frac{4 + 4}{52} = \frac{8}{52} = \frac{2}{13}$$

31. P(heart or spade)

$$= \frac{\text{number of hearts + number of spades}}{\text{total number of cards}}$$

$$= \frac{13 + 13}{52} = \frac{26}{52} = \frac{1}{2}$$

33. P(2, 3, 4, or 5)

$$= \frac{\text{number of 2's + number of 3's + number of 4's + number of 5's}}{\text{total number of cards}}$$

$$= \frac{4 + 4 + 4 + 4}{52} = \frac{16}{52} = \frac{4}{13}$$

35. P(red)

$$= \frac{\text{number of red marbles}}{\text{total marbles}}$$

$$= \frac{5}{25} = \frac{1}{5}$$

37. P(green)

$$= \frac{\text{number of green marbles}}{\text{total marbles}}$$

$$= \frac{9}{25}$$

39. P(not black)

$$= \frac{\text{number of nonblack marbles}}{\text{total marbles}}$$

$$= \frac{5 + 7 + 9}{25} = \frac{21}{25}$$

41. P(red or black)

$= \dfrac{\text{number of red + number of black}}{\text{total marbles}}$

$= \dfrac{5 + 4}{25} = \dfrac{9}{25}$

43. P(A)

$= \dfrac{\text{number of A's}}{\text{total letters in word}}$

$= \dfrac{5}{11}$

45. P(C or D)

$= \dfrac{\text{number of C's + number of D's}}{\text{total letters in word}}$

$= \dfrac{1 + 1}{11} = \dfrac{2}{11}$

Section 8.4

1. Owning a car and owning a truck are not mutually exclusive, since you can own a car and a truck at the same time.

3. Being married and being over 30 are not mutually exclusive, since you can be married and over the age of 30 at the same time.

5. Rolling a die once, getting a 4 and getting an odd number is mutually exclusive, since it is impossible to get a 4 and an odd number at the same time.

7. (a) P(sum is 2)

$= \dfrac{\text{number of ways to get a 2}}{\text{total number of possibilities}}$

$= \dfrac{1}{36}$

(b) P(sum is 4)

$= \dfrac{\text{number of ways to get a 4}}{\text{total number of possibilities}}$

$= \dfrac{3}{36} = \dfrac{1}{12}$

(c) P(sum is 5)

$= \dfrac{\text{number of ways to get a 5}}{\text{total number of possibilities}}$

$= \dfrac{4}{36} = \dfrac{1}{9}$

(d) P(sum is 6)

$= \dfrac{\text{number of ways to get a 6}}{\text{total number of possibilities}}$

$= \dfrac{5}{36}$

9. (a) P(sum is 9 or more)

$= \dfrac{\text{number of ways to get a 9, 10, 11, or 12}}{\text{total number of possibilities}}$

$= \dfrac{4 + 3 + 2 + 1}{36} = \dfrac{10}{36} = \dfrac{5}{18}$

(b) P(sum is less than 7)

$= \dfrac{\text{number of ways to get a 2, 3, 4, 5, or 6}}{\text{total number of possibilities}}$

$= \dfrac{1 + 2 + 3 + 4 + 5}{36} = \dfrac{15}{36} = \dfrac{5}{12}$

(c) P(sum is between 5 and 8)

$= \dfrac{\text{number of ways to get a 6 or 7}}{\text{total number of possibilities}}$

$= \dfrac{5 + 6}{36} = \dfrac{11}{36}$

11. (a) P(9 or 10)

$= \dfrac{\text{number of ways to get a 9 or 10}}{\text{total number of cards}}$

$= \dfrac{4 + 4}{52} = \dfrac{8}{52} = \dfrac{2}{13}$

(b) P(red or 3)

$$= \frac{\text{number of ways to get a red} + \text{number of ways to get a 3 - number of red 3's}}{\text{total number of cards}}$$

$$= \frac{26 + 4 - 2}{52} = \frac{28}{52} = \frac{7}{13}$$

(c) P(9 or black 10)

$$= \frac{\text{number of 9's + number of black 10's}}{\text{total number of cards}}$$

$$= \frac{4 + 2}{52} = \frac{6}{52} = \frac{3}{26}$$

(d) P(heart or black)

$$= \frac{\text{number of hearts + number of blacks}}{\text{total number of cards}}$$

$$= \frac{13 + 26}{52} = \frac{3}{4}$$

13. (a) P(brother or uncle)

$$= \frac{\text{number of brothers + number of uncles}}{\text{total number of relatives}}$$

$$= \frac{2 + 3}{10} = \frac{5}{10} = \frac{1}{2}$$

(b) P(brother or cousin)

$$= \frac{\text{number of brothers + number of cousins}}{\text{total number of relatives}}$$

$$= \frac{2 + 2}{10} = \frac{4}{10} = \frac{2}{5}$$

(c) P(brother or mother)

$$= \frac{\text{number of brothers + number of mothers}}{\text{total number of relatives}}$$

$$= \frac{2 + 1}{10} = \frac{3}{10}$$

15. There are 20 possibilities:
(1,2), (1,3), (1,4), (1,5), (3,5)
(2,3), (2,4), (2,5), (3,4), (4,5)
(2,1), (3,1), (4,1), (5,1), (5,3)
(3,2), (4,2), (5,2), (4,3), (5,4).

(a) P(sum of 9)

$$= \frac{\text{number of ways to get a 9}}{\text{total number of possibilities}}$$

$$= \frac{2}{20} = \frac{1}{10}$$

(b) P(sum if 5 or less)

$$= \frac{\text{number of ways to get a 5, 4, or 3}}{\text{total number of possibilities}}$$

$$= \frac{4 + 2 + 2}{20} = \frac{8}{20} = \frac{2}{5}$$

(c) P(first number 2 or sum is 2)

$$= \frac{\text{number with first number 2} + \text{number of sum of 6 - two on first and sum of 6}}{\text{total number of possibilities}}$$

$$= \frac{4 + 4 - 1}{20} = \frac{7}{20}$$

17. $P(E) = .26$, $P(F) = .41$, $P(E \cap F) = .17$

(a) $P(E \cup F) = P(E) + P(F) - P(E \cap F)$
$= .26 + .41 - .17 = .5$

Draw the Venn diagram showing probabilities. Insert $P(E \cap F) = .17$ and $P(E \cup F)' = 1 - P(E \cup F)$
$= 1 - .5$
$= .5.$

Note that $P(E \cup F)' = P(E' \cap F')$.

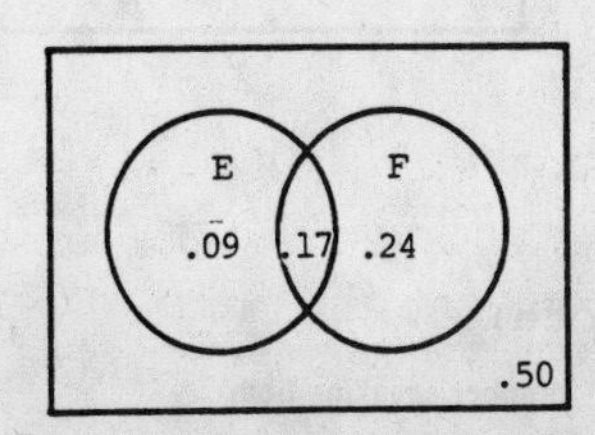

(b) $P(E' \cap F) = P(\text{not } E \text{ and } F)$
$= P(F) - P(E \cap F)$
$= .41 - .17 = .24$

See Venn diagram above.

(c) $P(E \cap F') = P(E \text{ and not } F)$
$= P(E) - P(E \cap F)$
$= .26 - .17 = .09$

See Venn diagram above.

(d) $P(E' \cup F')$
$= P(E') + P(F') - P(E' \cap F')$
$= 1 - P(E) + 1 - P(F) - P(E \cup F)'$
$= 1 - .26 + 1 - .41 - .5$
$= .83$

19. Draw Venn diagram showing numbers of students. Start with the innermost area: 8 spoke both languages. Since 45 spoke Spanish and 8 spoke both languages, 45 - 8 = 37 spoke Spanish and not Vietnamese. Likewise, 10 - 8 = 2 spoke Vietnamese and not Spanish. Of the 50 students, 50 - (8 + 37 + 2) = 3 spoke neither language. Use these numbers to solve the following.

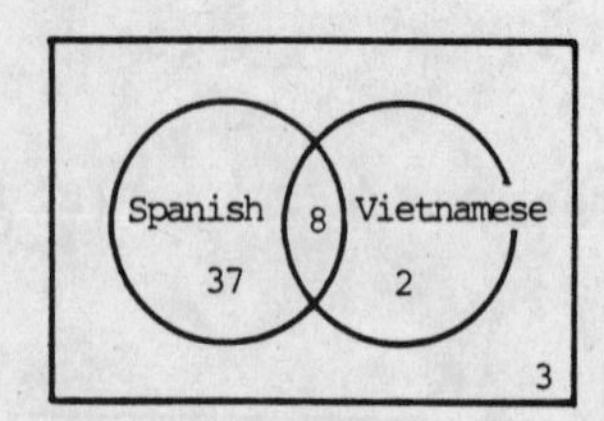

(a) P(both)

$$= \frac{\text{number speaking both}}{\text{total students}}$$

$$= \frac{8}{50} = \frac{4}{25}$$

(b) P(neither)

$$= \frac{\text{number speaking neither}}{\text{total students}}$$

$$= \frac{3}{50}$$

(c) P(only one)

$$= \frac{\text{number speaking only Spanish} + \text{number speaking only Vietnamese}}{\text{total students}}$$

$$= \frac{37 + 2}{50} = \frac{39}{50}$$

21. Complete a table to show the data.

	Men	Women	Total
Earn more than $40,000	13	3	16
Earn less than $40,000	62	52	114
Total	75	55	130

(a) P(woman earning less than $40,000)

$$= \frac{\text{women earning less than \$40,000}}{\text{total adults}}$$

$$= \frac{52}{130} = \frac{2}{5} = .4$$

(b) P(man earning more than $40,000)

$$= \frac{\text{men earning more than \$40,000}}{\text{total adults}}$$

$$= \frac{13}{130} = .1$$

(c) P(man or earning more than $40,000)

$$= \frac{\text{Number of men} + \text{number earning over \$40,000} - \text{men earning over \$40,000}}{\text{total adults}}$$

$$= \frac{75 + 16 - 13}{130} = \frac{78}{130} = \frac{3}{5} = .6$$

(d) P(woman or earning less than $40,000)

$$= \frac{\text{number of women + number earning less than \$40,000 - women earning less than \$40,000}}{\text{total adults}}$$

$$= \frac{55 + 114 - 52}{130} = \frac{117}{130} = .9$$

23.

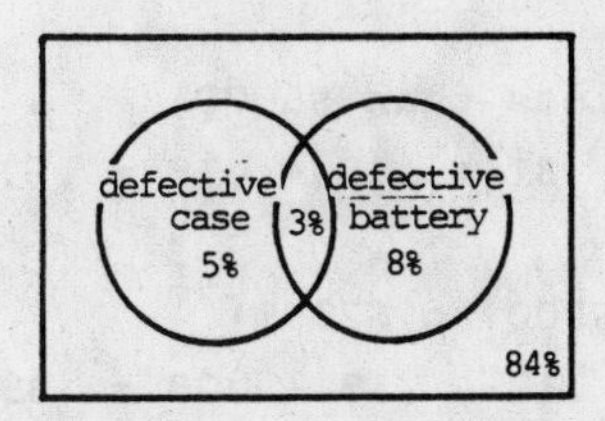

Since the case and batteries must be good, we need the percent outside the 2 circles, which is 84%.

To draw the Venn diagram for Exercises 25-29, find

$P(C) - P(M \cap C) = .049 - .042 = .007.$

Also, find the following.

$P(M \cup C) = P(M) + P(C) - P(M \cap C)$, substitute and solve for P(M)

$$.534 = P(M) + .049 - .042$$
$$P(M) = .534 - .049 + .042$$
$$P(M) = .527$$
$$P(M) - P(M \cap C) = .527 - .042 = .485$$
$$P(M \cup C)' = 1 - P(M \cup C)$$
$$= 1 - .534$$
$$= .466$$

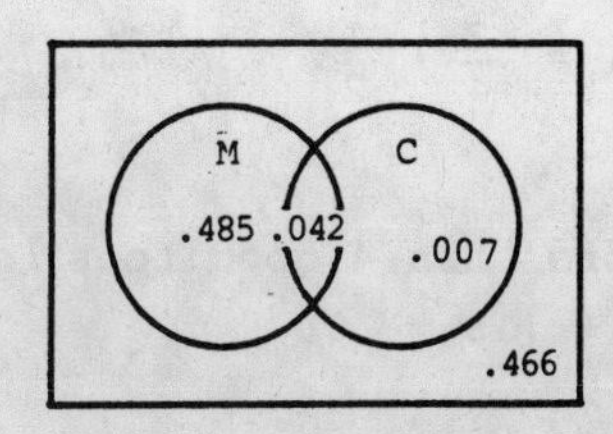

25. $P(C') = .485 + .466 = .951$

27. $P(M') = .007 + .466 = .473$

29. $P(C \cap M') = .007$
(inside C and outside M)

31. Number of ways to get a 5 = 1
Number of ways not to get a 5 = 5
Odds of getting a 5 = 1 to 5

33. Number of ways to get a 1, 2, 3 or 4 = 4
Number of ways not to get a 1, 2, 3, or 4 = 2
Odds of getting 1, 2, 3, or 4
= 4 to 2 or 2 to 1

35. **(a)** Probability of shooting below 75
= .01 + .08 + .15 + .28 = .52
Probability of not shooting below 75
= .48
Odds of shooting below 75
= .52 to .48 or 13 to 12

(b) Probability of shooting in 70's
= .50
Probability of not shooting in 70's = .50
Odds of shooting in the 70's
= .50 to .50 or 1 to 1

37. Number of ways not to get a white marble = 3 + 8 = 11
Number of ways to get a white = 4
Odds of not getting a white marble
= 11 to 4

39. Odds that it will rain are 4 to 7.
P(rain)

$$= \frac{4}{4 + 7} = \frac{4}{11}$$

41. No, because "odds against a direct hit are very low" means odds in favor of a direct hit are high. He should have said, "Odds in favor of a direct hit are very low."

43. Yes, this is possible since all probabilities are between 0 and 1 inclusive, and .92 + .03 + 0 + .02 + .03 = 1.

45. This is not possible since

$$\frac{1}{5} + \frac{1}{3} + \frac{1}{4} + \frac{1}{5} + \frac{1}{10} = \frac{13}{12} > 1.$$

47. This is not possible since $S_5 = -.3 < 0$.

49. Probability of winning if person watches TV for 23-50 hours per week is

$$\frac{14 + 96 + 327 + 165 + 55}{2000}$$

$$= \frac{657}{2000} = .3285.$$

51. Probability of winning if person has an income of at least $25,000 is

$$= \frac{18 + 31 + 73 + 32 + 85 + 60 + 88 + 160 + 52 + 327 + 165 + 55 + 189 + 100 + 12}{2000}$$

$$= \frac{1447}{2000} = .7235.$$

53. Probability of winning if person has an income less than $25,000 and watches at least 9 hours is

$$= \frac{15 + 8 + 14 + 120 + 19 + 28 + 96 + 232}{2000}$$

$$= \frac{532}{2000} = .266.$$

55. Probability of winning if person has an income of $0 - $24,999 and watches more than 30 hours is

$$= \frac{120 + 232}{2000}$$

$$= .176$$

57. **(a)** P($500 or more)
= .18 + .13 + .08 + .05 + .06 + .01
= .51

(b) P(less than $1000)
= .31 + .18 + .18 = .67

(c) P($500 to $2999)
= .18 + .13 + .08 = .39

(d) P($3000 or more)
= .05 + .06 + .01 = .12

59. **(a)** P(less than 240)
= .10 + .15 + .20 = .45

(b) P(220 or more)
= .20 + .26 + .29 = .75

(c) P(from 200 to 239)
= .15 + .20 = .35

(d) P(from 220 to 259)
= .220 + .26 = .46

61. Since red is dominant over white, the only combination to produce white is WW. Hence $P(\text{white}) = \frac{1}{4}$.

63. Pink is produced by RW or WR.

$P(\text{pink}) = \frac{2}{4} = \frac{1}{2}$.

65. P(no more than 4 good toes)
= .77 + .13 = .90

67. Answer will vary by the Monte Carlo method. To find the theoretical probability, the number of outcomes in the sample space is $2^5 = 32$. The possible outcomes of the event, 4 heads, are
(h, h, h, h, t), (h, h, h, t, h), (h, h, t, h, h), (h, t, h, h, h), and (t, h, h, h, h), which are 5 in number.
$P(\text{4 heads}) = \frac{5}{32} = .15625$

69. Answers will vary by the Monte Carlo method. To find the theoretical probability, find the number of possible outcomes in the sample space. On the first draw, 52 different cards can be drawn, on the second, 51, on the third, 50, and on the fourth, 49. So the total number of possible outcomes, in the sample space is
$$52 \cdot 51 \cdot 50 \cdot 49.$$
For the event, any 2 cards and then 2 kings, on the first draw, 50 different cards can be drawn because 2 kings must be left for the third and fourth draw, on the second, 49, on the third, 4 kings can be drawn, and on the fourth, since 1 king was drawn on the third, only 3 kings can be drawn. So the number of possible outcomes of the event is
$$50 \cdot 49 \cdot 4 \cdot 3.$$
P(any two cards, and then 2 kings)
$$= \frac{50 \cdot 49 \cdot 4 \cdot 3}{52 \cdot 51 \cdot 50 \cdot 49}$$
$$= \frac{12}{2652} = \frac{1}{221}$$
or about .004525

71. Answers will vary. The theoretical probability that at least 1 is right depends on the fact that if 1 watch is placed in the wrong box then at least 1 other watch is also wrong. Thus the sample space is
{all 8 right, 6 right and 2 wrong,
5 right and 3 wrong,
4 right and 4 wrong,
3 right and 5 wrong,
2 right and 6 wrong,
1 right and 7 wrong,
0 right and 8 wrong}.
By counting the possible outcomes,
$P(\text{at least 1 right}) = \frac{7}{8}$.

Section 8.5

1. There are 3 odd numbers, none of them is a 2.
$P(2 \mid \text{odd}) = \frac{0}{3} = 0.$

3. There is one 6, and it is even.
$P(\text{even} \mid 6) = \frac{1}{1} = 1.$

5. There are 6 doubles, 1 of which has a sum of 6.
$P(\text{sum of } 6 \mid \text{double}) = \frac{1}{6}.$

7. Since the first card is a heart, there are 51 cards remaining, 12 of them hearts.
$P(\text{second is heart} \mid \text{first is heart})$
$= \frac{12}{51} = \frac{4}{17}$

9. Since the first card is a jack, there are 51 cards remaining, 11 of them face cards.
$P(\text{second is face card} \mid \text{first is jack})$
$= \frac{11}{51}$

11. $P(\text{jack and } 10) = \dfrac{8 \cdot 4}{52 \cdot 51} = .012$

There are 8 possibilities for the first card (4 jacks and 4 tens), but for the second card there are only 4 (the 4 tens if a jack was picked or the 4 jacks if a 10 was picked).

13. $P(2 \text{ black cards}) = \dfrac{26 \cdot 25}{52 \cdot 51} = .245$

In the following tree diagrams the final probability is found by multiplying the probabilities on the pieces that make up each branch.

15. Complete a tree diagram.

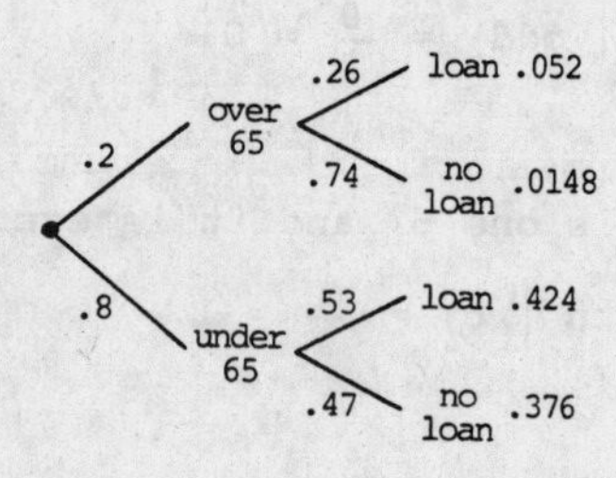

P(65 and has loan) = .052

17.

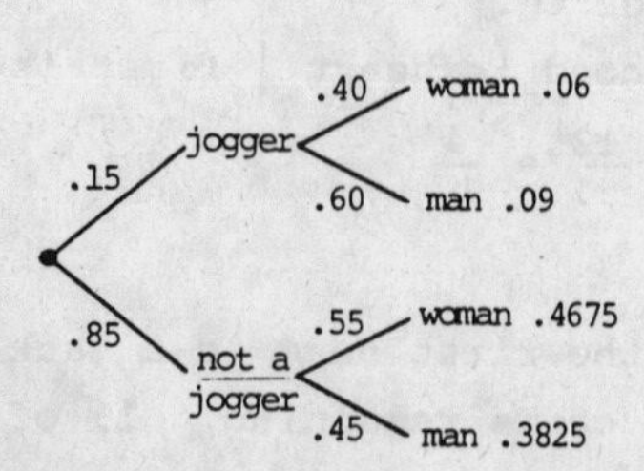

P(women jogger) = .06

19.

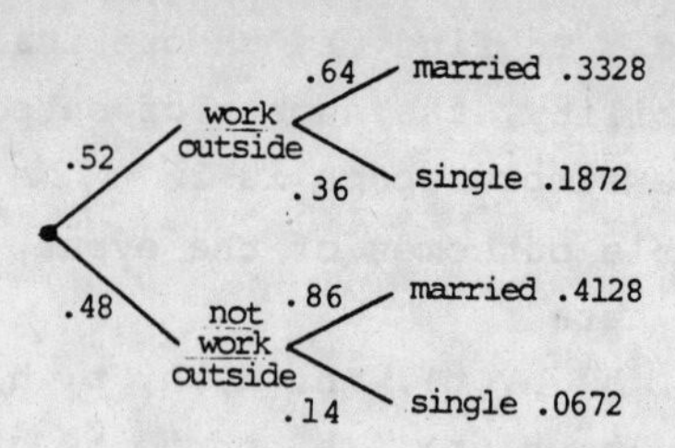

P(married) = .3328 + .4128 = .7456

21.

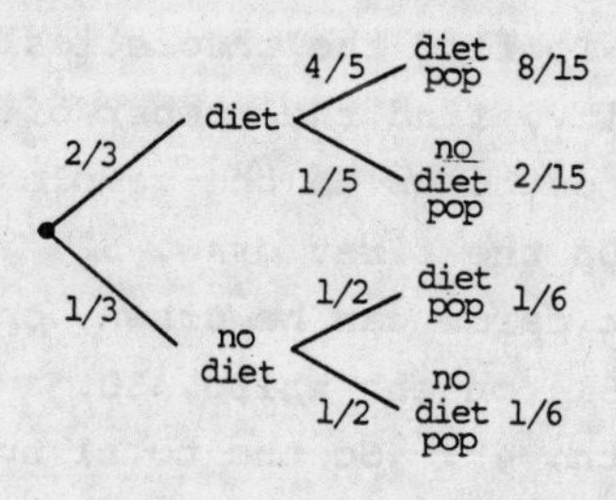

P(drink diet soft drink)

$= \dfrac{8}{15} + \dfrac{1}{6} = \dfrac{16 + 5}{30} = \dfrac{21}{30} = \dfrac{7}{10}$

23.

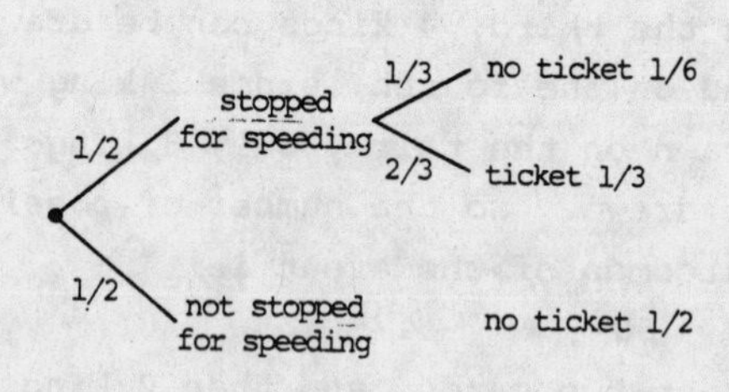

P(gets no ticket) $= \dfrac{1}{6} + \dfrac{1}{2} = \dfrac{2}{3}$

25. Since there are 2 evens and 3 odds, P(first is even and second is odd)

$= \dfrac{2}{5} \cdot \dfrac{3}{4} = \dfrac{6}{20} = \dfrac{3}{10}.$

27. Since there are 2 evens on the first draw and 1 even on the second, if another even is drawn on the first, P(both are even)

$= \frac{2}{5} \cdot \frac{1}{4} = \frac{2}{20} = \frac{1}{10}.$

29. Since there are 3 whites on the first draw and 2 whites on the second, P(both are white)

$= \frac{3}{7} \cdot \frac{2}{6} = \frac{6}{42} = \frac{1}{7}.$

31. Since the first is black, there are 6 marbles left, 3 of them black, 3 white.

P(second is white | first is black)

$= \frac{3}{6} = \frac{1}{2}$

33. P(one black and 1 white)

= P(black first, white second)

+ P(white first, black second)

$= \frac{4}{7} \cdot \frac{3}{6} + \frac{3}{7} \cdot \frac{4}{6}$

$= \frac{12}{42} + \frac{12}{42} = \frac{24}{42} = \frac{4}{7}$

35. $P(D' \mid C) = \frac{30}{80} = \frac{3}{8}$

37. $P(C' \mid D) = \frac{20}{70} = \frac{2}{7}$

39. Complete a table.

	Male	Female	Total
Beagle	1	2	3
Cocker Spaniel	1	0	1
Poodle	4	2	6
Total	6	4	10

$P(\text{beagle}) = \frac{3}{10}$

41. See the table in the solution for Exercise 39. There are 3 beagles, of which 1 is male.

$P(\text{male} \mid \text{beagle}) = \frac{1}{3}$

43. See the table in the solution for Exercise 39. There are 6 males, of which 4 are poodles.

$P(\text{poodle} \mid \text{male}) = \frac{4}{6} = \frac{2}{3}$

45.

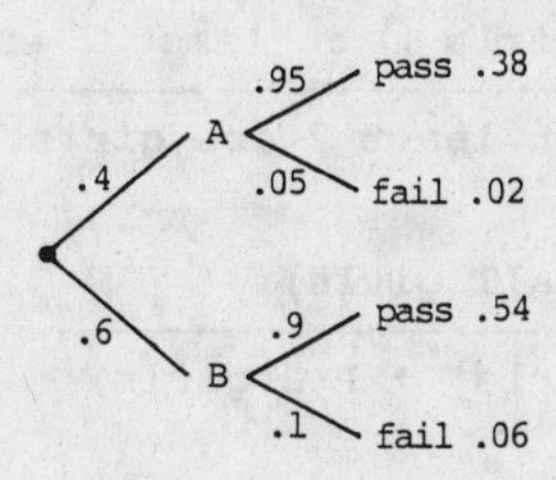

P(did not pass and came off line A)

= .02

47. There are 3 ways to get a red: RW, WR, RR. 2 of these combine a red and white: RW, WP.

$P(\text{combined gene} \mid \text{red}) = \frac{2}{3}$

The tree diagram used for Exercises 49 and 51 has 1/2 for each probability on a piece of a branch and 1/8 for each probability of a branch.

49. $P(\text{all girls} \mid \text{third is a girl})$

$$= \frac{P(\text{all girls} \cap \text{third is a girl})}{P(\text{third is a girl})}$$

$$= \frac{P(\text{all girls})}{P(\text{third is a girl})}$$

$$= \frac{\frac{1}{8}}{\frac{1}{2}} = \frac{1}{4}$$

51. $P(\text{all girls} \mid \text{at least 2 are girls})$

$$= \frac{P(\text{all girls} \cap \text{at least 2 are girls})}{P(\text{at least 2 are girls})}$$

$$= \frac{P(\text{all girls})}{P(\text{2 girls}) + P(\text{3 girls})}$$

$$= \frac{\frac{1}{8}}{\frac{3}{8} + \frac{1}{8}} = \frac{\frac{1}{8}}{\frac{1}{2}} = \frac{1}{4}$$

53. $P(M) = .527$ (directly from the chart)

55. $P(M \cap C) = P(M \text{ and } C) = .042$

57. $P(M \mid C) = \frac{P(M \cap C)}{P(C)} = \frac{.042}{.049} = \frac{6}{7}$

or .857

59. $P(C|M) = \frac{P(M \cap C)}{P(C)} = \frac{.042}{.527} = .0797$

$P(C) = .049$

Since $P(C|M) \neq P(C)$, they are dependent.

61. Complete the tree diagram.

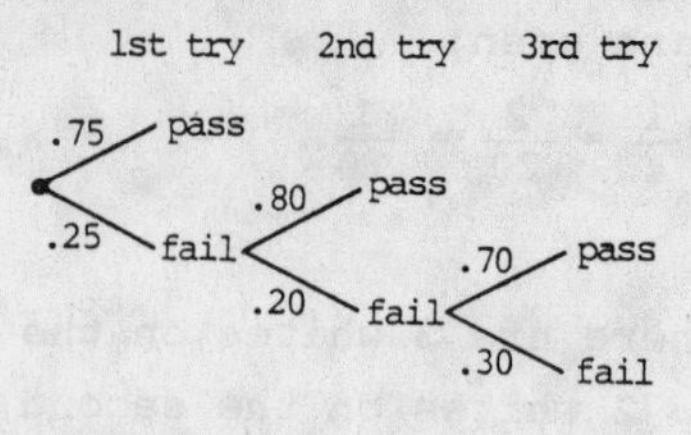

P(fails 1st and 2nd test)

$= P(\text{fails 1st}) \cdot P(\text{fails 2nd} | \text{fails 1st})$

$= (.25)(.20) = .05$

63. P(requires at least 2 tries)

= P(does not pass on 1st try)

= .25

65. Complete the tree diagram.

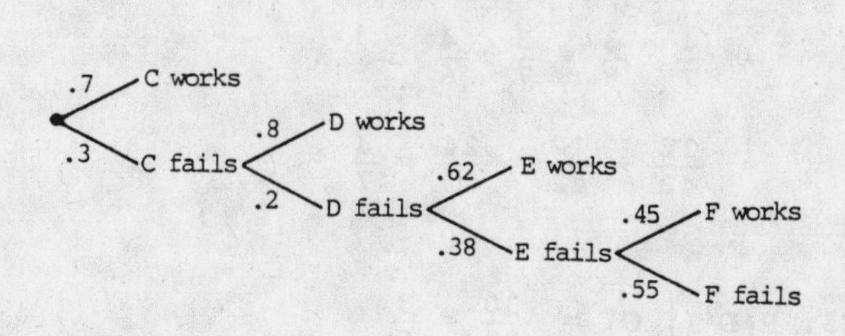

P(will take at least 2 and blood pressure reduced)

= P(C fails and D works)

+ P(C fails, D fails, and E works)

+ P(C fails, D fails, E fails, and F works)

$= .3(.8) + .3(.2)(.62)$

$+ .3(.2)(.38)(.45)$

$= .24 + .0372 + .01026$

$\approx .287$

67. Since the driver is selected at random from all drivers, the probabilities must be multiplied by the percent of drivers in each category of consumption of beer.
P(has an accident)
= P(no drinks)
+ P(1-2 beers) + P(3-5 beers)
+ P(more than 5 beers)
= .29(.003) + .37(.004)
+ .23(.011) + .11(.122)
= .0183

69. P(nondrinker injured in an accident)
= .29(.003)(.15) = .00013

71. P(less than 6 beer, has accident, and excapes injury)
= .29(.003)(.85) + .37(.004)(.82)
+ .23(.011)(.68)
= .00367

73. P(at least 1 hit) = 1 - P(no hit)
P(no hit) = P(no hits in first four)
= (1 - .32)(1 - .16)(1 - .08)
(1 - .04)
= .5045
P(at least 1 hit) = 1 - .5045 = .4955

75. P(hit in 1st 6 if 1st 3 are not hits)

$$= \frac{\text{P(hit in 1st 6 and 1st 3 are not hits)}}{\text{P(1st 3 are not hits)}}$$

P(hit in 1st 6 and 1st 3 are not hits)
= .68(.84)(.92)(.04)
+ (.68)(.84)(.92)(.96)(.02)
+ (.68)(.84)(.92)(.96)(.98)(.01)
= .036054
P(1st 3 are not hits
= .68(.84)(.92) = .525504
P(hit in 1st 6 if 1st 3 are not hits)

$$= \frac{.036054}{.525504} = .0686$$

77. The assumption of independence is not very realistic, since if weather is bad, many flights would be cancelled.

79. P(have computer service)
= 1 - P(no computer service)
= 1 - (.003)(.005) = 1 - .000015
= .999985
It is fairly realistic to assume independence because the chance of a failure of one computer does not usually depend on the failure of another.

Section 8.6

1.

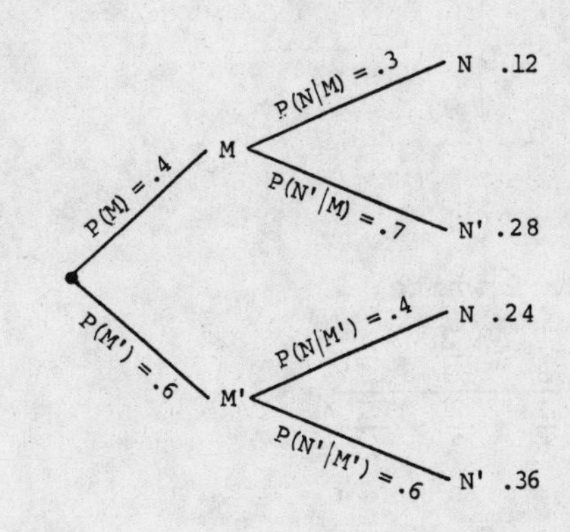

$$P(M|N) = \frac{.12}{.12 + .24} = \frac{.12}{.36} = \frac{1}{3}$$

3.

$P(R_1) = .05$
$P(R_2) = .6$
$P(R_3) = .35$
$P(Q|R_1) = .4$ Q .02
$P(Q'|R_1) = .6$ Q .03
$P(Q|R_2) = .3$ Q .18
$P(Q'|R_2) = .7$ Q .46
$P(Q|R_3) = .6$ Q .21
$P(Q'|R_3) = .4$ Q .14

$P(R_1|Q)$

$$= \frac{.02}{.02 + .18 + .21} = \frac{.02}{.41} = \frac{2}{41}$$

5. See art in Exercise 3.

$P(R_3|Q)$

$$= \frac{.21}{.02 + .18 + .21} = \frac{.21}{.41} = \frac{21}{41}$$

7.

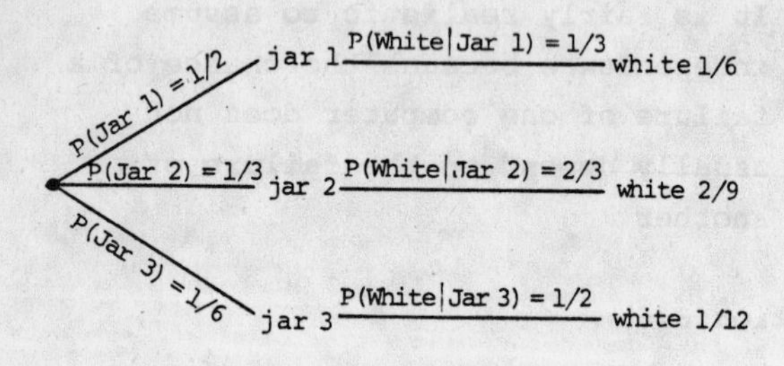

P(jar 2|white)

$$= \frac{\frac{2}{9}}{\frac{1}{6} + \frac{2}{9} + \frac{1}{12}} = \frac{8}{17}$$

9.

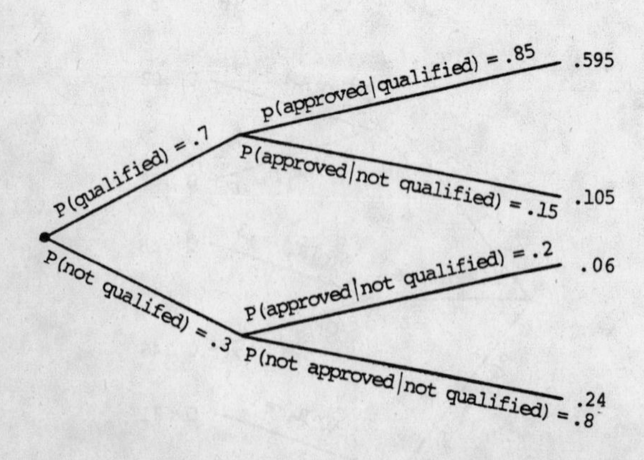

P(qualified | approved)

$$= \frac{.595}{.595 + .06} = \frac{119}{131} \approx .908$$

11.

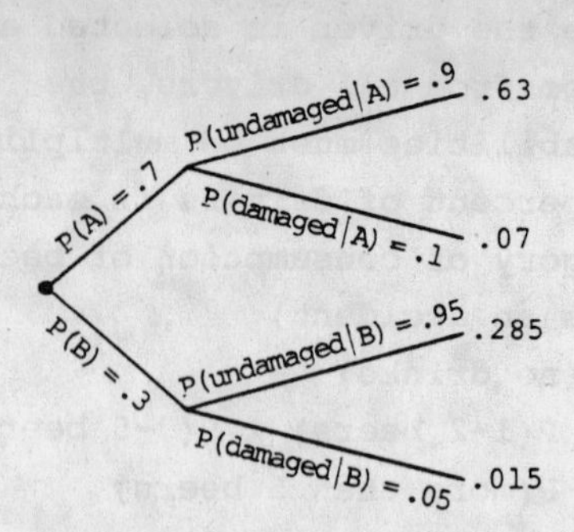

$$P(A \mid \text{damaged}) = \frac{.07}{.07 + .015} = \frac{14}{17} \approx .824$$

13.

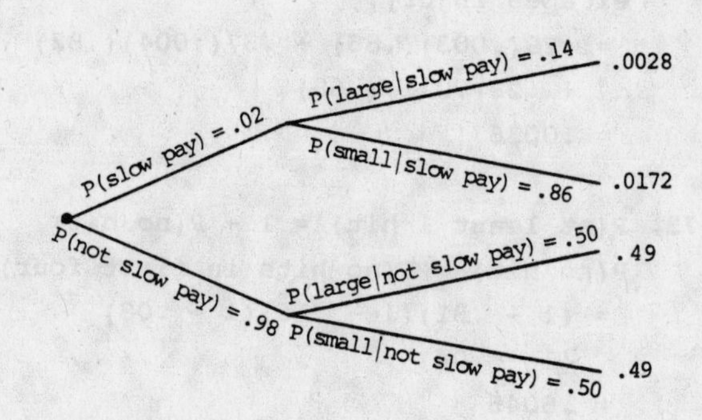

P(slow pay | large down payment)

$$= \frac{.0028}{.0028 + .49} = \frac{1}{176} \approx .006$$

15.

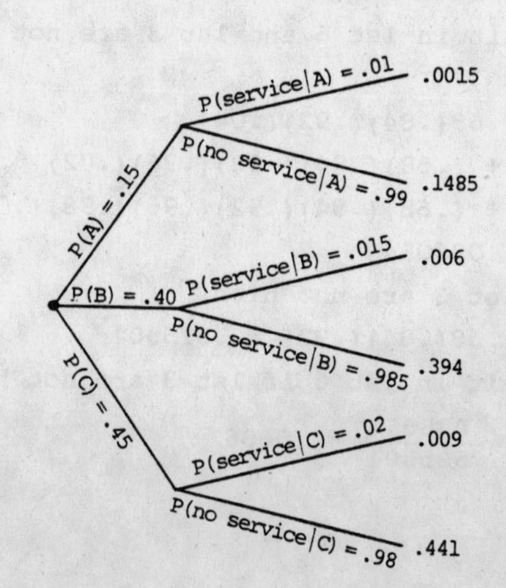

P(A | needs service)

$$= \frac{.0015}{.0015 + .006 + .009} = \frac{1}{11} \approx .091$$

17.

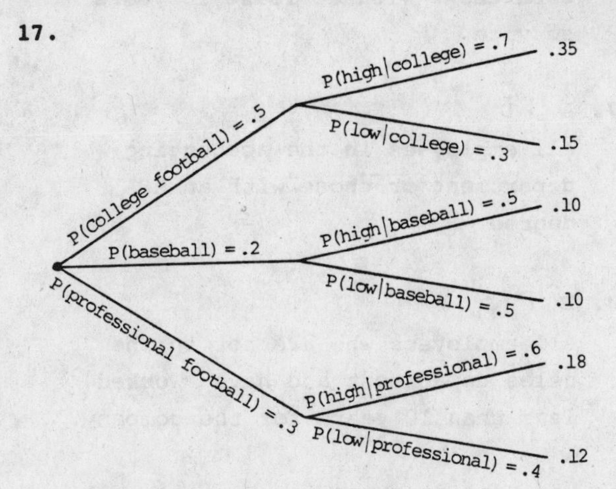

P(college | high rating)

$$= \frac{.35}{.35 + .10 + .18} = \frac{35}{63} = \frac{5}{9}$$

19.

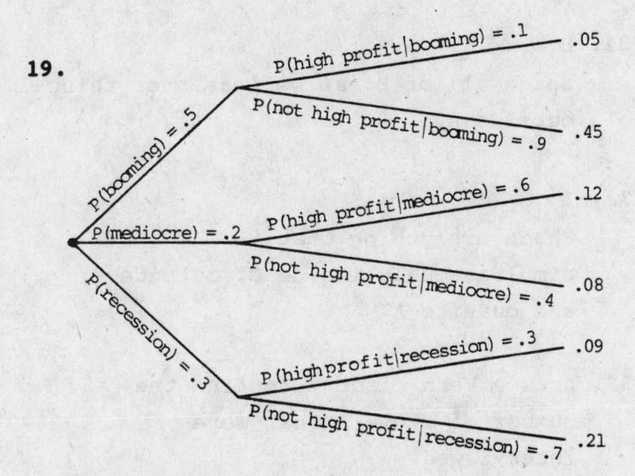

P(booming | high profit)

$$= \frac{.05}{.05 + .12 + .09} = \frac{.05}{.26} = \frac{5}{26}$$

21.

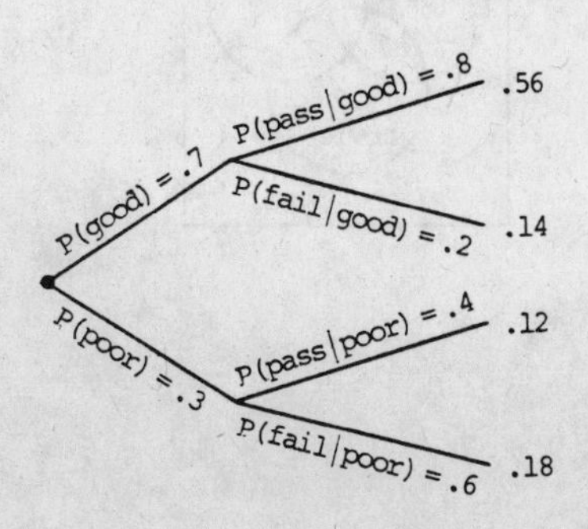

$$P(\text{Good} \mid \text{Pass}) = \frac{.56}{.56 + .12} = \frac{.56}{.68}$$

= .824 = 82.4%

23.

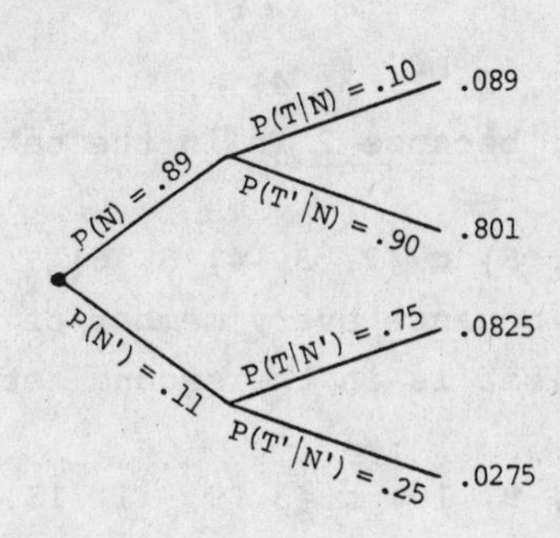

$$P(N \mid T) = \frac{.089}{.089 + .0825} = \frac{178}{343} \approx .519$$

25.

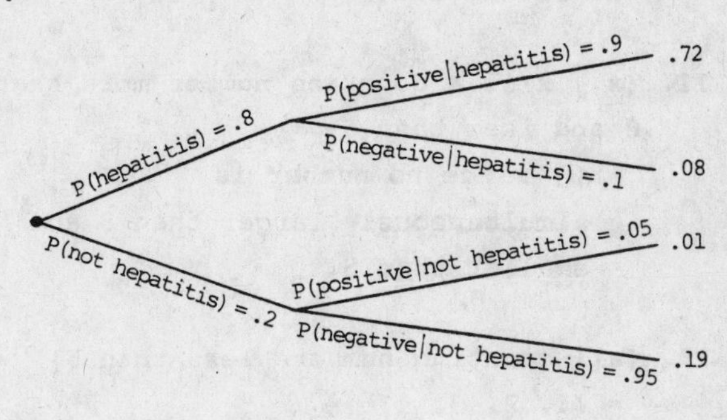

P(hepatitis | positive)

$$= \frac{.72}{.72 + .01} = \frac{72}{73} = .986$$

27.

$$\frac{(.185)(.07)}{(.128)(.04) + (.146)(.06) + (.185)(.07) + (.178)(.09) + (.139)(.12) + (.224)(.13)}$$

= .146

29.

$$\frac{(.11)(.48)}{(.11)(.48) + (.076)(.53) + (.376)(.68) + (.283)(.64) + (.155)(.74)}$$

= .082

Chapter 8 Review Exercises

1. $-6 \in \{8, 4, -3, -9, 6\}$
False, because -6 is not in the set.

3. $2 \notin \{0, 1, 2, 3, 4\}$
False, because 2 is in the set.

5. $\{3, 4, 5\} \subseteq \{2, 3, 4, 5, 6\}$
True, because every member of the first set is in the second set.

7. $\{3, 6, 9, 10\} \subseteq \{3, 9, 11, 13\}$
False, because 6 and 10 are not in the second set.

9. $\{2, 8\} \not\subseteq \{2, 4, 6, 8\}$
False, because 2 and 8 are both in the second set.

11. $\{x \mid x$ is a counting number more than 8 and less than $5\}$
$= \emptyset$, since no number is simultaneously larger than 8 and smaller than 5.

13. {all counting numbers less than 5}
$= \{1, 2, 3, 4\}$

15. The number of subsets of K is $2^4 = 16$.

17. $K' = \{a, b, e\}$

19. $K \cap R = \{c, d, g\}$

21. $(K \cap R)' = \{a, b, e, f\}$ since $K \cap R = \{c, d, g\}$

23. $\emptyset' = U$ *This is always the case*

25. $A \cap C$
All employees in the accounting department with at least 10 years' service.

27. $A \cup D$
All employees in the accounting department or those with an MBA degree.

29. $B' \cap C'$
All employees who are not in the sales department and have worked less than 10 years for the company.

See the Venn diagrams for Exercises 31 and 33 in the answer section of the textbook.

31. $B \cup A'$
Shade all of B, as well as everything outside A.

33. $A' \cap (B' \cap C)$
Shade everything that is simultaneously inside C, outside B, and outside A.

35. Draw a Venn diagram to find the number of viewers that were interviewed.

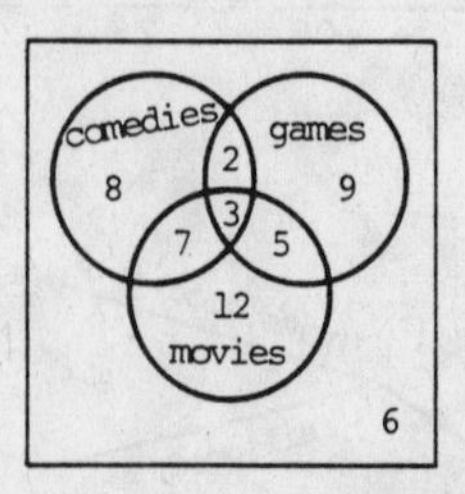

$8 + 2 + 9 + 7 + 3 + 5 + 12 + 6 = 52$

37. Use the diagram in Exercise 35 to find the number of viewers that watch only movies, or 12.

39. The sample space for a die is rolled is {1, 2, 3, 4, 5, 6}.

41. The sample space for the weights of a person measured in half pound, up to 300 is
{0, .5, 1, 1.5, 2, ..., 299, 299.5, 300}

43. The sample space is
{(3, R), (3, G), (5, R), (5, G), (7, R), (7, G), (9, R), (9, G), (11, R), (11, G)}

45. Event F, the second ball is green is
F = {(3, G), (5, G), (7, G), (9, G), (11, G)}.

47. A customer buys neither is
$E' \cap F'$.

49. There are 52 cards containing 13 hearts.

$P(\text{heart}) = \frac{13}{52} = \frac{1}{4}$

51. There are 52 cards containing 12 face cards

$P(\text{face card}) = \frac{12}{52} = \frac{3}{13}$

53. There are 4 queens, 2 of which are red.

$P(\text{red} \mid \text{queen}) = \frac{2}{4} = \frac{1}{2}$

55. There are 4 kings, each of which is a face card.

$P(\text{face card} \mid \text{king}) = \frac{4}{4} = 1$

57. There are 2 black jacks and 50 cards that are not black jacks. The odds are 2 to 50 or 1 to 25.

59. P(no more than 3 defective)
= .31 + .25 + .18 + .12 = .86

61. See the table in the answer section at the back of the textbook.

63. There are 4 possible combinations, but only 2 have a normal cell combined with a trait cell (N_1T_2, T_1N_2).

$P = \frac{2}{4} = \frac{1}{2}$

65. There are 36 possibilities, with 5 having a sum of 8:
(4,4), (3,5), (5,3), (2,6) and (6,2).

$P(8) = \frac{5}{36}$

67. P(at least 10)
= P(10) + P(11) + P(12)

$= \frac{3}{36} + \frac{2}{36} + \frac{1}{36} = \frac{6}{36} = \frac{1}{6}$

69. P(odd and greater than 8)
= P(9) + P(11)

$= \frac{4}{36} + \frac{2}{36} = \frac{6}{36} = \frac{1}{6}$

71. $P(7 \mid \text{one is a } 4) = \frac{2}{11}$, since there are 11 possibilities with 1 of the dice being 4 {(4,1), (1,4), (4,2), (2,4), (4,3), (3,4), (4,4), (5,4), (4,5), (6,4), (4,6)} with only (4,3) and (3,4) having a sum of 7.

73. $P(E \cup F) = P(E) + P(F) - P(E \cap F)$
= .51 + .37 - .22 = .66

75. Draw a Venn diagram.

E F
.29 .22 .15
.34

$$P(E' \cup F) = P(E') + P(F) - P(E' \cap F)$$
$$= (1 - .51) + .37 - .15$$
$$= .49 + .37 - .15 = .71$$

77. $P(N \cap S) = \dfrac{300}{1000} = \dfrac{3}{10}$ or .3

79. $P(N \mid S) = \dfrac{300}{750} = \dfrac{2}{5}$ or .4

81. $P(S \mid N') = \dfrac{450}{600} = \dfrac{3}{4}$ or .75

83. Draw a tree diagram.

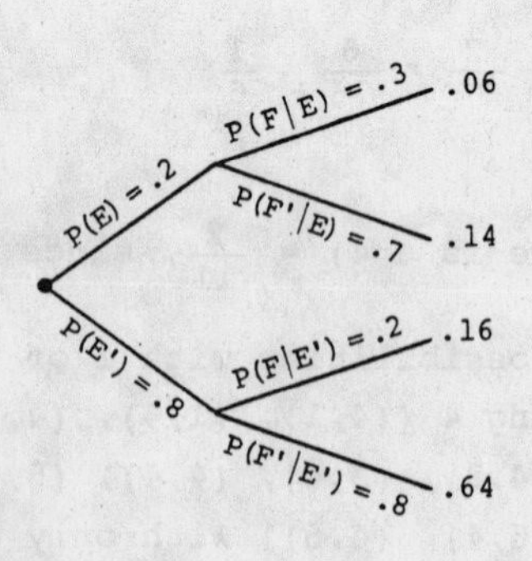

(a) $P(E \mid F) = \dfrac{.06}{.06+.16} = \dfrac{6}{22} = \dfrac{3}{11}$

(b) $P(E \mid F') = \dfrac{.14}{.14 + .64} = \dfrac{7}{39}$

85.

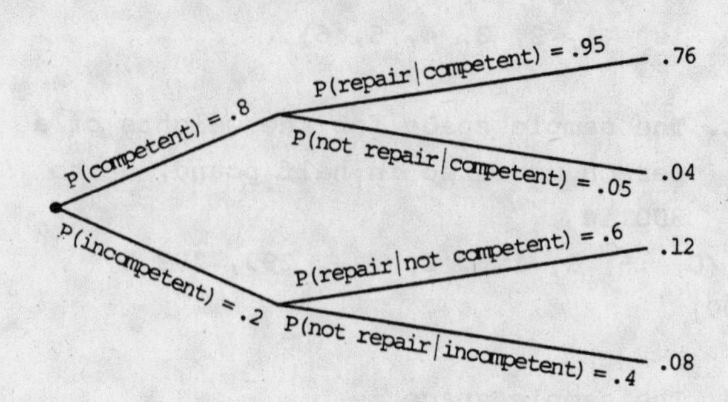

$P(\text{incompetent} \mid \text{repaired})$

$$= \frac{.12}{.76 + .12} = \frac{12}{88} = \frac{3}{22}$$

87. $P(\text{incompetent} \mid \text{not repaired})$

$$= \frac{.08}{.04 + .08} = \frac{8}{12} = \frac{2}{3}$$

89.

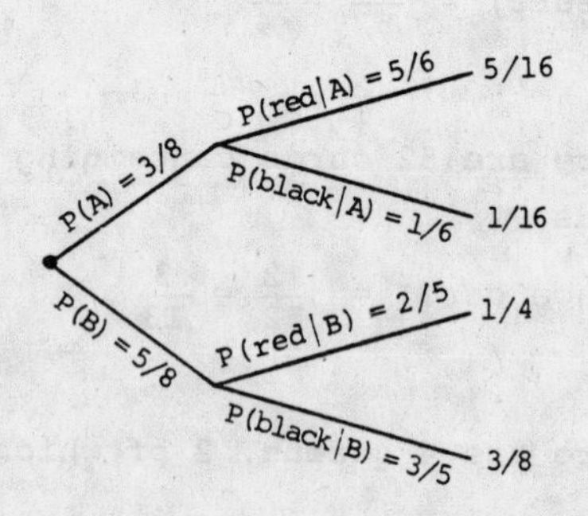

$$P(B \mid \text{red}) = \frac{\frac{1}{4}}{\frac{5}{16} + \frac{1}{4}} = \frac{4}{5 + 4} = \frac{4}{9}$$

Case 9 Exercises

1. P(producing)

$= \frac{1}{2000} = .0005$

2. P(not producing)

$1 - \frac{1}{2000} = .9995$

3. P(producing marketable drug) $= \frac{1}{2000}$

P(not producing a marketable drug)

$= \frac{1999}{2000}$

P(a products not producing)

$= (\frac{1999}{2000})^a$

4. $1 - (\frac{1999}{2000})^a$

5. Same as Exercise 3, except that with N scientists and c compounds per scientist, we have Nc different compounds.

$(\frac{1999}{2000})^{Nc}$

6. $1 - (\frac{1999}{2000})^{Nc}$

7. N = 100, c = 6, Nc = 100 · 6 = 600

$1 - (\frac{1999}{2000})^{600} = 1 - .741 = .259$

8. N = 25, c = 10, Nc = 25 · 10 = 250

$1 - (\frac{1999}{2000})^{250} = 1 - .882 = .118$

Case 10 Exercises

1. Last column of table:

$\frac{388}{543} = .715$

$\frac{186}{327} = .569$

$\frac{146}{356} = .410$

$\frac{97}{302} = .321$

$\frac{91}{336} = .271$

2. Since there are 100 yards on the playing field, there are 94 other possible values for n, from 6 to 99. Many are unlikely, but some, like "third and 10" is not uncommon. The outcomes listed do not exhaust the sample space.

Case 11 Exercises

1. $P(H_2|C_1) =$

$$\frac{P(C_1|H_2)P(H_2)}{P(C_1|H_1)P(H_1)+P(C_1|H_2)P(H_2)+P(C_1|H_3)P(H_3)}$$

$$= \frac{(.4)(.15)}{(.9)(.8)+(.4)(.15)+(.1)(.05)}$$

$$= \frac{.06}{.72+.06+.005} = .076$$

2. $P(H_1|C_2) =$

$$\frac{P(C_2|H_1)P(H_1)}{P(C_2|H_1)P(H_1)+P(C_2|H_2)P(H_2)+P(C_3|H_3)P(H_3)}$$

$$= \frac{(.2)(.8)}{(.2)(.8)+(.8)(.15)+(.3)(.05)}$$

$$= \frac{.16}{.16+.12+.015} = .542$$

3. $P(H_3|C_2) =$

$$\frac{P(C_2|H_3)P(H_3)}{P(C_2|H_1)P(H_1)+P(C_2|H_2)P(H_2)+P(C_3|H_3)P(H_3)}$$

$$= \frac{(.3)(.05)}{(.2)(.8)+(.8)(.15)+(.3)(.05)}$$

$$= \frac{.015}{.16+.12+.015} = .051$$

CHAPTER 9 FURTHER TOPICS IN PROBABILITY

Section 9.1

1. $P(4,2) = \frac{4!}{(4-2)!} = \frac{4!}{2!} = 4 \cdot 3 = 12$

3. $\binom{8}{3} = \frac{8!}{3!(8-3)!} = \frac{8!}{3!5!} = \frac{8 \cdot 7 \cdot 6}{3 \cdot 2 \cdot 1} = 56$

5. $P(8,1) = \frac{8!}{(8-1)!} = \frac{8!}{7!} = 8$

7. $4! = 4 \cdot 3 \cdot 2 \cdot 1 = 24$

9. $\binom{12}{5} = \frac{12!}{5!(12-5)!}$

$= \frac{12!}{5!7!} = \frac{12 \cdot 11 \cdot 10 \cdot 9 \cdot 8}{5 \cdot 4 \cdot 3 \cdot 2 \cdot 1} = 792$

11. $P(13,2) = \frac{13!}{(13-2)!} = \frac{13!}{11!} = 13 \cdot 12 = 156$

13. $P(25,5) = 6{,}375{,}600$

15. $P(14,5) = 240{,}240$

17. $\binom{21}{10} = 352{,}716$

19. $\binom{25}{16} = 2{,}042{,}975$

21. $5 \cdot 3 \cdot 2 = 30$ types of homes

23. **(a)** There are 2 possibilities for the first letter, K or W. There are 25 possibilities for the second letter, since no repetitions are allowed. Similarly, there are 24 possibilities for the third and 23 for the fourth.
$2 \cdot 25 \cdot 24 \cdot 23 = 27{,}600$ call letters

(b) There are 2 possibilities for the first letter, K or W. Since repetitions are allowed there are 26 possibilities for the other.
$2 \cdot 26 \cdot 26 \cdot 26 = 35{,}152$ call letters

(c) There are 2 possibilities for the first letter, K or W. There is 1 possibility for the last letter, R. Since no repetitions are allowed and 2 letters were already used, there are 24 possibilities for the second and 23 for third.
$2 \cdot 1 \cdot 24 \cdot 23 = 1104$ call letters

25. $3 \cdot 5 = 15$ first- and middle-name arrangements

27. There are 26 letters and 10 digits.
$26 \cdot 26 \cdot 26 \cdot 10 \cdot 10 \cdot 10 = 17{,}576{,}000$ license numbers

29. Number of arrangements for letters:
$26 \cdot 25 \cdot 24 = 15{,}600$
Number of arrangements for numbers:
$10 \cdot 9 \cdot 8 = 720$
Since either numbers or letters can come first, there are 2 possible arrangements of letters and numbers.
$(15600)(720)(2) = 22{,}464{,}000$ license numbers

31. $P(6, 6) = \frac{6!}{(6-6)!} = \frac{6!}{0!}$

$= \frac{720}{1} = 720$ ways

33. $P(6, 3) = \frac{6!}{(6-3)!} = \frac{6!}{3!}$

$= 6 \cdot 5 \cdot 4 = 120$ ways

35. $P(15, 3) = \frac{15!}{(15-3)!} = \frac{15!}{12!}$

$= 15 \cdot 14 \cdot 13 = 2730$ ways

37. $\binom{52}{2} = \frac{52!}{2!(52-2)!} = \frac{52!}{2!50!}$

$= \frac{52 \cdot 51}{2 \cdot 1} = 1326$ hands

39. $\binom{5}{2} = \frac{5!}{2!(5-2)!} = \frac{5!}{2!3!}$

$= \frac{5 \cdot 4}{2 \cdot 1} = 10$ combinations

41. "At least 2 good hitters" means 2 or 3 good hitters.
To get 2 good (and 1 bad):

$\binom{5}{2}\binom{4}{1} = 10 \cdot 4 = 40$

To get 3 good (and 0 bad):

$\binom{5}{3} = 10$

To choose at least 2 good hitters:
$40 + 10 = 50$ ways

43. This is a combinations problem.

$\binom{4}{2} = 6$ ways

45. This is a combinations problem.

(a) $\binom{8}{5} = 56$ committees

(b) $\binom{11}{5} = 462$ committees

(c) $\binom{8}{3}\binom{11}{2} = 56 \cdot 55 = 3080$ committees

(d) "No more than 3 women" means 3 women, 2 women, 1 woman, or no women.
3 women means there are 2 men.

$\binom{8}{2}\binom{11}{3} = 28 \cdot 165 = 4620$

2 women means there are 3 men.

$\binom{8}{3}\binom{11}{2} = 56 \cdot 55 = 3080$

1 woman means there are 4 men.

$\binom{8}{4}\binom{11}{1} = 70 \cdot 11 = 770$

0 women means there are 5 men.

$\binom{8}{5}\binom{11}{0} = 56 \cdot 1 = 56$

Total number of ways:
$4620 + 3080 + 770 + 56 = 8526$

47. This is a combinations problem.

(a) $\binom{12}{3} = 220$ ways

(b) $\binom{12}{9} = 220$ ways

49. Since 1 child will be left out, the problem is to find the number of different ways to seat 11 children. This is a permutations problem.
$P(12, 11) = 479{,}001{,}600$

51. This is a combinations problem.

(a) $\binom{9}{3} = 84$ delegations

(b) $\binom{5}{3} = 10$ delegations

(c) $\binom{5}{2}\binom{4}{1} = 10 \cdot 4 = 40$ delegations

(d) Since the mayor must be in the delegation, only 2 of the remaining 8 council members are to be selected.

$1 \cdot \binom{8}{2} = 1 \cdot 28 = 28$

53. This is a combinations problem.

(a) $\binom{7}{2}$ = 21 delegations

(b) Since 1 particular employee must go, only 1 employee from the remaining 6 must be chosen.

$\binom{6}{1}$ = 6 delegations

(c) At least 1 woman means 1 or 2 women.

2 women means 0 men.

$\binom{2}{2}\binom{5}{0} = 1 \cdot 1 = 1$

1 woman means 1 man.

$\binom{2}{1}\binom{5}{1} = 2 \cdot 5 = 10$

Total delegations: 1 + 10 = 11

55. Since the order makes a difference, this is a permutations problem.

P(10, 4) = 5040

57. This is a combinations problem.

(a) $\binom{5}{3}$ = 10 samples

(b) 0, since you are taking 3 red and only 1 red jelly bean exists.

(c) $\binom{3}{3}$ = 1 sample

(d) $\binom{5}{2}\binom{1}{1} = 10 \cdot 1 = 10$ samples

(e) $\binom{5}{2}\binom{3}{1} = 10 \cdot 3 = 30$ samples

(f) $\binom{3}{2}\binom{5}{1} = 3 \cdot 5 = 15$ samples

(g) 0, since you are picking 2 red and only 1 red jelly bean exists.

59. **(a)** initial

There are 7 total letters, 3 i's, 1 n, 1 t, 1 a, and 1 ℓ.

$\frac{7!}{3!1!1!1!1!} = 840$

(b) little

There are 6 letters, 2 ℓ's, 2 t's, 1 i, and 1 e.

$\frac{6!}{2!2!1!1!} = 180$

(c) decreed

There are 7 letters, 2 d's 3 e's, 1 c, and 1 r.

$\frac{7!}{2!3!1!1!} = 420$

61. **(a)** Total of 9 books:

P(9, 9) = 362,880 ways

(b) Since books of the same color are considered identical, the problem is to find the number of different arrangements of the three colors.

P(3, 3) = 6

(c) $\frac{9!}{4!3!2!} = 1260$

63. $\binom{52}{5}$ = 2,598,960

65. $\binom{52}{13} = 6.350135596 \times 10^{11}$ or

635,013,559,600

67. "At least 3 aces" means 3 or 4 aces.
3 aces, 10 cards that are not aces:

$\binom{4}{3}\binom{48}{10} = 4 \cdot 6540715896$
$= 26,162,863,584$

4 aces, 9 cards that are not aces:

$\binom{4}{4}\binom{48}{9} = 1 \cdot 1677106640$
$= 1,677,106,640$

Total:
26,162,863,584 + 1,677,106,640
= 27,839,970,224

69. No, since $26^3 = 17,576$ (the number of different 3-initial names) and the biologist needs 52,000 names. Actually, 4-initial names would do the biologist's job since
$26^4 = 456,976$.

71. **(a)** 3 numerals: $10^3 = 1000$
3 letters: $26^3 = 17,576$
License plates: $1000 \cdot 17,576$
$= 17,576,000$

(b) This year there are 12,000,000 licenses. Since the number increases by 10% or .10 each year, the total number of licenses each year is
$A = 12,000,000\ (1 + .10)^n$
where n is the number of years after this year. The problem is to find n when A is more than 17,576,000. Solve the inequality
$12,000,000(1 + .10)^n > 17,576,000.$

$$1.1^n > \frac{17,576,000}{12,000,000}$$

$$1.1^n > 1.46467$$

$$n > \frac{\ln 1.46467}{\ln 1.1}$$

$$n > 4.0041$$

The state will have no more license plates after about 4 years.

Section 9.2

1. The probability that 1 typewriter that is not defective will be drawn from 9 typewriters that contain 2 defectives is

$$\frac{\binom{7}{1}}{\binom{9}{1}} = \frac{7}{9} \approx .778.$$

3. 3 typewriters are drawn. The probability that none is defective is

$$\frac{\binom{7}{3}}{\binom{9}{3}} = \frac{35}{84} = \frac{5}{12} \approx .417.$$

5. 4 samples are chosen.

$$P(\text{no defectives}) = \frac{\binom{2}{0}\binom{10}{4}}{\binom{12}{4}}$$

$$= \frac{1(210)}{495} \approx .424$$

7. P(all red) =

$$\frac{\binom{6}{3}}{\binom{10}{3}} = \frac{20}{120} = \frac{1}{6} \approx .167$$

9. P(2 yellow and 1 red) =

$$\frac{\binom{4}{2}\binom{6}{1}}{\binom{10}{3}} = \frac{6(6)}{120} = \frac{3}{10} = .3$$

11. The number of 2-card hands that can be drawn from 52 cards is

$$\binom{52}{2} = 1326 \text{ hands.}$$

13. At least 1 ace means 1 ace with another card or 2 aces.

P(at least 1 ace)

= P(1 ace) + P(2 aces)

$$= \frac{\binom{4}{1}\binom{48}{1}}{\binom{15}{2}} + \frac{\binom{4}{2}}{1326}$$

$$= \frac{4\cdot 48}{1326} + \frac{6}{1326} = \frac{198}{1326}$$

$$= \frac{33}{221} \approx .149$$

15. There are 13 cards in each suit and 4 different suits.

P(2 of same suit)

$$= \frac{4\cdot\binom{13}{2}}{\binom{52}{2}} = \frac{312}{1326} = \frac{4}{17} \approx .235$$

17. There are 12 face cards and 40 non-face cards.

P(no face cards)

$$= \frac{\binom{40}{2}}{\binom{52}{2}} = \frac{780}{1326} = \frac{10}{17} \approx .588$$

19. Each time a letter is pulled out of 26 letters the probability a correct letter of "chuck" is pulled is 1/26. The probability of pulling out all 5 correct letters is

$$\left(\frac{1}{26}\right)^5 \approx 8.4 \times 10^{-8}.$$

21. Each time a letter is pulled, there are 26 possibilities. However, in order for each letter to differ, there is 1 less slip that is available of the 26.

$$\text{P(all different)} = \frac{26\cdot 25\cdot 24\cdot 23\cdot 22}{26^5}$$

$$= \frac{7,893,600}{11,881,376}$$

$$= \frac{18,975}{28,561} \approx .6644$$

23. There are 4 suits, and only 1 way to get a royal flush in each suit.

P(royal flush)

$$= \frac{1\cdot 4}{\binom{52}{5}} = \frac{4}{2,598,960}$$

$$= \frac{1}{649,740} \approx 1.539 \times 10^{-6}$$

25. There are 13 different denominations in each deck. There is only 1 way to pick 4 cards of the same value. The fifth card can be any of the remaining 48.

P(4 of a kind)

$$= \frac{13\cdot 1\cdot 48}{\binom{52}{5}} = \frac{624}{2,598,960}$$

$$= \frac{1}{4165} \approx .00024$$

27. Since there are 13 hearts in a deck and the hand must be all hearts, there is only 1 way to pick them. Furthermore, there are $\binom{52}{13}$ possible hands.

$$P(\text{all hearts}) = \frac{1}{\binom{52}{13}} = 1.57 \times 10^{-12}$$

29. Number of ways to get exactly 3 aces:

$\binom{4}{3}$

Number of ways to get exactly 3 kings:

$\binom{4}{3}$

Since a hand contains 13 cards, there are only 7 more out of the remaining 44 to pick. (They cannot be the remaining ace or king.) Number of ways to get remaining 7 cards: $\binom{44}{7}$

$$P\text{ (exactly 3 A's + 3 K's)} = \frac{\binom{4}{3}\binom{4}{3}\binom{44}{7}}{\binom{52}{13}}$$

31. **(a)** Number of ways to select 3 Hughes books:

$\binom{9}{3} = 84$

Number of ways to select 3 Morrison books:

$\binom{7}{3} = 35$

Number of ways to select 6 books out of 21:

$\binom{21}{6} = 54{,}264$

P(3 Hughes + 3 Morrison)

$= \frac{84 \cdot 35}{54{,}264} = .054$

(b) Number of ways to select 4 Baldwin books:

$\binom{5}{4} = 5$

Number of ways to select 2 books not by Baldwin:

$\binom{16}{2} = 120$

$P(\text{exactly 4 Baldwin}) = \frac{120 \cdot 5}{54{,}264} = .011$

(c) Number of ways to select 2 Hughes books:

$\binom{9}{2} = 36$

Number of ways to select 3 Baldwin books:

$\binom{5}{3} = 10$

Number of ways to select 1 Morrison book:

$\binom{7}{1} = 7$

P(2 Hughes, 3 Baldwin, 1 Morrison)

$= \frac{36 \cdot 10 \cdot 7}{54{,}264} = .046$

(d) "At least 4 Hughes books" means exactly 4 or exactly 5 or exactly 6 Hughes books

4 Hughes books with 2 books not by Hughes:

$\binom{9}{4}\binom{12}{2} = 126 \cdot 66 = 8316$

5 Hughes books with 1 book not by Hughes:

$\binom{9}{5}\binom{12}{1} = 126 \cdot 12 = 1512$

Exactly 6 Hughes books: $\binom{9}{6} = 84$

P(at least 4 Hughes)

$= \frac{8316 + 1512 + 84}{54,264} = .183$

(e) "Exactly 4 books by males" means exactly 2 are by females. Only 7 of the 21 are by females.

Number of ways to choose 4 books by males:

$\binom{14}{4} = 1001$

Number of ways to choose 4 books by females:

$\binom{7}{2} = 21$

P(exactly 4 males)

$= \frac{1001 \cdot 21}{54,264} = .387$

(f) "No more than 2 Baldwin books" means exactly 2 Baldwin books or 1 Baldwin book or no Baldwin book.

Number of ways to choose 2 Baldwin books with 4 books not by Baldwin:

$\binom{5}{2}\binom{16}{4} = 10(1820) = 18,200$

Number of ways to choose 1 Baldwin book and 5 others:

$\binom{5}{1}\binom{16}{5} = 5(4368) = 21,840$

Number of ways to choose 0 Baldwin books and 6 others:

$\binom{5}{0}\binom{17}{6} = 1(8008) = 8008$

P(no more than 2 Baldwins)

$= \frac{18,200 + 21,840 + 8008}{54,264} = .885$

33. (a) Number of ways to choose 3 detectors from box of 14:

$\binom{14}{3} = 364$

Number of ways to choose 3 detectors from 10 that are not defective:

$\binom{10}{3} = 120$

P(all 3 work) $= \frac{120}{364} = .3297$

(b) Number of ways to choose 2 detectors from box of 14:

$\binom{14}{2} = 91$

Number of ways to get 1 defective and 1 not defective:

$\binom{4}{1}\binom{10}{1} = 4 \cdot 10 = 40$

P(1 defective, 1 not) $= \frac{40}{91} = .4396$

(c) The box now has 12 detectors, 2 defective and 10 not defective.

Number of ways to choose 3 detectors:

$\binom{12}{3} = 220$

Number of ways to choose 3 good ones:

$\binom{10}{3} = 120$

P(all 3 work) $= \frac{120}{220} = .5455$

35. **(a)** Number of ways to choose 4 members:

$\binom{25}{4} = 12,650$

Number of ways to choose 4 Democrats:

$\binom{10}{4} = 210$

$$P(\text{all Democrats}) = \frac{210}{12,650} = .0166$$

(b) Number of ways to choose 4 Republicans:

$\binom{15}{4} = 1365$

$$P(\text{all Republicans}) = \frac{1365}{12,650} = .1079$$

(c) Number of ways to choose 2 Democrats:

$\binom{10}{2} = 45$

Number of ways to choose 2 Republicans:

$\binom{15}{2} = 105$

P(2 Democrats, 2 Republicans)

$$= \frac{45 \cdot 105}{12,650} = .3735$$

(d) Number of ways to choose 1 Democrat:

$\binom{10}{1} = 10$

Number of ways to choose 3 Republicans:

$\binom{15}{3} = 455$

P(1 Democrat, 3 Republicans)

$$= \frac{10 \cdot 455}{12,650} = .3597$$

(e) Since Rushby and McCullar were not chosen, there are now 8 Democrats to choose from and 23 members from which to choose.

Number of ways to choose 4 members:

$\binom{23}{4} = 8855$

Number of ways to choose 1 Democrat:

$\binom{8}{1} = 8$

Number of ways to choose 3 Republicans:
455 (from Part (d))
P(1 Democrat, 3 Republicans)

$$= \frac{8 \cdot 455}{8855} = .4111$$

37. Assume that the correct matching of the pictures is 1234. The number of ways to place 4 pictures, 4 at a time, is a permutation problem.
P(4, 4) = 24
Probably the easiest way to find how many possible ways no matches or 2 matches can be made is to list all 24 possibilities. Then mark those with no match N and those with exactly 2 matches M.

1234	2134M	3124	4123N
1243M	2143M	3142N	4132
1324M	2341N	3214M	4213
1342	2314	3241	4231M
1423	2413N	3412N	4312N
1432M	2431	3421N	4321N

There are 9 that have no match.

$$P(\text{no match}) = \frac{9}{24} = \frac{3}{8}$$

There are 6 that have exactly 2 matches.

$$P(\text{2 matches}) = \frac{6}{24} = \frac{1}{4}$$

39. Each person can pick 1 of 7 floors to exit. Let a "4-tuple" (a,b,c,d) be the indicator of where each person got off the elevator. Person 1 got off at floor a, person 2 got off at floor b, and so on. There are $7^4 = 2401$ possible 4-tuples. Since each person must pick a different floor, the number of possible 4-tuples that fulfill the requirement is P(7, 4) = 840.

$P = \frac{840}{2401} = \frac{120}{343} \approx .3498$

41. 4 items may be arranged (4-1)! = 3! = 6 ways.

43. 10 itmes may be arranged (10-1)! = 9! = 362,880 ways.

45. Number of ways to arrange 7 keys on ring:
(7 - 1)! = 6! = 720.
Since the 2 keys must be together (black and gold), consider them as a single key. 6 keys can be arranged (6 - 1)! = 5! = 120 ways.
Now these 2 keys can be arranged in 2 ways: black, gold or gold, black.

$P = \frac{2 \cdot 120}{720} = \frac{1}{3} = .333$

47. Prove by mathematical induction that n items can be arranged in a circle (n - 1)! ways. Show that the hypothesis is true for n = 2. (Basic step)

2 items can be arranged in 1 way.
(2 - 1)! = 1! = 1

Assume for n = k, that is k items can be arranged (k - 1)! ways (inductive hypothesis). Show the hypothesis is true for n = k + 1, that is, show that k + 1 items can be arranged (k + 1 - 1) = k! ways.

Remove 1 item from the k + 1 items leaving k items. By the inductive hypothesis, these k items can be arranged (k - 1)! ways. Now in each of these (k - 1)! ways insert the item that was removed.

There are k different positions into which the item can be placed. Thus, there are k·(k - 1)! ways = k! ways to place the k + 1 items.

49. Answers will vary, but each should be fairly close to the theoretical answers in Exercises 27-30.

Section 9.3

1. Probability of having a boy

$= 1 - \frac{1}{2} = \frac{1}{2}.$

$P(\text{exactly 2 girls}) = \binom{5}{2}\left(\frac{1}{2}\right)^2\left(\frac{1}{2}\right)^3 = \frac{5}{16}$

The first number counts the number of ways to pick 2 girls out of 5 children. The second number is the probability of having 2 girls. The third is the probability of having 3 boys.

3. $P(\text{no girls}) = \binom{5}{0}\left(\frac{1}{2}\right)^0\left(\frac{1}{2}\right)^5 = \frac{1}{32}$

The first number counts the number of ways to pick 0 girls out of 5 children. The second number is the probability of having 0 girls. The third is the probability of having 5 boys.

5. "At least 4 girls" means 4 girls or 5 girls.

$P(4\text{ girls}) = \binom{5}{4}\left(\frac{1}{2}\right)^4\left(\frac{1}{2}\right)^1 = \frac{5}{32}$

$P(5\text{ girls}) = \binom{5}{5}\left(\frac{1}{2}\right)^5\left(\frac{1}{2}\right)^0 = \frac{1}{32}$

P(at least 4 girls)

$= \frac{5}{32} + \frac{1}{32} = \frac{6}{32} = \frac{3}{16}$

7. "No more than 3 boys" means 0 boys or 1 boy or 2 boys or 3 boys.

$P(0\text{ boys}) = \binom{5}{0}\left(\frac{1}{2}\right)^0\left(\frac{1}{2}\right)^5 = \frac{1}{32}$

$P(1\text{ boys}) = \binom{5}{1}\left(\frac{1}{2}\right)^1\left(\frac{1}{2}\right)^4 = \frac{5}{32}$

$P(2\text{ boys}) = \binom{5}{2}\left(\frac{1}{2}\right)^2\left(\frac{1}{2}\right)^3 = \frac{10}{32}$

$P(3\text{ boys}) = \binom{5}{3}\left(\frac{1}{2}\right)^3\left(\frac{1}{2}\right)^2 = \frac{10}{32}$

P(no more than 3 boys)

$= \frac{1}{32} + \frac{5}{32} + \frac{10}{32} + \frac{10}{32} = \frac{26}{32} = \frac{13}{16}$

9. P(exactly 12 ones)

$= \binom{12}{12}\left(\frac{1}{6}\right)^{12}\left(\frac{5}{6}\right)^0 = \left(\frac{1}{6}\right)^{12} \approx .0000000005$

11. P(exactly 1 one)

$= \binom{12}{1}\left(\frac{1}{6}\right)^1\left(\frac{5}{6}\right)^{11} = .269$

13. "No more than 3 ones" means 0 ones, 1 one, 2 ones, or 3 ones.

$P(0\text{ ones}) = \binom{12}{0}\left(\frac{1}{6}\right)^6\left(\frac{5}{6}\right)^{12} = .1122$

P(1 one) = .2692 (from Exercise 11)

$P(2\text{ ones}) = \binom{12}{2}\left(\frac{1}{6}\right)^2\left(\frac{5}{6}\right)^{10} = .2961$

$P(3\text{ ones}) = \binom{12}{3}\left(\frac{1}{6}\right)^3\left(\frac{5}{6}\right)^9 = .1974$

P(no more than 3 ones)

$= .11216 + .269 + .29609 + .19740 \approx .875$

15. $P(\text{all heads}) = \binom{5}{5}\left(\frac{1}{2}\right)^5\left(\frac{1}{2}\right)^0 = \frac{1}{32}$

17. "No more than 3 heads" means 0 heads, 1 head, 2 heads, or 3 heads.

$P(0\text{ heads}) = \binom{5}{0}\left(\frac{1}{2}\right)^0\left(\frac{1}{2}\right)^5 = \frac{1}{32}$

$P(1\text{ head}) = \binom{5}{1}\left(\frac{1}{2}\right)^1\left(\frac{1}{2}\right)^4 = \frac{5}{32}$

$P(2\text{ heads}) = \binom{5}{2}\left(\frac{1}{2}\right)^2\left(\frac{1}{2}\right)^3 = \frac{10}{32}$

$P(3\text{ heads}) = \binom{5}{3}\left(\frac{1}{2}\right)^3\left(\frac{1}{2}\right)^2 = \frac{10}{32}$

P(no more than 3 heads)

$= \frac{1}{32} + \frac{5}{32} + \frac{10}{32} + \frac{10}{32} = \frac{26}{32} = \frac{13}{16}$

19. P(no defectives)

$= \binom{20}{0}(.05)^0(.95)^{20} = .358$

21. P(exactly 2 correct)

$= \binom{6}{2}\left(\frac{1}{5}\right)^2\left(\frac{4}{5}\right)^4 = .246$

23. "At least 4" means 4 or 5 or 6.

$P(4\text{ correct}) = \binom{6}{4}(.2)^4(.8)^2 = .0154$

$P(5\text{ correct}) = \binom{6}{5}(.2)^5(.8)^1 = .0015$

$P(6\text{ correct}) = \binom{6}{6}(.2)^6(.8)^0 = .0001$

P(at least 4 correct)
$= .0154 + .0015 + .0001 = .017$

25. Since 5 out of 50 clients lose their life savings, the probability of that happening is $\frac{5}{50} = \frac{1}{10}$. The probability of it not happening is $1 - \frac{1}{10} = \frac{9}{10}$.

P(exactly 1 out of 3)
$= \binom{3}{1}\left(\frac{1}{10}\right)^1\left(\frac{9}{10}\right)^2 = .243$

27. $P(\text{exactly } 5) = \binom{10}{5}(.2)^5(.8)^5 = .026$

Since 20% never had a Big Mac, 80% have had a Big Mac.

29. "4 or more have had a Big Mac" means 0, 1, 2, or 3 have not had a Big Mac.

$P(0) = \binom{10}{0}(.2)^0(.8)^{10} = .1074$

$P(1) = \binom{10}{1}(.2)^1(.8)^9 = .2684$

$P(2) = \binom{10}{2}(.2)^2(.8)^8 = .3020$

$P(3) = \binom{10}{3}(.2)^3(.8)^7 = .2013$

$.1074 + .2684 + .3020 + .2013 = .879$

31. Since 70% are cured, 30% are not.

$P(\text{exactly } 17) = \binom{20}{17}(.7)^{17}(.3)^3 = .072$

33. "At least 18" means 18 or 19 or 20.

$P(18) = \binom{20}{18}(.7)^{18}(.3)^2 = .0278$

$P(19) = \binom{20}{19}(.7)^{19}(.3)^1 = .0068$

$P(20) = \binom{20}{20}(.7)^{20}(.3)^0 = .0008$

P(at least 18)
$= .0278 + .0068 + .0008 = .035$

35. Since there are 5 choices for each question, there is a $\frac{1}{5}$ or 20% chance of a correct answer and $1 - \frac{1}{5} = \frac{4}{5}$ or 80% chance for an incorrect answer.

$P(\text{exactly } 7) = \binom{10}{7}(.2)^7(.8)^3 = .00079$

37. P(fewer than 8 correct)
= 1 - P(8 or more correct).
P(8 or more) = P(8) + P(9) + P(10).

$P(8) = \binom{10}{8}(.2)^8(.8)^2 = .000074$

$P(9) = \binom{10}{9}(.2)^9(.8)^1 = .000004$

$P(10) = \binom{10}{10}(.2)^{10}(.8)^0 = .000000102$

P(8 or more)
$= .000074 + .000004 + .000000102$
$= .000078$

P(fewer than 8 correct)
$= 1 - .000078 \approx .999922$

39. Since the probability of death is 20%, the probability of survival is 80%.

P(exactly 1 survives)
$= \binom{3}{1}(.8)^1(.2)^2 = .096$

41. No more than 1 survives means no one survives or 1 survives.

$P(0) = \binom{3}{0}(.8)^0(.2)^3 = .008$

$P(1) = .096$ (from Exercise 39)

P(no more than 1) = $.008 + .096 = .104$

43. P(exactly 3) = $\binom{6}{3}(.7)^3(.3)^3 = .185$

Since probability of recovery is 70% or .7, the probability of no recovery is 1 - .7 = .3.

45. "No more than 3 means" 0 or 1 or 2 or 3.

$P(0 \text{ recover}) = \binom{6}{0}(.7)^0(.3)^6 = .0007$

$P(1 \text{ recovers}) = \binom{6}{1}(.7)^1(.3)^5 = .0102$

$P(2 \text{ recover}) = \binom{6}{2}(.7)^2(.3)^4 = .0595$

$P(3 \text{ recover}) = \binom{6}{3}(.7)^3(.3)^3 = .1852$

P(no more than 3)
$= .0007 + .0102 + .0595 + .1852 = .256$

47. P(at least 1 mutation)
= 1 - P (no mutation)
P(no mutation)

$= \binom{10000}{0}(2.5\cdot10^{-7})^0(.99999975)^{10000}$

= .9975
P(at least 1 mutation)
= 1 - .9975 = .0025

49. **(a)** "At least 9" means 9 or 10.

$P(9) = \binom{10}{9}(.7)^9(.3)^1 = .1211$

$P(10) = \binom{10}{10}(.7)^{10}(.3)^0 = .0282$

P(at least 9) = .1211 + .0282 = .1493

(b) From the answer for part (a), the probability that a group of 10 has at least 9 with no cavities is .1493. Hence, the probability of less than 9 with no cavities is .8507.

P(at least 1 group) = 1 - P(no group)
P(no group)

$= \binom{15}{0}(.1493)^0(.8507)^{15} = .0884$

P(at least 1 group)
= 1 - .0884 = .9116

51. **(a)** The probability of not getting the flu is .8.

The probability of getting the flu is 1 - .8 = .2
P(exactly 10)

$= \binom{134}{10}(.2)^{10}(.8)^{124} = .000036$

(b) "No more than 10" means 0, 1, 2, 3, 4, 5, 6, 7, 8, 9, or 10.

$P(0) = \binom{134}{0}(.2)^0(.8)^{134} = 1.0329 \times 10^{-13}$

$P(1) = \binom{134}{1}(.2)^1(.8)^{133}$

$= 3.4602149 \times 10^{-2}$

$P(2) = \binom{134}{2}(.2)^2(.8)^{132}$

$= 5.7526073 \times 10^{-11}$

$P(3) = \binom{134}{3}(.2)^3(.8)^{131}$

$= 6.327868 \times 10^{-10}$

$P(4) = \binom{134}{4}(.2)^4(.8)^{136}$

$= 5.1809419 \times 10^{-9}$

$P(5) = \binom{134}{5}(.2)^5(.8)^{129}$

$= 3.3676123 \times 10^{-8}$

$P(6) = \binom{134}{6}(.2)^6(.8)^{128} = .0000002$

$P(7) = \binom{134}{7}(.2)^7(.8)^{127} = .0000008$

$P(8) = \binom{134}{8}(.2)^8(.8)^{126} = .0000033$

$P(9) = \binom{134}{9}(.2)^9(.8)^{125} = .0000115$

P(10) = .0000359 (from part (a))
The sum of the preceding is .000052.

(c) With n = 134, x = 0,
P(none get the flu)

$= \binom{134}{0}(.2)^0(.8)^{134} = 1.0329 \times 10^{-13}$

which is very small, so small in fact that we can call it 0.

53. **(a)** The probability of a defective item is .05.
The probability of a good item is 1 - .05 = .95.
With n = 75, x = 5, P(exactly 5 defectives)

$= \binom{75}{5}(.05)^5(.95)^{70} = .148774.$

(b) With n = 75, x = 0,

$P(\text{no defectives}) = \binom{75}{0}(.05)^0(.95)^{75} = .021344.$

(c) P(at least 1 item)
= 1 - P(0 items)
P(0 items) from part (b)
1 - .021344 = .978656

Section 9.4

1. $\begin{bmatrix} \frac{2}{3} & \frac{1}{3} \end{bmatrix}$ is a probability vector since

$\frac{2}{3} + \frac{1}{3} = 1.$

3. [0 1] is a probability vector since 0 + 1 = 1.

5. [.4 .2 0] is not a probability vector since $.4 + .2 + 0 = .6 \neq 1$.

7. $\begin{bmatrix} .5 & 0 \\ 0 & .5 \end{bmatrix}$ is not a transition matrix

since the sum of row 1 is
$.5 + 0 = .5 \neq 1$.

9. $\begin{bmatrix} \frac{1}{4} & \frac{3}{4} \\ \frac{1}{2} & \frac{1}{2} \end{bmatrix}$ is a transition matrix since

(a) it is square (2 × 2), (b) all entries are between 0 and 1 inclusive, and (c) sum of row 1 is

$\frac{1}{4} + \frac{3}{4} = 1$ and sum of row 2 is

$\frac{1}{2} + \frac{1}{2} = 1.$

11. $\begin{bmatrix} \frac{1}{3} & \frac{1}{3} & \frac{1}{3} \\ 0 & 1 & 0 \\ \frac{1}{2} & 0 & \frac{1}{2} \end{bmatrix}$ is a transition matrix since

(a) it is square (3 × 3), (b) all entries are between 0 and 1 inclusive, and (c) sum of row 1 is $\frac{1}{3} + \frac{1}{3} + \frac{1}{3} = 1$ sum of row 2 is $0 + 1 + 0 = 1$, and sum of row 3 is $\frac{1}{2} + 0 + \frac{1}{2} = 1$.

13. $\begin{bmatrix} \frac{1}{3} & \frac{1}{2} & 1 \\ \frac{1}{3} & 0 & 0 \\ \frac{1}{3} & \frac{1}{2} & 0 \end{bmatrix}$ is not a transition matrix

since the sum of row 1 is $\frac{1}{3} + \frac{1}{2} + 1 = \frac{11}{6} \neq 1$.

15. The transition matrix is regular since $\begin{bmatrix} .2 & .8 \\ .9 & .1 \end{bmatrix}\begin{bmatrix} .2 & .8 \\ .9 & .1 \end{bmatrix} = \begin{bmatrix} .76 & .24 \\ .27 & .73 \end{bmatrix}$ and all entries are positive.

17. The square of the matrix is

$$\begin{bmatrix} 1 & 0 \\ .6 & .4 \end{bmatrix}\begin{bmatrix} 1 & 0 \\ .6 & .4 \end{bmatrix} = \begin{bmatrix} 1 & 0 \\ .84 & .16 \end{bmatrix}.$$

The cube of the matrix is

$$\begin{bmatrix} 1 & 0 \\ .84 & .16 \end{bmatrix}\begin{bmatrix} 1 & 0 \\ .6 & .4 \end{bmatrix} = \begin{bmatrix} 1 & 0 \\ .936 & .064 \end{bmatrix}.$$

The matrix to the fourth power is

$$\begin{bmatrix} 1 & 0 \\ .84 & .16 \end{bmatrix}\begin{bmatrix} 1 & 0 \\ .84 & .16 \end{bmatrix} = \begin{bmatrix} 1 & 0 \\ .9744 & .0256 \end{bmatrix}.$$

Notice that any power of the transition matrix yields a 0 in the first row and second column. Therefore, it is not regular.

19. The square of the matrix is

$$\begin{bmatrix} 0 & 1 & 0 \\ .4 & .2 & .4 \\ 1 & 0 & 0 \end{bmatrix}\begin{bmatrix} 0 & 1 & 0 \\ .4 & .2 & .4 \\ 1 & 0 & 0 \end{bmatrix} = \begin{bmatrix} .4 & .2 & .4 \\ .48 & .44 & .08 \\ 0 & 1 & 0 \end{bmatrix}.$$

The cube of the matrix is

$$\begin{bmatrix} .4 & .2 & .4 \\ .48 & .44 & .08 \\ 0 & 1 & 0 \end{bmatrix}\begin{bmatrix} 0 & 1 & 0 \\ .4 & .2 & .4 \\ 1 & 0 & 0 \end{bmatrix} = \begin{bmatrix} .48 & .44 & .08 \\ .256 & .568 & .176 \\ .4 & .2 & .4 \end{bmatrix}.$$

Notice that in the cube of the matrix, all entries are positive. Thus, it is a regular matrix.

21. $[v_1 \ v_2]\begin{bmatrix} \frac{1}{4} & \frac{3}{4} \\ \frac{1}{2} & \frac{1}{2} \end{bmatrix} = \left[\frac{1}{4}v_1 + \frac{1}{2}v_2 \quad \frac{3}{4}v_1 + \frac{1}{2}v_2\right]$

$\frac{1}{4}v_1 + \frac{1}{2}v_2 = v_1$ and $\frac{3}{4}v_1 + \frac{1}{2}v_2 = v_2$

$-\frac{3}{4}v_1 + \frac{1}{2}v_2 = 0$ $\qquad \frac{3}{4}v_1 + -\frac{1}{2}v_2 = 0$

Since $v_1 + v_2 = 1$, $v_1 = 1 - v_2$

Substituting into either statement gives

$$-\frac{3}{4}(1 - v_2) + \frac{1}{2}v_2 = 0$$

$$-\frac{3}{4} + \frac{3}{4}v_2 + \frac{1}{2}v_2 = 0$$

$$-\frac{3}{4} + \frac{5}{4}v_2 = 0$$

$$-3 + 5v_2 = 0$$

$$5v_2 = 3$$

$$v_2 = \frac{3}{5}$$

$$v_1 = 1 - \frac{3}{5} = \frac{2}{5}$$

The equilibrium vector is $\left[\frac{2}{5} \ \frac{3}{5}\right]$.

23. $[v_1\ v_2]\begin{bmatrix} .3 & .7 \\ .4 & .6 \end{bmatrix} = [.3v_1 + .4v_2 \quad .7v_1 + .6v_2]$

$v_1 = .3v_1 + .4v_2$

$0 = -.7v_1 + .4v_2$ and

$v_2 = .7v_1 + .6v_2$

$0 = .7v_1 - .4v_2$

Since $v_1 + v_2 = 1$, $v_1 = 1 - v_2$.

Substituting into the first equation gives

$0 = -.7(1 - v_2) + .4v_2$

$0 = -.7 + .7v_2 + .4v_2$

$0 = -.7 + 1.1v_2$

$v_2 = .636$ or $\frac{7}{11}$.

$v_1 = 1 - \frac{7}{11} = \frac{4}{11}$

$\begin{bmatrix} \frac{4}{11} & \frac{7}{11} \end{bmatrix}$

25. $[v_1\ v_2\ v_3]\begin{bmatrix} .1 & .1 & .8 \\ .4 & .4 & .2 \\ .1 & .2 & .7 \end{bmatrix} =$

$[.1v_1+.4v_2+.1v_3 \quad .1v_1+.4v_2+.2v_3 \quad .8v_1+.2v_2+.7v_3]$

$v_1 = .1v_1 + .4v_2 + .1v_3$

$0 = -.9v_1 + .4v_2 + .1v_3$ *(1)*

$v_2 = .1v_1 + .4v_2 + .2v_3$

$0 = .1v_1 - .6v_2 + .2v_3$ *(2)*

$v_3 = .8v_1 + .2v_2 + .7v_3$

$0 = .8v_1 + .2v_2 - .3v_3$ *(3)*

Since $v_1 + v_2 + v_3 = 1$, $v_1 = 1 - v_2 - v_3$.

Substituting into the first 2 equations gives

$0 = -.9(1 - v_2 - v_3) + .4v_2 + .1v_3$

$0 = -.9 + .9v_2 + .9v_3 + .4v_2 + .1v_3$

$.9 = 1.3v_2 + 1.0v_3$ *(4)*

$0 = -.1(1 - v_2 - v_3) - .6v_2 + .2v_3$

$0 = .1 - .1v_2 - .1v_3 - .6v_2 + .2v_3$

$-.1 = -7v_2 + .1v_3$ *(5)*

Solve equation 4 for v_3.

$.9 - 1.3v_2 = v_3$

Substitute into equation 5.

$-.1 = -.7v_2 + .1(.9 - 1.3v_2)$

$-.1 = -.7v_2 + .09 - .13v_2$

$-.19 = -.83v_2$

$\frac{19}{83} = v_2$

$v_3 = .9 - (1.3)\frac{19}{83} = \frac{50}{83}$

$v_1 = 1 - \frac{19}{83} - \frac{50}{83} = \frac{14}{83}$

$\begin{bmatrix} \frac{14}{83} & \frac{19}{83} & \frac{50}{83} \end{bmatrix}$ or $[.169 \quad .229 \quad .602]$

27. Let V be the probability vector $[v_1\ v_2\ v_3]$.

$[v_1\ v_2\ v_3]\begin{bmatrix} .25 & .35 & .4 \\ .1 & .3 & .6 \\ .55 & .4 & .05 \end{bmatrix} = [v_1\ v_2\ v_3]$

$.25v_1 + .1v_2 + .55v_3 = v_1$

$.35v_1 + .3v_2 + .4v_3 = v_2$

$.4v_1 + .6v_2 + .05v_3 = v_3$

Simplify these equations to get the dependent system

$-.75v_1 + .1v_2 + .55v_3 = 0$

$.35v_1 - .7v_2 + .4v_3 = 0$

$.4v_1 + .6v_2 - .95v_3 = 0$

Since V is a probability vector, $v_1 + v_2 + v_3 = 1$.

Solving this system we obtain

$v_1 = \frac{170}{563}$, $v_2 = \frac{197}{563}$, $v_3 = \frac{196}{563}$

or $v_1 = .302$, $v_2 = .350$, $v_3 = .348$.

Thus, the equilibrium vector is

$\begin{bmatrix} \frac{170}{563} & \frac{197}{563} & \frac{196}{563} \end{bmatrix}$ or $[.302 \quad .350 \quad .348]$.

29.

	Works	Doesn't work
Works	.95	.05
Doesn't work	.7	.3

$$[v_1 \ \ v_2]\begin{bmatrix} .95 & .05 \\ .7 & .3 \end{bmatrix}$$

$= [.95v_1 + .7v_2 \quad .05v_1 + .3v_2]$

$.95v_1 + .7v_2 = v_1$

$-.05v_1 + .7v_2 = 0$ *(1)*

$.05v_1 + .3v_2 = v_2$

$.05v_1 - .7v_2 = 0$ *(2)*

Since $v_1 + v_2 = 1$, $v_1 = 1 - v_2$.

Substitute this value into equation 1.

$-.05(1 - v_2) + .7v_2 = 0$

$-.05 + .05v_2 + .7v_2 = 0$

$.75v_2 = .05$

$v_2 = \frac{1}{15}$

$v_1 = 1 - \frac{1}{15} = \frac{14}{15}$

In the long run, the assembly line will run $\frac{14}{15}$ of the time.

31. (a)

	Red	White
Red	.75	.25
White	.50	.50

(b) [.72 .25]

(c) The square of the transition matrix is

$$\begin{bmatrix} .75 & .25 \\ .50 & .50 \end{bmatrix}\begin{bmatrix} .75 & .25 \\ .50 & .50 \end{bmatrix} = \begin{bmatrix} .6875 & .3125 \\ .625 & .375 \end{bmatrix}.$$

The transition matrix to the fourth power is

$$\begin{bmatrix} .6875 & .3125 \\ .625 & .375 \end{bmatrix}\begin{bmatrix} .6875 & .3125 \\ .625 & .375 \end{bmatrix} = \begin{bmatrix} .6680 & .3320 \\ .6641 & .3359 \end{bmatrix}$$

$$[.75 \ \ .25]\begin{bmatrix} .6680 & .3320 \\ .6641 & .3359 \end{bmatrix} = [.6670 \ \ .3330].$$

(d) $[v_1 \ \ v_2]\begin{bmatrix} .75 & .25 \\ .50 & .50 \end{bmatrix}$

$= [.75v_1 + .50v_2 \quad .25v_1 + .50v_2]$

$v_1 = .75v_1 + .50v_2$

$0 = -.25v_1 + .50v_2$ *(1)*

$v_2 = .25v_1 + .50v_2$

$0 = .25v_1 - .50v_2$ *(2)*

Since $v_1 + v_2 = 1$, $v_1 = 1 - v_2$.

Substitute into equation 1.

$0 = -.25(1 - v_2) + .50v_2$

$0 = -.25 + .25v_2 + .50v_2$

$.25 = .75v_2$

$\frac{1}{3} = v_2$

$v_1 = 1 - \frac{1}{3} = \frac{2}{3}$

$$\begin{bmatrix} \frac{2}{3} & \frac{1}{3} \end{bmatrix}$$

33. The transition matrix is

	Low	Medium	High
Low	.5	.4	.1
Medium	.25	.45	.3
High	.05	.4	.55

$$[v_1 \ \ v_2 \ \ v_3]\begin{bmatrix} .5 & .4 & .1 \\ .25 & .45 & .3 \\ .05 & .4 & .55 \end{bmatrix} = [v_1 \ \ v_2 \ \ v_3]$$

$.5v_1 + .25v_2 + .05v_3 = v_1$

$.4v_1 + .45v_2 + .4v_3 = v_2$

$.1v_1 + .3v_2 + .55v_3 = v_3$

Also, $v_1 + v_2 + v_3 = 1$.

Solving this system, we obtain

$v_1 = \frac{51}{209}$, $v_2 = \frac{88}{209}$, $v_3 = \frac{70}{209}$.

The equilibrium vector is

$$\begin{bmatrix} \frac{51}{209} & \frac{88}{209} & \frac{70}{209} \end{bmatrix}.$$

Thus, the long-range trends for the proportions of low, medium, and high producers are 51/209, 88/209, and 70/209, respectively.

35. The transition matrix is

$$\begin{bmatrix} .85 & .10 & .05 \\ .15 & .75 & .10 \\ .10 & .30 & .60 \end{bmatrix}.$$

The square of the transition matrix is

$$\begin{bmatrix} .85 & .10 & .05 \\ .15 & .75 & .10 \\ .10 & .30 & .60 \end{bmatrix}\begin{bmatrix} .85 & .10 & .05 \\ .15 & .75 & .10 \\ .10 & .30 & .60 \end{bmatrix}$$

$$= \begin{bmatrix} .7425 & .175 & .0825 \\ .25 & .6075 & .21 \\ .19 & .415 & .44 \end{bmatrix}.$$

The cube of the transition matrix is

$$\begin{bmatrix} .85 & .10 & .05 \\ .15 & .75 & .10 \\ .10 & .30 & .60 \end{bmatrix}\begin{bmatrix} .7425 & .175 & .0825 \\ .25 & .6075 & .21 \\ .19 & .415 & .44 \end{bmatrix}$$

$$= \begin{bmatrix} .665625 & .23025 & .113125 \\ .317875 & .523375 & .213875 \\ .26325 & .44875 & .33525 \end{bmatrix}.$$

(a) $[50{,}000 \quad 0 \quad 0]\begin{bmatrix} .85 & .10 & .05 \\ .15 & .75 & .10 \\ .10 & .30 & .60 \end{bmatrix}$

$= [42{,}500 \quad 5000 \quad 2500]$

The numbers in the groups after 1 year are 42,500, 5000, and 2500.

(b) $[50{,}000 \quad 0 \quad 0]\begin{bmatrix} .7425 & .175 & .0825 \\ .25 & .6075 & .21 \\ .19 & .415 & .44 \end{bmatrix}$

$= [37{,}125 \quad 8750 \quad 4125]$

The numbers in the groups after 2 years are 37,125, 8750, and 4125.

(c)

$$[50{,}000 \quad 0 \quad 0]\begin{bmatrix} .665625 & .23025 & .113125 \\ .317875 & .523375 & .213875 \\ .26325 & .44875 & .33525 \end{bmatrix}$$

$= [33{,}281 \quad 11{,}513 \quad 5206]$

The numbers in the groups after 3 years are 33,281, 11,513, and 5206.

(d) The system of equations is

$$.85x_1 + .15x_2 + .10x_3 = x_1$$
$$.10x_1 + .75x_2 + .30x_3 = x_2$$
$$.05x_1 + .10x_2 + .60x_3 = x_3$$

After collecting like terms, we have the matrix

$$\left[\begin{array}{ccc|c} -.15 & .15 & .10 & 0 \\ .10 & -.25 & .30 & 0 \\ .05 & .10 & -.40 & 0 \end{array}\right].$$

Clear the first column.

$$\left[\begin{array}{ccc|c} 1 & -1 & -\frac{2}{3} & 0 \\ 0 & -\frac{3}{20} & \frac{11}{30} & 0 \\ 0 & \frac{3}{20} & -\frac{11}{30} & 0 \end{array}\right]$$

Clear the second column.

$$\left[\begin{array}{ccc|c} 1 & 0 & -\frac{28}{9} & 0 \\ 0 & 1 & -\frac{22}{9} & 0 \\ 0 & 0 & 0 & 0 \end{array}\right]$$

Hence, $x_1 = \frac{28}{9}x_3$ and

$x_2 = \frac{22}{9}x_3$,

which we substitute into $x_1 + x_2 + x_3 = 1$.

The result is

$$\frac{28}{9}x_3 + \frac{22}{9}x_3 + \frac{9}{9}x_3 = 1$$

$$\frac{59}{9}x_3 = 1$$

$$x_3 = \frac{9}{59}.$$

Then $x_2 = \frac{22}{59}$, $x_1 = \frac{28}{59}$, and

$$V = \left[\frac{28}{59} \ \ \frac{22}{59} \ \ \frac{9}{59}\right] \text{ or } [.475 \ \ .373 \ \ .152]$$

In the long run, the probabilities of no accidents, one accident, and more than one accident are .475, .373, and .152, respectively.

37.

		Child: Have	Child: Have not
Father	Have	.95	.05
	Have not	.1	.9

is the transition matrix.

$$[v_1 \ \ v_2]\begin{bmatrix} .95 & .05 \\ .1 & .9 \end{bmatrix}$$

$= [.95v_1 + .1v_2 \ \ .05v_1 + .9v_2]$

Since $v_1 + v_2 = 1$, $v_1 = 1 - v_2$.

Also, $v_1 = .95v_1 + .1v_2$ *(1)*

and $v_2 = .05v_1 + .9v_2$. *(2)*

Substituting into (1) and putting (1) into standard form gives

$0 = -.05v_1 + .1v_2$

$0 = -.05(1 - v_2) + .1v_2$

$0 = -.05 + .15v_2$

$\frac{1}{3} = v_2$

$v_2 = 1 - \frac{1}{3} = \frac{2}{3}.$

Thus, $\frac{2}{3}$ will have the defect.

39. (a)

	Single	Multiple
Single	.90	.10
Multiple	.05	.95

(b) [.75 .25]

(c) $[.75 \ \ .25]\begin{bmatrix} .90 & .10 \\ .05 & .95 \end{bmatrix}$

$= [.6875 \ \ .3125]$

$= [68.8\% \ \ 31.3\%]$

(d) $[v_1 \ \ v_2]\begin{bmatrix} .90 & .10 \\ .05 & .95 \end{bmatrix}$

$= [.9v_1 + .05v_2 \ \ .1v_1 + .95v_2]$

$v_1 = .9v_1 + .05v_2$

$0 = -.1v_1 + .05v_2$ *(1)*

$v_2 = .1v_1 + .95v_2$

$0 = .1v_1 - .05v_2$ *(2)*

Since $v_1 + v_2 = 1$, $v_1 = 1 - v_2$.

Substitute into (1).

$0 = -1(1 - v_2) + .05v_2$

$0 = -.1 + .1v_2 + .05v_2$

$.1 = .15v_2$

$\frac{2}{3} = v_2$

$v_1 = 1 - \frac{2}{3} = \frac{1}{3}.$

The long-range prediction is

$\left[\frac{1}{3} \ \ \frac{2}{3}\right].$

41.

	H	S	U
Humanities	.35	.2	.45
Science	.15	.5	.35
Undecided	.5	.3	.2

Let $[v_1 \ \ v_2 \ \ v_3]$ be a probability vector. Then

$$[v_1 \ \ v_2 \ \ v_3]\begin{bmatrix} .35 & .2 & .45 \\ .15 & .5 & .35 \\ .5 & .2 & .3 \end{bmatrix} = [v_1 \ \ v_2 \ \ v_3].$$

We have the system

$$.35v_1 + .15v_2 + .5v_3 = v_1$$
$$.2v_1 + .5v_2 + .3v_3 = v_2$$
$$.45v_1 + .35v_2 + .2v_3 = v_3$$
$$v_1 + v_2 + v_3 = 0,$$

which is equivalent to the system

$$-.65v_1 + .15v_2 + .5v_3 = 0$$
$$.2v_1 - .5v_2 + .3v_3 = 0$$
$$.45v_1 + .35v_2 - .8v_3 = 0$$
$$v_1 + v_2 + v_3 = 1.$$

To solve this system, we use the augmented matrix

$$\left[\begin{array}{ccc|c} -.65 & .15 & .5 & 0 \\ .2 & -.5 & .3 & 0 \\ .45 & .35 & -.8 & 0 \\ 1 & 1 & 1 & 1 \end{array}\right].$$

Solving by the Gauss-Jordan method, we obtain the matrix

$$\left[\begin{array}{ccc|c} 1 & 0 & 0 & .333 \\ 0 & 1 & 0 & .333 \\ 0 & 0 & 1 & .333 \\ 0 & 0 & 0 & 0 \end{array}\right].$$

Thus, $[v_1 \ \ v_2 \ \ v_3]$

$$= \left[\frac{1}{3} \ \ \frac{1}{3} \ \ \frac{1}{3}\right]$$

43. Using a computer program for raising a matrix to a power, we can generate the following. Let T = the given matrix.

$$T^2 = \begin{bmatrix} .2 & .15 & .17 & .19 & .29 \\ .16 & .2 & .15 & .18 & .31 \\ .19 & .14 & .24 & .21 & .22 \\ .16 & .19 & .16 & .2 & .29 \\ .16 & .19 & .14 & .17 & .34 \end{bmatrix}$$

$$T^3 = \begin{bmatrix} .17 & .178 & .171 & .191 & .29 \\ .171 & .178 & .161 & .185 & .305 \\ .18 & .163 & .191 & .197 & .269 \\ .175 & .174 & .164 & .187 & .3 \\ .167 & .184 & .158 & .182 & .309 \end{bmatrix}$$

$$T^4 = \begin{bmatrix} .1731 & .175 & .1683 & .188 & .2956 \\ .1709 & .1781 & .1654 & .1866 & .299 \\ .1748 & .1718 & .1753 & .1911 & .287 \\ .1712 & .1775 & .1667 & .1875 & .2971 \\ .1706 & .1785 & .1641 & .1858 & .301 \end{bmatrix}$$

$$T^5 = \begin{bmatrix} .17193 & .17643 & .1678 & .18775 & .29609 \\ .17167 & .17689 & .16671 & .18719 & .29754 \\ .17293 & .17488 & .17007 & .18878 & .29334 \\ .17192 & .17654 & .16713 & .18741 & .297 \\ .17142 & .17726 & .16629 & .18696 & .29807 \end{bmatrix}$$

The probability that state 2 changes to state 4 after 5 repetitions of the experiment would be the entry in the second row, fourth column of T^5, .18719.

45. **(a)** Find the fifth power of the transition matrix. The entry in row 1, column 4 is .847723 or about 85%, which represents the percent of employees never in the program that completed it after the program was offered 5 times.

(b) Let T^4 represent the fourth power of the transition matrix.

$[.5 \ .5 \ 0 \ 0] \ T^4$

$= [.032 \ .0998125 \ .0895625 \ .778625]$

47. The equilibrium vector is

[.219086 .191532 .252352 .178091 .158938]

For various powers of the transition matrix, consult Exercise 43.

Section 9.5

See the histograms for Part (b) of Exercises 1-5 in the answers at the back of the textbook.

1. **(a)**

Number germinated	0	1	2	3	4	5
Probability	0	0	.1	.3	.4	.2

3. **(a)**

Number of bullseyes	0	1	2	3	4	5	6
Probability	0	.04	0	.16	.40	.32	.08

5. **(a)**

Number with disease	0	1	2	3	4	5
Probability	.15	.25	.3	.15	.1	.05

7. Let x be the number of heads observed. Then x can take on 0, 1, 2 3, 4 as values. The probabilities follow.

$P(x=0) = \binom{4}{0}\left(\frac{1}{2}\right)^0\left(\frac{1}{2}\right)^4 = \frac{1}{16}$

$P(x=1) = \binom{4}{1}\left(\frac{1}{2}\right)\left(\frac{1}{2}\right)^3 = \frac{4}{16} = \frac{1}{4}$

$P(x=2) = \binom{4}{2}\left(\frac{1}{2}\right)^2\left(\frac{1}{2}\right)^2 = \frac{6}{16} = \frac{3}{8}$

$P(x=3) = \binom{4}{3}\left(\frac{1}{2}\right)^3\left(\frac{1}{2}\right) = \frac{4}{16} = \frac{1}{4}$

$P(x=4) = \binom{4}{4}\left(\frac{1}{2}\right)^4\left(\frac{1}{2}\right)^0 = \frac{1}{16}$

No. of Heads	0	1	2	3	4
Probability	$\frac{1}{16}$	$\frac{1}{4}$	$\frac{3}{8}$	$\frac{1}{4}$	$\frac{1}{16}$

9. Let x be the number of aces drawn. Then x can take on values 0, 1, 2, 3. The probabilities follow.

$P(x = 0) = \binom{3}{0}\left(\frac{48}{52}\right)\left(\frac{47}{51}\right)\left(\frac{46}{50}\right)$

$= \frac{4324}{5525} \approx .783$

$P(x = 1) = \binom{3}{1}\left(\frac{4}{52}\right)\left(\frac{48}{51}\right)\left(\frac{47}{50}\right)$

$= \frac{1128}{5525} \approx .204$

$P(x = 2) = \binom{3}{2}\left(\frac{4}{52}\right)\left(\frac{3}{51}\right)\left(\frac{48}{50}\right)$

$= \frac{75}{5525} \approx .013$

$P(x = 3) = \binom{3}{3}\left(\frac{4}{52}\right)\left(\frac{3}{51}\right)\left(\frac{2}{50}\right)$

$= \frac{1}{5525} \approx .0002$

No. of aces	0	1	2	3
Probability	.783	.204	.013	.0002

11. Let x be the number of hits the batter gets in the next 4 tries. Then x can take on the values 0, 1, 2, 3, 4.

$P(x = 0) = \binom{4}{0}(.290)^0(.710)^4 \approx .254$

$P(x = 1) = \binom{4}{1}(.290)^1(.710)^3 \approx .415$

$P(x = 2) = \binom{4}{2}(.290)^2(.710)^2 \approx .254$

$P(x = 3) = \binom{4}{3}(.290)^3(.710)^1 \approx .069$

$P(x = 4) = \binom{4}{4}(.290)^4(.710)^0 \approx .007$

No. of hits	0	1	2	3	4
Probability	.254	.415	.254	.069	.007

See the histograms for Exercises 13-17 in the answers at the back of the textbook.

13. Sketch the histogram and shade the bars for x = 0, x = 1, and x = 2.

15. Sketch the histogram and shade the bar for x = 1.

17. Sketch the histogram and shade the bars for x = 2 and x = 3.

Use $E(x) = x_1p_1 + x_2p_2 + x_3p_3 + \ldots + x_np_n$ for Exercises 19-25.

19. $E(x) = 2(.1) + 3(.4) + 4(.3) + 5(.2)$
$= 3.6$

21. $E(z) = 9(.14) + 12(.22) + 15(.36)$
$+ 18(.18) + 21(.10)$
$= 14.64$

23. $E(x) = 1(.2) + 2(.3) + 3(.1) + 4(.4)$
$= 2.7$

25. $E(x) = 6(.1) + 12(.2) + 18(.4)$
$+ 24(.2) + 30(.1)$
$= 18$

27. The probability distribution of x, which stands for the person's net winnings, follows.

x	\$99	\$39	-\$1
P(x)	$\frac{1}{500} = .002$	$\frac{2}{500} = .004$	$\frac{497}{500} = .994$

$E(x) = 99(.002) + 39(.004)$
$+ (-1)\ (.994)$
$\approx -.64$

The expected winnings are -\$.64. This is not a fair game.

29. The number of possible samples is

$$\binom{7}{3} = \frac{7!}{3!4!} = \frac{7 \cdot 6 \cdot 5}{3 \cdot 2 \cdot 1} = 35.$$

The number of samples containing no yellows and, therefore, 3 whites is

$\binom{4}{3} = 4$, so the probability of drawing a sample containing no yellows is $\frac{4}{35}$.

The number of samples containing 1 yellow and, therefore, 2 whites is

$\binom{3}{1}\binom{4}{2} = 3 \cdot 6 = 18$, so the probability of drawing a sample containing 1 yellow is $\frac{18}{35}$. Similarly, the probability of drawing a sample containing 2 yellows is

$$\frac{\binom{3}{2}\binom{4}{1}}{\binom{7}{3}} = \frac{12}{35},$$

and the probability of drawing a sample containing 3 yellows is

$$\frac{\binom{3}{3}}{\binom{7}{3}} = \frac{1}{35}.$$

Let x denote the number of yellow marbles drawn. Below is the probability distribution of x.

x	0	1	2	3
P(x)	$\frac{4}{35}$	$\frac{18}{35}$	$\frac{12}{35}$	$\frac{1}{35}$

$$E(x) = 0\left(\frac{4}{35}\right) + 1\left(\frac{18}{35}\right) + 2\left(\frac{12}{35}\right) + 3\left(\frac{1}{35}\right)$$
$$= \frac{45}{35} = \frac{9}{7} \approx 1.3$$

31. The probability that the delegation contains no liberals and 3 conservatives is

$$\frac{\binom{5}{0}\binom{4}{3}}{\binom{9}{3}} = \frac{1 \cdot 4}{84} = \frac{4}{84}.$$

(Since the calculation of expected value involves adding fractions, it will be easier if the fractions are not reduced, so there will be a common denominator.)

Similarly, use combinations to calculate the remaining probabilities for the probability distribution.

(a) Let x be the number of liberals on the delegation. The probability distribution of x follows.

x	0	1	2	3
P(x)	$\frac{4}{84}$	$\frac{30}{84}$	$\frac{40}{84}$	$\frac{10}{84}$

$$E(x) = 0\left(\frac{4}{84}\right) + 1\left(\frac{30}{84}\right) + 2\left(\frac{40}{84}\right) + 3\left(\frac{10}{84}\right)$$

$$= \frac{140}{84} = \frac{5}{3}$$

$$\approx 1.67$$

(b) Let y be the number of conservatives on the committee. The probability distribution of y follows.

y	0	1	2	3
P(y)	$\frac{10}{84}$	$\frac{40}{84}$	$\frac{30}{84}$	$\frac{4}{84}$

$$E(y) = 0\left(\frac{10}{84}\right) + 1\left(\frac{40}{84}\right) + 2\left(\frac{30}{84}\right) + 3\left(\frac{4}{84}\right)$$

$$= \frac{112}{84} = \frac{4}{3} \approx 1.33$$

33. Let x represent the number of junior members on the committee. Use combinations to find the probabilities of 0, 1, 2, and 3 junior members.
The probability distribution of x is

x	0	1	2	3
P(x)	$\frac{57}{203}$	$\frac{95}{203}$	$\frac{45}{203}$	$\frac{6}{203}$

$$E(y) = 0\left(\frac{57}{203}\right) + 1\left(\frac{95}{203}\right) + 2\left(\frac{45}{203}\right) + 3\left(\frac{6}{203}\right)$$

$$= 1$$

35. The probability of drawing 2 diamonds is

$$\frac{\binom{13}{2}}{\binom{52}{2}} = \frac{78}{1326},$$

and the probability of not drawing 2 diamonds is

$$1 - \frac{78}{1326} = \frac{1248}{1326}.$$

Let x denote your net winnings. Then

$$E(x) = 4.5\left(\frac{78}{1326}\right) + (-.5)\left(\frac{1248}{1326}\right)$$

$$= -\frac{273}{1326} \approx -\$.21$$

or -21¢.

The game is not fair since your expected winnings are not zero.

37. $P(\text{even}) = \frac{18}{37}$, $P(\text{noneven}) = \frac{19}{37}$

If an even number comes up, you win \$1. Otherwise, you lose \$1 (win -\$1.) If x denotes your winnings, then

$$E(x) = \frac{18}{37} - \frac{19}{37} = -\frac{1}{37} \approx -2.7¢.$$

39. $P(\text{your number comes up}) = \frac{20}{80} = \frac{1}{4}$

P(your number does not come up)

$$= \frac{60}{80} = \frac{3}{4}$$

If your number comes up, you win \$2.20. Otherwise, you lose \$1 (win -\$1). If x denotes your winnings, then

$$E(x) = 2.20\left(\frac{1}{4}\right) - 1\left(\frac{3}{4}\right)$$

$$= .55 - .75$$

$$= -\$.20 = -20¢.$$

41.

	Possible results	
Result of toss	H	H
Call	H	T
Caller wins?	Yes	No
Probability	1/2	1/2

(a) Yes, since the probability of Donna matching is still 1/2.

(b) Since Donna will match with probability 1, if she calls heads, her expected gain is

$$1(40¢) = 40¢.$$

(c) Since Donna will not match with probability 1, if she calls tails, her expected gain is

$$1(-40¢) = -40¢.$$

43. First compute the amount of money the company can expect to pay out for each kind of policy. The sum of these amounts will be the total the company can expect to pay out. For a single $10,000 policy,

Outcome	Pay $10,000	Don't pay $10,000
Probability	.001	.999

$$E(\text{payoff}) = 10{,}000(.001) + 0(.999) = \$10$$

For all 100 such policies, the company can expect to pay out

$$100(10) = \$1000.$$

For a single $5000 policy,

$$E(\text{payoff}) = 5000(.001) + 0(.999) = \$5.$$

For all 500 such policies, the company can expect to pay out

$$500(5) = \$2500.$$

Similarly, for all 1000 policies of $1000, the company can expect to pay out

$$1000(1) = \$1000.$$

Thus, the total amount the company can expect to pay out is

$$\$1000 + \$2500 + \$1000 = \$4500.$$

45. The probability of a light snowfall is 1 - .4 = .6, so

$$E(x) = .4(160) + .6(90) = 118 \text{ guests}.$$

47.

$$E(x) = 0(.01) + 1(.05) + 2(.15) + 3(.26) + 4(.33) + 5(.14) + 6(.06) = 3.51$$

49. (a) If the two players are equally skilled, the old pro's expected winnings are

$$\frac{1}{2}(80{,}000) + \frac{1}{2}(20{,}000) = \$50{,}000.$$

(b) If the pro's chance of winning is 3/4, then his expected winnings are

$$\frac{3}{4}(80{,}000) + \frac{1}{4}(20{,}000) = \$65{,}000.$$

51. At any one restaurant your expected winnings are

$$E(x) = 100{,}000\left(\frac{1}{176{,}402{,}500}\right) + 25{,}000\left(\frac{1}{39{,}200{,}556}\right) + 5000\left(\frac{1}{17{,}640{,}250}\right) + 1000\left(\frac{1}{1{,}568{,}022}\right) + 100\left(\frac{1}{282{,}244}\right) + 5\left(\frac{1}{7056}\right) + 1\left(\frac{1}{588}\right) = .00488.$$

Going to 25 restaurants gives you expected earnings of 25(.00488) = .122. Since you spent $1, you lose 87.8¢ on the average.

Section 9.6

1. **(a)** Buy speculative; she thinks the market will go up.

   (b) Buy blue-chip; she thinks the market will go down.

   (c) If there is a .7 probability the market will go up, there is a .3 probability it will go down. Find her expected profit for each strategy.
   Blue-chip:
   (.7)(25,000) + (.3)(18,000)
   = $22,900
   Speculative:
   (.7)(30,000) + (.3)(11,000)
   = $24,300
   She should buy speculative; her expected profit is $24,300.

   (d) Find her expected profit for each strategy.
   Blue-chip:
   (.2)(25,000) + (.8)(18,000)
   = $19,400
   Speculative:
   (.2)(30,000) + (.8)(11,000)
   = $14,800
   She should buy blue-chip.

3. **(a)** Set up in the stadium; she doesn't think it will rain.

   (b) Set up in the gym; the worst that can happen is a profit of $1000.

   (c) If there is a .6 probability of rain, there is a .4 probability of no rain. Find her expected profit for each strategy.
   Stadium: .6(-1550) + .4(1500) = -$330
   Gym: .6(1000) + .4(1000) = $1000
   Both: .6(750) + .4(1400) = $1010
   She should set up both; her expected profit is $1010.

5. **(a)**

	Better	Not better
Market	50,000	-25,000
Don't market	-40,000	-10,000

   (b) Find the expected profit under the 2 strategies.
   Market product:
   (.4)(50,000) + (.6)(-25,000)
   = $5000
   Don't market:
   (-.4)(-40,000) + (.6)(-10,000)
   = -$22,000
   They should market the product.

7. **(a)**

	Strike	No strike
Bid $30,000	-5500	4500
Bid $40,000	4500	0

   (b) Find his expected earnings under each strategy.
   Bid $30,000:
   (.6)(-5500) + (.4)(4500) = -$1500
   Bid $40,000:
   (.6)(4500) + (.4)(0) = $2700
   He should bid $40,000.

9. Find the expected utility under each strategy.
   Jobs:
   (.35)(25) + (.65)(-10) = 2.25
   Environment:
   (.35)(-15) + (.65)(30) = 14.25
   She should emphasize the environment. The expected utility of this strategy is 14.25.

Chapter 9 Review Exercises

1. There are 6 taxicabs, order makes a difference.

$$P(6, 6) = \frac{6!}{(6-6)!} = \frac{6!}{0!} = \frac{720}{1} = 720.$$

3. There are 12 oranges from which a sample of 3 of them is taken.

$$\binom{12}{3} = \frac{12!}{3!(12-3)!} = \frac{12!}{3!9!} = 220$$

5. With the 5 pictures, a certain picture must be first. The other 4 may be arranged in any way.

$$P(4, 4) = \frac{4!}{(4-4)!} = 24$$

7. There are 7 choices in the first department, 5 in the second, and 4 for the third. The total number of choices is $7 \cdot 5 \cdot 4 = 140$.

9. There are 4 black balls, 11 balls in all, and 3 balls must be drawn.
Number of ways to pick 3 balls:
$C(11, 3) = 165$
Number of ways to pick 3 black balls:
$C(4, 3) = 4$

$$P(\text{all black}) = \frac{4}{165}$$

11. From Exercise 9, there are 165 ways to pick 3 balls.
Number of ways to pick 2 black balls from 4 black balls:
$C(4, 2) = 6$
Number of ways to pick 1 green ball from 5 green balls.
$C(5, 1) = 5$
Total ways: $6 \cdot 5 = 30$

$$P(\text{2 black, 1 green}) = \frac{30}{165} = \frac{2}{11}$$

13. From Exercise 9, there are 165 ways to pick 3 balls.
Number of ways to pick 2 green balls from 5 green balls:
$C(5, 2) = 10$
Number of ways to pick 1 blue ball from 2 blue balls:
$C(2, 1) = 2$
Total ways: $10 \cdot 2 = 20$

$$P(\text{2 green, 1 blue}) = \frac{20}{165} = \frac{4}{33}$$

15. Number of ways to pick 2 cards from a deck of 52:
$C(52, 2) = 1326$
Number of ways to pick 2 red cards from the 26 red cards in a deck:
$C(26, 2) = 325$

$$P(\text{both red}) = \frac{325}{1326} = \frac{25}{102}$$

17. From Exercise 15, total number of ways to pick 2 cards is 1326.
"At least 1 spade" means 1 spade or 2 spades.
Number of ways to pick 2 spades:
$C(13, 2) = 78$
Number of ways to pick 1 spade and 1 other card:
$C(13, 1) \cdot C(39, 1) = 13 \cdot 39 = 507$
P(at least 1 spade) =

$$\frac{78 + 507}{1326} = \frac{585}{1326} = \frac{15}{34}$$

19. Use binomial trials with $n = 6$ and

$$P(\text{girl}) = \frac{1}{2}$$

$$P(\text{exactly } 3) = \binom{6}{3}\left(\frac{1}{2}\right)^3\left(\frac{1}{2}\right)^{6-3} = \frac{20}{64} = \frac{5}{16}$$

21. "At least 4 girls" means 4, 5, or 6 girls. Use binomial trials.

$P(\text{4 girls}) = \binom{6}{4}\left(\frac{1}{2}\right)^4\left(\frac{1}{2}\right)^{6-4} = \frac{15}{64}$

$P(\text{5 girls}) = \binom{6}{5}\left(\frac{1}{2}\right)^5\left(\frac{1}{2}\right)^{6-5} = \frac{6}{64}$

$P(\text{6 girls}) = \binom{6}{6}\left(\frac{1}{2}\right)^6\left(\frac{1}{2}\right)^{6-6} = \frac{1}{64}$

$P(\text{at least 4}) = \frac{15}{64} + \frac{6}{64} + \frac{1}{64} = \frac{22}{64} = \frac{11}{32}$

23. Use binomial trials with n = 20, p = .01.
P(exactly 4 defectives)
$= \binom{20}{4}(.01)^4(.99)^{20-4} = .00004$

25. "No more than 4" means 0, 1, 2, 3, or 4 defectives.
P(0 defectives)
$= \binom{20}{0}(.01)^0(.99)^{20} = .81791$

P(1 defective)
$= \binom{20}{1}(.01)^1(.99)^{19} = .16523$

P(2 defectives)
$= \binom{20}{2}(.01)^2(.99)^{18} = .01586$

P(3 defectives)
$= \binom{20}{3}(.01)^3(.99)^{17} = .00096$

P(4 defectives)
$= \binom{20}{4}(.01)^4(.99)^{16} = .00004$

P(no more than 4)
= .81791 + .16523 + .01586
+ .00096 + .0004
≈ .1.0000

27. This is a transition matrix since (a) it is a square matrix, (b) all entries are between 0 and 1 inclusive, and (c) the sum of any row is 1.

29. This is a transition matrix since (a) it is a square matrix, (b) all entries are between 0 and 1 inclusive, and (c) the sum of any row is 1.

31. **(a)** Multiply the transition matrix by the vector [.35 .65].

$[.35\ \ .65]\begin{bmatrix} .8 & .2 \\ .4 & .6 \end{bmatrix} = [.54\ \ .46]$

(b) Cube the transition matrix first.

$\begin{bmatrix} .8 & .2 \\ .4 & .6 \end{bmatrix}\begin{bmatrix} .8 & .2 \\ .4 & .6 \end{bmatrix}\begin{bmatrix} .8 & .2 \\ .4 & .6 \end{bmatrix}$

$= \begin{bmatrix} .72 & .28 \\ .56 & .44 \end{bmatrix}\begin{bmatrix} .8 & .2 \\ .4 & .6 \end{bmatrix}$

$= \begin{bmatrix} .688 & .312 \\ .624 & .376 \end{bmatrix}$

Multiply the result by the vector [.35 .65].

$[.35\ \ .65]\begin{bmatrix} .688 & .312 \\ .624 & .376 \end{bmatrix} = [.6464\ \ .3536]$

(c) $[v_1 \ \ v_2]\begin{bmatrix} .8 & .2 \\ .4 & .6 \end{bmatrix} = [.8v_1 + .4v_2 \ \ \ .2v_1 + .6v_2]$

$$.8v_1 + .4v_2 = v_1$$
$$-.2v_1 + .4v_2 = 0$$

Since $v_1 + v_2 = 1$,

$$v_1 = 1 - v_2.$$
$$-.2(1 - v_2) + .4v_2 = 0$$
$$-.2 + .2v_2 + .4v_2 = 0$$
$$-.2 + .6v_2 = 0$$
$$.6v_2 = .2$$
$$v_2 = \frac{1}{3}$$
$$v_1 = 1 - \frac{1}{3} = \frac{2}{3}$$

Dogkins long-range share is 2/3 of the market.

33. (a)

$$[.2 \ \ .55 \ \ .25]\begin{bmatrix} .3 & .5 & .2 \\ .2 & .6 & .2 \\ .1 & .5 & .4 \end{bmatrix} = [.195 \ \ .555 \ \ .25]$$

(b) For 2 generations, square the transition matrix.

$$\begin{bmatrix} .3 & .5 & .2 \\ .2 & .6 & .2 \\ .1 & .5 & .4 \end{bmatrix}\begin{bmatrix} .3 & .5 & .2 \\ .2 & .6 & .2 \\ .1 & .5 & .4 \end{bmatrix} = \begin{bmatrix} .21 & .55 & .24 \\ .20 & .56 & .24 \\ .17 & .55 & .28 \end{bmatrix}$$

Now multiply by the vector $[.2 \ \ .55 \ \ .25]$

$$[.2 \ \ .55 \ \ .25]\begin{bmatrix} .21 & .55 & .24 \\ .20 & .56 & .24 \\ .17 & .55 & .28 \end{bmatrix}$$

$$= [.1945 \ \ .5555 \ \ .2500].$$

(c) $[v_1 \ \ v_2 \ \ v_3]\begin{bmatrix} .3 & .5 & .2 \\ .2 & .6 & .2 \\ .1 & .5 & .4 \end{bmatrix} =$

$[.3v_1 + .2v_2 + .1v_3 \ \ \ .5v_1 + .6v_2 + .5v_3 \ \ \ .2v_1 + .2v_2 + .4v_3]$

$$.3v_1 + .2v_2 + .1v_3 = v_1$$
$$.5v_1 + .6v_2 + .5v_3 = v_2$$
$$.2v_1 + .2v_2 + .4v_3 = v_3$$

$$-.7v_1 + .2v_2 + .1v_3 = 0$$
$$.5v_1 - .4v_2 + .5v_3 = 0$$
$$.2v_1 + .2v_2 - .6v_3 = 0$$

Solve for v_1, v_2, and v_3.
Since $v_1 + v_2 + v_3 = 1$, $v_1 = 1 - v_2 - v_3$.

$$-.7(1 - v_2 - v_3) + .2v_2 + .1v_3 = 0$$
$$.5(1 - v_2 - v_3) - .4v_2 + .5v_3 = 0$$
$$.9v_2 + .8v_3 = .7$$
$$-.9v_2 = -.5$$
$$v^2 = \frac{5}{9}$$
$$.9\left(\frac{5}{9}\right) + .8v_3 = .7$$
$$.5 + .8v_3 = .7$$
$$v_3 = \frac{1}{4}$$
$$v_1 = 1 - \frac{5}{9} - \frac{1}{4}$$
$$= 1 - \frac{20}{36} - \frac{9}{36} = \frac{7}{36}$$

The long-range prediction is

$$\left[\frac{7}{36} \ \ \frac{5}{9} \ \ \frac{1}{4}\right] \text{ or } [.1944 \ \ .5556 \ \ .2500]$$

See the histograms for Parts (b), Exercises 35 and 37 in the answer section at the back of the textbook.

35. To find the probability distribution, first add the frequencies.

$1 + 0 + 2 + 5 + 8 + 4 + 3 = 23$

The probabilities are found by dividing the frequency by 23.

Number	8	9	10	11	12	13	14
Probability	$\frac{1}{23}$	0	$\frac{2}{23}$	$\frac{5}{23}$	$\frac{8}{23}$	$\frac{4}{23}$	$\frac{3}{23}$

37. **(a)** When a pair of dice are rolled, the sums range from 2 to 12. There are 36 different rolls of the dice. To find the probabilities, divide the number of ways to get a particular sum by 36.

Number	2	3	4	5	6	7
Probability	$\frac{1}{36}$	$\frac{1}{18}$	$\frac{1}{12}$	$\frac{1}{9}$	$\frac{5}{36}$	$\frac{1}{6}$

Number	8	9	10	11	12
Probability	$\frac{5}{36}$	$\frac{1}{9}$	$\frac{1}{12}$	$\frac{1}{18}$	$\frac{1}{36}$

39. Add the probabilities in each shaded region.

$.3 + .2 + .1 = .6$

41. If you roll a 6, you win $8 - 6 = \$2$.
If you roll a 5, you win $8 - 7 = \$1$.
If you roll anything else, you win $4 - 6 = \$2$.

The probability of a 6 is $\frac{1}{6}$.

The probability of a 5 is $\frac{1}{6}$.

The probability of anything else $= 1 - \frac{1}{6} - \frac{1}{6} = \frac{4}{6}$.

The expected value is

$$E(x) = 2\left(\frac{1}{6}\right) + 1\left(\frac{1}{6}\right) + (-2)\left(\frac{4}{6}\right)$$

$$= \frac{2}{6} + \frac{1}{6} - \frac{8}{6} = \frac{-5}{6} = -83\frac{1}{3} \text{ cents}$$

This is not a fair game.

43. The probability of having a girl is $\frac{1}{2}$.

The probability of not having a girl is $1 - \frac{1}{2} = \frac{1}{2}$.

Number	Probability
0	$\binom{5}{0}\left(\frac{1}{2}\right)^0\left(\frac{1}{2}\right)^5 = \frac{1}{32}$
1	$\binom{5}{1}\left(\frac{1}{2}\right)^1\left(\frac{1}{2}\right)^4 = \frac{5}{32}$
2	$\binom{5}{2}\left(\frac{1}{2}\right)^2\left(\frac{1}{2}\right)^3 = \frac{10}{32}$
3	$\binom{5}{3}\left(\frac{1}{2}\right)^3\left(\frac{1}{2}\right)^2 = \frac{10}{32}$
4	$\binom{5}{4}\left(\frac{1}{2}\right)^4\left(\frac{1}{2}\right)^1 = \frac{5}{32}$
5	$\binom{5}{5}\left(\frac{1}{2}\right)^5\left(\frac{1}{2}\right)^0 = \frac{1}{32}$

$$E(x) = 0\left(\frac{1}{32}\right) + 1\left(\frac{5}{32}\right) + 2\left(\frac{10}{32}\right) + 3\left(\frac{10}{32}\right) + 4\left(\frac{5}{32}\right) + 5\left(\frac{1}{32}\right)$$

$$= \frac{80}{32} = 2.5$$

45. If you win the contest you win $\$1000 - .30 - .01 = \$999.69$.
If you lose, you lose $-\$.31$.

The probability of winning is $\frac{1}{4000}$.

The probability of losing is

$1 - \frac{1}{4000} = \frac{3999}{4000}$.

$$E(x) = \left(\frac{1}{4000}\right)(999.69) + \left(\frac{3999}{4000}\right)(-.31)$$

$$= -.06 \text{ or } -6¢.$$

47. Probability of getting 3 clubs

$$= \frac{\binom{13}{3}}{\binom{52}{3}} = \frac{286}{22100} = .0129$$

Probability of not getting 3 clubs = 1 - .0129 = .9871.
Let x = the amount to pay for the game.
If you win, you get 100 - x.
If you lose, you get -x. The expected value must be 0.

$$(100-x)(.0129)+(-x)(.9871) = 0$$
$$1.29-.0129x-.9871x = 0$$
$$1.29 = x$$

You must pay $1.29.

49. **(a)** Since the candidate is an optimist, look for the biggest value in the matrix, which is 5000. Hence, she should oppose it.

(b) A pessimistic candidate wants to find the best of the worst things than can happen. If she favors, the worse is -4000. If she waffles the worse is -500. If she opposes, the worse is 0. Since the best of these is 0, she should oppose.

(c) Since there is a 40% chance the candidate favors the plant and a 35% that he will waffle, the chance he will oppose is
1 - .4 - .35 = .25.
Expected gain if she favors:
0(.4) + (-1000)(.35) + (-4000)(.25)
= -1350
Expected gain if she waffles:
1000(.4) + 0(.35) + (-500)(.25) = 275
Expected gain if she opposes:
5000(.4) + 2000(.35) + 0(.25) = 2700
She should oppose and get 2700 additional votes.

(d) Now the opponent has 0 probability of favoring, .7 of waffling and .3 of opposing.
Expected gain if she favors:
0(0) + (.7)(-1000) + .3(-4000)
= -1900
Expected gain if she waffles:
0(1000) + .7(0) + .3(-500) = -150
Expected gain if she opposes:
0(5000) + .7(2000) + .3(0) = 1400
She should oppose and get 1400 additional votes.

Case 12 Exercises

1. **(a)** $C(M_0) = NL[1-(1-P_1)(1-P_2)(1-P_3)]$
$= 3(54)[1-(1-.09)(1-.24)(1-.17)]$
$= 162[1-(.91)(.76)(.83)]$
$= \$69.01$

(b) $C(M_2) = H_2+NL[1-(1-P_1)(1-P_3)]$
$= 40+3(54)[1-(1-.09)(1-.17)]$
$= 40+162[1-(.91)(.83)]$
$= \$79.64$

(c) $C(M_3) = H_3+NL[1-(1-P_1)(1-P_2)]$
$= 90+3(54)[1-(1-.09)(1-.24)]$
$= 9+162[1-(.91)(.76)]$
$= \$58.96$

(d) $C(M_{12}) = H_1+H_2+NL[1-(1-P_3)]$
$= 15+40+3(54)[1-(1-.17)]$
$= 55+162[1-.83)]$
$= \$82.54$

(e) $C(M_{13}) = H_1+H_3+NL[1-(1-P_2)]$
$= 15+9+3(54)[1-(1-.24)]$
$= 24+162[.24]$
$= \$62.88$

(f) $C(M_{123}) = H_1 + H_2 + H_3$

$= 15 + 40 + 9$

$= \$64$

2. Of all the answers in Exercise 1, \$58.96 is the least. Hence, the lowest expected value comes from stocking only part 3 on the truck.

3. The probabilities need not add to 1 since P_1, P_2, and P_3 are not the only events in the sample space.

4. $\binom{n}{0} + \binom{n}{1} + \binom{n}{2} + \ldots + \binom{n}{n}$ different policies would be needed.

CHAPTER 10 INTRODUCTION TO STATISTICS

Section 10.1

1.(a) The bar for the 10-19 age group has a height equivalent to about 17.5%.

(b) The bar for the 60-69 age group has a height equivalent to about 8%.

(c) The bar with the tallest height is the bar for the 20-29 age group. Thus the group with the largest percent of the population is the 20-29 age group.

See the graphs for Exercises 3-7 in the answers at the back of the textbook.

3. Find the size of each interval:

$$\frac{149 - 4}{6} = \frac{145}{6} \approx 24.2$$

A convenient size of each interval is 25. Start the lowest interval at 0.

(a)

Class	Tally	Frequency
0-24	////	4
25-49	///	3
50-74	//////	6
75-99	///	3
100-124	/////	5
125-149	/////////	9

(b)

Interval	Frequency	Cumulative Frequency
0-24	4	4
25-49	3	7
50-74	6	13
75-99	3	16
100-124	5	21
125-149	9	30

(c) Draw bars whose heights correspond to the frequencies and whose widths correspond to the class interval sizes.

(d) The frequency polygon is formed by joining the midpoints of the tops of the histogram bars with straight-line segments.

(e) Plot points using ordered pairs, each with the end of each interval as the first component and the cumulative frequency for that interval as the second.

5. The size of each interval:

$$\frac{110 - 72}{9} \approx 4.2$$

A convenient size is 5. Start the lowest interval at 70.

(a)

Class	Tally	Frequency
70-74	//	2
75-79	/	1
80-84	///	3
85-89	//	2
90-94	//////	6
95-99	/////	5
100-104	//////	6
105-109	////	4
110-114	//	2

(b)

Interval	Frequency	Cumulative Frequency
70-74	2	2
75-79	1	3
80-84	3	6
85-89	2	8
90-94	6	14
95-99	5	19
100-104	6	25
105-109	4	29
110-114	2	31

(c) Draw bars whose heights correspond to the frequency and whose widths correspond to the class interval sizes.

(d) The frequency polygon is formed by joining the midpoints of the tops of the histogram bars with straight-line segments.

(e) Plot points using the end of each interval and the cumulative frequency for that interval.

7. Draw bars whose heights correspond to the frequency and whose widths correspond to the class interval sizes. The frequency polygon can be formed by joining the midpoints of the tops of the histogram bars with straight-line segments.

9. The number of heads in 5 tosses (0, 1, 2, 3, 4) will be plotted along the horizontal axis. The frequency of occurrence of each will be plotted along the vertical axis. Draw a bar centered over each number of heads with a height equal to the frequency of occurrence. Each student's graph will be different.

11. Plot the number of heads along the horizontal axis. Draw a bar centered over the number of heads with a height equal to the frequency of occurrence. See the graph in the answer section at the back of the textbook.

13. Use the histogram obtained in Exercise 12. Connect the midpoints of the tops of the histogram bars with straight-line segments. Each student's graph will be different.

15. Plot the numbers of red cards 0-5 along the horizontal axis. Plot the frequencies 0-325 along the vertical axis. Points of the frequency polygon are the graphs of ordered pairs with number of cards as the first component and frequency as the second component. Connect the points with straight-line segments. See the graph in the answer section at the back of the textbook.

17. Count the number of occurrences of each letter for the frequency. Divide this number by the total number of letters in the paragraph, which is 242, to find the percent. The results obtained are given in the table.

Letter	Frequency	Percent
E	41	16.9
T	22	9.1
N	17	7.0
R	17	7.0
I	16	6.6
A	15	6.2
H	14	5.8
S	14	5.8
L	12	5.0
C	11	4.5
O	10	4.1
F	9	3.7
G	9	3.7
U	9	3.7
D	5	2.1
Y	5	2.1
M	4	1.7
P	4	1.7
W	4	1.7
Q	3	1.2
V	1	.4
B,J,K,X,Z	0	0

Section 10.2

1. $\overline{x} = \dfrac{8 + 10 + 16 + 21 + 25}{5}$

$\overline{x} = \dfrac{80}{5} = 16$

3. $\overline{x} = \dfrac{130 + 141 + 149 + 152}{8}$

$+ \dfrac{158 + 163 + 139 + 170}{8}$

$\overline{x} = \dfrac{1202}{8} \approx 150.3$

5. $\bar{x} = \dfrac{21{,}900 + 22{,}850 + 24{,}930}{6} + \dfrac{29{,}710 + 28{,}340 + 40{,}000}{6}$

$\bar{x} = \dfrac{167{,}730}{6} = 27{,}955$

7. $\bar{x} = \dfrac{9.4 + 11.3 + 10.5 + 7.4 + 9.1}{10} + \dfrac{8.4 + 9.7 + 5.2 + 1.1 + 4.7}{10}$

$\bar{x} = \dfrac{76.8}{10} = 7.68 \approx 7.7$

9. $\bar{x} = \dfrac{.06 + .04 + .05 + .08 + .03}{10} + \dfrac{.14 + .18 + .29 + .07 + .01}{10}$

$\bar{x} = \dfrac{.95}{10} = .095 \approx .1$

11.

x	f	xf
3	4	12
5	2	10
9	1	9
12	3	36
Total	10	67

$\bar{x} = \dfrac{67}{10} = 6.7$

13.

x	f	xf
12	4	48
13	2	26
15	5	75
19	3	57
22	1	22
23	5	115
Total	20	343

$\bar{x} = \dfrac{343}{20} = 17.2$

15.

x	f	xf
104	6	624
112	14	1568
115	21	2415
119	13	1547
123	22	2706
127	6	762
132	9	1188
Total	91	10,810

$\bar{x} = \dfrac{10{,}810}{91} = 118.8$

17. The numbers are already in numerical order. There are 7 numbers. The median is the middle number. 7 divided by 2 is 3.5. Adding .5 gives 4, so count 4 entries: The fourth number is 51.

19. The numbers are already in numerical order. There are 6 numbers, so 6 divided by 2 is 3. The median is the sum of the third and fourth number divided by 2.

$\text{Median} = \dfrac{125 + 135}{2} = \dfrac{260}{2} = 130.$

21. The data, arranged in numerical order is 21, 32, 38, 46, 49, 53, 58, 72, 97. There are 9 numbers, so the median is the fifth number or 49.

23. The data arranged in numerical order is 525, 542, 551, 559, 565, 576, 578, 590. There are 8 numbers, so the median is the mean of the fourth and fifth numbers.

$\text{Median} = \dfrac{559 + 565}{2} = \dfrac{1124}{2}$

$= 562.$

25. The data arranged in numerical order is 3.4, 9.1, 27.6, 28.4, 29.8, 32.1, 47.6, 59.8. There are 8 numbers, so the median is the mean of the fourth and fifth numbers or

$$\text{Median} = \frac{28.4 + 29.8}{2} = \frac{58.2}{2}$$
$$= 29.1.$$

27. The number 9 occurs most (twice) and therefore, 9 is the mode.

28. The mode is 64 as it occurs three times.

31. There are two modes, 68 and 74.

33. No numbers occur more than once. So, there are no modes.

35. The number 6.1 occurs most (twice) and therefore, it is the mode.

37.

Interval	Midpoint, x	f	xf
0-24	12	4	48
25-49	37	3	111
50-74	62	6	372
75-99	87	3	261
100-124	112	5	560
125-149	137	9	1233
Total		30	2585

$$\bar{x} = \frac{2585}{30} = 86.16 \approx 86.2$$

Modal class = 125-149, since it contains the most data values.

39.

Interval	Midpoint, x	f	xf
30-39	34.5	1	34.5
40-49	44.5	6	267.0
50-59	54.5	13	708.5
60-69	64.5	22	1419.0
70-79	74.5	17	1266.5
80-89	84.5	13	1098.5
90-99	94.5	8	756.0
Total		80	5550.0

$$\bar{x} = \frac{5550}{80} = 69.4$$

Modal class is 60-69.

41. The mean for the maximum temperature, $\bar{x}$, is

$$\bar{x} = \frac{39 + 39 + 44 + 50 + 60 + 69}{12} + \frac{79 + 78 + 70 + 51 + 47 + 40}{12}$$

$$\bar{x} = \frac{666}{12} = 55.5° \text{ F.}$$

43. The mean price per bushel is given by

$$\bar{x} = \frac{2.70 + 2.30 + 2.95 + 3.80 + 3.90}{10} + \frac{3.60 + 3.55 + 3.50 + 3.35 + 3.20}{10}$$

$$\bar{x} = \frac{32.85}{10} = 3.285 \approx \$3.29.$$

45.

$$\text{Mean} = \frac{-1 + 0 + (-3) + 7 + 1}{10} + \frac{1 + 5 + 4 + 1 + 4}{10}$$
$$= \frac{19}{10} \approx 1.9$$

In order, the data are -3, -1, 0, 1, 1, 1, 4, 4, 5, 7. The median is the middle value or the average of the fifth and sixth value or

$$\frac{1 + 1}{2} = 1.$$

The mode is 1 since it occurs most often (3 times).

47.

$$\text{Mean} = \frac{20 + 15 + 18 + 22}{12} + \frac{10 + 12 + 16 + 17}{12} + \frac{19 + 21 + 23 + 13}{12}$$
$$= \frac{206}{12} = 17.167 \approx 17.2$$

The numbers 17 and 18 are the middle two numbers, that is, the sixth and seventh, when the data are arranged in ascending order, and so the median is

$$\frac{17 + 18}{2} = 17.5.$$

No number occurs more than once and, thus, there is no mode.

Section 10.3

1. Range = 12 - 6 = 6

$$\bar{x} = \frac{6 + 8 + 9 + 10 + 12}{5} = \frac{45}{5} = 9.$$

Number	Deviation from mean	Square of deviation
6	-3	9
8	-1	1
9	0	0
10	1	1
12	3	9
Total		20

$$s = \sqrt{\frac{20}{5 - 1}} = \sqrt{5} = 2.24$$

3. Range = 18 - 6 = 12

$$\bar{x} = \frac{7 + 6 + 12 + 14 + 18 + 15}{6} = \frac{72}{6} = 12$$

Number	Deviation from mean	Square of deviation
7	-5	25
6	-6	36
12	0	0
14	2	4
18	6	36
15	3	9
Total		110

$$s = \sqrt{\frac{110}{6 - 1}} = \sqrt{\frac{110}{5}} = \sqrt{22} \approx 4.7$$

5. Range = 82 - 29 = 53

$$\bar{x} = \frac{42 + 38 + 29 + 74}{7} + \frac{82 + 71 + 35}{7} = \frac{371}{7} = 53.$$

Number	Deviation from mean	Square of deviation
42	-11	121
38	-15	225
29	-24	576
74	21	441
82	29	841
71	18	324
35	-18	324
Total		2852

$$s = \sqrt{\frac{2852}{7 - 1}} = \sqrt{\frac{2852}{6}} = \sqrt{475.33} \approx 21.8.$$

7. Range: 287 - 241 = 46

$\bar{x} = 256$

Number	Deviation from mean	Square of deviation
241	-15	225
248	-8	64
251	-5	25
257	1	1
252	-4	16
287	31	961
Total		1292

$$s = \sqrt{\frac{1292}{6 - 1}} \approx 16.1$$

9. Range: 27 - 3 = 24

$\bar{x} = 14$

Number	Deviation from mean	Square of deviation
3	-11	121
7	-7	49
4	-10	100
12	-2	4
15	1	1
18	4	16
19	5	25
27	13	169
24	10	100
11	-3	9
Total		594

$$s = \sqrt{\frac{594}{10 - 1}} \approx 8.1$$

11. Range = 51 - 21 = 30

$\overline{x} = 34.8$

Number	Deviation from mean	Square of deviation
21	-13.8	190.44
28	-6.8	46.24
32	-2.8	7.84
42	7.2	51.84
51	16.2	262.44
Total		558.80

$$s = \sqrt{\frac{558.80}{5 - 1}} \approx 11.8$$

13.

Interval	Midpoint	f	$x - \overline{x}$	$(x - \overline{x})^2$	$f(x - \overline{x})^2$
0-24	12	4	-74.2	5505.64	22,022.56
25-49	37	3	-49.2	2420.64	7261.92
50-74	62	6	-24.2	585.64	3513.84
75-99	87	3	.8	.64	1.92
100-124	112	5	25.8	665.64	3328.20
125-149	137	9	50.8	2580.64	23,225.76
Total		30			59,354.20

$\overline{x} = 86.2$ (see Section 10.2, Exercise 37)

$$s = \sqrt{\frac{59,354.20}{30 - 1}} = 45.2$$

15.

Interval	Midpoint	f	$f\overline{x}$	$x - \overline{x}$	$(x - \overline{x})^2$	$f(x - \overline{x})^2$
70-74	72	2	144	-22.9	524.41	1048.82
75-79	77	1	77	-17.9	320.41	320.41
80-84	82	3	246	-12.9	166.41	499.23
85-89	87	2	174	-7.9	62.41	124.82
90-94	92	6	552	-2.9	8.41	50.46
95-99	97	5	485	2.1	4.41	22.05
100-104	102	6	612	7.1	50.41	302.46
105-109	107	4	428	12.1	146.41	585.64
110-114	112	2	224	17.1	292.41	584.84
Total		31	2942			3538.73

$$\overline{x} = \frac{2942}{31} = 94.9$$

$$s = \sqrt{\frac{3538.73}{30}} \approx 10.9$$

17. k = 2. Thus,

$$1 - \left(\frac{1}{2^2}\right) = 1 - \frac{1}{4} = \frac{3}{4}.$$

So, 3/4 of the data lie within 2 standard deviations from the mean.

19. k = 5. Thus,

$$1 - \left(\frac{1}{5^2}\right) = 1 - \frac{1}{25} = \frac{24}{25}.$$

So, 24/25 of the data lie within 5 standard deviations from the mean.

21. Since s = 6, $\frac{32 - 50}{6} = -3$

and $\frac{68 - 50}{6} = 3.$

So 32 and 68 both lie within 2 standard deviations from the mean.

(Go to top of right column.)

Thus,

$$k = 3 \text{ and } 1 - \frac{1}{k^2} = 1 - \frac{1}{3^2}$$

$$= 1 - \frac{1}{9} = \frac{8}{9} = 88.9\%.$$

So at least 88.9% of the numbers are between 32 and 68.

23. $\frac{20 - 50}{6} = -5 \qquad \frac{80 - 50}{6} = 5$

So, $1 - \frac{1}{k^2} = 1 - \frac{1}{5^2}$

$$= 1 - \frac{1}{25} = \frac{24}{25} - 96\%.$$

So at least 96% of the data lie between 20 and 80.

25. From Exercise 21, 88.9% of the data lie between 32 and 68. So, 100 - 88.9 = 11.1% of the data are less then 32 or more than 68.

27. Forever Power

x	f	fx	$x - \overline{x}$	$(x - \overline{x})^2$	$f(x - \overline{x})^2$
20	1	20	-6.2	38.44	38.44
22	2	44	-4.2	17.64	35.28
25	1	25	-1.2	1.44	1.44
26	1	26	- .2	.04	.04
27	2	54	.8	.64	1.28
28	1	28	1.8	3.24	3.24
30	1	30	3.8	14.44	14.44
35	1	35	8.8	77.44	77.44
Total	10	262			171.60

$$\overline{x} = \frac{262}{10} = 26.2, \; s = \sqrt{\frac{171.6}{9}} \approx 4.4$$

Brand X

x	f	fx	$x - \bar{x}$	$(x - \bar{x})^2$	$f(x - \bar{x})^2$
15	1	15	-10.5	110.25	110.25
18	1	18	-7.5	56.25	56.25
19	1	19	-6.5	42.25	42.25
23	1	23	-2.5	6.25	6.25
25	2	50	- .5	.25	.50
28	1	28	2.5	6.25	6.25
30	1	30	4.5	20.25	20.25
34	1	34	8.5	72.25	72.25
38	1	38	12.5	156.25	156.25
Total	10	255			470.50

$\bar{x} = \frac{255}{10} = 25.5$, $s = \sqrt{\frac{470.5}{9}} \approx 7.2$.

(a) Forever Power (s of 4.4 as opposed to s of 7.2)

(b) Forever Power (26.2 as opposed to 25.5)

29.(a) For the test cities,

$$\bar{x} = \frac{18 + 15 + 7 + 10}{4} = 12.5.$$

(b) For the four control cities,

$$\bar{x} = \frac{1 + (-8) + (-5) + 0}{4} = -3.0.$$

(c) Test cities

x	$x - \bar{x}$	$(x - \bar{x})^2$
18	5.5	30.25
15	2.5	6.25
7	-5.5	30.25
10	-2.5	6.25
Total		73.00

$$s = \sqrt{\frac{73.00}{4 - 1}} \approx 4.9$$

(d) Control cities

x	$x - \bar{x}$	$(x - \bar{x})^2$
1	4	16
-8	-5	25
-5	-2	4
0	3	9
Total		54

$$s = \sqrt{\frac{54}{3}} \approx 4.2.$$

(e) $12.5 - (-3) = 15.5$

(f) $15.5 - 7.95 = 7.55$

$15.5 + 7.95 = 23.45$

Exercises 31 and 33 were found by calculator.

31. $\overline{x} = \frac{43.58}{24} = 1.8158$

$s = \sqrt{\frac{4.55618}{23}} = .4451$

33. $\overline{x} = \frac{154.5}{21} = 7.3571$

$s = \sqrt{\frac{.35142}{20}} = .1326$

Section 10.4

1. 2.50 standard deviations means that $z = 2.50$.
From Table 8, for $z = 2.50$ the entry is .4938. So, the percent of area is 49.38%.

3. From Table 8, the entry for $z = .45$ is .1736, so the percent of area between the mean and .45 is 17.36%.

5. From Table 8, the entry for $z = 1.71$ is .4564. (The minus sign indicates that the area is to the left of the mean.) So the percent of area is 45.64%.

7. From Table 8, the entry for $z = 3.11$ is .4991, so the percent of area is 49.91%.

9. The entry corresponding to $z = 1.41$ is .4207 and the entry for $z = 2.83$ is .4977. So the area between these values is
$.4977 - .4207 = .0770$ or 7.7%.

11. For $z - 2.48$, the entry is .4934 and for $z = .05$, the entry is .0239. The area between these values is $.4934 - .0199 = .4735$ or 47.35%.

13. For $z = 3.11$, the area is .4991 and for $z = 1.44$, the area is .4251. since $z = -3.11$ is negative, the area .4991 is to the left of the mean and the other area is to the right of the mean. Therefore, the area between the two values is
$.4991 + .4251 = .9241$ or 92.42%.

15. See Exercise 13 for details.
$z = .42$ corresponds to the area .1628 and, therefore, the area between the two values is
$2 \cdot .1628 = .3256$ or 32.56%.

17. The mean divides the area in half or .5. The area from the mean to z standard deviations is
$.5 - .05 = .45$. Using Table 8 backwards, we get the z-score corresponding to the area .45 as 1.65 (approximately).

19. The area from the mean to z standard deviations is
$.5 - .15 = .35$. The z-score from the table is 1.04. As the area is to the left of the mean, $z = -1.04$.

21. Since the average life is 500 hours, the total number of bulbs that will last at least 500 hours corresponds to $.5 = 50\%$ of the total area to the right of the mean. So, 50% of 10,000, or 5000 bulbs will last at least 500 hours.

23. $z = \frac{x - \mu}{\sigma} = \frac{650 - 500}{100} = 1.5$

This z-score corresponds to 43.32% of the total number of bulbs. So,
(.4332)(10,000) = 4332
bulbs will last between 500 and 650 hours.

25. For x = 780,

$z = \frac{780 - 500}{100} = 2.8.$

The area corresponding to this z-score is .4974. For x = 650,

$z = \frac{650 - 500}{100} = 1.5$

and the area corresponding to this z-score is .4332. The area between these values is .4974 - .4332 = .0642. So, 6.42% of 10,000 = 642 bulbs will last between 650 and 780 hours.

27. $z = \frac{740 - 500}{100} = 2.4.$

The area from mean to 2.4 standard deviations (to the right) is .4918 and
.5 + .4918 = .9918 = 99.18%.
Therefore,
99.18% of 10,000 = 9918
bulbs will last 740 hours.

29. $z = \frac{790 - 500}{100} = 2.9$

The area from the mean to this z-value is .4981. So, the area to the right of the z-value is .5 - .4981 = .0019.
Thus,
.0019 · 10,000 = 19
bulbs will last more than 790 hours.

31. We are given σ = .5 and μ = 16.5. A box is underweight if it weighs less than 16 ounces. We will find the area below the x value of 16.

$z = \frac{16.5 - 16}{.5} = 1.$

The area from the mean to this z-value is .3413. So,
.5 - .3413 = .1587 or 15.87%
of the boxes will be underweight.

33. See Exercise 31 for details.

$z = \frac{16.5 - 16}{.2} = 2.5$

The area corresponding to this z-value is .4938. So,
.5 - .4938 = .0062 or .62%
of the boxes will weigh less than 16 ounces.

35. We are given mean μ = 1850 and standard deviation σ = 150. For x = 1700,

$z = \frac{1850 - 1700}{150} = 1.$

The corresponding area is .3413. So,
.3413 + .5 = .8413 or 84.13%
of the chickens will weigh more than 1700 grams.

37. $z_1 = \frac{1850 - 1750}{150} = .67$

and the corresponding area is .2486. $z_2 = \frac{1900 - 1850}{150} = .33$

and the area is .1293. So,
.2486 + .1293 = .3779 or 37.79%
of the chickens will weigh between 1750 and 1900 grams.

39. $z = \frac{1850 - 1550}{150} = 2$

The area is .4773. So,
.5 - .4773 = .0227 or 2.27%
of the chickens will weigh less than 1500 grams.

41. We are given z = 2.5. The area corresponding to this z-value is .4938 from the mean. So,
.5 + .4938 = .9938 or 99.38%
of the population will receive an adequate amount of vitamins.

43. The recommended daily allowance is
159 + (2.5 · 12) = 189 units.

45. $\mu = .25$, $\sigma = .02$, and $x = .3$.

$z = \frac{.3 - .25}{.02} = 2.5$, Area = .4938.

Note that the x-value, .3, is greater than the mean .25 and hence, the probability that the diameter is greater than .2 is
.5 - .4938 = .0062.

47. $\mu = 32.2$, $\sigma = 1.2$, and $x = .3$.

$z = \frac{32 - 32.2}{1.2} = -1.7$,

Area = .0675.
The probability that the carton will contain less than 32 ounces is
.5 - .0675 = .4325.

49. $\mu = 52.25$ and $\sigma = 15.50$.
Of the 50% of the customers, 25% will be to the right of the mean and the other 25% will be to the left of the mean. The z-score corresponding to the area .25 is .67. Therefore, the largest amount spent is
52.25 + (.67 · 15.50) = $62.64
and the smallest amount spent is
52.25 - (.67 · 15.50) = $41.86.

51. $\mu = 12.3$, $\sigma = 4.1$, and $x = 18$.

$z = \frac{18 - 12.3}{4.1} = 1.39$

The area is .4177. Therefore, the probability that the fish is longer than 18 inches is
.5 - .4177 = .0823.

53. We are given $\mu = 74$ and $\sigma = 6$. Let us find the z-score corresponding to the grade A. The area is .5 - .08 = .42 and the z-value from Table 8 is 1.41. Therefore, to get an A one must obtain
74 + (1.41 x 6) = 82 points.

55. For a grade of C, the area from the mean is .5 - .23 = .27 and the z-value is .74. So, for a C grade, one must obtain
74 - (.74 · 6) = 70 points.

57. Find the z-value for $x = 4\frac{15}{16}$,

$\mu = 7.5$, and $\sigma = \frac{21}{16}$.

$$z = \frac{4\frac{15}{16} - 7.5}{\frac{21}{16}} = -1.95.$$

The area between the mean and z = -1.95 is .4744. So the area to the left of z = -1.95 is
.5 - .4744 = .0256.
This is the probability that a newborn baby will weigh less than 4 pounds 15 ounces.

59. The value of z so that 95% (.95) of the area is to the left of z is found by looking up the area

$.95 - .5 = .45$

in the body of the table. The z-value is 1.65. So,

$x = \mu + \sigma z = 35 + 7(1.65) \approx 47.$

So to be at work by 8:00 A.M. (with a 95% chance), the commuter should leave at 7:13 A.M. (47 minutes before 8).

Section 10.5

1.(a) The probability that 1 comes up on the die is

$p = \frac{1}{6}.$

Find the probability distribution if n = 6.

$P(x) = \binom{n}{x} p^x (1 - p)^{(n-x)}$

$P(0) = \binom{6}{0}\left(\frac{1}{6}\right)^0\left[1 - \left(\frac{1}{6}\right)\right]^{6-0}$

$= 1(1)(.3349)$

$\approx .335$

$P(1) = \binom{6}{1}\left(\frac{1}{6}\right)^1\left[1 - \left(\frac{1}{6}\right)\right]^{6-1}$

$= 6\left(\frac{1}{6}\right)(.4019)$

$\approx .402$

$P(2) = \binom{6}{2}\left(\frac{1}{6}\right)^2\left[1 - \left(\frac{1}{6}\right)\right]^4$

$= 15\left(\frac{1}{36}\right)\left(\frac{5}{6}\right)^4$

$\approx .201$

Similarly we get the following probability distribution.

x	0	1	2	3	4	5	6
P(x)	.335	.402	.201	.054	.008	.001	.000

(b) $\mu = np = 6\left(\frac{1}{6}\right) = 1$

(c) $\sigma = \sqrt{np(1 - p)}$

$= \sqrt{6\left(\frac{1}{6}\right)\left[1 - \left(\frac{1}{6}\right)\right]}$

$= \sqrt{\frac{5}{6}} \approx .91$

3. p - .02, 1 - p = .98, and n = 3

(a)

x	P(x)
0	$\binom{3}{0}(.02)^0(.98)^3 \approx .941$
1	$\binom{3}{1}(.02)^1(.98)^2 \approx .058$
2	$\binom{3}{2}(.02)^2(.98)^1 \approx .001$
3	$\binom{3}{3}(.02)^3(.98)^0 \approx .000$

(b) $\mu = np = (.02)(3) = .06$

(c) $\sigma = \sqrt{np(1 - p)} = \sqrt{(3)(.02)(.98)}$

$\sigma \approx .24$

5. p = .7, 1 - p = .3 and n = 4

(a)

x	P(x)
0	$\binom{4}{0}(.7)^0(.3)^4 = .0081$
1	$\binom{4}{1}(.7)^1(.3)^3 = .0756$
2	$\binom{4}{2}(.7)^2(.3)^2 = .2646$
3	$\binom{4}{3}(.7)^3(.3)^1 \approx .4116$
4	$\binom{4}{4}(.7)^4(.3)^0 \approx .2401$

Note that the sum of the probabilities in the distribution is 1, which should be the case in any binomial distribution.

(b) $\mu = np = (4)(.7) = 2.8$

(c) $\sigma = \sqrt{np(1-p)} = \sqrt{4(.7)(.3)} \approx .92$

7. $p = .025$ and $n = 500$

$\mu = np = 500(.025) = 12.5$

$\sigma = \sqrt{n(p)(1-p)}$

$\sigma = \sqrt{500(.025)(.975)} = 3.49$

9. $n = 64$, $p = .8$

$\mu = np$

$= 64(.8)$

$= 51.2$

$\sigma = \sqrt{n(p)(1-p)}$

$\sigma = \sqrt{16 \cdot \frac{1}{2}(1 - \frac{1}{2})} = 2$

11. The probability of a head coming up is

$p = \frac{1}{2}$ and $n = 16$.

Since we are using a normal curve approximation,

$\mu = np$

$\mu = 16(\frac{1}{2})$

$= 8$

$\sigma = \sqrt{n(p)(1-p)}$

$\sigma = \sqrt{16 \cdot \frac{1}{2}(1 - \frac{1}{2})} = 2.$

The required probability is the area of the region corresponding to the x-values of 7.5 and 8.5.

$z = \frac{x - \mu}{\sigma} = \frac{8.5 - 8}{2} = .25$

and from Table 8, the corresponding area is .0987. Therefore, the total area is

$2 \cdot .0987 = .1974.$

This is the probability of getting exactly 8 heads. (Note that a direct calculation using binomial distribution gives

$P(8) = \binom{16}{8}(\frac{1}{2})^8 (1 - \frac{1}{2})^8 = .1964.$)

13. Here, $p = \frac{1}{2}$ and hence, $\mu = 8$ and $\sigma = 2$. See Exercise 11 for details. We need to find the areas corresponding to the two x-values 9.5 and 10.5.

$z_1 = \frac{10.5 - 8}{2} = 1.25$

The area is .3944.

$z_2 = \frac{9.5 - 8}{2} = .75$

The area is .2734. Therefore, the probability of getting exactly 10 tails is $.3944 - .2734 = .1210$.

15. $n = 1000$, $p = \frac{1}{2}$

$\mu = np = 1000\left(\frac{1}{2}\right) = 500$

$\sigma = \sqrt{np(1-p)} = \sqrt{100(\frac{1}{2})(\frac{1}{2})} = 15.8$

We need to find the area of the region corresponding to the x-values of 499.5 and 500.5.

$z = \frac{500.5 - 500}{15.8} = .03$

The area is .0120. Since the region is symmetric about the mean $\mu = 500$, the probability (area) is $2(.0120) = .0240$.

17. $p = \frac{1}{2}$, $\mu = 500$, and $\sigma = 15.8$.

The required area is to the right of the x-value 479.5.

$z = \frac{479.5 - 500}{15.8} = -1.3$

The area is .4032. This area is to the left of the mean. So, the required area (probability) is $.4032 + .5 = .9032$.

19. We need to find the probability for 517 heads or less. $\mu = 500$ and $\sigma = 15.8$ (see Exercise 15). For $x = 517.5$,

$$z = \frac{517.5 - 500}{15.8} = 1.11.$$

The area is .3665, which is from the mean $\mu = 500$ to $x = 517.5$. So the probability is
$.5 + .3665 = .8665$.

21. The probability of a five is $p = \frac{1}{6}$ and $n = 120$.

$$\mu = np = 120\left(\frac{1}{6}\right) = 20$$

$$\sigma = \sqrt{120\left(\frac{1}{6}\right)\left(\frac{5}{6}\right)} = 4.08$$

$$z = \frac{20.5 - 20}{4.08} = .12$$

The area is .0478. So, the area between the x-values 19.5 and 20.5 is
$2(.0478) = .0956$.
This is the required probability.

23. The probability of a three is $p = \frac{1}{6}$.

$$z_1 = \frac{16.5 - 20}{4.08} = -.86$$

The area is .3051.

$$z_2 = \frac{17.5 - 20}{4.08} = -.61$$

The area is .2291. So, the probability of 17 threes is
$.3051 - .2291 = .0760$.

25. $p = \frac{1}{6}$ and the needed area is to the right of $x = 18.5$.

$$z = \frac{18.5 - 20}{4.08} = -.37$$

The area is .1443. So, for more than 18 threes, the probability is
$.1443 + .5 = .6443$.

27. $p = 2\% = .02$ and $n = 10{,}000$
$\mu = np = 10{,}000(.02) = 200$
$\sigma = \sqrt{10{,}000 \cdot .02 \cdot .98} = 14$
Since 169 or fewer were defective, the area is to the left of $x = 169.5$.

$$z = \frac{169.5 - 200}{14} = -2.18$$

The area from $x = 169.5$ to the mean is .4854. So, the area to the left of $x = 169.5$ is
$.5 - .4854 = .0146$.

29. $p = 80\% = .8$ and $n = 25$.
$\mu = 25$ and $\sigma = 2$. See Exercise 15 for details.

$$z = \frac{20.5 - 20}{2} = .25$$

The area is .0987. Similarly, the area corresponding to $x = 19.5$ is .0987. Therefore, the probability is
$2(.0987) = .1974$.

31. "All are cured" is the same as "exactly 25 are cured." $p = .8$, $n = 25$, $\mu = 20$, and $\sigma = 2$.

$$z = \frac{x - \mu}{\sigma} = \frac{24.5 - 20}{2} = 2.25$$

The area is .4878. For $x - 25.5$,

$$z = \frac{25.5 - 20}{2} = 2.75.$$

The area is .4970. So the probability is
$.4970 - .4878 = .0092$.

(Note that all cured is also the same as 25 or more are cured. By this interpretation, the probability is
$.5 - .4878 = .0122$
which is close to .0092.)

33. For $x = 12.5$,

$$z = \frac{12.5 - 20}{2} = -3.75$$

The area is .4999. (This area is from $x = 12.5$ to the mean $\mu = 20$.) So, the probability is
$.5 - .4999 = .0001$.

35. Exercise 51

$\mu = np = (134)(.8) = 107.2$

$\sigma = \sqrt{np(1 - p)} = \sqrt{(134)(.8)(.2)}$
$= 4.63$

(a) $z = \dfrac{9.5 - 107.2}{4.63} = -21.1$

$z = \dfrac{10.5 - 107.2}{4.63} = -20.89$

The probability that exactly 10 get the flu is equal to the area between $z = 21.1$ and -20.89, which is approximately .0001.

(b) The probability that no more than 10 get the flu is equal to the area to the left of $z = -20.89$ or approximately .0002.

(c) $z = \dfrac{0 - 107.2}{4.63} = -23.3$

The probability that none get the flu is the area to the left of $z = -23.3$ or approximately 0.

Exercise 52

$\mu = np = (53)(.042) = 2.226$

$\sigma = \sqrt{np(1 - p)} = \sqrt{(53)(.042)(.958)}$
$= 1.4603$

(a) $z = \dfrac{4.5 - 2.226}{1.4603} = 1.56$

$z = \dfrac{5.5 - 2.226}{1.4603} = 2.24$

The probability that exactly 5 are color blind is the area between $z = 1.56$ and $z = 2.24$ or
$.4875 - .4406 = .0469$.

(b) The probability that no more than 5 are color blind is the area to the left of $z = 2.24$ or
$.5 + .4875 = .9875$.

(c) The probability that at least 1 is color blind $(x = 1)$ is found by determining the area to the right of the z-value that corresponds to $x = .5$.

$z = \dfrac{.5 - 2.226}{1.4603} = -1.18$

So the area from $x = .5$ to μ is .3810. The area from μ to the right is .5000. Therefore, the total area is
$.3810 + .5000 = .8810$,
which is the probability that at least 1 is color blind.

Exercise 53

$\mu = (75)(.05) = 3.75$

$\sigma = \sqrt{(75)(.05)(.95)} = 1.89$

(a) $z = \frac{4.5 - 3.75}{1.89} \approx .40$

$z = \frac{5.5 - 3.75}{1.89} \approx .93$

The probability of exactly 5 defectives is the area between z = .40 and z = .93 or
.3238 - .1554 = .1684.

(b) The probability of no defectives (x = 0) is found by determining the area under the standard normal curve between the z-values corresponding to x = -.5 and x = .5.

$z = \frac{-.5 - 3.75}{1.89} = -2.25$

$z = \frac{.5 - 3.75}{1.89} = -1.72$

Area between z = -2.25 and z = -1.72 is .4878 - .4573 = .0305. This is the probability of no defectives.

(c) The probability of at least 1 defective is the area to the right of the z-value that corresponds to x = .5.

$z = \frac{.5 - 3.75}{1.89} = -1.72$

The area from x = .5 to μ is .4573 and from μ to the right is .5000. Therefore, the probability is
.4573 + .5000 = .9573.

Exercise 54

(a) $\mu = (58)(.7) = 40.6$

$\sigma = \sqrt{(58)(.7)(.3)} = 3.49$

To find the probability that 58 liked the product, find the area under the standard normal curve between the z-values corresponding to x = 57.5 and x = 58.5.

$z = \frac{57.5 - 40.16}{3.49} = 4.97$

$z = \frac{58.5 - 40.16}{3.49} = 5.25$

Area between 4.97 and 5.25 is
.5000 - .5000 = 0.
So the probability that 58 liked the product is 0.

(b) Find the area under the standard normal curve between the z-values corresponding to x = 27.5 and x = 30.5.

$z = \frac{27.5 - 40.6}{3.49} = -3.75$

$z = \frac{30.5 - 40.6}{3.49} = -2.89$

The area between z = -3.75 and z = -2.89 is
.4999 - .4981 = .0018.

Chapter 10 Review Exercises

1.(a)

Interval	Frequency
450-474	5
475-499	6
500-524	5
525-549	2
550-574	2

(b) and (c) See the graphs on the answer section at the back of the textbook.

3. $\bar{x} = \frac{41 + 60 + 67 + 68 + 72}{10}$

$+ \frac{74 + 78 + 83 + 90 + 97}{10}$

$= \frac{730}{10} = 73$

5.

Midpoint	f	xf	$(x - \bar{x})^2$	$f(x - \bar{x})^2$
14.5	6	87	416.16	2496.96
24.5	12	294	108.16	1297.92
34.5	14	483	.16	2.24
44.5	10	445	92.16	921.60
54.5	8	436	384.16	3073.28
Total	50	1745		7792.00

$\bar{x} = \frac{1745}{50} = 34.9.$

7. The median is the fourth number, 44. The mode is the most frequent number, 46.

9. The modal class is 30-39. This class has the most data.

11. Range = 32 - 14 = 18

x	$x - \bar{x}$	$(x - \bar{x})^2$
14	-16	36
17	-3	9
18	-2	4
19	-1	1
32	12	144
Total		194

$\bar{x} = \frac{100}{20}, \ \sigma = \sqrt{\frac{194}{5 - 1}} \approx 7.0.$

13. $s = \sqrt{\frac{\sum (x - \bar{x})^2 f}{n - 1}} = \sqrt{\frac{7792}{50 - 1}} \approx 12.6$

15.(a) Stock I

$\bar{x} = \frac{11 + (-1) + 14}{3} = 8\%$

$s = \sqrt{\frac{(11 - 8)^2 + (-1 - 8)^2 + (14 - 8)^2}{2}}$

$\approx 7.9\%$

Stock II

$\bar{x} = \frac{9 + 5 + 10}{3} = 8\%$

$s = \sqrt{\frac{(9 - 8)^2 + (5 - 8)^2 + (10 - 8)^2}{2}}$

$\approx 2.6\%$

(b) Stock II is preferable because the standard deviation is less, that is, the annual returns fluctuate less, compared to stock I.

17.(a) $\bar{x} = 28, \ s = 4$

Since $\frac{36 - 28}{4} = 2$ and

$\frac{20 - 28}{4} = -2$, 20 and 36 are within 2 standard deviations from the mean. Thus, k = 2 and

$1 - \frac{1}{2^2} = \frac{3}{4} = 75\%.$

Hence, 75% of the distribution is between 20 and 36.

(b) Since $\frac{32.8 - 28}{4} = 1.2$ and

$\frac{23.2 - 28}{4} = -1.2$, 32.8 and 23.2 are within 1.2 standard deviations of the mean. So, k = 1.2 and

$1 - \frac{1}{1.2^2} = .306 = 30.6\%.$

Thus 30.6% lies within these numbers so
100 - 30.6% = 69.4%
is less than 23.2 or greater than 32.8.

19. From Table 8, the area is .3980.

21. For z = .41, the area is .1591. So, the area to the left of z = .41 is .5 - .1591 = .3409.

23. $\mu = 100$, $\sigma = 15$. We need to find the area to the right of x = 130.

$$z = \frac{130 - 100}{15} = 2$$

The area is .4772. The area to the right of

x = 130 is .5 - .4773 = .0227.
Therefore, 2.27% of the people score more than 130.

25. For x = 115,

$$z = \frac{115 - 100}{15} = 1.$$

The area is .3413. Since 85 and 115 are equidistant from the mean, the area is
2(.3413) = .6826.
So, 68.26% score between 85 and 115.

27. p = .005, 1 - p = .995 and n = 4

x	P(x)
0	$\binom{4}{0}(.005)^0(.995)^4 = .9801$
1	$\binom{4}{1}(.005)^1(.995)^3 = .0197$
2	$\binom{4}{2}(.005)^2(.995)^2 = .0001485$
3	$\binom{4}{3}(.005)^3(.995)^1 = .0000004975$
4	$\binom{4}{4}(.005)^4(.995)^0 = 6 \times 10^{-10}$

$\mu = np = 4(.005) = .02$ and

$s = \sqrt{4(.005)(.995)} = .141$

29. Use the normal curve approximation with $\mu = np = (50)(.21) = 10.5$ and

$$\sigma = \sqrt{np(1 - p)}$$
$$= \sqrt{(50)(.21)(.79)} = 2.88$$

(a) Find the area between z-values corresponding to x = 7.5 and x = 8.5. The z-values are

$$z = \frac{7.5 - 10.5}{2.88} = -1.04$$

$$z = \frac{8.5 - 10.5}{2.88} = -.69.$$

So the area between z = -1.04 and z = -.69 is .3508 - .2549 = .0959.

(b) Find the probability that 2 or less go bankrupt.

$$z = \frac{2.5 - 10.5}{2.88} = -2.78$$

The area is .4973. So the area to the left of z = -2.78 is .5 - .4973 = .0027. So the probability that no more than 2 go bankrupt is .0027.

31. We assume that the time spent on commuting can be approximated by a normal curve. Let us find the percent of people commuting 35 minutes or less a day.

$\mu = 42$ and $\sigma = 12$

$$z = \frac{35 - 42}{12} = -.58$$

The area is .2910. So, the area to the left of z = -.58 is
.5 - .2190 = .2810
and 28.10% of people commute no more than 35 minutes.

33. $\mu = 42$, x = 38, and x = 60
For x = 60,

$$z = \frac{60 - 42}{12} = 1.5$$

The area is .4332.
For x = 38,

$$z = \frac{38 - 42}{12} = -.33.$$

The area is .1293. The total area is .4332 + .1293 = .5625 and so, 56.25% of people commute between 38 and 60 minutes per day.

35. $p = 6\% = .06$ and $n = 500$.
The problem satisfies the hypotheses of a binomial distribution, so the normal curve approximation can be used.
$\mu = np = 500(.06) = 30$ and
$\sigma = \sqrt{np(1 - p)}$
$= \sqrt{500(.06)(.94)} = 5.3$.
We need to find the area between two x-values, $x = 29.5$ and $x = 30.5$.
$z = \frac{30.5 - 30}{5.3} = .09$
The area is .0359. Since the two x-values are symmetric about the mean, the area is
$2(.0359) = .0718$.

37. $p = .98$, $1 - p = .02$, and $n = 100$.
$x = 95\%$ of $100 = 95$
$P(x) = \binom{100}{95}(.98)^{95}(.02)^{5}$

39. $P = .98$, $1 - p = .02$, and $n = 100$.
The probability that 90 or more flies are killed is
$\binom{100}{90}(.98)^{90}(.02)^{10}$
$+ \binom{100}{91}(.98)^{91}(.02)^{9}$
$+ \ldots$
$+ \binom{100}{100}(.98)^{100}(.02)^{0}$

Case 13 Exercises

1. $P = S(F + L) + 1.64\ \sqrt{2S(F + L)}$
when $L = 4$, $S = 12$, $F = 1$.
$P = 12(1 + 4) + 1.64\ \sqrt{2(12)(5)}$
$P = 12(5) + 1.64\ \sqrt{120}$
$P = 77.9$
So the level of inventory at which reordering should occur is 77.9.
If level is 85, do not reorder. If level is 50, reorder.

2. $P = S(F + L) + 1.64\sqrt{2S(F + L)}$
when $L = 5$, $S = 3$, $F = 4$.
$P = 3(4 + 5) + 1.64\sqrt{2(3)(9)}$
$P = 3(9) = 1.64\sqrt{54}$
$P = 39.05$
So, the level of inventory at which reordering should occur is 39.05. If inventory level is 50, do not reorder. If inventory level is 30, reorder.

CHAPTER 11 DIFFERENTIAL CALCULUS

Section 11.1

1. **(a)** By reading the graph as x gets closer to 3 from the left or the right, f(x) gets closer to 3, so

$$\lim_{x\to 3} f(x) = 3$$

(b) As x gets closer to -1.5 from the left or right, f(x) gets closer to 0, so

$$\lim_{x\to -1.5} f(x) = 0.$$

3. **(a)** By reading the graph, as x gets closer to -2 from the left, f(x) approaches -1. As x gets closer to -2 from the right, f(x) approaches -1/2. Since these two values of f(x) are not equal,

$\lim_{x\to -2} f(x)$ does not exist.

(b) By reading the graph, as x gets closer to 1 from the left or right, f(x) gets closer to -1/2 so

$$\lim_{x\to 1} f(x) = -\frac{1}{2}.$$

5. **(a)** By reading the graph, as x gets closer to 0 from the left or right, f(x) gets closer to 2, so

$$\lim_{x\to 0} f(x) = 2.$$

(b) By reading the graph, as x gets closer to -1 from the right, f(x) increases without bound, and from the left, f(x) decreases without bound. Since f(x) does not approach any fixed number, f(x) has no limit as x approaches -1 from both the right and left, so

$\lim_{x\to -1} f(x)$ does not exist.

7. **(a)** By reading the graph, as x gets closer to 1 from the left or right, g(x) gets closer to 1, so

$$\lim_{x\to 1} g(x) = 1.$$

(b) By reading the graph, as x gets closer to -1 from the left or right, g(x) gets closer to -1, so

$$\lim_{x\to 0} g(x) = 1.$$

9. **(a)** By reading the graph, as x increases without bound, g(x) continues to increase. Since g(x) does not approach any fixed number, g(x) has no limit as x approaches ∞. So,

$\lim_{x\to \infty} g(x)$ does not exist.

(b) By reading the graph, as x gets closer to 0 from the left or right, g(x) gets closer to 1. Thus,

$$\lim_{x\to 0} g(x) = 1.$$

11. By calculating the value of $\sqrt{x + 1}$ for each x, entering this into the calculator, and then using the ln key, the following values are obtained.

x	2.9	2.99	2.999	2.9999	3.0001	3.001	3.01	3.1
x + 1	3.9	3.99	3.999	3.9999	4.0001	4.001	4.01	4.1
$\sqrt{x + 1}$	1.9748	1.9975	1.9997	1.99998	2.0000	2.0002	2.0025	2.0248
$\ln\sqrt{x + 1}$	.6805	.6919	.6930	.6931	.6932	.6933	.6944	.7055

An estimate of $\lim_{x \to 3} \ln\sqrt{x + 1}$ is .693.

13.

x	-1.1	-1.01	-1.001	-1.0001	-.9999	-.999	-.99	-.9
3x + 2	-1.3	-1.03	-1.003	-1.0003	-.9997	-.997	-.97	-.7
3x + 2	.2725	.3570	.3668	.3678	.3680	.3690	.3791	.4966

An estimate of $\lim_{x \to -1} e^{3x+2}$ is .368.

15. $\lim_{x \to -1} (4x^3 - x^2 + 3x - 1)$

$= 4(-1)^3 - (-1)^2 + 3(-1) - 1$
$= 4(-1) - (1) - 3 - 1$
$= -4 - 1 - 3 - 1$
$= -9$

17. $\lim_{x \to -1} \frac{2x + 1}{3x - 4} = \frac{2(-2) + 1}{3(-2) - 4}$

$= \frac{-4 + 1}{-6 - 4} = \frac{-3}{-10} = \frac{3}{10}$

19. $\lim_{x \to 2} \frac{-4x^3 + 6x - 8}{3x^2 + 7x - 2}$

$= \frac{-4(2)^2 + 6(2) - 8}{3(2)^2 + 7(2) - 2}$

$= \frac{-16 + 12 - 8}{12 + 14 - 2} = \frac{-12}{24}$

$= -\frac{1}{2}$

21. $\lim_{x \to -2} \frac{x^2 - 4}{x + 2}$

$= \lim_{x \to -2} \frac{(x + 2)(x - 2)}{x + 2}$

$= \lim_{x \to -2} (x - 2)$

$= (-2) - 2 = -4$

23. $\lim_{x \to 5} \frac{x^2 - 3x - 10}{x - 5}$

$= \lim_{x \to 5} \frac{(x - 5)(x + 2)}{x - 5}$

$= \lim_{x \to 5} (x + 2)$

$= (5) + 2 = 7$

25. $\lim\limits_{x \to 3} \dfrac{x^2 + 5x + 6}{x^2 + 2x - 3}$

$= \lim\limits_{x \to -2} \dfrac{(x + 1)(x + 2)}{(x - 3)(x + 2)}$

$= \lim\limits_{x \to -2} \dfrac{x + 1}{x - 3} = \dfrac{-2 + 1}{-2 - 3}$

$= \dfrac{-1}{-5} = \dfrac{1}{5}$

27. $\lim\limits_{x \to -3} \dfrac{x^2 + 5x + 6}{x^2 + 2x - 3}$

$= \lim\limits_{x \to -3} \dfrac{(x + 3)(x + 2)}{(x + 3)(x - 1)}$

$= \lim\limits_{x \to -3} \dfrac{x + 2}{x - 1} = \dfrac{-3 + 2}{-3 - 1}$

$= \dfrac{-1}{-4} = \dfrac{1}{4}$

29. $\lim\limits_{x \to -3} \dfrac{(x + 3)(x - 3)(x + 4)}{(x + 8)(x + 3)(x - 4)}$

$= \lim\limits_{x \to -3} \dfrac{(x - 3)(x + 4)}{(x + 8)(x - 4)}$

$= \dfrac{(-3 - 3)(-3 + 4)}{(-3 + 8)(-3 - 4)} = \dfrac{(-6)(1)}{(5)(-7)}$

$= \dfrac{-6}{-35} = \dfrac{6}{35}$

31. $\lim\limits_{x \to 3} \sqrt{x^2 - 5} \quad = \sqrt{\lim\limits_{x \to 3} (x^2 - 5)}$

$= \sqrt{(3)^2 - 5}$

$= \sqrt{9 - 5} = \sqrt{4} = 2$

33. $\lim\limits_{x \to -2} \dfrac{3x}{(x + 2)^3}$

As $x \to -2$, $3x \to -6$.

As $x \to -2$, $(x + 2)^3 \to 0$.

So the denominator gets close to 0

and $\dfrac{3x}{(x + 2)^3}$ increases without bound.

Thus, $\lim\limits_{x \to -2} \dfrac{3x}{(x + 2)^3}$ does not exist.

35. $\lim\limits_{x \to 0} \dfrac{-x^5 - 9x^3 + 8x^2}{5x}$

$= \lim\limits_{x \to 0} \dfrac{x(-x^4 - 9x^2 + 8x}{5x}$

$= \lim\limits_{x \to 0} \dfrac{-x^4 - 9x^2 + 8x}{5}$

$= \dfrac{-(0)^4 - 9(0)^2 + 8(0)}{5}$

$= \dfrac{0}{5} = 0$

37. $\lim\limits_{x \to 0} \dfrac{\dfrac{-1}{x + 2} + \dfrac{1}{2}}{x}$

$= \lim\limits_{x \to 0} \dfrac{\dfrac{-2 + (x + 2)}{2(x + 2)}}{x}$

$= \lim\limits_{x \to 0} \dfrac{\dfrac{x}{2(x + 2)}}{x}$

$= \lim\limits_{x \to 0} \dfrac{x}{2x(x + 2)}$

$= \lim\limits_{x \to 0} \dfrac{1}{2(x + 2)}$

$= \dfrac{1}{2(0 + 2)} = \dfrac{1}{4}$

39. $\lim_{x\to 36} \frac{\sqrt{x} - 6}{x - 36}$

$= \lim_{x\to 36} \frac{\sqrt{x} - 6}{x - 36} \cdot \frac{\sqrt{x} + 6}{\sqrt{x} + 6}$

$= \lim_{x\to 36} \frac{x - 36}{(x - 36)(\sqrt{x} + 6)}$

$= \lim_{x\to 36} \frac{1}{\sqrt{x} + 6}$

$= \frac{1}{\sqrt{36} + 6} = \frac{1}{6 + 6} = \frac{1}{12}$

41. $\lim_{x\to 8} \frac{\sqrt{x} - \sqrt{8}}{x - 8}$

$= \lim_{x\to 8} \frac{\sqrt{x} - \sqrt{8}}{x - 8} \cdot \frac{\sqrt{x} + \sqrt{8}}{\sqrt{x} + \sqrt{8}}$

$= \lim_{x\to 8} \frac{x - 8}{(x - 8)(\sqrt{x} + \sqrt{8})}$

(Go to top of right column.)

$= \lim_{x\to 8} \frac{1}{\sqrt{x} + \sqrt{8}}$

$= \frac{1}{\sqrt{8} + \sqrt{8}} = \frac{1}{2\sqrt{8}}$

$= \frac{1}{2(2\sqrt{2})}$

$= \frac{1}{4\sqrt{2}}$, or $\frac{\sqrt{2}}{8}$

43. $\lim_{x\to 2} \frac{(x - 2)^2}{|x - 2|} = \lim_{x\to 2} (x - 2) \cdot \frac{x - 2}{|x - 2|}$

$= \lim_{x\to 2} (x - 2) \lim_{x\to 2} \frac{x - 2}{|x - 2|}$

$= (0)(1) = 0$

45. $\lim_{x\to 3} \frac{\ln x - \ln 3}{x - 3}$

x	2.9	2.99	2.999	2.9999	3.0001	3.001	3.01	3.1
ln x	1.0647	1.0953	1.09827	1.09857	1.0986	1.0989	1.1019	1.1314
ln x - ln 3	-.0339	-.00334	-.000333	-.000033	.00003	.0003	.0033	.0328
$\frac{\ln x - \ln 3}{x - 3}$	.3390	.33389	.3333	.3333	.3333	.3332	.3328	.3279

So the limit is .3333 or 1/3.

47. $\lim\limits_{x \to 0} x \ln|x|$

x	-.1	-.01	-.001	-.0001	.0001	.001	.01	.1
$|x|$	.1	.01	.001	.0001	.0001	.001	.01	.1
$\ln|x|$	-2.3026	-4.6051	-6.9078	-9.2103	-9.2103	-6.9078	-4.6051	-2.3026
$x \ln|x|$	.23026	.046051	.00691	.0009	-.0009	-.00691	-.046051	-.23026

So $\lim\limits_{x \to 0} x \ln|x| = 0$

49. $\lim\limits_{x \to \infty} \dfrac{5x}{3x - 1}$

$$= \lim_{x \to \infty} \frac{5}{3 - \frac{1}{x}}$$

$$= \frac{\lim\limits_{x \to \infty}}{\lim\limits_{x \to \infty} 3 - \lim\limits_{x \to \infty} \frac{1}{x}}$$

$$= \frac{5}{3 - 0} = \frac{5}{3}$$

51. $\lim\limits_{x \to \infty} \dfrac{8x + 2}{2x - 5}$

$$= \lim_{x \to \infty} \frac{8 + \frac{2}{x}}{2 - \frac{5}{x}}$$

$$= \frac{\lim\limits_{x \to \infty} (8) + \lim\limits_{x \to \infty} (\frac{2}{x})}{\lim\limits_{x \to \infty} (2) - \lim\limits_{x \to \infty} (\frac{5}{x})}$$

$$= \frac{8 + 0}{2 - 0}$$

$$= \frac{8}{2} = 4$$

53. $\lim\limits_{x \to \infty} \dfrac{x^2 + 2x - 5}{3x^2 + 2}$

$$= \lim_{x \to \infty} \frac{1 - \frac{2}{x} - \frac{5}{x^2}}{3 + \frac{2}{x^2}}$$

$$= \frac{\lim\limits_{x \to \infty} 1 - \lim\limits_{x \to \infty} \frac{2}{x} - \lim\limits_{x \to \infty} \frac{5}{x^2}}{\lim\limits_{x \to \infty} 3 + \lim\limits_{x \to \infty} \frac{2}{x^2}}$$

$$= \frac{1 - 0 - 0}{3 + 0} = \frac{1}{3}$$

55. $\lim\limits_{x \to \infty} \dfrac{2x^2 + 11x - 10}{5x^3 + 3x^2 + 2x}$

$$= \lim_{x \to \infty} \frac{\frac{2}{x} + \frac{11}{x^2} - \frac{10}{x^3}}{5 + \frac{3}{x} + \frac{2}{x^2}}$$

$$= \frac{\lim\limits_{x \to \infty} \frac{2}{x} + \lim\limits_{x \to \infty} \frac{11}{x^2} - \lim\limits_{x \to \infty} \frac{10}{x^3}}{\lim\limits_{x \to \infty} 5 + \lim\limits_{x \to \infty} \frac{3}{x} + \lim\limits_{x \to \infty} \frac{2}{x^2}}$$

$$= \frac{0 + 0 + 0}{5 + 0 + 0} = \frac{0}{5} = 0$$

57. $\lim_{x\to\infty} \frac{2x^2 - 1}{3x^4 + 2}$

$= \lim_{x\to\infty} \frac{\frac{2}{x^2} - \frac{1}{x^4}}{3 + \frac{2}{x^4}}$

$= \frac{\lim_{x\to\infty} \frac{2}{x^2} - \lim_{x\to\infty} \frac{1}{x^4}}{\lim_{x\to\infty} 3 + \lim_{x\to\infty} \frac{2}{x^4}} = \frac{0 - 0}{3 + 0}$

$= \frac{0}{3} = 0$

59. $c(x) = 15{,}000 + 6x$

$\overline{c}(x) = \frac{c(x)}{x} = \frac{15{,}000 + 6x}{x}$

(a) $\overline{c}(1000) = \frac{15{,}000 + 6(1000)}{1000}$

$= \frac{15{,}000 + 6000}{1000}$

$= \frac{21{,}000}{1000}$

$= 21$

(b) $\overline{c}(100{,}000)$

$= \frac{15{,}000 + 6(100{,}000)}{100{,}000}$

$= \frac{15{,}000 + 600{,}000}{100{,}000}$

$= \frac{615{,}000}{100{,}000}$

$= 6.15$

(c) $\lim_{x\to 10{,}000} \overline{c}(x)$

$= \lim_{x\to 10{,}000} \frac{15{,}000 + 6x}{x}$

$= \frac{15{,}000 + 6(10{,}000)}{10{,}000}$

$= \frac{15{,}000 + 60{,}000}{10{,}000}$

$= \frac{75{,}000}{10{,}000}$

$= 7.5$

(d) $\lim_{x\to\infty} \overline{c}(x)$

$= \lim_{x\to\infty} \frac{15{,}000 + 6x}{x}$

$= \lim_{x\to\infty} \frac{\frac{15{,}000}{x} + 6}{1}$

$= \frac{\lim_{x\to\infty} \frac{15{,}000}{x} + \lim_{x\to\infty} 6}{\lim_{x\to\infty} 1}$

$= \frac{0 + 6}{1}$

$= 6$

61. (a) $\lim_{t\to 12} G(t)$

As t approaches 12, the value of G(t) for the corresponding point on the graph approaches 3.
Thus, $\lim_{t\to 12} G(t) = 3$

(b) $\lim_{t\to 16} G(t)$

As t approaches 16 from the right, G(t) gets closer to 1.5. As t approaches 16 from the left, G(t) gets closer to 2. Since G(t) does not get closer to a single real number, $\lim_{t\to 16} G(t)$ does not exist.

(c) G(16) is the value of function G(t) when t = 16 as shown by the solid dot on the graph.
So G(16) = 2.

(d) The tipping point occurs at the break in the graph or when t = 16 months.

63. $p_n = \frac{1}{2} + (p_0 - \frac{1}{2})(1 - 2p)^n$ with

$p_0 = .7$ and $p = .2$

(a) $p_2 = \frac{1}{2} + (.7 - \frac{1}{2})[1 - 2(.2)]^2$

$= .5 + (.2)(1 - .4)^2$
$= .5 + (.2)(.6)^2$
$= .5 + (.2)(.36)$
$= .5 + .072$
$\approx .57$

Thus, the probability that the congressman will vote "yes" on the second roll call is about .57.

(b) $p_4 = \frac{1}{2} + (.7 - \frac{1}{2})(1 - 2(.2))^4$

$= .5 + (.2)(1 - .4)^4$
$= .5 + (.2)(.6)^4$
$= .5 + .02592$
$\approx .53$

The probability on the fourth vote is about .53.

(c) $p_8 = p_8 = \frac{1}{2} + (.7 - \frac{1}{2})(1 - .2(.2))^8$

$= .5 + (.2)(1 - .4)^8$
$= .5 + (.2)(.6)^8$
$= .5 + .0034$
$\approx .50$

The probability on the eighth vote is about .50.

(d) As the number of roll calls increases ($n \to \infty$), p_n approaches .5.

So $\lim\limits_{n\to\infty} p_n = .5$.

This means that, as the number of roll calls increases, the probability the congressman will vote "yes" gets close to .5 but is never less than .5.

65. $\lim\limits_{x\to 5.2} \frac{x^4 - 6x^3 - 10x + 5}{4x^3 + 6x^2 - 12x - 6}$

$= \frac{5.2^4 - 6(5.2)^3 - 10(5.2) + 5}{4(5.2)^3 + 6(5.2)^2 - 12(5.2) - 6}$

$= \frac{731.1616 - 843.648 - 52 + 5}{562.432 + 162.24 - 62.4 - 6}$

$= \frac{-159.4864}{656.272}$

$= -.24302$

67. $\lim\limits_{x\to 1.89} \frac{3.1x^3 + 5.2\sqrt{x}}{-2x^3 + 12.3x - 2\sqrt{x}}$

$= \frac{3.1(1.89) + 5.2\sqrt{1.89}}{-2(1.89)^3 + 12.3(1.89) - 2\sqrt{1.89}}$

$= \frac{28.0778}{6.9949}$

$= 4.0140$

Section 11.2

1. $f(x) = x^2 + 2x$ between $x = 0$ and $x = 3$

Average rate of change

$= \frac{f(3) - f(0)}{3 - 0} = \frac{15 - 0}{3} = 5$

3. $f(x) = 2x^3 - 4x^2 + 6x$ between $x = -1$ and $x = 1$

Average rate of change

$= \frac{f(1) - f(-1)}{1 - (-1)}$

$= \frac{4 - (-12)}{1 + 1} = \frac{4 + 12}{2}$

$= \frac{16}{2} = 8$

5. $f(x) = \sqrt{x}$ between $x = 1$ and $x = 4$

Average rate of change

$= \frac{f(4) - f(1)}{4 - 1}$

$= \frac{2 - 1}{4 - 1} = \frac{1}{3}$

(Go to top of right column.)

7. $f(x) = \frac{1}{x - 1}$ between $x = -2$ and $x = 0$

Average rate of change

$= \frac{f(0) - f(-2)}{0 - (-2)} = \frac{-1 - (-\frac{1}{3})}{2}$

$= \frac{-1 + \frac{1}{3}}{2} = \frac{\frac{-2}{3}}{2} = -\frac{1}{3}$

9. $s(t) = 2t^2$

Time Interval	Average Speed
$t = 5$ to $t = 5.1$	$\frac{s(5.1) - s(5)}{5.1 - 5} = \frac{52.02 - 50}{.1} = 20.2$
$t = 5$ to $t = 5.01$	$\frac{s(5.01) - s(5)}{5.01 - 5} = \frac{50.2002 - 50}{.01} = 20.02$
$t = 5$ to $t = 5.001$	$\frac{s(5.001) - s(5)}{5.001 - 5} = \frac{50.020002 - 50}{.001} = 20.0002$

This chart suggests that the exact speed at $t = 5$ is 20 feet per second.

11. $s(t) = 2t^2$

Time Interval	Average Speed
$t = 25$ to $t = 25.1$	$\frac{s(25.1) - s(25)}{25.1 - 25} = \frac{1260.02 - 1250}{.1} = 100.2$
$t = 25$ to $t = 25.01$	$\frac{s(25.01) - s(25)}{25.01 - 25} = \frac{1251.0002 - 1250}{.01} = 100.02$
$t = 25$ to $t = 25.001$	$\frac{s(25.001) - s(25)}{25.001 - 25} = \frac{1250.100002 - 1250}{.001} = 100.0002$

This chart shows that the exact speed at $t = 5$ is 100 feet per second.

13. $s(t) = 2t^2$

Average speed

$= \frac{s(30) - s(0)}{30 - 0} = \frac{1800 - 0}{30}$

= 60 feet per second

15. $s(t) = t^2 + 5t + 2$

For $t = 1$, the instantaneous velocity

is $\lim_{h \to 0} \frac{s(1 + h) - s(1)}{h}$

(Go to middle of right column.)

$s(1 + h) = (1 + h)^2 + 5(1 + h) + 2$

$= 1 + 2h + h^2 + 5 + 5h + 2$

$= h^2 + 7h + 8$

$s(1) = (1)^2 + 5(1) + 2 = 8$

So $s(1 + h) - s(1) = (h^2 + 7h + 8) - 8$

$= h^2 + 7h$

and the instantaneous velocity at $t = 1$ is

$\lim_{h \to 0} \frac{h^2 + 7h}{h} = \lim_{h \to 0} \frac{h(h + 7)}{h}$

$\lim_{h \to 0} (h + 7) = 7$ feet per second.

17. Let S(x) be the function representing sales in thousands of dollars of x thousand catalogs.

(a) $S(20) = 40$

$S(10) = 30$

Average rate of change

$= \frac{S(20) - S(10)}{20 - 10}$

$= \frac{40 - 30}{10}$

$= 1$

As catalog distribution changes from 10,000 to 20,000, sales will have an average increase of 1000. Thus, on the average, sales will increase $1000 for each additional 1000 catalogs distributed.

(b) $S(40) = 50$

$S(10) = 30$

Average rate of change

$= \frac{S(40) - S(10)}{40 - 10}$

$= \frac{50 - 30}{30}$

$= \frac{20}{30}$

$= \frac{2}{3}$

As catalog distribution changes from 10,000 to 40,000, on the average, sales will incease $667 for each additional 1000 catalogs distributed.

(c) $S(30) = 45$

$S(20) = 40$

Average rate of change

$= \frac{S(30) - S(20)}{30 - 20}$

$= \frac{45 - 40}{10}$

$= \frac{5}{10}$

$= \frac{1}{2}$

As catalog distribution changes from 20,000 to 30,000, on the average, sales will increase $600 for each additional 1000 catalcgs distributed.

(d) $S(40) = 50$

$S(30) = 45$

Average rate of change

$= \frac{S(40) - S(30)}{40 - 30}$

$= \frac{50 - 45}{10}$

$= \frac{5}{10}$

$= \frac{1}{2}$

As catalog distribution changes from 30,000 to 40,000, on the average, sales will increase $400 for each additional 1000 catalogs distributed. As more catalogs are distributed, overall sales increase at a smaller and smaller rate.

19. Estimates may vary.

(a) $R(1935) = 120$

$R(1870) = 0$ *Use solid line*

Average rate of change

$= \frac{R(1935) - R(1870)}{1935 - 1870} = \frac{120 - 0}{65} = \frac{120}{65}$

≈ 1.8 per year

(b) $R(1975) = 610$

$R(1945) = 20$

Average rate of change

$= \frac{R(1975) - R(1945)}{1975 - 1945} = \frac{610 - 20}{30} = \frac{590}{30}$

≈ 19.67 per year

(c) R(1985) = 205

R(1975) = 610

Average rate of change

$= \frac{R(1985) - R(1975)}{1985 - 1975} = \frac{205 - 610}{10}$

≈ -40 per year

(d) R(1970) = 190

R(1945) = 20 *Use dotted line*

Average rate of change

$= \frac{R(1970) - R(1945)}{1970 - 1945} = \frac{190 - 20}{25}$

≈ 6.8 per year

(e) R(1980) = 180

R(1970) = 190

Average rate of change

$= \frac{R(1980) - R(1970)}{1980 - 1970} = \frac{180 - 190}{10}$

≈ -1 per year

(f) R(1985) ≈ 85

R(1980) ≈ 180

Average rate of change

$= \frac{R(1985) - R(1980)}{1985 - 1980} = \frac{85 - 180}{5}$

≈ -19 per year

21. (a) Let v = the average velocity.

From 0 second to 6 seconds,

$v = \frac{14 - 0}{6 - 0}$.

$v = \frac{7}{3}$ feet per second.

(b) From 2 seconds to 10 seconds,

$v = \frac{18 - 10}{10 - 2}$

= 1 foot per second.

(c) From 2 seconds to 12 seconds,

$v = \frac{30 - 10}{12 - 2}$

= 2 feet per second.

(d) From 6 seconds to 10 seconds,

$v = \frac{18 - 14}{10 - 6}$

= 1 foot per second.

(e) From 6 seconds to 18 seconds,

$v = \frac{36 - 14}{18 - 6}$

$= \frac{11}{6}$ feet per second.

(f) From 12 seconds to 20 seconds,

$v = \frac{40 - 30}{20 - 12}$

$= \frac{5}{4}$ feet per second.

Since the average velocity is positive in each case, the car is always moving in the same direction (forward).

23. (a) S(1989) = 6.5

S(1985) = 6

Average rate of change

$= \frac{S(1989) - S(1985)}{1989 - 1985} = \frac{6.5 - 6}{4} = \frac{.5}{4}$

= .125%

(b) S(1989) = 6.5

S(1986) = 6.25

Average rate of change

$= \frac{S(1989) - S(1985)}{1989 - 1985} = \frac{6.5 - 6}{3} = \frac{.5}{3}$

≈ .1%

(c) S(1989) = 6.5

S(1987) = 6.0

Average rate of change

$= \frac{S(1989) - S(1987)}{1989 - 1987} = \frac{6.5 - 6.0}{2} = \frac{.5}{2}$

≈ .25%

The average rate of change of market share is increasing in each period.

25. **(a)** $\frac{8000 - 5000}{1987 - 1981} = \frac{3000}{6}$

$= \$500$ per year

(Go to top of right column.)

(b) $\frac{7500 - 6000}{1986 - 1982} = \frac{15}{4}$

$= \$375$ per year

27.

h	1	.1	.01	.001	.0001
4 + h	5	4.1	4.01	4.001	4.0001
P(4 + h)	31	19.21	18.1102	18.011	18.0011
P(4)	18	18	18	18	18
P(4 + h) - P(4)	13	1.12	.110212	.0109959	.00110817
$\frac{P(4+h) - P(4)}{h}$	13	11.2	11.0212	10.9959	11.0817

By the table, it appears that

$$\lim_{h\to 0} \frac{P(4 + h) - P(4)}{h} = 11.$$

By the rule for limits, we have

$$\lim_{h\to 0} \frac{P(4 + h) - P(4)}{h}$$

$$= \lim_{h\to 0} \frac{2(4 + h)^2 - 5(4 + h) + 6 - 18}{h}$$

$$= \lim_{h\to 0} \frac{2h^2 + 11h + 18 - 18}{h}$$

$$= \lim_{h\to 0} \frac{2h^2 + 11h}{h}$$

$$= \lim_{h\to 0} 2h + 11$$

$$= 11$$

29. $N(p) = 80 - 5p^2,\ 1 \le p \le 4$

(a) The average rate of change of demand is

$N(3) - N(2)$

$= (80 - 45) - (80 - 20)$

$= -25$ boxes per dollar.

(Go to middle of right column.)

(b) The instantaneous rate of change when p is 2 is

$$\lim_{h\to 0} \frac{N(2 + h) - N(2)}{h}$$

$$= \lim_{h\to 0} \frac{80 - 5(2 + h)^2 - [80 - 5(2)^2]}{h}$$

$$= \lim_{h\to 0} \frac{80 - 20 - 20h - 5h^2 - (80 - 20)}{h}$$

$$= \lim_{h\to 0} \frac{-5h^2 - 20h}{h}$$

$= -20$ boxes per dollar.

(c) The instantaneous rate of change when p is 3 is

$$\lim_{h\to 0} \frac{80 - 5(3 + h)^2 - [80 - 5(3)^2]}{h}$$

$$= \lim_{h\to 0} \frac{80 - 45 - 30h - 5h^2 - 80 + 45}{h}$$

$$= \lim_{h\to 0} \frac{-30h - 5h^2}{h}$$

$= -30$ boxes per dollar.

31. $R(t) = -.03t^2 + 15$

(a) The instantaneous rate of change where $t = 5$ is

$$\lim_{h \to 0} \frac{-.03(5+h)^2 + 15 - [-.03(5)^2 + 15]}{h}$$

$$= \lim_{h \to 0} \frac{-.75 - .3h + .03h^2 + 15 + .75 - 15}{h}$$

$$= \lim_{h \to 0} \frac{-.3h + .03h^2}{h}$$

$= -.3$ word per minute.

(b) At $t = 15$, the instantaneous rate of change is

$$\lim_{h \to 0} \frac{-.03(15+h)^2 + 15 - [-.03(15)^2 + 15]}{h}$$

$$= \lim_{h \to 0} \left(\frac{-.03(225 + 30h + h^2)}{h} + \frac{15 + .03(225) - 15}{h} \right)$$

$$= \lim_{h \to 0} \frac{-.9h + h^2}{h}$$

$= -.9$ word per minute.

Section 11.3

1. $f(x) = -4x^2 + 11x$; $x = -2$

$$\frac{f(x+h) - f(x)}{h}$$

$$= \frac{[-4(x+h)^2 + 11(x+h)] - (-4x^2 + 11x)]}{h}$$

$$= \frac{[-4(x^2 + 2hx + h^2) + 11x + 11h]}{h} - \frac{(-4x^2 + 11x)}{h}$$

$$= \frac{(-4x^2 - 8hx - 4h^2 + 11x + 11h)}{h} - \frac{(-4x^2 + 11x)}{h}$$

$$= \frac{-8hx - 4h^2 + 11h}{h}$$

$$= -8x - 4h + 11$$

$$\lim_{h \to 0} (-8x - 4h + 11) = -8x + 11$$

$$= f'(x)$$

$f'(-2) = -8(-2) + 11 = 27$ is the slope of tangent line at $x = -2$.

3. $f(x) = \frac{-2}{x}$; $x = 4$

$$\frac{f(x+h) - f(x)}{h}$$

$$= \frac{\frac{-2}{x+h} - \left(\frac{-2}{x}\right)}{h}$$

$$= \frac{\frac{-2x + 2(x+h)}{(x+h)x}}{h}$$

$$= \frac{-2x + 2x + 2h}{h(x+h)(x)}$$

$$= \frac{2h}{h(x+h)x}$$

$$= \frac{2}{(x+h)x}$$

$$\lim_{h \to 0} \frac{2}{(x+h)x} = \frac{2}{x^2}$$

$$= f'(x)$$

$f'(4) = \frac{2}{4^2} = \frac{1}{8}$ is the slope of the tangent line at $x = -4$.

5. $f(x) = \sqrt{x}$; $x = 16$

$$\frac{f(x+h) - f(x)}{h}$$

$$= \frac{\sqrt{x+h} - \sqrt{x}}{h} \cdot \frac{\sqrt{x+h} + \sqrt{x}}{\sqrt{x+h} + \sqrt{x}}$$

$$= \frac{(x+h) - x}{h(\sqrt{x+h} + \sqrt{x})}$$

$$= \frac{1}{\sqrt{x+h} + \sqrt{x}}$$

$$\lim_{h \to 0} \frac{1}{\sqrt{x+h} + \sqrt{x}} = \frac{1}{\sqrt{x} + \sqrt{x}}$$

$$= \frac{1}{2\sqrt{x}} = f'(x)$$

$f'(16) = \frac{1}{2\sqrt{16}} = \frac{1}{8}$ is the slope of the tangent line at $x = 16$.

7. $f(x) = x^2 + 2x;\ x = 3$

$$\frac{f(x+h) - f(x)}{h}$$

$$= \frac{[(x+h)^2 + 2(x+h)] - (x^2 + 2x)}{h}$$

$$= \frac{(x^2 + 2hx + h^2 + 2x + 2h) - (x^2 + 2x)}{h}$$

$$= \frac{2hx + h^2 + 2h}{h}$$

$$= 2x + h + 2$$

$$\lim_{h \to 0} (2x + h + 2) = 2x + 2$$

$$= f'(x)$$

$f'(3) = 2(3) + 2 = 8$ is the slope of the tangent line at $x = 3$. Now use $m = 8$ and $(3, 15)$ in the point-slope form to write the equation of the tangent line.

$$y - 15 = 8(x - 3)$$
$$y = 8x - 9$$

9. $f(x) = \frac{5}{x};\ x = 2$

$$\frac{f(x+h) - f(x)}{h}$$

$$= \frac{\frac{5}{x+h} - \frac{5}{x}}{h}$$

$$= \frac{\frac{5x - 5(x+h)}{(x+h)x}}{h}$$

$$= \frac{5x - 5x - 5h}{h(x+h)(x)}$$

$$= \frac{-5h}{h(x+h)(x)}$$

$$= \frac{-5}{(x+h)(x)}$$

$$\lim_{h \to 0} \frac{-5}{(x+h)(x)} = \frac{-5}{x^2}$$

$$= f'(x)$$

$f'(2) = \frac{-5}{2^2} = -\frac{5}{4}$ is the slope of the tangent line at $x = 2$.

Now use $m = -\frac{5}{4}$ and $(2, \frac{5}{2})$ in the point-slope form.

$$y - \frac{5}{2} = -\frac{5}{4}(x - 2)$$

$$y - \frac{5}{2} = \frac{5}{2} = -\frac{5}{4}x + \frac{10}{4}$$

$$y = -\frac{5}{4}x + 5$$

$$5x + 4y = 20$$

11. $f(x) = 4\sqrt{x};\ x = 9$

$$\frac{f(x+h) - f(x)}{h}$$

$$= \frac{4\sqrt{x+h} - 4\sqrt{x}}{h} \cdot \frac{4\sqrt{x+h} + 4\sqrt{x}}{4\sqrt{x+h} + 4\sqrt{x}}$$

$$= \frac{16(x+h) - 16x}{h(4\sqrt{x+h} + 4\sqrt{x})}$$

$$= \frac{16h}{4h(\sqrt{x+h} + \sqrt{x})}$$

$$= \frac{4}{\sqrt{x+h} + \sqrt{x}}$$

$$\lim_{h \to 0} \frac{4}{(\sqrt{x+h} + \sqrt{x}} = \frac{4}{2\sqrt{x}}$$

$$= \frac{2}{\sqrt{x}}$$

$$= f'(x)$$

$f'(9) = \frac{2}{\sqrt{9}} = \frac{2}{3}$ is the slope of the tangent line at $x = 9$.

Now use $m = \frac{2}{3}$ and $(9, 12)$ in the point-slope form.

$y - 12 = \frac{2}{3}(x - 9)$

$y = \frac{2}{3}x + 6$

$3y = 2x + 18$

For Exercises 13-17, choose any two convenient points on each of the tangent lines.

13. Using the points (5, 3) and (6, 8), we have

$m = \frac{5 - 3}{6 - 5}$

$= 2.$

15. Using the points (-2,2) and (3, 3), we have

$m = \frac{3 - 2}{3 - (-2)}$

$= \frac{1}{5}.$

17. Using the points (-3, -3) and (0, -3), we have

$m = \frac{-3 - (-3)}{0 - 3}$

$= 0.$

19. $f(x) = -4x^2 + 11x$

$\frac{f(x + h) - f(x)}{h}$

$= \frac{[-4(x + h)^2 + 11(x + h)] - (-4x^2 + 11x)}{h}$

$= \frac{(-4x^2 - 8xh - 4h^2 + 11x + 11h) + 4x^2 - 11x}{h}$

$= \frac{-8xh - 4h^2 + 11h}{h} = -8x - 4h + 11$

$\lim_{h \to 0} (-8x - 4h + 11) = -8x + 11$

$= f'(x)$

$f'(2) = -8(2) + 11 = -5$

$f'(0) = -8(0) + 11 = 11$

$f'(-3) = -8(-3) + 11 = 35$

21. $f(x) = 8x + 6$

$\frac{f(x + h) - f(x)}{h}$

$= \frac{[8(x + h) + 6] - (8x + 6)}{h}$

$= \frac{(8x + 8h + 6) - (8x + 6)}{h}$

$= \frac{8h}{h}$

$= 8$

$\lim_{h \to 0} 8 = 8 = f'(x)$

$f'(2) = 8;\ f'(0) = 8;\ f(-3) = 8$

23. $f(x) = x^3 + 3x$

$\frac{f(x + h) - f(x)}{h}$

$= \frac{[(x + h)^3 + 3(x + h)] - (x^3 + 3x)}{h}$

$= \frac{x^3 + 3x^2h + 3xh^2 + h^3 + 3x + 3h}{h}$

$= \frac{3x^2h + 3xh^2 + h^3 + 3h}{h}$

$= 3x^2 + 3xh + h^2 + 3$

$\lim_{h \to 0} (3x^2 + 3xh + h^2 + 3)$

$= 3^2 = 3 = f'(x)$

$f'(2) = 3(2)^2 + 3 = 15$

$f'(0) = 3(0)^2 + 3 = 3$

$f'(-3) = 3(-3)^2 + 3 = 30$

25. $f(x) = -\frac{2}{x}$

$$\frac{f(x+h) - f(x)}{h}$$

$$= \frac{\frac{-2}{x+h} - \left(\frac{-2}{x}\right)}{h}$$

$$= \frac{\frac{-2x + 2(x+h)}{(x+h)x}}{h}$$

$$= \frac{2h}{h(x+h)(x)}$$

$$= \frac{2}{(x+h)(x)}$$

$$\lim_{h \to 0} \frac{2}{(x+h)(x)} = \frac{2}{x^2}$$

$$= f'(x)$$

$f'(2) = \frac{2}{2^2} = \frac{1}{2}$

$f'(0) = \frac{2}{0^2}$ *Not defined*

The derivative does not exist at $x = 0$.

$f'(3) = \frac{2}{3^2} = \frac{2}{9}$

27. $f(x) = \frac{4}{x-1}$

$$\frac{f(x+h) - f(x)}{h}$$

$$= \frac{\frac{4}{x+h-1} - \frac{4}{x-1}}{h}$$

$$= \frac{\frac{4(x-1) - 4(x+h-1)}{(x+h-1)(x-1)}}{h}$$

$$= \frac{-4h}{h(x-1+h)(x-1)}$$

$$= \frac{-4}{(x-1+h)(x-1)}$$

$$\lim_{h \to 0} \frac{-4}{(x-1+h)(x-1)}$$

$$= \frac{-4}{(x-1)^2} = f'(x)$$

$f'(2) = \frac{-4}{(2-1)^2} = -4$

$f'(0) = \frac{-4}{(0-1)^2} = -4$

$f'(-3) = \frac{-4}{(-3-1)^2} = \frac{-4}{16} = -\frac{1}{4}$

29. $f(x) = \sqrt{x}$

$$\frac{f(x+h) - f(x)}{h}$$

$$= \frac{\sqrt{x+h} - \sqrt{x}}{h} \cdot \frac{\sqrt{x+h} + \sqrt{x}}{\sqrt{x+h} + \sqrt{x}}$$

$$= \frac{(x+h) - x}{h(\sqrt{x+h} + \sqrt{x})}$$

$$= \frac{h}{h(\sqrt{x+h} + \sqrt{x})}$$

$$= \frac{1}{\sqrt{x+h} + \sqrt{x}}$$

$$\lim_{h \to 0} \frac{1}{\sqrt{x+h} + \sqrt{x}} = \frac{1}{2\sqrt{x}} = f'(x)$$

$f'(2) = \frac{1}{2\sqrt{2}}$

$f'(0) = \frac{1}{2\sqrt{0}}$ does not exist.

$f'(-3) = \frac{1}{2\sqrt{-3}}$ does not exist.

31. At $x = 0$, the graph of $f(x)$ has a sharp point. Therefore, there is no derivative for $x = 0$.

33. For $x = \pm 6$, the graph of $f(x)$ has sharp points. Therefore, there is no derivative for $x = 6$ or $x = -6$.

35. For $x = -3$ and $x = 0$, the tangent to the graph of $f(x)$ is vertical. For $x = 2$, the function $f(x)$ does not exist. For $x = 3$ and $x = 5$, the graph of $f(x)$ has sharp points. Therefore, no derivative exists for $x = -3$, $x = 0$, $x = 2$, $x = 3$, and $x = 5$.

37.(a) Since the derivative is positive at $x = 10$, the profit is increasing at $x = 10$.

(b) Since the derivative is 0 at $x = 12$, the profit is constant at $x = 12$.

(c) Since the derivative is negative at $x = 15$, the profit is decreasing at $x = 15$.

39. $R(p) = 20p = \frac{p^2}{500}$

(a) The marginal revenue is $R'(p)$, so find $R'(p)$.

$$\frac{R(p + h) - R(p)}{h}$$

$$= \frac{20(p + h) - \frac{(p + h)^2}{500} - 20p - \frac{p^2}{500}}{h}$$

$$= \frac{20p + 20h - \frac{p^2 + 2ph + h^2}{500} - 20p - \frac{p^2}{500}}{h}$$

$$= \frac{20h - \frac{2ph + h^2}{500}}{h}$$

$$= 20 - \frac{2p + h}{500}$$

$$\lim_{h \to 0} \frac{R(p + h) - R(p)}{h}$$

$$= \lim_{h \to 0} 20 - \frac{2p + h}{500}$$

$$= 20 - \frac{2p}{500} = R'(p)$$

At $p = 1000$,

$$R'(p) = \frac{10{,}000 - 2(1000)}{500}$$

$= \$16$ per table.

(b) The actual revenue is $R(1001) - R(1000)$

$$= 20(1001) - \frac{1001^2}{500} - \left[20(1000) - \frac{1000^2}{500}\right]$$

$= 18{,}015.998 - 18{,}000$

$= \$15.998$ or $\$16$.

(c) The marginal revenue found in part (a) approximates the actual revenue from the sale of the 1001st item found in part (b).

41. Rate of change of demand is given by the derivative of the demand function $D'(x)$ where $D(x) = -2x^2 + 4x + 6$.

$$\frac{D(x + h) - D(x)}{h}$$

$$= \frac{[-2(x + h)^2 + 4(x + h) + 6]}{h} - \frac{[-2x^2 + 4x + 6]}{h}$$

$$= \frac{-2x^2 - 4xh - 2h^2 + 4x}{h} + \frac{4h - 6 + 2x^2 - 4x - 6}{h}$$

$$= \frac{-4xh - 2h^2 + 4h}{h}$$

$$= \frac{h(-4x - 2h + 4)}{h}$$

$$= -4x - 2h + 4$$

$$\lim_{h \to 0} (-4x - 2h + 4) = -4x + 4 = D'(x)$$

(a) $D'(3) = -4(3) + 4 = -12 + 4 = -8$

(b) $D'(6) = -4(6) + 4 = -24 + 4 = -20$

43. $B(t) = 1000 + 50t - 5t^2$

B is millions of bacteria.

t is hours.

$$\frac{B(t+h) - B(t)}{h}$$

$$= \frac{[1000 + 50(t+h) - 5(t+h)^2]}{h} - \frac{(1000 + 50t - 5t^2)}{h}$$

$$= \frac{[1000 + 50t + 50h - 5(t^2 + 2ht + h^2)]}{h} - \frac{1000 + 50t - 5t^2}{h}$$

$$= \frac{1000 + 50t + 50h - 5t^2 - 10ht - 5h^2}{h} - \frac{1000 + 50t - 5t^2}{h}$$

$$= \frac{50h - 10ht - 5h^2}{h} = 50 - 10t - 5h$$

$$\lim_{h \to 0} (50 - 10t + 5h) = 50 - 10t = B'(t)$$

(a) $B'(2) = 50 - 10(2) = 30$

(b) $B'(3) = 50 - 10(3) = 20$

(c) $B'(4) = 50 - 10(4) = 10$

(d) $B'(5) = 50 - 10(5) = 0$

(e) $B'(6) = 50 - 10(6) = -10$

(f) $B'(8) = 50 - 10(8) = -30$

45. Slope of tangent line to graph at first point is found by finding two points on tangent line.

$(x_1, y_1) = (1000. 13.5)$

$(x_2, y_2) = (0, 18.5)$

$$m = \frac{18.5 - 13.5}{0 - 1000} = \frac{5}{-1000} = -.005$$

At second point,

$(x_1, y_1) = (1000. 13.5)$

$(x_2, y_2) = (2000, 21.5)$.

$$m = \frac{21.5 - 13.5}{2000 - 1000} = \frac{8}{1000} = .008$$

At third point,

$(x_1, y_1) = (5000, 20)$

$(x_2, y_2) = (3000, 22.5)$.

$$m = \frac{22.5 - 20}{3000 - 5000} = \frac{2.5}{-2000}$$

$$= -.00125$$

47. $f(x) = \dfrac{x^2 + 4.9}{3x^2 - 1.7x}$

With computer software,

$$f'(x) = \frac{(2x)(3x^2 - 1.7x)}{(3x^2 - 1.7x)^2} - \frac{(6x - 1.7)(x^2 + 4.9)}{(3x^2 - 1.7x)^2}$$

If x is replace by 4.9,

$f'(4.9) = -.0435$.

49. $f(x) = \sqrt{x^3 + 4.8}$

With computer software,

$$f'(x) = \frac{3x^2}{2\sqrt{x^3 + 4.8}}.$$

If x is replace by 8.17,

$f'(8.17) = 4.2687$.

51. With computer software,

$$f'(x) = \frac{-1.4}{\sqrt{x}(\sqrt{x} + 1.2)^2}.$$

If x is replaced by 2.65,

$f'(2.65) \approx -.10754$

Section 11.4

1. $f(x) = 9x^2 - 8x + 4$

$f'(x) = 2 \cdot 9x - 8 + 0 = 18x - 8$

3. $y = 10x^3 - 9x^2 + 6x$

$y' = 3 \cdot 10x^2 - 2 \cdot 9x + 6$

$= 30x^2 - 18x + 6$

5. $y = x^4 - 5x^3 + 9x^2 + 5$

$= 4x^3 - 3 \cdot 5x^2 + 2 \cdot 9x$

$= 4x^3 - 15x^2 + 18x$

7. $f(x) = 6x^{1.5} - 4x^{.5}$

$f'(x) = 1.5(6)x^{1.5-1} - .5(4)x^{-.5}$

$= 9x^{.5} - 2x^{-.5}$ or $9x^{.5} - \dfrac{2}{x^{.5}}$

9. $y = -15x^{3.2} + 2x^{1.9}$

$y' = 3.2(-15)\ x^{3.2-1} + 1.9(2)x^{1.9-1}$

$= -48x^{2.2} + 3.8x^{.9}$

11. $y = 24t^{3/2} + 4t^{1/2}$

$y' = \dfrac{3}{2}(24)(t^{1/2}) + \dfrac{1}{2}(4)(t^{-1/2})$

$= 36t^{1/2} + 2t^{-1/2}$ or $36t^{1/2} + \dfrac{2}{t^{1/2}}$

13. $y = 8\sqrt{x} + 6x^{3/4}$

$= 8x^{1/2} + 6x^{3/4}$

$y' = \dfrac{1}{2}(8)(x^{-1/2}) + \dfrac{3}{4}(6)(x^{-1/4})$

$= 4x^{-1/2} + \dfrac{9}{2}x^{-1/4}$ or $\dfrac{4}{x^{1/2}} + \dfrac{9}{2x^{1/4}}$

15. $g(x) = 6x^{-5} - x^{-1}$

$g'(x) = -5(6)(x^{-6}) - (-1)(x^{-2})$

$= -30x^{-6} + x^{-2}$ or $\dfrac{-30}{x^6} + \dfrac{1}{x^2}$

17. $y = -4x^{-2} + 3x^{-3}$

$y' = -2(-4)(x^{-3}) + (-3)(3x^{-4})$

$= 8x^{-3} - 9x^{-4}$ or $\dfrac{8}{x^3} - \dfrac{9}{x^4}$

19. $y = 10x^{-2} + 3x^{-4} - 6x$

$y' = -2(10)(x^{-3}) + (-4)(3x^{-5}) - 6$

$= -20x^{-3} - 12x^{-5} - 6$

or $\dfrac{-20}{x^3} - \dfrac{12}{x^5} - 6$

21. $f(t) = \dfrac{6}{t} - \dfrac{8}{t^2}$

$= 6t^{-1} - 8t^{-2}$

$f'(t) = (-1)(6t^{-2}) - (-2)(8t^{-3})$

$= -6t^{-2} + 16t^{-3}$

or $\dfrac{-6}{t^2} + \dfrac{16}{t^3}$

23. $y = \dfrac{9 - 8x + 2x^3}{x^4}$

$y = \dfrac{9}{x^4} - \dfrac{8}{x^3} + \dfrac{2}{x}$

$= 9x^{-4} - 8x^{-3} + 2x^{-1}$

$y' = (-4)(9x^{-5}) - (-3)(8x^{-4}) + (-1)(2x^{-2})$

$= -36x^{-5} + 24x^{-4} - 2x^{-2}$

or $\dfrac{-36}{x^5} + \dfrac{24}{x^4} - \dfrac{2}{x^2}$

25. $f(x) = 12x^{-1/2} - 3x^{1/2}$

$f'(x) = -\dfrac{1}{2}(12)(x^{-3/2}) - \dfrac{1}{2}(3)(x^{-1/2})$

$= -6x^{-3/2} - \dfrac{3}{2}x^{-1/2}$

or $\dfrac{-6}{x^{3/2}} - \dfrac{3}{2x^{1/2}}$

27. $p(x) = -10x^{-1/2} + 8x^{-3/2}$

$p'(x) = \left(-\dfrac{1}{2}\right)(-10)(x^{-3/2}) + \left(-\dfrac{3}{2}\right)(8x^{-5/2})$

$= 5x^{-3/2} - 12x^{-5/2}$

or $\dfrac{5}{x^{3/2}} - \dfrac{12}{x^{5/2}}$

29. $y = \frac{6}{4\sqrt{x}} = 6x^{-1/4}$

$y' = -\frac{1}{4}(6)(x^{-5/4})$

$= -\frac{3}{2}x^{-5/4}$ or $\frac{-3}{2x^{5/4}}$

31. $y = \frac{-5t}{3\sqrt{t^2}}$

$= -5t(t^{-2/3})$

$y = -5t^{1/3}$

$y' = \frac{1}{3}(-5)(t^{-2/3})$

$= \frac{-5}{3}t^{-2/3}$ or $\frac{-5}{3t^{2/3}}$

33. $y = 8x^{-5} - 9x^{-4}$

$\frac{dy}{dx} = -5(8)(x^{-6}) - (-4)(9)(x^{-5})$

$= -40x^{-6} + 36x^{-5}$

or $\frac{-40}{x^6} + \frac{36}{x^5}$

35. $D_x(9x^{-1/2} + \frac{2}{x^{3/2}})$

$= D_x(9x^{-1/2} + 2x^{-3/2})$

$= -\frac{1}{2}(9)(x^{-3/2}) - \frac{3}{2}(2)(x^{-5/2})$

$= -\frac{9}{2}x^{-3/2} - 3x^{-5/2}$

or $\frac{-9}{2x^{3/2}} - \frac{3}{x^{5/2}}$

37. $f(x) = 6x^2 - 4x$

$f'(x) = 12x - 4$

$f'(-2) = 12(-2) - 4$

$= -24 - 4$

$= -28$

39. $f(t) = 2\sqrt{t} - \frac{3}{\sqrt{t}}$

$= 2t^{1/2} - 3t^{-1/2}$

$f'(t) = \frac{1}{2}(2)t^{-1/2} - (-\frac{1}{2})3t^{-3/2}$

$= t^{-1/2} + \frac{3}{2}t^{-3/2}$

$f'(4) = (4)^{-1/2} + \frac{3}{2}(4)^{-3/2}$

$= \frac{1}{2} + \frac{3}{2}\left(\frac{1}{2}\right)^3$

$= \frac{1}{2} + \frac{3}{2}\left(\frac{1}{8}\right) = \frac{11}{16}$

41. $y = x^4 - 5x^3 + 2;\ x = 2$

$y' = 4x^3 - 15x^2$

$y'(2) = 4(2)^3 - 15(2)^2$

$= -28$ is the slope of tangent line.

Use $(2, -22)$ to obtain the equation.

$y - (-22) = -28(x - 2)$

$y = -28x + 34$

$28x + y = 34$

43. $y = 3x^{3/2} - 2x^{1/2};\ x = 9$

$y' = \frac{3}{2}(3)x^{1/2} - \left(\frac{1}{2}\right)2x^{-1/2}$

$= \frac{9}{2}x^{1/2} - x^{-1/2}$

$y'(9) = \frac{9}{2}(3) - \frac{1}{3} = \frac{79}{6}$

45. $y = 5x^{-1} - 2x^{-2};\ x = 3$

$y' = 5(-1)x^{-2} - (-2)(2)x^{-3}$

$= -5x^{-2} + 4x^{-3}$

$y'(3) = -5(3)^{-2} = 4(3)^{-3}$

$= \frac{-5}{9} + \frac{4}{27} = -\frac{11}{27}$

47. Demand is $x = 5000 - 100p$

$$p = \frac{5000 - x}{100}$$

$$R(x) = x\left(\frac{5000 - x}{100}\right)$$

$$= \frac{5000x - x^2}{100}$$

$$R'(x) = \frac{5000 - 2x}{100}$$

(a) $R'(1000) = \frac{5000 - 2000}{100}$

$$= 30$$

(b) $R'(2500) = \frac{5000 - 2(2500)}{100}$

$$= 0$$

(c) $R'(3000) = \frac{5000 - 2(3000)}{100}$

$$= -10$$

49. $P(x) = \frac{-1000}{x} + 1000$

$$= -1000x^{-1} + 1000$$

$$P'(x) = (-1)(-1000)(x^{-2}) = 1000x^{-2}$$

$$P'(10) = 1000(10)^{-2} = \frac{1000}{100}$$

$$= 10$$

51. $P(t) = \frac{100}{t}$

(a) $P(1) = \frac{100}{1} = 100$

(b) $P(100) = \frac{100}{100} = 1$

(c) $P(t) = \frac{100}{t} = 100t^{-1}$

$$P'(t) = 100(-1t^{-2})$$

$$= -100t^{-2}$$

$$= \frac{-100}{t^2}$$

$$P'(100) = \frac{-100}{(100)^2}$$

$$= \frac{-1}{100}$$

$$= -.01$$

The percent of acid is decreasing at the rate of .01 per day after 100 days.

53. $G(x) = -.2x^2 + 450$

$$\frac{dG}{dx} = -2(.2)x = -.4x$$

(a) $\frac{dG}{dx}(10) = -.4(10) = -4$

When 10 units of insulin are injected, the blood sugar level is decreasing at a rate of 4 points per unit of insulin.

(b) $\frac{dG}{dx}(25) = -.4(25)$

$$= -10$$

When 10 units of insulin are injected, the blood sugar level is decreasing at a rate of 10 points per unit of insulin.

55. $M(t) = 4t^{3/2} + 2t^{1/2}$

(a) $M(16) = 4 \cdot 16^{3/2} + 2 \cdot 16^{1/2}$

$$= 256 + 8$$

$$= 264$$

(b) $M(25) = 4 \cdot 25^{3/2} + 2 \cdot 25^{1/2}$

$$= 500 + 10$$

$$= 510$$

(c) $M'(T) = \frac{3}{2} \cdot 4t^{1/2} + \frac{1}{2} \cdot 2t^{-1/2}$

$$= 6t^{1/2} + t^{-1/2}$$

$$M'(16) = 6 \cdot 16^{1/2} + 16^{-1/2}$$

$$= 24 + \frac{1}{4}$$

$$= \frac{97}{4} \text{ Or } 24.25$$

(d) The number of matings per hour is increasing by about 24.25 matings at a temperature of 16°C.

57. $C(x) = 100 + 8x - x^2 + 4x^3$

Marginal Cost

$= C'(x) = 0 + 8 - 2x + 4(3x^2)$

$= 8 - 2x + 12x^2$

(a) $C'(0) = 8 - 2(0) + 12(0)^2$

$= 8 - 0 + 0 = \$8$

(b) $C'(4) = 8 - 2(4) + 12(4)^2$

$= 8 - 8 + 12(16)$

$= 8 - 8 + 192 = \$192$

(c) $C'(6) = 8 - 2(6) + 12(6)^2$

$= 8 - 12 + 12(36)$

$= 8 - 12 + 432 = \$428$

(d) $C'(8) = 8 - 2(8) + 12(8)^2$

$= 8 - 16 + 12(64)$

$= 8 - 16 + 768 = \$760$

59. $s(t) = 6t + 5$

(a) $v(t) = s'(t) = 6$

(b) $v(0) = 6$

$v(5) = 6$

$v(10) = 6$

61. $s(t) = 11t^2 + 4t + 2$

(a) $v(t) = s'(t) = 22t + 4$

(b) $v(0) = 22(0) + 4 = 4$

$v(5) = 22(5) + 4 = 114$

$v(10) = 22(10) + 4 = 224$

63. $s(t) = 4t^3 + 8t^2$

(a) $v(t) = s'(t) = 12t^2 + 16t$

(b) $v(0) = 12(0)^2 + 16(0) = 0$

$v(5) = 12(5)^2 + 16(5) = 380$

$v(10) = 12(10)^2 + 16(10)$

$= 1360$

65. $s(t) = -16t^2 + 144$

velocity $= s'(t)$

$= -32t$

(a) $s'(1) = -32$ feet per second

$s'(2) = -32 \cdot 2$

$= -64$ feet per second

(b) The rock will hit the ground when $s(t) = 0$.

$-16t^2 + 144 = 0$

$t = \sqrt{\frac{144}{16}}$

$= 3$ seconds

(c) The velocity at impact is the velocity at 3 seconds.

$v = -32 \cdot 3$

$= -96$ feet per second

67. $(p + q)^n = p^n + np^{n-1} + \frac{n(n-1)}{2} p^{n-2}q^2 + \ldots + q^n$

(a) $(x + h)^n = x^n + nx^{n-1}h + \frac{n(n-1)}{2}x^{n-2}h^2 + \ldots + h^n$

(b) $\frac{(x+h)^n - x^n}{h}$

$$= \frac{x^n + nx^{n-1}h + \frac{n(n-1)}{2}x^{n-2}h^2 + \ldots + h^n - x^n}{h}$$

$$= nx^{n-1} + \frac{n(n-1)}{2}x^{n-2}h + \ldots + h^{n-1}$$

(c) $y' = \lim_{h \to 0} \frac{(x+h)^n - x^n}{h}$

$$= \lim_{h \to 0} \left(nx^{n-1} + \frac{n(n-1)}{2} x^{n-2}h + \ldots\right)$$

$= nx^{n-1}$

Section 11.5

1. $y = (2x - 5)(x + 4)$

$y' = (2x - 5)(1) + (x + 4)(2)$

$= 2x - 5 + 2x + 8$

$= 4x + 3$

3. $f(x) = (8x - 2)(3x + 9)$

$f'(x) = (8x - 2)(3) + (3x + 9((8)$

$= 24x - 6 + 24x + 72$

$= 48x + 66$

5. $y = (3x^2 + 2)(2x - 1)$

$y' = (3x^2 + 2)(2) + (2x - 1)(6x)$

$= 6x^2 + 4 + 12x^2 - 6x$

$= 18x^2 - 6x + 4$

7. $y = (x^2 + x)(3x - 5)$

$y' = (x^2 + x)(3) + (3x - 5)(2x + 1)$

$= 3x^2 + 3x + 6x^2 - 7x - 5$

$= 9x^2 - 4x - 5$

9. $y = (9x^4 + 7x)(x^2 - 1)$

$y' = (9x^4 + 7x)(2x)$

$+ (x^2 - 1)(36x^3 + 7)$

$= 18x^5 + 14x^2 + 36x^5 + 7x^2 - 36x^3 - 7$

$= 54x^5 - 36x^3 + 21x^2 - 7$

11. $y = (2x - 5)^2 = (2x - 5)(2x - 5)$

$y' = (2x -5)(2) + (2x - 5)(2)$

$= 4x - 10 + 4x - 10$

$= 4(2x - 5)$ or $8x - 20$

13. $k(t) = (x^2 - 1)^2$

$= (x^2 - 1)(x^2 - 1)$

$k'(t) = (x^2 - 1)(2x) + (x^2 - 1)(2x)$

$= 2x^3 - 2x + 2x^3 - 2x$

$= 4x^3 - 4x$ or $4x(x^2 - 1)$

15. $y = (x + 1)\ (\sqrt{x} + 2)$

$= (x + 1)(\ x^{1/2} + 2)$

$y' = (x + 1)\left(\frac{1}{2}x^{-1/2}\right) + (x^{1/2} + 2)(1)$

$= \frac{1}{2}x^{1/2} + \frac{1}{2}x^{-1/2} + x^{1/2} + 2$

$= \frac{3}{2}x^{1/2} + \frac{1}{2}x^{-1/2} + 2$

or $\frac{3\sqrt{x}}{2} + \frac{1}{2\sqrt{x}} + 2$

17. $g(x) = (5\sqrt{x} - 1)(2\sqrt{x} + 1)$

$= (5x^{1/2} - 1)(2x^{1/2} + 1)$

$g'(x) = (5x^{1/2} - 1)(x^{-1/2})$

$+ (2x^{1/2} + 1)\left(\frac{5}{2}x^{-1/2}\right)$

$= 5 - x^{-1/2} + 5 + \frac{5}{2}x^{-1/2}$

$= 10 + \frac{3}{2}x^{-1/2}$ or $10 + \frac{3}{2\sqrt{x}}$

19. $y = \frac{x + 1}{2x - 1}$

$y' = \frac{(2x - 1)(1) - (x + 1)(2)}{(2x - 1)^2}$

$= \frac{2x - 1 - 2x - 2}{(2x - 1)^2}$

$= \frac{-3}{(2x - 1)^2}$

21. $f(x) = \frac{7x + 1}{3x + 8}$

$f'(x) = \frac{(3x + 8)(7) - (7x + 1)(3)}{(3x + 8)^2}$

$= \frac{21x + 56 - 21x - 3}{(3x + 8)^2}$

$= \frac{53}{(3x + 8)^2}$

23. $y = \frac{2}{3x - 5}$

$y' = \frac{(3x - 5)(0) - 2(3)}{(3x - 5)^2}$

$= \frac{-6}{(3x - 5)^2}$

25. $y = \frac{5 - 3x}{4 + x}$

$$y' = \frac{(4 + x)(-3) - (5 - 3x)(1)}{(4 + x)^2}$$

$$= \frac{-12 - 3x - 5 + 3x}{(4 + x)^2}$$

$$= \frac{-17}{(4 + x)^2}$$

27. $f(t) = \frac{t^2 + t}{t - 1}$

$$f'(t) = \frac{(t - 1)(2t + 1) - (t^2 + t)(1)}{(t - 1)^2}$$

$$= \frac{2t^2 - t - 1 - t^2 - t}{(t - 1)^2}$$

$$= \frac{t^2 - 2t - 1}{(t - 1)^2}$$

29. $y = \frac{x - 2}{x^2 + 1}$

$$y' = \frac{(x^2 + 1)(1) - (x - 2)(2x)}{(x^2 + 1)^2}$$

$$= \frac{x^2 + 1 - 2x^2 + 4x}{(x^2 + 1)^2}$$

$$= \frac{-x^2 + 4x + 1}{(x^2 + 1)^2}$$

31. $g(x) = \frac{3x^2 + x}{2x^3 - 1}$

$$g'(x) = \frac{(2x^3 - 1)(6x + 1)}{(2x^3 - 1)^2} - \frac{(3x^2 + x)(6x^2)}{(2x^3 - 1)^2}$$

$$= \frac{12x^4 + 2x^3 - 6x - 1 - 18x^4 - 6x^3}{(2x^3 - 1)^2}$$

$$= \frac{-6x^4 - 4x^3 - 6x - 1}{(2x^3 - 1)^2}$$

33. $g(x) = \frac{x^2 - 4x + 2}{x + 3}$

$$g'(x) = \frac{(x + 3)(2x - 4)}{(x + 3)^2} - \frac{(x^2 - 4x + 2)(1)}{(x + 3)^2}$$

$$= \frac{2x^2 + 2x - 12 - x^2 + 4x - 2}{(x + 3)^2}$$

$$= \frac{x^2 + 6x - 14}{(x + 3)^2}$$

35. $p(t) = \frac{\sqrt{t}}{t - 1}$

$$= \frac{t^{1/2}}{t - 1}$$

$$p'(t) = \frac{(t - 1)\left(\frac{1}{2}t^{-1/2}\right)}{(t - 1)^2} - \frac{t^{1/2}(1)}{(t - 1)^2}$$

$$= \frac{\frac{1}{2}t^{1/2} - \frac{1}{2}t^{-1/2} - t^{1/2}}{(t - 1)^2}$$

$$= \frac{\frac{-1}{2}t^{1/2} - \frac{1}{2}t^{-1/2}}{(t - 1)^2}$$

$$= \frac{-\frac{\sqrt{t}}{2} - \frac{1}{2\sqrt{t}}}{(t - 1)^2} \text{ or } \frac{-t - 1}{2\sqrt{t}(t - 1)^2}$$

37. $y = \frac{5x + 6}{\sqrt{x}} = \frac{5x + 6}{x^{1/2}}$

$$y' = \frac{(x^{1/2})(5) - (5x + 6)\left(\frac{1}{2}x^{-1/2}\right)}{x}$$

$$= \frac{5x^{1/2} - \frac{5}{2}x^{1/2} - 3x^{-1/2}}{x}$$

$$= \frac{\frac{5}{2}x^{1/2} - 3x^{-1/2}}{x}$$

$$= \frac{\frac{5\sqrt{x}}{2} - \frac{3}{\sqrt{x}}}{x} \text{ or } \frac{5x - 6}{2x\sqrt{x}}$$

39. $f(p) = \frac{(2p + 3)(4p - 1)}{3p + 2}$

$$f'(p) = \frac{(3p + 2)D_p[(2p + 3)(4p - 1)]}{(3p + 2)^2} - \frac{(2p + 3)(4p - 1)D_p(3p + 2)}{(3p + 2)^2}$$

$$= \frac{(3p + 2)[(2)(4p - 1) + (2p + 3)(4)]}{(3p + 2)^2} - \frac{(2p + 3)(4p - 1)(3)}{(3p + 2)^2}$$

$$= \frac{(3p + 2)(8p - 2 + 8p + 12)}{(3p + 2)^2} - \frac{(8p^2 + 10p - 3)(3)}{(3p + 2)^2}$$

$$= \frac{24p^2 + 16p - 6p - 4 + 24p^2}{(3p + 2)^2} + \frac{16p + 36p + 24 - 24p^2 - 30p + 9}{(3p + 2)^2}$$

$$= \frac{24p^2 + 32p + 29}{(3p + 2)^2}$$

41. $g(x) = \frac{x^3 + 1}{(2x + 1)(5x + 2)}$

$$g'(x) = \frac{(2x + 1)(5x + 2)D_x(x^3 + 1)}{[(2x + 1)(5x + 2)]^2} - \frac{(x^3 + 1)D_x[(2x + 1)(5x + 2)]}{[(2x + 1)(5x +2)]^2}$$

$$= \frac{(2x + 1)(5x +2)(3x^2)}{(2x + 1)^2(5x + 2)^2} + \frac{-(x^3 + 1)[2(5x + 2) + (2x + 1)(5)]}{(2x + 1)^2(5x + 2)^2}$$

$$= \frac{(10x^2 + 5x + 4x + 2)(3x^2)}{(2x + 1)^2(5x + 2)^2} - \frac{(x^3 + 1)(10x + 4 + 10x + 5)}{(2x + 1)^2(5x + 2)^2}$$

$$= \frac{30x^4 + 15x^3 - 12x^3 + 6x^2}{(2x + 1)^2(5x + 2)^2} + \frac{-10x^4 - 4x^3 - 10x^4 - 5x^3 - 10x - 4 - 10x - 5}{(2x + 1)^2(5x + 2)^2}$$

$$= \frac{10x^4 + 18x^3 + 6x^2 - 20x - 9}{(2x + 1)^2(5x + 2)^2}$$

43. $C(x) = (3x + 2)(3x + 4)$
$= 9x^2 + 18x + 8$

The average cost is $\overline{C}(x) = \frac{C(x)}{x}$.

(a) 10 units

$$C(10) = \frac{C(10)}{10} = \frac{(3 \cdot 10 + 2)(3 \cdot 10 + 4)}{10} = \frac{(32)(34)}{10} = \frac{1088}{10} = 108.80$$

(b) 20 units

$$\overline{C}(20) = \frac{(3 \cdot 20 + 2)(3 \cdot 20 + 4)}{20} = \frac{(62)(64)}{20} = \frac{3968}{20} = 198.40$$

(c) x units

$$\overline{C}(x) = \frac{9x^2 + 18x + 8}{x} = 9x + 18 + \frac{8}{x}$$

(d) The marginal average cost function is

$$\frac{d\overline{C}}{dx} = 9 - \frac{8}{x^2}$$

45. $M(d) = \frac{200d}{3d + 10}$

(a) $M'(d)$

$$= \frac{(3d + 10)D_d(200d) - (200d)D_t(3d + 10)}{(3d + 10)^2}$$

$$= \frac{(3d + 10)(200) - (200d)(3)}{(3d + 10)^2}$$

$$= \frac{600d + 2000 - 600d}{(3d + 10)^2}$$

$$= \frac{2000}{(3d + 10)^2}$$

(b) $M'(2) = \frac{2000}{(3 \cdot 2 + 10)^2}$

$$= \frac{2000}{(16)^2}$$

$$= \frac{2000}{256}$$

$$= 7.8125$$

$M'(5) = \frac{2000}{(3 \cdot 5 + 10)^2}$

$$= \frac{2000}{(25)^2}$$

$$= \frac{2000}{625}$$

$$= 3.2$$

The new employee can assemble 7.8125 additional bicycles per day after 2 days of training and 3.2 additional bicycles after 5 days of training.

47. $s(x) = \frac{x}{m + nx}$; m and n are constants.

(a) $s'(x) = \frac{(m + nx)(1) - x(n)}{(m + nx)^2}$

$$= \frac{m + nx - nx}{(m + nx)^2}$$

$$= \frac{m}{(m + nx)^2}$$

(b) $x = 50$, $m = 10$, $n = 3$

$s'(50) = \frac{m}{(m + 50n)^2}$

$$= \frac{10}{[10 + 50(3)]^2}$$

$$= \frac{1}{2560} \approx .000391$$

(c) The amount of muscle contraction is increasing by .000391 millimeters when the concentration of the drug is 50 milliliters.

Section 11.6

For Exercises 1-5, $f(x) = 4x^2 - 2x$ and $g(x) + 8x + 1$.

1. $f[g(x)] = 4(8x + 1)^2 - (8x + 1)$

$f(g(2)] = 4(17)^2 - 2(17)$

$= 1122$

3. $g[f(2)] = 8(4x^2 - 2x) + 1$

$= 8(12) + 1$

$= 97$

5. $f[g(k)] = 4(8k + 1)^2 - 2(8k + 1)$

$= 4(64k^2 + 16k + 1) - 16k - 2$

$= 256k^2 + 48k + 2$

7. $f(x) = 8x + 12$; $g(x) = 3x - 1$

$f[g(x)] = 8(3x - 1) + 12$

$= 24x + 4$

$g[f(x)] = 3(8x + 12) - 1$

$= 24x + 35$

9. $f(x) = -x^3 + 2$; $g(x) = 4x$

$f[g(x)] = -(4x)^3 + 2$

$= -64x^3 + 2$

$g[f(x)] = 4(-x^3 + 2)$

$= -4x^3 + 8$

11. $f(x) = \frac{1}{x};\ g(x) = x^2$

$f[g(x)] = \frac{1}{x^2}$

$g[f(x)] = \left(\frac{1}{x}\right)^2$

$= \frac{1}{x^2}$

13. $f(x) = \sqrt{x + 2};\ g(x) = 8x^2 - 6$

$f[g(x)] = \sqrt{8x^2 - 6 + 2}$

$= \sqrt{8x^2 - 4}$

$g[f(x)] = 8(\sqrt{x + 2})^2 - 6$

$= 8x + 16 - 6$

$= 8x + 10$

15. $y = (3x - 7)^{1/3}$

If $f(x) = x^{1/3}$ and $g(x) = 3x - 7$, then

$y = f[g(x)] = (3x - 7)^{1/3}$.

17. $y = \sqrt{9 - 4x}$

If $f(x) = \sqrt{x}$ and $g(x) = 9 - 4x$, then

$y = f[g(x)] = \sqrt{9 - 4x}$.

19. $y = \frac{\sqrt{x} + 3}{\sqrt{x} - 3}$

If $f(x) = \frac{x + 3}{x - 3}$ and $g(x) = \sqrt{x}$, then

$y = f[g(x)] = \frac{\sqrt{x} + 3}{\sqrt{x} - 3}$.

21. $y = (x^{1/2} - 3)^2 + (x^{1/2} - 3) + 5$

If $f(x) = x^2 + x + 5$ and

$g(x) = x^{1/2} - 3$, then

$y = f[g(x)] = (x^{1/2} - 3)^2 + (x^{1/2} - 3) + 5$.

23. $y = (2x + 9)^2$

$y' = 2(2x + 9)(2)$

$= 4(2x + 9)$

25. $y = 6(5x - 1)^3$

$y' = 6(3)(5x - 1)^2(5)$

$= 90(5x - 1)^2$

27. $y = -2(12x^2 + 4)^3$

$y' = (-2)(3)(12x^2 + 4)^2(24x)$

$= -144x\ (12x^2 + 4)^2$

29. $y = 9(x^2 + 5x)^4$

$y' = 9(4)(x^2 + 5x)^3(2x + 5)$

$= 36(2x + 5)(x^2 + 5x)^3$

31. $y = 12(2x + 5)^{3/2}$

$y' = 12(2x + 5)^{1/2}\ (2)$

$= 36(2x + 5)^{1/2}$

33. $y = -7(4x^2 + 9x)^{3/2}$

$y' = -7\left(\frac{3}{2}\right)(4x^2 + 9x)^{1/2}(8x + 9)$

$= -\frac{21}{2}(8x + 9)(4x^2 + 9x)^{1/2}$

35. $y = 8\sqrt{4x + 7} = 8(4x + 7)^{1/2}$

$y' = 8\left(\frac{1}{2}\right)(4x + 7)^{-1/2}(4)$

$= 16(4x + 7)^{-1/2}$

$= \frac{16}{(4x + 7)^{1/2}}$ or $\frac{16}{\sqrt{4x + 7}}$

37. $y = -2\sqrt{x^2 + 4x} = -2(x^2 + 4x)^{1/2}$

$y' = -2\left(\frac{1}{2}\right)(x^2 + 4x)^{-1/2}(2x + 4)$

$= -(2x + 4)\ (x^2 + 4x)^{-1/2}$

or $\frac{-(2x + 4)}{(x^2 + 4x)^{1/2}}$

39. $y = 4x(2x + 3)^2$

$y' = 4x(2)(2x + 3)(2) + (2x + 3)^2(4)$

$= 16x(2x + 3) + 4(2x + 3)^2$

$= 4(2x + 3)[4x + (2x + 3)]$

$= 4(2x + 3)(6x + 3)$

$= 12(2x + 3)(2x + 1)$

41. $y = (x + 2)(x - 1)^2$

$y' = (x + 2)(2)(x - 1)(1) + (x - 1)^2(1)$

$= 2(x + 2)(x - 1) + (x - 1)^2$

This answer may be simplified:

$(x - 1)[2(x + 2) + (x - 1)]$

$= (x - 1)(2x + 4 + x - 1)$

$= (x - 1)(3x + 3)$

$= 3(x - 1)(x + 1).$

43. $y = 5(x + 3)^2(2x - 1)^5$

$y' = 5(x + 3)^2(5)(2x - 1)^4(2) + (5)(2x - 1)^5(2)(x + 3)$

$= 50(x + 3)^2(2x - 1)^4 + 10(2x - 1)^5(x + 3)$

$= 10(x + 3)(2x - 1)^4[5(x + 3) + (2x - 1)]$

$= 10(x + 3)(2x - 1)^4(7x + 14)$

$= 70(x + 3)(2x - 1)^4(x + 2)$

45. $y = (3x + 1)^3 \sqrt{x} = (3x + 1)^3 \ (x)^{1/2}$

$y' = 3(3x + 1)^2(3)(x)^{1/2} + (3x + 1)^3\left(\frac{1}{2}\right)(x)^{-1/2}$

$= 9x^{1/2}(3x + 1)^2 + \frac{1}{2} \cdot \frac{1}{x^{1/2}}(3x + 1)^3$

$= 9\sqrt{x}(3x + 1)^2 + \frac{(3x + 1)^3}{2\sqrt{x}}$

$= (3x + 1)^2x^{-1/2}\left[9x + \frac{3x + 1}{2}\right]$

$= (3x + 1)^2x^{-1/2}\left[\frac{18x + 3x + 1}{2}\right]$

$= (3x + 1)^2x^{-1/2}\left(\frac{21x + 1}{2}\right)$

$= \frac{(3x + 1)^2(21x + 1)}{2\sqrt{x}}$

47. $y = \frac{1}{(x - 4)^2} = (x - 4)^{-2}$

$y' = (-2)(x - 4)^{-3}$

$= \frac{-2}{(x - 4)^3}$

49. $y = \frac{(4x + 3)^2}{2x - 1}$

$y' = \frac{(2x - 1)(2)(4x + 3)(4) - (4x + 3)^2(2)}{(2x - 1)^2}$

$= \frac{8(2x - 1)(4x + 3) - 2(4x + 3)^2}{(2x - 1)^2}$

$= \frac{(4x + 3)[8(2x - 1) - 2(4x + 3)]}{(2x - 1)^2}$

$= \frac{(4x + 3)(16x - 8 - 8x - 6)}{(2x - 1)^2}$

$= \frac{(4x + 3)(8x - 14)}{(2x - 1)^2}$

$= \frac{2(4x + 3)(4x - 7)}{(2x - 1)^2}$

51. $y = \frac{x^2 + 4x}{(5x + 2)^3}$

$y' = \frac{(2x + 4)(5x + 2)^3}{(5x + 2)^6} - \frac{3(5x + 2)^2(5)(x^2 + 4x)}{(5x + 2)^6}$

$= \frac{(2x + 4)(5x + 2)^3}{(5x + 2)^6} - \frac{15(5x + 2)^2(x^2 + 4x)}{(5x + 2)^6}$

$= \frac{(5x + 2)^2}{(5x + 2)^6} \cdot [(2x + 4)(5x + 12) - 15(x^2 + 4x)]$

$= \frac{(10x^2 + 20x + 4x + 8 - 15x^2 - 60x)}{(5x + 2)^4}$

$= \frac{-5x^2 - 36x + 8}{(5x + 2)^4}$

53. $y = (x^{1/2} + 1)(x^{1/2} - 1)^{1/2}$

$y' = (x^{1/2} + 1)\left(\frac{1}{2}\right)(x^{1/2} - 1)^{-1/2}\left(\frac{1}{2}x^{-1/2}\right) + (x^{1/2} - 1)^{1/2}\left(\frac{1}{2}x^{-1/2}\right)$

$$= \frac{x^{-1/2}(x^{1/2}+1)(x^{1/2}-1)^{-1/2}}{4}$$

$$+ \frac{x^{-1/2}(x^{1/2}-1)^{-1/2}}{2}$$

$$= \frac{x^{-1/2}(x^{1/2}-1)^{-1/2}}{2} \cdot \left[\frac{x^{1/2}+1}{2} + (x^{1/2}-1)\right]$$

$$= \frac{x^{-1/2}(x^{1/2}-1)^{-1/2}(3x^{1/2}-1)}{4}$$

or $\frac{3x^{1/2}-1}{4x^{1/2}(x^{1/2}-1)^{1/2}}$

55. $D(p) = \frac{-p^2}{100} + 500;\ p(c) = 2c - 10$

The demand in terms of the cost is

$D(c) = D[p(c)]$

$$= \frac{-(2c-10)^2}{100} + 500$$

$$= \frac{-4(c-5)^2}{100} + 500$$

$$= \frac{-c^2+10c-25}{25} + 500$$

57. $A = 1500\left(1 + \frac{r}{36,500}\right)^{1825}$

dA/dr is the rate of change of A with respect to r.

$$\frac{dA}{dr} = 1500(1825)\left(1 + \frac{r}{36,500}\right)^{1824}\left(\frac{1}{36,500}\right)$$

$$= 75\left(\frac{36,500 + r}{36,500}\right)^{1824}$$

(a) For r = 6%,

$$\frac{dA}{dr} = \$101.22.$$

(b) For r = 8%,

$$\frac{dA}{dr} = \$111.86.$$

(c) For r = 9%,

$$\frac{dA}{dr} = \$117.59.$$

59. $V = \frac{6000}{1 + .3t + .1t^2}$

The rate of change of the value is

$$V' = \frac{(-1)(6000)(.3 + .2t)}{(1 + .3t + .1t^2)^2}$$

(a) 2 years after purchase the rate of decrease in the value is

$$V' = \frac{-6000(.3 + .4)}{(1 + .6 + .4)^2}$$

$$= \frac{-4200}{4}$$

$$= -\$1050.$$

(b) 4 years after purchase:

$$V' = \frac{-6000(.3 + .8)}{(1 + 1.2 + 1.6)^2}$$

$$= \frac{-6600}{14.44}$$

$$= -\$457.06.$$

61. Demand function is $p = 300/x^{1/3}$; $x = 8n$

The marginal revenue product is

$$\frac{dR}{dn} = \left(p + x\frac{dp}{dx}\right)\frac{dx}{dn}$$

$$= [300x^{-1/3} + 8n(-100x^{-4/3})]8.$$

If n = 8, then x = 64 and

$$\frac{dR}{dn} = \left[\frac{300}{4} + 64\left(\frac{-100}{256}\right)\right]8$$

$= 50(8)$

$= \$400$ per additional worker.

63. $P(x) = 2x^2 + 1;\ x = f(a) = 3a + 2$

$$P[f(a)] = 2(3a + 2)^2 + 1$$
$$= 2(9a^2 + 12a + 4) + 1$$
$$= 18a^2 + 24a + 9$$

65. $r(t) = 2t;\ A(r) = \pi r^2$

$$A[r(t)] = \pi(2t)^2$$
$$= 4\pi t^2$$

$A = 4\pi t^2$ gives the area of the pollution in terms of the time since the pollutants were first emitted.

67. $C(t) = \frac{1}{2}(2t + 1)^{-1/2}$

$C'(t) = \frac{1}{2}\left(-\frac{1}{2}\right)(2t + 1)^{-3/2}(2)$

$= -\frac{1}{2}(2t + 1)^{-3/2}$

(a) $C'(0) = -\frac{1}{2}[2(0) + 1]^{-3/2}$

$= -\frac{1}{2} = -.5$

(b) $C'(4) = -\frac{1}{2}[2(4) + 1]^{-3/2}$

$= -\frac{1}{2}(9)^{-3/2}$

$= -\frac{1}{2}\left(\frac{1}{27}\right)$

$= -\frac{1}{54} \approx -.02$

(c) $C'(6) = -\frac{1}{2}[2(6) + 1]^{-3/2}$

$= -\frac{1}{2}(13)^{-3/2} \approx -.011$

(d) $C'(7.5) = -\frac{1}{2}[2(7.5) + 1]^{-3/2}$

$= -\frac{1}{2}(16)^{-3/2}$

$= -\frac{1}{2}\left(\frac{1}{64}\right)$

$= -\frac{1}{128} \approx -.008$

Section 11.7

In Exercises 1-17, use

$\frac{d}{dx}e^{g(x)} = g'(x)e^{g(x)}$.

1. $y = e^{4x}$

$g(x) = 4x$

$g'(x) = 4$

$y' = 4e^{4x}$

3. $y = -6e^{-2x}$

$g(x) = -2x$

$g'(x) = -2$

$y' = -6[-2e^{-2x}]$

$= 12e^{-2x}$

5. $y = -8e^{2x}$

$y' = -8[2e^{2x}]$

$= -16e^{2x}$

7. $y = -16e^{x+1}$

$g(x) = x + 1$

$g'(x) = 1$

$y' = -16[(1)e^{x+1}]$

$= -16e^{x+1}$

9. $y = e^{x^2}$

$g(x) = x^2$

$g'(x) = 2x$

$y' = 2xe^{x^2}$

11. $y = 3e^{2x^2}$

$y' = 3[4xe^{2x^2}]$

$= 12xe^{2x^2}$

13. $y = 4e^{2x^2-4}$

$y' = 4[(4x)e^{2x^2-4}]$

$= 16xe^{2x^2-4}$

15. $y = xe^x$

$y' = xe^x + e^x$ *Use product rule*

$= e^x(x + 1)$

17. $y = (x - 3)^2e^{2x}$

$y' = (x - 3)^2(2)e^{2x}(2)(x - 3)$

Product rule

$= 2(x - 3)^2e^{2x} + 2(x - 3)e^{2x}$

$= 2(x - 3)e^{2x}[(x - 3) + 1]$

$= 2(x - 3)(x - 2)e^{2x}$

In Exercises 19-29, use

$$\frac{d}{dx}\ln|g(x)| = \frac{g'(x)}{g(x)}.$$

19. $y = \ln(3 - x)$

$g(x) = 3 - x$

$g'(x) = -1.$

$y' = \frac{-1}{3 - x}$ or $\frac{1}{x - 3}$

21. $y = \ln(2x^2 - 7x)$

$g(x) = 2x^2 - 7x$

$g'(x) = 4x - 7$

$y' = \frac{4x - 7}{2x^2 - 7x}$

23. $y = \ln\sqrt{x + 5}$

$g(x) = \sqrt{x + 5}$

$= (x + 5)^{1/2}$

$g'(x) = \frac{1}{2}(x + 5)^{-1/2}$

$= \frac{1}{2\sqrt{x + 5}}$

$y' = \frac{\frac{1}{2\sqrt{x + 5}}}{\sqrt{x + 5}}$

$= \frac{1}{2(\sqrt{x + 5})^2}$

$= \frac{1}{2(x + 5)}$

25. $y = \ln(2x + 7)(3x - 2)$

$g(x) = (2x + 7)(3x - 2)$

$g'(x) = (2)(3x - 2) + (2x + 7)(3)$

$= 6x - 4 + 6x + 21$

$= 12x + 17$

$y' = \frac{12x + 17}{(2x + 7)(3x - 2)}$

27. $y = \ln\left(\frac{5x - 1}{2x + 4}\right)$

$g(x) = \frac{5x - 1}{2x + 4}$

$g'(x) = \frac{(5)(2x + 4) - (2)(5x - 1)}{(2x + 4)^2}$

$= \frac{10x + 20 - 10x + 2}{(2x + 4)^2}$

$= \frac{22}{(2x + 4)^2}$

$y' = \frac{22}{(2x + 4)^2} \div \frac{5x - 1}{(2x + 4)}$

$= \frac{22}{(2x + 4)^2} \cdot \frac{2x + 4}{5x - 1}$

$= \frac{22}{(2x + 4)(5x - 1)}$

29. $y = \ln(x^4 + 5x^2)^{3/2}$

$g(x) = (x^4 + 5x^2)^{3/2}$

$g'(x) = \frac{3}{2}(x^4 + 5x^2)^{1/2}(4x^3 + 10x)$

$= \frac{3}{2} \cdot (x^2)^{1/2}(x^2 + 5)^{1/2}(2x)(2x^2 + 5)$

$= \frac{3}{2} \cdot (x)(2x)(x^2 + 5)^{1/2}(2x + 5)$

$= 3x^2(x^2 + 5)^{1/2}(2x^2 + 5)$

$y' = \frac{3x^2(x^2 + 5)^{1/2}(2x^2 + 5)}{(x^4 + 5x^2)^{3/2}}$

$= \frac{3x^2(x^2 + 5)^{1/2}(2x^2 + 5)}{(x^2)^{3/2}(x^2 + 5)^{3/2}}$

$= \frac{3x^2(x^2 + 5)^{1/2}(2x^2 + 5)}{x^3(x^2 + 5)^{3/2}}$

$= \frac{3(2x^2 + 5)}{x(x^2 + 5)}$

31. $y = -3x\ln(x + 2)$

$y' = -3xD_x[\ln(x + 2)]$

$+ \ln(x + 2)D_x(-3x)$

$= -3x\left(\frac{1}{x + 2}\right) + \ln(x + 2)(-3)$

$= \frac{-3x}{x + 2} - 3\ln(x + 2)$

33. $y = x^2 \ln|x|$

$y' = x^2 D_x(\ln |x|) + \ln |x| \cdot D_x(x)^2$

$= x^2\left(\frac{1}{x}\right) + \ln |x| \cdot 2x$

$= x + 2x \ln |x|$

35. $y = \frac{x^2}{e^x}$

$y' = \frac{e^x(2x) - x^2e^x}{e^{2x}}$ *Quotient rule*

$= \frac{xe^x(2 - x)}{e^{2x}}$

$= \frac{x(2 - x)}{e^x}$

37. $y = (2x^3 - 1) \ln |x|$

$y' = (6x^2)\ln |x| + (2x^3 - 1) \frac{1}{x}$

$= 6x^2 \ln |x| + \frac{2x^3 - 1}{x}$

$= \frac{6x^3\ln |x| + 2x^3 - 1}{x}$

39. $y = \frac{\ln |x|}{4x + 7}$

$y' = \frac{(4x + 7)D_x(\ln |x|)}{(4x + 7)^2} - \frac{\ln |x| \cdot D_x(4x + 7)}{(4x + 7)^2}$

$= \frac{(4x + 7)\left(\frac{1}{x}\right) - \ln |x| \cdot (4)}{(4x + 7)^2}$

$= \frac{\frac{4x + 7}{x} - 4 \ln |x|}{(4x + 7)^2}$

$= \frac{4x + 7 - 4x \ln |x|}{x(4x + 7)^2}$

41. $y = \frac{\ln |x|}{e^x}$

$y' = \frac{\frac{1}{x}e^x - e^x \ln |x|}{(e^x)^2}$

$= \frac{\frac{e^x}{x} - e^x \ln |x|}{(e^x)^2}$

$= \frac{e^x - xe^x \ln |x|}{x(e^x)^2}$

$= \frac{e^x(1 - x \ln |x|)}{x(e^x)^2}$

$= \frac{1 - x \ln |x|}{xe^x}$

43. $y = \frac{3x^2}{\ln |x|}$

$y' = \frac{\ln |x| D_x(3x^2) - 3x^2 D_x(\ln |x|)}{(\ln |x|)^2}$

$= \frac{\ln |x| (6x) - 3x^2\left(\frac{1}{x}\right)}{(\ln |x|)^2}$

$= \frac{6x \ln |x| - 3x}{(\ln |x|)^2}$

45. $y = [\ln(x + 1)]^4$

$y' = 4[\ln(x + 1)]^3 \cdot D_x[\ln(x + 1)]$

$= 4[\ln(x + 1)]^3 \left(\frac{1}{x + 1}\right)$

$= \frac{4[\ln(x + 1)]^3}{x + 1}$

47. $y = \frac{e^x}{\ln |x|}$

$y' = \frac{(\ln |x|)e^x - e^x(1/x)}{(\ln |x|)^2}$

$= \frac{xe^x \ln |x| - e^x}{x(\ln |x|)^2}$ *Quotient rule*

$= \frac{e^x(x \ln |x| - 1)}{x(\ln |x|)^2}$

49. $y = \frac{e^x + e^{-x}}{x}$

$y' = \frac{x(e^x - e^{-x}) - (e^x + e^{-x})}{x^2}$

Quotient rule

$= \frac{xe^x - xe^{-x} - e^x - e^{-x}}{x^2}$

$= \frac{x(e^x - e^{-x}) - (e^x + e^{-x})}{x^2}$

51. $y = e^{x^2} \ln x$

$y' = e^{x^2}\left(\frac{1}{x}\right) + (\ln x)(2x)e^{x^2}$

$= \frac{e^{x^2}}{x} + 2xe^{x^2} \ln x$

53. $y = \frac{5000}{1 + 10e^{.4x}}$

$y' = \frac{(1 + 10e^{.4x}) \cdot 0 - 5000[0 + 10(.4)e^{.4x}]}{(1 + 10e^{.4x})^2}$

$= \frac{-20{,}000e^{.4x}}{(1 + 10e^{.4x})^2}$

55. $y = \frac{10{,}000}{9 + 4e^{-.2x}}$

$y' = \frac{(9 + 4e^{-.2x}) \cdot 0}{(9 + 4e^{-.2x})^2}$

$- \frac{10{,}000[0 + 4(-.2)e^{-.2x}]}{(9 + 4e^{-.2x})^2}$

$= \frac{8000e^{-.2x}}{(9 + 4e^{-.2x})^2}$

57. $y = \ln(\ln |x|)$

$y' = \frac{1}{\ln |x|} \cdot \frac{1}{x}$

$= \frac{1}{x \ln |x|}$

59. $p = 100 + \frac{50}{\ln x}$

(a) $R = px$

$R = 100x + \frac{50x}{\ln x}$

Marginal revenue is

$\frac{dR}{dx} = 100 + \frac{50(\ln x - 1)}{(\ln x)^2}.$

(b) Revenue from 8001st item is dR/dx for x = 8.

$100 + \frac{50(\ln 8 - 1)}{(\ln 8)^2} = \$112.48$

61. $C(x) = 100x + 100$

(a) $C'(x) = 100$

(b) $P(x) = R(x) - C(x)$

$= x\left(100 + \frac{50}{\ln x}\right) - (100x + 100)$

$P(x) = 100x + \frac{50x}{\ln x} - 100x - 100$

$P(x) = \frac{50x}{\ln x} - 100$

(c) $P'(x) = \frac{50\left(\ln x - \frac{x}{x}\right)}{(\ln x)^2} = \frac{50(\ln x - 1)}{(\ln x)^2}$

$P'(8) = \frac{50(\ln 8 - 1)}{(\ln 8)^2} \approx 12.48$

So additional profit is about \$12.48.

63. $S(t) = 100 - 90e^{-.3t}$

$S'(t) = -90(-.3)e^{-.3t}$

$= 27e^{-.3t}$

(a) $S'(1) = 27e^{-.3(1)}$

$= 27e^{-.3}$

≈ 20

(b) $S'(10) = 27e^{-.3(10)}$

$= 27e^{-3}$

≈ 1.34

65. $y = 100e^{-.03045t}$

(a) For $t = 0$,

$y = 100e^{-.03045(0)}$

$= 100e^0$

$= 100\%$.

(b) For $t = 2$,

$y = 100e^{-.03045(2)}$

$= 100e^{-.0609}$

$\approx 94.1\%$.

(c) For $t = 4$,

$y = 100e^{-.03045(4)}$

$\approx 88.5\%$.

(d) For $t = 6$,

$y = 100e^{-.03045(6)}$

$\approx 83.3\%$.

(e) $y' = 100(-.03045)e^{-.03045t}$

$= -3.045e^{-.03045t}$

For $t = 0$,

$y' = -3.045e^{-.03045(0)}$

$\approx -3.05\%$.

(f) For $t = 2$,

$y' = -3.045e^{-.03045(2)}$

$\approx -2.87\%$.

(g) For $t = 4$,

$y' = -3.045e^{-.03045(4)}$

$\approx -2.70\%$.

(h) For $t = 6$,

$y' = -3.045e^{-.03045(6)}$

$\approx -2.54\%$.

67. $P(t) = 1000e^{.2t}$

$P'(t) = (.2)(1000)e^{.2t}$

$= 200e^{.2t}$

For $t = 2$,

$P'(t) = P'(2) = 200e^{.2(2)}$

$= 200e^{.4}$

$= 298.$

For $t = 8$,

$P'(t) = P'(8) = 200e^{.2(8)}$

$= 200e^{1.6}$

$= 990.$

69. $M(t) = (e^{.1t} + 1) \ln \sqrt{t}$

(a) $M(15) = (e^{.1(15)} + 1) \ln \sqrt{15}$

$= (4.48 + 1)(1.354)$

$= 7.42$

(b) $M(25) = (e^{.1(25)} + 1) \ln \sqrt{25}$

$= (13.1825)(1.609)$

$= 21.2$

(c) $M'(t) = (.1)(e^{.1t}\ln\sqrt{t} + (e^{.1t} + 1)\left(\frac{1}{2t}\right)$

$M'(25) + (.1)e^{.1(15)}\ln\sqrt{15}$

$+ (e^{.1(15)} + 1)\left(\frac{1}{2(15)}\right)$

$= .79$

71. $P(x) = .04e^{-4x}$

(a) $P(.5) = .04e^{-4(.5)}$

$= .04e^{-2}$

$\approx .005$

(b) $P(1) = .04e^{-4(1)}$

$= .04e^{-4}$

$\approx .0007$

(c) $P(2) = .04e^{-4(2)}$

$= .04e^{-8}$

$\approx .000013$

$P'(x) = .04(-4)e^{-4x} = -.16e^{4x}$

(d) $P'(5) = -.16e^{-4(.5)}$

$= -.16e^{-2}$

$\approx -.022$

(e) $P'(1) = -.16e^{-4(1)}$

$= -.16e^{-4}$

$\approx -.0029$

(f) $P'(2) = -.16e^{-4(2)}$

$= -.16e^{-8}$

$\approx -.000054$

Section 11.8

1. The open circle on the graph at $x = -1$ means that $f(-1)$ does not exist. Because of this, part (a) of the definition fails and the function is discontinuous at $x = -1$.

3. As $g(x)$ approaches -1 from the right, $g(x)$ is 3. As $g(x)$ approaches -1 from the left, $g(x)$ is 1. Since the values of the function are not equal, the function is discontinuous at $x = -1$.

5. $f(x)$ is not defined at $x = 0$ and the open circle at $x = 2$ mean $f(2)$ does not exist. So the function is discontinuous at $x = 0$ and $x = 2$.

7. The limit of $f(x)$ as x approaches 1 is -2. The value of $f(x)$ at $x = 1$ is 2 (solid circle). Since the limit of $f(x)$ and the value of $f(x)$ are not equal, $f(x)$ is discontinuous at $x = 1$.

9. $f(x) = \frac{2}{x - 3}$; $x = 0$, $x = 3$

Since f is a rational function, it will not be continuous for any x for which $x - 3 = 0$. Therefore, f is not continuous for $x = 3$, but is continuous at all other points, including $x = 0$.

11. $g(x) = \frac{1}{x(x - 2)}$; $x = 0$, $x = 2$, $x = 4$

Since g is a rational function, it will not be continuous for any x for which $x(x - 2) = 0$. Therefore, g is not continuous for $x = 0$ or $x = 2$, but is continuous at all other points, including $x = 4$.

13. $h(x) = \frac{1 + x}{(x - 3)(x + 1)}$; $x = 0$, $x = 3$, $x = -1$

Since h is a rational function, it will not be continuous for any x for which $(x - 3)(x + 1) = 0$. Therefore, h is not continuous for $x = 3$ or $x = -1$, but is continuous at all other points, including $x = 0$.

15. $k(x) = \frac{5 + x}{2 + x}$; $x = 0$, $x = -2$, $x = -5$

Since k is a rational function, it will not be continuous for any x for which $2 + x = 0$. Therefore, k is not continuous for $x = -2$, but is continuous at all other points, including $x = 0$ and $x = -5$.

17. $g(x) = \frac{x^2 - 4}{x - 2}$; $x = 0$, $x = 2$, $x = -2$

Since g is a rational function, it will not be continuous for any x for which $x - 2 = 0$. Therefore, g is not continuous for $x = 2$, but is continuous at all other points, including $x = 0$ and $x = -2$.

19. $p(x) = x^2 - 4x + 11$; $x = 0$, $x = 2$, $x = -1$

Since p is a polynomial function it is continuous for all real values of x, including $x = 0$, $x = 2$, and $x = -1$.

21. $p(x) = \frac{|x + 2|}{x + 2}$; $x = -2$, $x = 0$, $x = 2$

Since p is a rational function, it is not continuous at $x = -2$, but is continuous at all other points, including $x = 0$ and $x = 2$.

23. $F(x) = 1.20x$ if $0 < x \le 100$
$F(x) = 1.0x$ if $x > 100$

(a) $F(80) = 1.20(80) = \$96$

(b) $F(150) = 1.0(150) = \$150$

(c) See the graph in the answer section at the back of the textbook.

(d) By reading the graph of part (c), F is discontinuous at $x = 100$.

25. Total cost in dollars:
$C(t) = 30t$, if $1 \le t \le 5$ or
$C(t) = 30(5) = 150$, if $t = 6$ or $t = 7$
$C(t) = 150 + 30(t - 7)$ if $8 \le t \le 12$
$A(t) = \frac{C(t)}{t}$

(a) $A(4) = \frac{30(4)}{4} = \$30$

(b) $A(5) = \frac{3(5)}{5} = \$30$

(c) $A(6) = \frac{150}{6} = \$25$

(d) $A(7) = \frac{150}{7} \approx \$21.43$

(e) $A(8) = \frac{150 + 30(8 - 7)}{8}$

$= \frac{180}{8} = \$22.50$

(f) $\lim_{t \to 5} A(t) = 30$ because as t approaches 5 from the left, A(t) approaches 30 (think of the graph for $t = 1, 2, \ldots, 5$), and t approaches 5 from the right (think of the graph for $5 < t < 6$), A(t) approaches \$30.

(g) $\lim_{t \to 6} A(t)$ does not exist because as t approaches 6 from the left (think of the graph for $5 < t < 6$), A(t) approaches 30, but as t approaches 6 from the right (think of the graph for $6 < t < 7$), A(t) approaches \$25.

27. The function is discontinuous at values of x where breaks or gaps occur. These values are $x = 6$, $x = 8$, $x = 12$, and $x = 16$.

29. Since neither -4 or 6 are included, the interval is $(-4, 6)$.

31. Since -2 is included, the interval is $[-\infty, -2]$.

33. Since both -11 and 9 are included the interval is $[-11, 9]$.

35. Since neither -6 or -2 are included, the interval is $(-6, -2)$.

37. Since -4 is not included, the interval is $(-4, \infty)$.

39. Since 0 is not included, the interval is $(-\infty, 0)$.

41. Points of discontinuity occur at $x = -3$, $x = 0$, and $x = 3$.
So the interval $(-3, 0)$ does not contain one of the discontinuities, the function is continuous on $(-3, 0)$.
The interval $(0, 3)$ does not contain one of the discontinuities, so the function is continuous on $(0, 3)$.
The interval $(0, 4)$ contains the discontinuity $x = 3$ so the function is discontinuous on $(0, 4)$.

43. The discontinuities occur at $x = 0$ and $x = 6$. $(-12, 6)$ contains a discontinuity so the function is discontinuous on $(-12, 6)$.
$(0, 6)$ does not contain a discontinuity, so the function is continuous on $(0, 6)$.
$(6, 12)$ does not contain a discontinuity, so the function is continuous on $(6, 12)$.

Chapter 11 Review Exercises

1. The graph of $f(x)$ approaches 4 from the left and from the right of $x = -3$.

$\lim_{x \to -3} f(x) = 4$

3. The graph of $f(x)$ approaches 4 from the left of $x \to \infty$, so

$\lim_{x \to \infty} f(x) = -3$

5. $\lim_{x \to -1} (2x^2 + 3x + 5)$

$= \lim_{x \to -1} 2x^2 + \lim_{x \to -1} 3x + \lim_{x \to -1} 5$

$= (\lim_{x \to -1} 2)(\lim_{x \to -1} x) + (\lim_{x \to -1} 3)(\lim_{x \to -1} x) + \lim_{x \to -1} 5$

$= 2(-1)^2 + 3(-1)\ 5$
$= 4$

7. $\lim_{x \to 6} \dfrac{2x + 5}{x - 3}$

$$= \frac{\lim_{x \to 6} (2x + 5)}{\lim_{x \to 6} (x - 3)}$$

$$= \frac{\lim_{x \to 6} 2x + \lim_{x \to 6} 5}{\lim_{x \to 6} x + \lim_{x \to 6} (-3)}$$

$$= \frac{\left(\lim_{x \to 6} 2\right)\left(\lim_{x \to 6} x\right) + \lim_{x \to 6} 5}{\lim_{x \to 6} x + \lim_{x \to 6} (-3)}$$

$$= \frac{2(6) + 5}{6 + (-3)}$$

$$= \frac{17}{3}$$

9. $\lim_{x \to 4} \dfrac{x^2 - 16}{x - 4}$

$$= \lim_{x \to 4} \frac{(x - 4)(x + 4)}{x - 4}$$

$= \lim_{x \to 4} (x + 4) = \lim_{x \to 4} x + \lim_{x \to 4} 4$

$= 4 + 4$
$= 8$

11. $\lim\limits_{x\to-4} \dfrac{2x^2 + 3x - 20}{x + 4}$

$= \lim\limits_{x\to-4} \dfrac{(2x - 5)(x + 4)}{x + 4}$

$= \lim\limits_{x\to-4} (2x - 5)$

$= \lim\limits_{x\to-4} 2x + \lim\limits_{x\to-4} -5$

$= \left(\lim\limits_{x\to-4} 2\right)\left(\lim\limits_{x\to-4} x\right) + \lim\limits_{x\to-4} -5$

$= 2(-8) - 5$

$= -13$

13. $\lim\limits_{x\to9} \dfrac{\sqrt{x} - 3}{x - 9}$

$= \lim\limits_{x\to9} \dfrac{\sqrt{x} - 3}{x - 9} \cdot \dfrac{\sqrt{x} + 3}{\sqrt{x} + 3}$

$= \lim\limits_{x\to9} \dfrac{x - 9}{(x - 9)(\sqrt{x} + 3)}$

$= \lim\limits_{x\to9} \dfrac{1}{\sqrt{x} + 3}$

$= \dfrac{\lim\limits_{x\to9} 1}{\lim\limits_{x\to9} (\sqrt{x} + 3)}$

$= \dfrac{\lim\limits_{x\to9} 1}{\lim\limits_{x\to9} \sqrt{x} + \lim\limits_{x\to9} 3}$

$= \dfrac{1}{\sqrt{9} + 3}$

$= \dfrac{1}{6}$

15. $\lim\limits_{x\to\infty} \dfrac{x^2 + 5}{5x^2 - 1}$

$= \lim\limits_{x\to\infty} \dfrac{\dfrac{x^2 + 5}{x^2}}{\dfrac{5x^2 - 1}{x^2}}$

$= \lim\limits_{x\to\infty} \dfrac{1 + \dfrac{5}{x^2}}{5 - \dfrac{1}{x^2}}$

$= \dfrac{\lim\limits_{x\to\infty} 1 + \lim\limits_{x\to\infty} \dfrac{5}{x^2}}{\lim\limits_{x\to\infty} 5 + \lim\limits_{x\to\infty} -\dfrac{1}{x^2}}$

$= \dfrac{1 + 0}{5 + 0}$

$= \dfrac{1}{5}$

17. $\lim\limits_{x\to\infty} \left(\dfrac{3}{4} + \dfrac{2}{x} - \dfrac{5}{x^2}\right)$

$= \lim\limits_{x\to\infty} \dfrac{3}{4} = \lim\limits_{x\to\infty} \dfrac{2}{x} + \lim\limits_{x\to\infty} \dfrac{-5}{x^2}$

$= \dfrac{3}{4} + 0 + 0$

$= \dfrac{3}{4}$

19. Average rate of change of f, from $x = 0$ to $x = 4$.

By reading graph,

$\dfrac{1 - 0}{4 - 0} = \dfrac{1}{4}$.

21. Average rate of change of f, from $x = 2$ to $x = 4$.

By reading graph,

$$\frac{1 - 4}{4 - 2} = -\frac{3}{2}$$

23. $y = 6x^2 + 2,\ x = 1,\ x = 4$

$y(4) = 6(4)^2 + 2 = 98$

$y(1) = 6(1)^2 + 2 = 8$

Average rate of change

$$= \frac{98 - 8}{4 - 1} = \frac{90}{3} = 30$$

25. $y = \frac{-6}{3x - 5},\ x = 4,\ x = 9$

$$y(9) = \frac{-6}{3(9) - 5} = \frac{-6}{22} = -\frac{3}{11}$$

$$y(4) = \frac{-6}{3(4) - 5} = -\frac{6}{7}$$

Average rate of change

$$= \frac{\frac{-3}{11} - \left(-\frac{6}{7}\right)}{9 - 4}$$

$$= \frac{\frac{-21 + 66}{77}}{5}$$

$$= \frac{45}{5(77)} = \frac{9}{77}$$

27. $y = 4x + 3$

$$y' = \lim_{h \to 0} \frac{y(x + h) - y(x)}{h}$$

$$= \lim_{h \to 0} \frac{[4(x + h) + 3] - [4x + 3]}{h}$$

$$= \lim_{h \to 0} \frac{4x + 4h + 3 - 4x - 3}{h}$$

$$= \lim_{h \to 0} \frac{4h}{h} = \lim_{h \to 0} 4$$

$$= 4$$

29. $y = -x^3 + 7x$

$$y' = \lim_{h \to 0} \frac{y(x + h) - y(x)}{h}$$

$$= \lim_{h \to 0} \left(\frac{[-(x + h)^3 + 7(x + h)]}{h} - \frac{[-x^3 + 7x]}{h}\right)$$

$$= \lim_{h \to 0} \left(\frac{-x^3 - 2x^2h - xh^2 - hx^2 - 2xh^2}{h} + \frac{-h^3 + 7x + 7h + x^3 - 7x}{h}\right)$$

$$= \lim_{h \to 0} \left(\frac{-2x^2h - xh^2 - hx^2}{h} - \frac{2xh^2 - h^3 + 7h}{h}\right)$$

$$= \lim_{h \to 0} -2x^2 - xh - x^2 - 2xh - h^2 + 7$$

$$= -3x^2 + 7$$

31. $y = x^2 = 6x$, tangent at $x = 2$.

$y' = 2x - 6 = m$

$y'(2) = 2(2) - 6 = -2 = m$

Use $(2, -8)$ and point-slope form.

$y - (-8) = -2(x - 2)$

$y + 8 = -2x + 4$

$y + 2x = -4$

33. $y = \frac{3}{x - 1}$, tangent at $x = -1$.

$$y = \frac{3}{x - 1} = 3(x - 1)^{-1}$$

$$y' = 3(-1)(x - 1)^{-2}(1)$$

$$= -3(x - 1)^{-2}$$

$$y'(-1) = -3(-1 - 1)^{-2} = -\frac{3}{4} = m$$

Use (-1, -3/2) and point-slope form.

$$y - \left(\frac{-3}{2}\right) = -\frac{3}{4}(x - 1(-1))$$

$$y + \frac{3}{2} = -\frac{3}{4}(x + 1)$$

$$y = -\frac{3}{4}x - \frac{9}{4}$$

$$3x + 4y = -9$$

35. $y = (3x^2 - 5x)(2x)$, tangent at $x = -1$.

$$y' = (3x^2 - 5x)(2) + (2x)(6x - 5)$$

$$y'(-1) = [3(-1)^2 - 5(-1)](2) + 2(-1)(6(-1) - 5)$$

$$= 38 = m$$

Use (-1, -16) and point-slope form.

$$y - (-16) = 38(x - (-1))$$

$$y + 16 = 38x + 38$$

$$y = 38x + 22$$

37. $y = \sqrt{6x - 2}$, tangent at $x = 3$.

First find y'. Since $y = (6x - 2)^{1/2}$, the chain rule gives

$$y' = \frac{1}{2}(6x - 2)^{-1/2}(6)$$

or $y' = 3(6x - 2)^{-1/2}$.

Find the slope of the tangent by replacing x with 3. This gives

$$m = 3(6 \cdot 3 - 2)^{-1/2}$$

$$= 3(16)^{-1/2}$$

$$= \frac{3}{16^{1/2}}$$

$$= \frac{3}{4}.$$

Find the value of y corresponding to $x = 3$:

$$y = \sqrt{6x - 2} = \sqrt{6 \cdot 3 - 2} = \sqrt{16} = 4$$

The answer is the line through (3, 4) with slope 3/4. The equation of this line is

$$y - 4 = \frac{3}{4}(x - 3)$$

$$4y - 16 = 3x - 9$$

$$4y = 3x + 7.$$

39. $C(x) = \begin{cases} 1.50x \text{ for } 0 < x \le 125 \\ 1.35x \text{ for } x < 125 \end{cases}$

(a) $C(100) = 1.50(100) = \$150$

(b) $C(125) = 1.50(125) = \$187.50$

(c) $C(140) = 1.35(140) = \$189$

(d) See the graph in the answer section at the back of your textbook.

(e) By reading the graph, C(x) is discontinuous at x = $125.

41. $y = 5x^2 - 7x - 9$

$$y' = 2(5)x - 7$$

$$= 10x - 7$$

43. $y = 6x^{7/3}$

$$y' = \frac{7}{3}(6)x^{4/3}$$

$$= \frac{42}{3}x^{4/3}$$

$$= 14x^{4/3}$$

45. $f(x) = x^{-3} + \sqrt{x}$

$$= x^{-3} + x^{1/2}$$

$$f'(x) = -3x^{-4} + \left(\frac{1}{2}\right)x^{-1/2}$$

$$\text{or } \frac{-3}{x^4} + \frac{1}{2x^{1/2}}$$

47. $y = (3t^2 + 7)(t^3 - t)$

$y' = (3t^2 + 7)(3t^2 - 1) + (t^3 - t)(6t)$

$= 9t^4 + 18t^2 - 7 + 6t^4 - 6t^2$

$= 15t^4 + 12t^2 - 7$

49. $y = 4\sqrt{x}(2x - 3)$

$= 4x^{1/2}(2x - 3)$

$y' = 4x^{1/2}(2) + (2x - 3)(4)\left(\frac{1}{2}\right)(x^{-1/2})$

$= 8x^{1/2} + (4x - 6)x^{-1/2}$

$= 8x^{1/2} + 4x^{1/2} - 6x^{-1/2}$

$= 12^{1/2} - 6x^{-1/2}$

or $12x^{1/2} - \frac{6}{x^{1/2}}$

51. $g(t) = -3t^{1/3}(5t + 7)$

$g'(t) = (-3t^{1/3})(5)$

$+ (5t + 7)\left(\frac{1}{3}\right)(-3t^{-2/3})$

$= -15t^{1/3} + (5t + 7)(-t^{-2/3})$

$= -15t^{1/3} + (-5t^{1/3}) - 7t^{-2/3}$

$= -20t^{1/3} - 7t^{-2/3}$

or $-20t^{1/3} - \frac{7}{t^{2/3}}$

53. $y = 12x^{-3/4}(3x + 5)$

$y' = (12x^{-3/4})(3)$

$+ (3x + 5)(12)\left(-\frac{3}{4}\right)(x^{-7/4})$

$= 36x^{-3/4} + (3x + 5)(-9)x^{-7/4}$

$= 36x^{-3/4} + (-27)x^{-3/4} + (-45)x^{-7/4}$

$= 9x^{-3/4} - 45x^{-7/4}$

or $\frac{9}{x^{3/4}} - \frac{45}{x^{7/4}}$

55. $k(x) = \frac{3x}{x + 5}$

$k'(x) = \frac{(x + 5)(3) - (3x)(1)}{(x + 5)^2}$

$= \frac{3x + 15 - 3x}{(x + 5)^2}$

$= \frac{15}{(x + 5)^2}$

57. $y = \frac{\sqrt{x} - 1}{x + 2}$

$y' = \frac{(x + 2)\left(\frac{1}{2}x^{-1/2}\right) - (\sqrt{x} - 1)1}{(x + 2)^2}$

$= \left(\frac{\frac{1}{2}x^{1/2} + x^{-1/2} - \sqrt{x} + 1}{(x + 2)^2}\right) \cdot \left(\frac{2}{2}\right)$

$= \frac{x^{1/2} + 2x^{-1/2} - 2\sqrt{x} + 2}{2(x + 2)^2} \cdot \frac{x^{1/2}}{x^{1/2}}$

$= \frac{x + 2 - 2x + 2x^{1/2}}{2x^{1/2}(x + 2)^2}$

$= \frac{2 - x + 2x^{1/2}}{2x^{1/2}(x + 2)^2}$

59. $y = \frac{x^2 - x + 1}{x - 1}$

$y' = \frac{(x - 1)(2x - 1) - (x^2 - x + 1)(1)}{(x - 1)^2}$

$= \frac{2x^2 - 3x + 1 - x^2 + x - 1}{(x - 1)^2}$

$= \frac{x^2 - 2x}{(x - 1)^2}$

61. $f(x) = (3x - 2)^4$

$f'(x) = 4(3x - 2)^3(3)$

$= 12(3x - 2)^3$

63. $y = \sqrt{2t - 5}$

$= (2t - 5)^{1/2}$

$y' = \frac{1}{2}(2t - 5)^{-1/2}(2)$

$= (2t - 5)^{-1/2}$ or $\frac{1}{(2t - 5)^{1/2}}$

65. $y = 3x(2x + 1)^3$

$y' = 3x(3)(2x + 1)^2(2) + (2x + 1)^3(3)$

$= (18x)(2x + 1)^2 + 3(2x + 1)^3$

$= [18x + 3(2x + 1)](2x + 1)^2$

$= (24x + 3)(2x + 1)^2$

$= 3(8x + 1)(2x + 1)^2$

67. $r(t) = \frac{5t^2 - 7t}{(3t + 1)^3}$

$r'(t) = \frac{(3t + 1)^3(10 - 7)}{(3t + 1)^6}$

$- \frac{(5t^2 - 7t)3(3t + 1)^2(3)}{(3t + 1)^6}$

$= \frac{(3t + 1)^3(10t - 7)}{(3t + 1)^6}$

$- \frac{9(5t^2 - 7t)(3t + 1)^2}{(3t + 1)^6}$

$= \frac{(3t + 1)(10 - 7) - 9(5t^2 - 7t)}{(3t + 1)^4}$

$= \frac{30t^2 - 11t - 7 - 45t^2 + 63t}{(3t + 1)^4}$

$= \frac{-15t + 52t - 7}{(3t + 1)^4}$

69. $y = \frac{x^2 + 3x - 10}{x - 2}$

$y' = \frac{(x - 2)(2x + 3) - (x^2 + 3x - 10)(1)}{(x - 2)^2}$

$= \frac{2x^2 - x - 6 - x^2 - 3x + 10}{(x - 2)^2}$

$= \frac{x^2 - 4x + 4}{(x - 2)^2}$

$= \frac{(x - 2)^2}{(x - 2)^2}$

$= 1$

71. $y = -6e^{2x}$

$y' = -6(2)e^{2x} = -12e^{2x}$

73. $y = e^{-2x^3}$

$y' = -6x^2e^{-2x^3}$

75. $y = 5x \cdot e^{2x}$

$y' = 5x(2)e^{2x} + e^{2x}(5)$ *Product rule*

$= 10xe^{2x} + 5e^{2x}$

$= 5e^{2x}(2x + 1)$

77. $y = \ln(2 + x^2)$

$y' = \frac{2x}{2 + x^2}$

79. $y = \frac{\ln(3x)}{x - 3}$

$y' = \frac{(x - 3)\left(\frac{1}{3x}\right)(3) - (\ln(3x))1}{(x - 3)^2}$

Quotient rule

$= \frac{\frac{x - 3}{x} - \ln(3x)}{(x - 3)^2}$

$= \frac{x - 3 - x\ln(3x)}{x(x - 3)^2}$

81. $D_x\left(\frac{\sqrt{x}+1}{\sqrt{x}-1}\right)$

$= D_x\left(\frac{x^{1/2}+1}{x^{1/2}-1}\right)$

$= \frac{(x^{1/2}-1)\left(\frac{1}{2}x^{-1/2}\right) - (x^{1/2}+1)\left(\frac{1}{2}x^{-1/2}\right)}{(x^{1/2}-1)^2}$

$= \frac{(x^{1/2}-1-x^{1/2}-1)\left(\frac{1}{2}x^{-1/2}\right)}{(x^{1/2}-1)^2}$

$= \frac{-2\left(\frac{1}{2}\right)x^{-1/2}}{(x^{1/2}-1)^2}$

$= \frac{-1}{x^{1/2}(x^{1/2}-1)^2}$

83. $y = \sqrt{t^{1/2}+t} = (t^{1/2}+t)^{1/2}$

$\frac{dy}{dt} = \frac{1}{2}(t^{1/2}+t)^{-1/2}\left(\frac{1}{2}t^{-1/2}+1\right)$

$= \frac{\left(\frac{1}{2}t^{-1/2}+1\right)(2t^{1/2})}{2(t^{1/2}+t)^{1/2}(2t^{1/2})}$

$= \frac{1+2t^{1/2}}{4t^{1/2}(t^{1/2}+t)^{1/2}}$

85. $f(x) = \frac{\sqrt{8+x}}{x+1} = \frac{(8+x)^{1/2}}{x+1}$

$f'(x) = \frac{(x+1)\left(\frac{1}{2}\right)(8+x)^{-1/2}(1)}{(x+1)^2}$

$- \frac{(8+x)^{1/2}(1)}{(x+1)^2}$

$= \frac{\frac{1}{2}(x+1)(8+x)^{-1/2} - (8+x)^{1/2}}{(x+1)^2}$

$f(1) = \frac{\frac{1}{2}(1+1)(8+1)^{-1/2} - (8+1)^{1/2}}{(1+1)^2}$

$= \frac{\frac{1}{2}(2)(9)^{-1/2} - (9)^{1/2}}{2^2}$

$= \frac{\frac{1}{3}-3}{4} = \frac{-\frac{8}{3}}{4} = -\frac{2}{3}$

87. Because the graph shows no breaks in the function for any of the open intervals (-3, 0), (-3, 3), (0, 4), the function is continuous on each of these intervals.

89. $f(x) = \frac{6x+1}{2x-3}$; $x = \frac{3}{2}$, $x = -\frac{1}{6}$, $x = 0$

Since $f(x)$ is a rational function, it will not be continuous for any x for which $2x - 3 = 0$. Then if $x = \frac{3}{2}$, $f(x)$ will not be continuous, but it is continuous at all other points, including $x = -1/6$ and $x = 0$.

91. $f(x) = \frac{-5}{3x(2x-1)}$

$x = -5$, $x = 0$, $x = -1/3$, $x = 1/2$

Since $f(x)$ is a rational function, it will not be continuous for any x for which $3x(2x - 1) = 0$. Set $3x(2x - 1) = 0$ and solve for x. If $x = 0$ or $x = 1/2$, $f(x)$ will not be continuous but it is continuous at all other points, including $x = -5$ and $x = -1/3$.

93. $f(x) = \dfrac{x - 6}{x + 5}$

$x = 6$, $x = -5$, $x = 0$

Since $f(x)$ is a rational function, it will not be continuous for any x for which $x + 5 = 0$. Then, if $x = -5$, $f(x)$ will not be continuous, but it is continuous at all other points, including $x = 6$ and $x = 0$.

95. $f(x) = x^2 + 3x - 4$; $x = 1$, $x = -4$, $x = 0$

Since $f(x)$ is a polynomial function, it is continuous at all points, including $x = 1$, $x = -4$, and $x = 0$.

97. $P(x) = 15x + 25x^2$

$P'(x) = 15 + 50x$ is marginal profit.

(a) $P'(6) = 15 + 50(6) = 315$

(b) $P'(20) = 15 + 50(20) = 1015$

(c) $P'(30) = 15 + 50(30) = 1515$

99. $S(x) = 1000 + 50\sqrt{x} + 10x$

$= 1000 + 50x^{1/2} + 10x$

$\dfrac{dS}{dx} = 50\left(\dfrac{1}{2}\right)x^{-1/2} + 10$

$= 25x^{-1/2} + 10$

(a) $\dfrac{dS}{dx}(9) = \dfrac{25}{\sqrt{9}} + 10$

$= \dfrac{25}{3} + 10$

$= \dfrac{55}{3}$ *or* $18\dfrac{1}{3}$

(Sales will increase by 55 million dollars when 3 thousand more dollars are spent on research)

(b) $\dfrac{dS}{dx}(16) = \dfrac{25}{\sqrt{16}} + 10$

$= \dfrac{25}{4} + 10$

$= \dfrac{65}{4}$ or $16\dfrac{1}{4}$

(Sales will increase by 65 million dollars when 4 thousand more dollars are spent on research.)

(c) $\dfrac{dS}{dx}(25) = \dfrac{25}{\sqrt{25} + 10}$

$= 5 + 10$

$= 15$

(Sales will increase by 15 million dollars when 1 thousand more dollars are spent on research.)

101. $T(x) = \dfrac{1000 + 50x}{x + 1}$

$T'(x) = \dfrac{(x + 1)(50) - (1000 + 50x)(1)}{(x + 1)^2}$

$= \dfrac{-950x}{(x + 1)^2}$

(a) $T'(9) = \dfrac{-950x}{(9 + 1)^2}$

$= \dfrac{-950}{100}$

$= -9.5$

(Cost will decrease by \$9500 for the next \$100 spent on training.)

(b) $T'(19) = \dfrac{-950}{(19 + 1)^2}$

$= \dfrac{-950}{400}$

$= -\dfrac{19}{8} = -2.375$

(Cost will decrease by \$2375 for the next \$100 spent on training.)

CHAPTER 12 APPLICATIONS OF THE DERIVATIVE

Section 12.1

1. The lowest point on the graph has coordinates (1, -4). So the relative minimum of -4 occurs at $x = 1$.

3. The highest point on the graph has coordinates (-2, 3). So a relative maximum of 3 occurs at $x = -2$.

5. A relatively high point on the graph occurs at (-4, 3) and a relatively low point on the graph occurs at (-2, 1). So a relative maximum of 3 occurs at $x = -4$ and a relative minimum of 1 occurs at $x = -2$.

7. A relatively high point on the graph occurs at (-4, 3) and relatively low points occur at (-7, -2) and (-2, -2). So a relative maximum of 3 occurs at $x = -4$ and a relative minimum of -2 occurs at $x = -7$ and $x = -2$.

9. $f(x) = x^2 + 12x - 8$

$f'(x) = 2x + 12 = 0$ when

$2(x + 6) = 0$

$x = -6.$

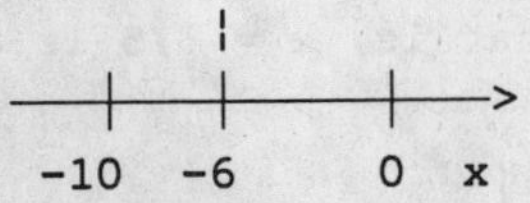

-10 -6 0 x

Test $f'(x)$ at -10 and 0.

$f'(-10) = -8 < 0$

$f'(0) = 12 > 0$

This shows that $f(x)$ is decreasing on $(-\infty, -6)$ and increasing on $(-6, \infty)$. By the first derivative test, $f(x)$ has a relative minimum at 6.

$f(-6) = (-6)^2 + 12(-6) - 8$

$= 36 - 72 - 8$

$= -44$

Relative minimum of -44 at -6

11. $f(x) = x^3 + 6x^2 + 9x - 8$

$f'(x) = 3x^2 + 12x + 9 = 0$ when

$3(x^2 + 4x + 3) = 0$

$3(x + 3)(x + 1) = 0$

$x = -1$ or $x = -3.$

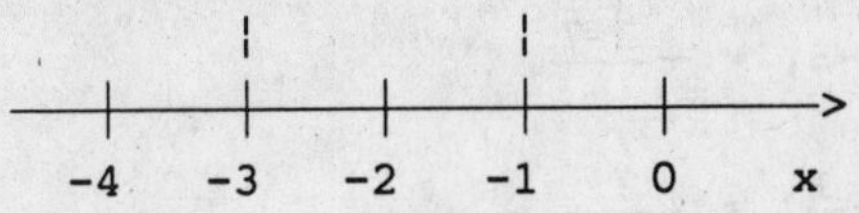

-4 -3 -2 -1 0 x

Test $f'(x)$ at -4, -2, and 0.

$f'(-4) = 3(-4 + 3)(-4 + 1)$

$= 3(-1)(-3) = 9 > 0$

$f'(-2) = 3(-2 + 3)(-2 + 1)$

$= 3(1)(-1) = -3 < 0$

$f'(0) = 3(0 + 3)(0 + 1)$

$= 3(3)(1) = 9 > 0$

By the first derivative test, $f(x)$ has a relative maximum at $x = -3$ and a relative minimum at $x = -1$.
Since $f(-3) = -8$, $f(x)$ has a relative maximum of -8 at -3.
Since $f(-1) = -12$, $f(x)$ has a relative minimum of -12 at -1.

13. $f(x) = -\frac{4}{3}x^3 - \frac{21}{2}x^2 - 5x + 8$

$f'(x) = -4x^2 - 21x - 5 = 0$

$4x^2 + 21x + 5 = 0$

$(4x + 1)(x + 5) = 0$

$x = -5$ or $x = -\frac{1}{4}$

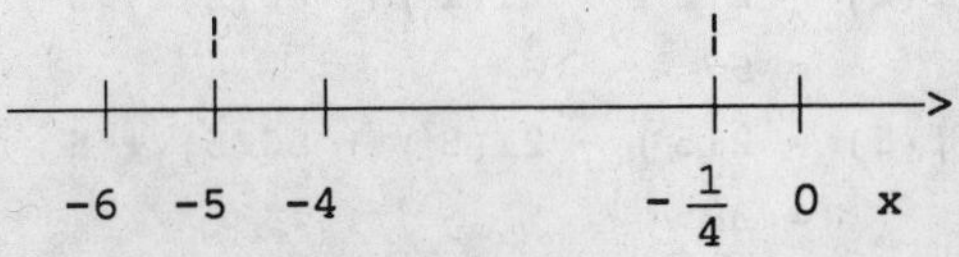

-6 -5 -4 $-\frac{1}{4}$ 0 x

Test $f'(x)$ at -6, -4, and 0.

$f'(-6) = -23 < 0$

$f'(-4) = 15 > 0$

$f'(0) = -5 < 0$

$f'(x)$ changes from negative to positive as x increases through -5. $f'(x)$ changes from positive to negative as x increases through -1/4. By the first derivative test, $f(x)$ has a relative minimum at -5 and a relative maximum at -1/4.

$$f(-5) = -\frac{377}{6}$$

$$f\left(-\frac{1}{4}\right) = \frac{827}{96}$$

Relative maximum of 827/96 at -1/4
Relative minimum of -377/6 at -5

15. $f(x) = 2x^3 - 21x^2 + 60x + 5$

$f'(x) = 6x^2 - 42x + 60 = 0$ when

$$6x^2 - 42x + 60 = 0$$
$$x^2 - 7x + 10 = 0$$
$$(x - 5)(x - 2) = 0$$
$$x = 5 \quad \text{or} \quad x = 2$$

0 2 3 5 6 x

Test $f'(x)$ at 0, 3, and 6.

$f'(0) = 60 > 0$

$f'(3) = -12 < 0$

$f'(6) = 24 > 0$

$f'(x)$ changes from positive to negative as x increases through 2. $f'(x)$ changes from negative to positive as x increases through 5.

$f(2) = 2(2^3) - 21(2^2) + 60(2) + 5$
$= 57$

$f(5) = 2(5^3) - 21(5^2) + 60(5) + 5$
$= 30$

Relative maximum of 57 at 2
Relative minimum of 30 at 5

17. $f(x) = x^4 - 18x^2 - 4$

$f'(x) = 4x^3 - 36x = 0$ when

$$4x^3 - 36x = 0$$
$$4x(x^2 - 9) = 0$$
$$4x(x + 3)(x - 3) = 0$$
$$x = 0 \quad \text{or} \quad x = -3 \quad \text{or} \quad x = 3.$$

-4 -3 -1 0 1 3 4 x

Test $f'(x)$ at -4, -1, 1, and 4.

$f'(-4) = 4(-4)^3 - 36(-4) = -122 < 0$

$f'(-1) = -4 + 36 = 32 > 0$

$f(1) = 4 - 36 = -32 < 0$

$f(4) = 4(4)^3 - 36(4) = 112 > 0$

$f'(x)$ is negative on $(-\infty, -3)$ and $(0, 3)$ and $f'(x)$ is positive on $(-3, 0)$ and $(3, \infty)$.

$f(-3) = (-3)^4 - 18(-3)^2 - 4$
$= -85$

$f(0) = -4$

$f(3) = 3^4 - 18(3^2) - 4$
$= -85$

Relative maximum of -4 at 0
Relative minimum of -85 at 3 and -3

19. $f(x) = -(8 - 5x)^{2/3}$

$$f'(x) = -\frac{2}{3}(8 - 5x)^{-1/3}(-5)$$

$$= \frac{10}{3(8 - 5x)^{1/3}}$$

$f'(x)$ is never zero but fails to exist if $8 - 5x = 0$, that is, $x = 8/5$ is a critical point.

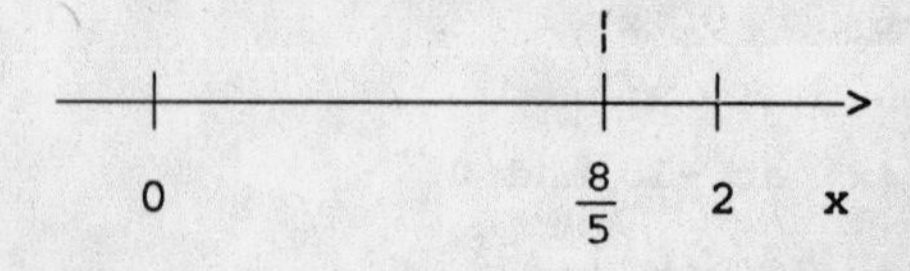

Test $f'(x) = 0$ and $x = 2$.

$$f'(0) = \frac{10}{3(8)^{1/3}} = \frac{5}{3} > 0$$

$$f'(2) = \frac{10}{3(8 - 10)^{1/3}} \approx -2.6 < 0$$

$f'(x)$ is positive on $(-\infty, 8/5)$ and negative on $(8/5, \infty)$.

$$f\left(\frac{8}{5}\right) = -[8 - 5\left(\frac{8}{5}\right)]^{2/3} = 0$$

Since 8/5 is in the domain of f(x), there is a relative maximum of 0 at 8/5.

21. $f(x) = x - \frac{1}{x}$

$f'(x) = 1 + \frac{1}{x^2}$ is never zero, but fails to exist at x = 0.

Since f(x) also fails to exist at x = 0, there are no critical numbers and no relative extrema.

23. $f(x) = \frac{x^2}{x^2 + 1}$

$$f'(x) = \frac{(x^2 + 1)2x - x^2(2x)}{(x^2 + 1)^2}$$

$$= \frac{2x}{(x^2 + 1)^2} = 0 \text{ when}$$

$$x = 0.$$

-1 0 1 x

Test f′(x) at x = -1 and x = 1.

$$f'(-1) = -\frac{1}{2} < 0$$

$$f'(1) = \frac{1}{2} > 0$$

f′(x) is negative on (-∞, 0) and positive on (0, ∞).

$$f(0) = 0$$

Relative minimum of 0 at 0

25. $f(x) = -xe^x$

$$f'(x) = (-1)(e^x) + (-x)(e^x)$$

$$= -e^x - xe^x$$

$$= -e^x(1 + x) = 0 \text{ when}$$

$-e^x = 0$ or $1 + x = 0$

No solution $\quad x = -1.$

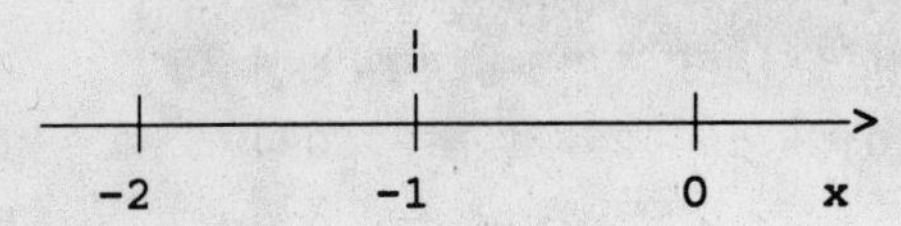

Test f′(x) at x = -2 and x = 0.

$$f'(-2) = -e^{-2}(1 - 2) = -e^{-2}(-1)$$

$$= 135 > 0$$

$$f'(0) = -e^0(1) = -1 < 0$$

f′(x) is positive on (-∞, -1) and negative on (-1, ∞).

$$f(-1) = -(-1)e^{-1} = \frac{1}{e}$$

Relative maximum of 1/e at -1

27. $f(x) = e^x + e^{-x}$

$$f'(x) = e^x - e^{-x}$$

$$= e^x(1 - e^{-2x}) = 0$$

$e^x = 0$ or $1 - e^{-2x} = 0$

No solution $\quad e^{-2x} = 1$

$$-2x = 0$$

$$x = 0$$

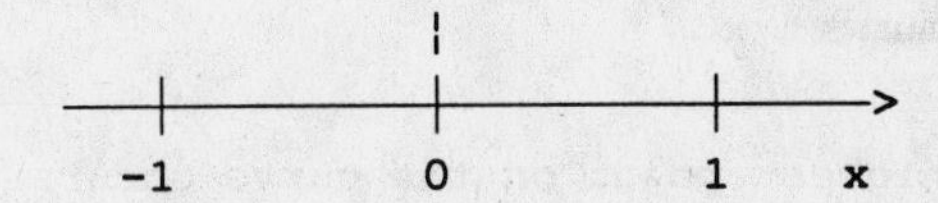

Test f′(x) at x = -1 and x = 1.

$$f'(-1) = e^{-1} - e^1 = \frac{1}{e} - e = -2.35$$

$$f'(1) = e^1 - e^{-1} = e - \frac{1}{e} = 2.35$$

f′(x) is negative on (-∞, 0) and positive on (0, ∞).

$$f(0) = e^0 + e^0 = 2$$

Relative minimum of 2 at 0

29. $f(x) = x \ln x.$

$$f'(x) = (1)\ln x + x\left(\frac{1}{x}\right)$$

$$= \ln x + 1 = 0$$

$$\ln x = -1$$

$$x = e^{-1} = \frac{1}{e} = .3678$$

Test $f'(x)$ at $x = .2$ and $x = 1$

$f'(0) = \ln .2 + 1 = -.6094$

$f'(1) = \ln 1 + 1 = 1$

$f'(x)$ is negative on $(-\infty, \frac{1}{e})$ and positive on $(\frac{1}{e}, \infty)$.

$$f\left(\frac{1}{e}\right) = \frac{1}{e} \ln\left(\frac{1}{e}\right) = \frac{1}{e}(-1) = -\frac{1}{e}$$

Relative minimum of $-1/e$ at $1/e$
By symmetry and absolute value, there is a relative maximum of $1/e$ at $-1/e$.

31. The lowest point on the curve occurs at x_1. $f(x_1)$ is the absolute minimum. The curve does not have a high point so there is no absolute maximum.

33. The highest point on the curve occurs at x_1. $f(x_1)$ is an absolute maximum at x_1. The lowest point on the curve occurs at x_2. $f(x_2)$ is an absolute minimum at x_2.

35. The highest point on the curve occurs at x_2. $f(x_2)$ is an absolute maximum at x_2. The lowest point on the curve occurs at x_3. $f(x_3)$ is an absolute minimum at x_3.

37. $f(x) = 5 - 8x - 4x^2$; $[-5, 1]$

$f'(x) = -8 - 8x = 0$ when

$-8(x + 1) = 0$

$x = -1.$

Test endpoints and critical number -1.

$f(-5) = -55$ Absolute minimum

$f(-1) = 9$ Absolute maximum

$f(1) = -7$

Absolute maximum of 9 at -1
Absolute minimum of -55 at -5

39. $f(x) = x^3 - 3x^2 - 24x + 5$; $[-3, 6]$

$f'(x) = 3x^2 - 6x - 24 = 0$ when

$3(x^2 - 2x - 8) = 0$

$3(x + 2)(x - 4) = 0$

Test endpoints and critical numbers in $f(x)$.

$f(-3) = 23$

$f(-2) = 32$

$f(4) = -75$

$f(6) = -31$

Absolute maximum of 32 at -2
Absolute minimum of -75 at 4

41. $f(x) = \frac{1}{3}x^3 - \frac{1}{2}x^2 - 6x + 3$; $[-4, 4]$

$f'(x) = x^2 - x - 6 = 0$ when

$(x + 2)(x - 3) = 0$

$x = -2$ or $x = 3.$

$f(-4) = -\frac{7}{3}$

$f(-2) = \frac{31}{3}$

$f(3) = -\frac{21}{2}$

$f(4) = -\frac{23}{3}$

Absolute maximum of $31/3$ at -2
Absolute minimum of $-21/2$ at 3

43. $f(x) = x^4 - 32x^2 - 7$; $[-5, 6]$

$f'(x) = 4x^3 - 64x = 0$ when

$4x(x^2 - 16) = 0$

$4x(x - 4)(x + 4) = 0$

$x = 0$ or $x = 4$ or $x = -4.$

$f(-5) = -182$

$f(-4) = -263$

$f(0) = -7$

$f(4) = -263$

$f(6) = 137$

Absolute maximum of 137 at 6
Absolute minimum of -263 at -4 and 4

45. $f(x) = \frac{8 + x}{8 - x}$; [4, 6]

$$f'(x) = \frac{(8 - x)(1) - (8 + x)(-1)}{(8 - x)^2}$$

$$= \frac{16}{(8 - x)^2}$$

$f'(x)$ is never zero, but fails to exist if $x = 8$. However, 8 is not in the given interval so it is of no interest. There are no critical points in the interval.

$f(4) = 3$

$f(6) = 7$

Absolute maximum of 7 at 6

Absolute minimum of 3 at 4

47. $f(x) = \frac{x}{x^2 + 2}$; [0, 4]

($f'(x)$ is defined for all x.)

$$f'(x) = \frac{(x^2 + 2)1 - x(2x)}{(x^2 + 2)^2}$$

$$= \frac{-x^2 + 2}{(x^2 + 2)^2} = 0$$

$$-x^2 + 2 = 0$$

$$x^2 = 2$$

$x = \sqrt{2}$ or $x = -\sqrt{2}$ but $-\sqrt{2}$ is not in [0, 4].

$f(0) = 0$

$f(\sqrt{2}) = \frac{\sqrt{2}}{4}$

$f(4) = \frac{4}{18} = \frac{2}{9}$

$\sqrt{2} > 1$ so $\frac{\sqrt{2}}{4} > \frac{1}{4} = \frac{9}{36}$ and

$\frac{2}{9} = \frac{8}{36}$, so $\frac{\sqrt{2}}{4} > \frac{2}{9}$.

Absolute maximum of $\sqrt{2}/4$ at $\sqrt{2}$

Absolute minimum of 0 at 0

49. $f(x) = (x^2 + 18)^{2/3}$; [-3,3]

$$f'(x) = \frac{2}{3}(x^2 + 18)^{-1/3}(2x)$$

$$= \frac{4x}{(x^2 + 18)^{1/3}}$$

This derivative always exists, and is 0 when

$$\frac{4x}{(x^2 + 18)^{1/3}} = 0$$

$$4x = 0$$

$$x = 0.$$

$f(0) = 18^{2/3} \approx 6.87$

$f(-3) = (9 + 18)^{2/3} = 9$

$f(3) = 9$

Absolute maximum of 9 at -3 and 3

Absolute minimum of about 6.87 at 0

51. $f(x) = \frac{1}{\sqrt{x^2 + 1}}$; [-1, 1]

$$f'(x) = \frac{(x + 1)(0) - 1\left(\frac{1}{2}\right)(x^2 + 1)^{-1/2}(2x)}{x^2 + 1}$$

$$= \frac{-x(x^2 + 1)^{-1/2}}{x^2 + 1}$$

$$= -x(x^2 + 1)^{-1/2}$$

$$= \frac{-x}{\sqrt{x^2 + 1}} = 0 \text{ when}$$

$$x = 0.$$

$f'(x)$ is defined for all x.

$f(-1) = \frac{1}{\sqrt{2}}$

$f(0) = 1$

$f(1) = \frac{1}{\sqrt{2}}$

Absolute maximum of 1 at 0

Absolute minimum of $1/\sqrt{2}$ at -1 and 1

53. $P(x) = 80 + 108x - x^3$

(a) $P'(x) = 108 - 3x^2 = 0$

$108 = 3x^2$

$36 = x^2$

$6 = x$

Use test points $x = 5$ and $x = 7$.

$P'(5) = 33 \quad P'(7) = -39$

$P'(x)$ is positive on $(-\infty, 6)$ and negative on $(6, \infty)$. So $x = 6$ is the location of a relative maximum. Thus, an expenditure of $6 leads to a maximum profit.

(b) $P(6) = 80 + 108(6) - (6)^3$

$= 512$

The maximum profit is $512.

55. $C(x) = x^2 + 200x + 100$

Since x is given in hundreds, the domain is [1, 12].

Average cost per unit is

$$\overline{C}(x) = \frac{C(x)}{x} = \frac{x^2 + 200x + 100}{x}$$

$$= x + 200 + \frac{100}{x}.$$

$$\overline{C}'(x) = 1 - \frac{100}{x^2} = 0 \text{ when}$$

$$1 = \frac{100}{x^2}$$

$$x^2 = 100$$

$$x = 10.$$

Test for minimum.

$$\overline{C}'(5) = 1 - \frac{100}{25} = -3 < 0$$

$$\overline{C}'(15) = 1 - \frac{100}{225} = 1 - \frac{4}{9} = \frac{5}{9} > 0$$

The average cost per unit is as small as possible when $x = 10$, or when 1000 manuals are produced. If the student produces 1000 manuals, his cost will be

$C(10) = 10^2 + 200(10) + 100 = \$2200.$

Thus, in order to make a profit, for each copy the student will have to charge at least

$$\frac{2200}{1000} = 2.2,$$

or more than $2.20.

57. $f(x) = \dfrac{x^2 + 36}{2x}$, $1 \le x \le 12$

$$f'(x) = \frac{2x(2x) - (x^2 + 36)2}{4x^2}$$

$$= \frac{4x^2 - 2x^2 - 72}{4x^2}$$

$$= \frac{2x^2 - 72}{4x^2}$$

$$= \frac{2(x^2 - 36)}{4x^2}$$

$$= \frac{(x + 6)(x - 6)}{2x^2}$$

$f'(x) = 0$ when $x = 6$ and $x = -6$.

Only 6 is in the interval $1 \le x \le -12$.

Test for maximum or minimum.

$$f'(5) = \frac{(11)(-1)}{50} < 0$$

$$f'(7) = \frac{(13)(1)}{98} > 0$$

So the minimum occurs at $x = 6$, when

$$f(x) = \frac{6^2 + 36}{12}$$

$$= 6.$$

The selenium is reduced to a minimum at 6 months when 6% of the soil is selenium.

59. $M(x) = -\dfrac{1}{45}x^2 + 2x - 20,$

$30 \le x \le 65$

$$M'(x) = -\frac{1}{45}(2x) + 2$$

$$= -\frac{2x}{45} + 2$$

When $M'(x) = 0$,

$$-\frac{2x}{45} + 2 = 0$$

$$2 = \frac{2x}{45}$$

$$45 = x.$$

$$M(30) = 20$$

$$M(45) = 25$$

$$M(65) \approx 16.1$$

Absolute maximum is 25 miles per gallon and absolute minimum is about 16.1 per gallon.

61. $B(x) = x^3 - 7x^2 - 160x + 1800$

(a) $B'(x) = 3x^2 - 14x - 160$

When $B'(x) = 0$

$$3x^2 - 14x - 160 = 0$$

$$(3x + 16)(x - 10) = 0$$

$3x + 16 = 0$ or $x - 10 = 0$

$x = -16/3$ or $x = 10$

Since x must be positive, 10 is the only critical number of interest. Test 0 and 250.

$B'(0) = -160$ and $B'(20) = 760$

So 10 units of oxygen produces the minimum microbe concentration.

(b) $B(10) = (10)^3 - 7(10)^2 - 160(10) + 1800$

$B(10) = 500$

So the minimum microbe concentration is 500 units.

63. $f(x) = x^3 - 9x^2 + 18x - 7$

Interval: [0, 1.5]

x	f(x)
0	-7
.1	-5.289
.2	-3.752
.3	-2.383
.4	-1.176
.5	-.0125
.6	.776
.7	1.533
.8	2.152
.9	2.639
1.0	3.000
1.1	3.241
1.2	3.368
1.3	3.387
1.4	3.304
1.5	3.125

Absolute maximum of about 3.4 at about 1.3

Absolute minimum of -7 at 0

65. $f(x) = x^3 + 6x^2 - 6x + 3$

Interval: [0, 1]

x	f(x)
0	3
.1	2.461
.2	2.048
.3	1.767
.4	1.624
.5	1.625
.6	1.776
.7	2.083
.8	2.552
.9	3.189
1.0	4

Absolute maximum on [0, 1] is 4 at 1
Absolute minimum on [0, 1] is about 1.6 at about .4 or .5

Interval: [-5, -4]

x	f(x)
-5	58
-4.9	58.811
-4.8	59.448
-4.7	59.917
-4.6	60.224
-4.5	60.375
-4.4	60.376
-4.3	60.223
-4.2	59.952
-4.1	59.539
-4	59

Absolute maximum on [-5, -4] is about 60.4 at -4.5 or -4.4
Absolute minimum on [-5, -4] is about 58 at -5

Section 12.2

1. $f(x) = 3x^3 - 4x + 5$

$f'(x) = 9x^2 - 4$

$f''(x) = 18x$

Replace x, in turn, with 0, 2, and -3 in 18x.

$f''(0) = 0$

$f''(2) = 36$

$f''(-3) = -54$

3. $f(x) = 3x^4 - 5x^3 + 2x^2$

$f'(x) = 12x^3 - 15x^2 + 4x$

$f''(x) = 36x^2 - 30x + 4$

Using a calculator, replace x, in turn, with 0, 2, -3 in $f''(x)$.

$f''(0) = 4$

$f''(2) = 88$

$f''(-3) = 418$

5. $f(x) = (x + 4)^3$

$f'(x) = 3(x + 4)^2(1)$

$= 3(x + 4)^2$

$f''(x) = 6(x + 4)(1)$

$= 6(x + 4)$

$f''(0) = 24$

$f''(2) = 36$

$f''(-3) = 6$

7. $f(x) = \frac{2x + 1}{x - 2}$

$$f'(x) = \frac{(x - 2)(2) - (2x + 1)(1)}{(x - 2)^2}$$

$$= \frac{-5}{(x - 2)^2}$$

$$f''(x) = \frac{(x - 2)^2(0) - (-5)(2)(x - 2)(1)}{(x - 2)^4}$$

$$= \frac{10}{(x - 2)^3}$$

$f''(0) = -\frac{5}{4}$

$f''(2)$ does not exist.

$f''(-3) = -\frac{2}{25}$

9. $f(x) = \frac{x^2}{1 + x}$

$f'(x) = \frac{(1 + x)(2x) - x^2(1)}{(1 + x)^2}$

$= \frac{2x + x^2}{(1 + x)^2}$

$f''(x)$

$= \frac{(1 + x)^2(2 + 2x) - (2x + x^2)(2)(1 + x)}{(1 + x)^4}$

$= \frac{(1 + x)(2 + 2x) - (2x + x^2)(2)}{(1 + x)^3}$

$= \frac{2}{(1 + x)^3}$

$f''(0) = 2$

$f''(2) = \frac{2}{27}$

$f''(-3) = -\frac{1}{4}$

11. $f(x) = \sqrt{x + 4} = (x + 4)^{1/2}$

$f'(x) = \frac{1}{2}(x + 4)^{-1/2}$

$f''(x) = \frac{1}{2}(-\frac{1}{2})(x + 4)^{-3/2}$

$= \frac{-1}{4}(x + 4)^{-3/2}$

$= \frac{-1}{4(x + 4)^{-3/2}}$

Replacing x, in turn, with 0, 2, and -3 gives

$f''(0) = \frac{-1}{4(0 + 4)^{3/2}} = \frac{-1}{4(4)^{3/2}}$

$= \frac{-1}{4(8)}$

$= -\frac{1}{32}$

$f''(2) = \frac{-1}{4(2 + 4)^{3/2}}$

$= \frac{-1}{4(6)^{3/2}} \approx -.0170$

$f''(-3) = \frac{-1}{4(-3 + 4)^{3/2}} = \frac{-1}{4(1)^{3/2}}$

$= -\frac{1}{4}.$

13. $f(x) = 5x^{3/5}$

$f'(x) = 3x^{-2/5}$

$f''(x) = -\frac{6}{5}x^{-7/5}$ or $\frac{-6}{5x^{7/5}}$

$f''(0)$ does not exist.

$f''(2) = -\frac{6}{5}(2^{-7/5}) = \frac{-6}{5(2^{7/5})}$

$\approx -.4547$

$f''(-3) = -\frac{6}{5}(-3)^{-7/5} = \frac{-6}{5(-3)^{7/5}}$

$\approx .2578$

15. $f(x) = 2e^x$

$f'(x) = 2e^x$

$f''(x) = 2e^x$

$f''(0) = 2e^0 = 2(1) = 2$

$f''(2) = 2e^2$

$f''(-3) = 2e^{-3} = \frac{2}{e^3}$

17. $f(x) = -x^4 + 2x^2 + 8$

$f'(x) = -4x^3 + 4x$

$f''(x) = -12x^2 + 4$

$f'''(x) = -24x$

$f^{(4)}(x) = -24$

19.
$$f(x) = 4x^5 + 6x^4 - x^2 + 2$$
$$f'(x) = 20x^4 + 24x^3 - 2x$$
$$f''(x) = 80x^3 + 72x^2 - 2$$
$$f'''(x) = 240x^2 + 144x$$
$$f^{(4)}(x) = 480x + 144$$

21.
$$f(x) = \frac{x-1}{x+2}$$
$$f'(x) = \frac{(x+2)-(x-1)}{(x+2)^2}$$
$$= \frac{3}{(x+2)^2}$$
$$f''(x) = \frac{-3(2)(x+2)}{(x+2)^4}$$
$$= \frac{-6}{(x+2)^3}$$
$$f'''(x) = \frac{(-6)(-3)(x+2)^2}{(x+2)^6}$$
$$= \frac{18}{(x+2)^4} \text{ or } 18(x+2)^{-4}$$
$$f^{(4)}(x) = \frac{-18(4)(x+2)^3}{(x+2)^8}$$
$$= \frac{-72}{(x+2)^5}$$
$$\text{or } -72(x+2)^{-5}$$

23.
$$f(x) = 5e^{2x}$$
$$f'(x) = 10e^{2x}$$
$$f''(x) = 20e^{2x}$$
$$f'''(x) = 40e^{2x}$$
$$f^{(4)}(x) = 80e^{2x}$$

25.
$$f(x) = \ln|x|$$
$$f'(x) = \frac{1}{x} = x^{-1}$$
$$f''(x) = -x^{-2}$$
$$f'''(x) = 2x^{-3} = \frac{2}{x^3}$$
$$f^{(4)}(x) = -6x^{-4} = \frac{-6}{x^4}$$

27.
$$f(x) = x \ln|x|$$
$$f'(x) = (1)\ln|x| + x\left(\frac{1}{x}\right)$$
$$= \ln|x| + 1$$
$$f''(x) = \frac{1}{x} = x^{-1}$$
$$f'''(x) = -x^{-2} = \frac{-1}{x^2}$$
$$f^{(4)}(x) = 2x^{-3} = \frac{2}{x^3}$$

29.
$$f(x) = x^2 + 6x - 3$$
$$f'(x) = 2x + 6$$

Solve $f'(x) = 0$.
$$2x + 6 = 0$$
$$2x = -6$$
$$x = -3$$

Use the second derivative test.
$$f''(x) = 2$$

$f''(x) > 0$ for all x.

So $x = -3$ gives a relative minimum.

31.
$$f(x) = 12 - 8x + 4x^2$$
$$f'(x) = -8 + 8x$$

Solve $f'(x) = 0$.
$$-8 + 8x = 0$$
$$8x = 8$$
$$x = 1$$

Use the second derivative test.
$$f''(x) = 8$$

$f''(x) > 0$ for all x.

So $x = 1$ gives a relative minimum.

33.
$$f(x) = x^3 + 2x^2 - 5$$
$$f'(x) = 3x^2 + 4x$$

Solve $f'(x) = 0$.
$$3x^2 + 4x = 0$$
$$x(3x + 4) = 0$$
$$x = 0 \quad \text{or} \quad 3x + 4 = 0$$
$$x = -4/3$$

Use the second derivative test.

$f''(x) = 6x + 4$

$f''(0) = 4 > 0$

$f''(-4/3) = 6(-4/3) + 4 = -4 < 0$

So $x = 0$ gives a relative minimum, and $x = -4/3$ gives a relative maximum.

35. $f(x) = -4x^3 - 15x^2 + 18x + 5$

$f'(x) = -12x^2 - 30x + 18$

Solve $f'(x) = 0$

$-12x^2 - 30x + 18 = 0$

$x^2 + 5x - 3 = 0$

$(2x - 1)(x + 3) = 0$

$2x - 1 = 0$ or $x + 3 = 0$

$x = 1/2$ or $x = -3$

Use the second derivative test.

$f''(x) = -24x - 30$

$f''(1/2) = -24(1/2) - 30 = -12 - 30$

$= -42 < 0$

$f''(-3) = -24(-3) - 30 = 72 - 30$

$= 42 > 0$

So $x = 1/2$ gives a relative maximum, and $x = -3$ gives a relative minimum.

37. $f(x) = \frac{2}{3}x^3 + \frac{1}{2}x^2 - x - \frac{1}{4}$

$f''(x) = 2x^2 + x - 1$

Solve $f''(x) = 0$.

$2x^2 + x - 1 = 0$

$(2x - 1)(x + 1) = 0$

$2x - 1 = 0$ or $x + 1 = 0$

$x = \frac{1}{2}$ or $x = -1$

Use the second derivative test.

$f''(x) = 4x + 1$

$f''(1/2) = 4(1/2) + 1 = 3 > 0$

$f''(-1) = 4(-1) + 1 = -4 + 1 = -3 < 0$

So $x = 1/2$ gives a relative minimum and $x = -1$ gives a relative maximum.

39. $f(x) = -2x^3 + 1$

$f'(x) = -6x^2$

Solve $f'(x) = 0$

$-6x^2 = 0$

$x = 0.$

Use the second derivative test.

$f''(x) = -12x$ *Test*

$f''(x) = 0$ *fails*

Use the first derivative test.

$f'(-1) = -6(-1)^2 = -6 < 0$

$f'(1) = -6(1)^2 = -6 < 0$

$f'(x)$ is negative on $(-\infty, 0)$ and on $(0, \infty)$.

So $x = 0$ is a critical value but there is neither a relative maximum or minimum.

41. $f(x) = x^4 - 8x^2$

$f'(x) = 4x^3 - 16x$

Solve $f'(x) = 0$.

$4x^3 - 16x = 0$

$4x(x^2 - 4) = 0$

$4x = 0$ or $x^2 - 4 = 0$

$x^2 = 4$

$x = 0$ or $x = 2$ or $x = -2$

Use the second derivative test.

$f''(x) = 12x^2 - 16$

$f''(0) = -16$

$f''(2) = 48 - 16 = 32$

$f''(-2) = 48 - 16 = 32$

So $x = 0$ gives a relative maximum and $x = 2$ and $x = -2$ give a relative minimum.

43. $f(x) = x + \frac{3}{x}$

$= x + 3x^{-1}$

$f'(x) = 1 - 3x^{-2}$

Solve $f'(x) = 0$.

$$1 - 3x^{-2} = 0$$

$$1 - \frac{3}{x^2} = 0$$

$$1 = \frac{3}{x^2}$$

$$x^2 = 3$$

$$x = \pm\sqrt{3}$$

Use the second derivative test.

$$f''(x) = 6x^{-3} = \frac{6}{x^3}$$

$$f''(\sqrt{3}) = \frac{6}{(\sqrt{3})^3} = \frac{6}{3\sqrt{3}} = \frac{2}{\sqrt{3}} > 0$$

$$f''(-\sqrt{3}) = \frac{6}{(-\sqrt{3})^3} = \frac{6}{-3\sqrt{3}} = \frac{-2}{\sqrt{3}} < 0$$

So $x = \sqrt{3}$ gives a relative minimum, and $x = -\sqrt{3}$ gives a relative maximum.

45. $$f(x) = \frac{x^2 + 9}{2x}$$

$$f'(x) = \frac{(2x)(2x) - (2)(x^2 + 9)}{(2x)^2}$$

$$= \frac{4x^2 - 2x^2 - 18}{4x^2}$$

$$= \frac{2x^2 - 18}{4x^2}$$

Solve $f'(x) = 0$.

$$\frac{2x^2 - 18}{4x^2} = 0$$

$$2x^2 - 18 = 0$$

$$x^2 = 9$$

$$x = 3 \quad \text{or} \quad x = -3$$

Use the second derivative test.

$$f''(x) = \frac{(4x)(4x^2) - (8x)(2x^2 - 18)}{(4x^2)^2}$$

$$= \frac{144x}{16x^4} = \frac{9}{x^3}$$

$$f''(3) = \frac{9}{(3)^3} = \frac{9}{27} = \frac{1}{3} > 0$$

$$f''(-3) = \frac{9}{(-3)^3} = -\frac{9}{27} = -\frac{1}{3}$$

So $x = 3$ gives a relative minimum and $x = -3$ gives a relative maximum.

47. $$f(x) = \frac{2 - x}{2 + x}$$

$$f'(x) = \frac{(-1)(2 + x) - (1)(2 - x)}{(2 + x)^2}$$

$$= \frac{-2 - x - 2 + x}{(2 + x)^2}$$

$$= \frac{-4}{(2 + x)^2}$$

$f'(x)$ does not equal 0 for any value. $f'(x)$ does not exist for $x = -2$ but $f(x)$ does not exist for $x = -2$. So there are no critical values and, thus, no relative maxima or minima.

49. $s(t) = 8t^2 + 4t$

$v(t) = s'(t) = 16t + 4$

$a(t) = s''(t) = 16$

$v(0) = 4$ cm/sec

$v(4) = 68$ cm/sec

$a(0) = 16$ cm/sec^2

$a(4) = 16$ cm/sec^2

51. $s(t) = -5t^3 - 8t^2 + 6t - 3$

$v(t) = s'(t) = -15t^2 - 16t + 6$

$a(t) = s''(t) = -30t - 16$

$v(0) = 6$ cm/sec

$v(4) = -298$ cm/sec

$a(0) = -16$ cm/sec^2

$a(4) = -136$ cm/sec^2

53. $$s(t) = \frac{-2}{3t + 4}$$

$$v(t) = s'(t) = \frac{-(-2)(3)}{(3t + 4)^2}$$

$$= \frac{6}{(3t + 4)^2} \quad \text{or} \quad 6(3t + 4)^{-2}$$

$$a(t) = s''(t) = \frac{-6(2)(3t + 4)^3}{(3t + 4)^4}$$

$$= \frac{-36}{(3t + 4)^3} \text{ or } -36(3t + 4)^{-3}$$

$$v(0) = \frac{3}{8} \text{ cm/sec}$$

$$v(4) = \frac{3}{128} \text{ cm/sec}$$

$$a(0) = -\frac{9}{16} \text{ cm/sec}^2$$

$$a(4) = -\frac{9}{1024} \text{ cm/sec}^2$$

55. $s(t) = -16t^2$

$v(t) = s'(t) = -32(t)$

(a) $s'(3) = -32(3) = -96$ ft/sec

(b) $s'(5) = -32(5) = -160$ ft/sec

(c) $s'(8) = -32(8) = -256$ ft/sec

(d) $a(t) = s''(t) = -32$ ft/sec^2

57. $p(x) = 1000 - 12x + 10x^2 - x^3$

$p'(x) = -12 + 20x - 3x^2$

Set $p'(x) = 0$.

$-3x^2 + 20x - 12 = 0$

$3x^2 - 20x + 12 = 0$

$(3x - 2)(x - 6) = 0$

$3x - 2 = 0$ or $x - 6 = 0$

$x = \frac{2}{3}$ or $x = 6$

Use the second derivative test.

$p'(x) = 20 - 6x$

$p'(\frac{2}{3}) = 20 - 6(\frac{2}{3}) = 20 - 4 = 16$

$p''(6) = 20 - 6(6) = 20 - 36 = -16$

So $x = 6$ is the number of units that leads to a maximum profit.

$p(6) = 1000 - 12(6) + 10(6)^2 - (6)^3$

$= 1072$

The maximum profit is \$1072.

59. $M(x) = 50 - 32x + 14x^2 - x^3$

$M'(x) = -32 + 28x - 3x^2$

Solve $M'(x) = 0$.

$-3x^2 + 28x - 32 = 0$

$3x^2 - 28x + 32 = 0$

$(3x - 4)(x - 8) = 0$

$3x - 4 = 0$ or $x - 8 = 0$

$x = \frac{4}{3}$ or $x = 8$

Use the second derivative test.

$M''(x) = 28 - 6x$

$M'''\left(\frac{4}{3}\right) = 28 - 6\left(\frac{4}{3}\right) = 28 - 8 = 20$

$M'''(8) = 28 - 6(8) = 28 - 48 = -20$

So $x = 8$ inches produces the maximum value of $M(x)$.

61. $K(x) = \dfrac{3x}{x^2 + 4}$

(a) $K'(x)$

$$= \frac{3(x^2 + 4) - (2x)(3x)}{(x^2 + 4)^2}$$

$$= \frac{-3x^2 + 12}{(x^2 + 4)^2} = 0$$

$-3x^2 + 12 = 0$

$x^2 = 4$

$x = 2$ or $x = -2$

Discard -2 since $x \geq 0$

2 is a critical number.

Test 1 and 3 in $K'(x)$:

$K'(1) = \frac{9}{25}$

$K'(3) = \frac{-15}{(13)^2}$

$K'(x)$ changes from positive to negative as x increases through 2. So a relative maximum occurs at $x = 2$ hours.

(b) $K(2) = \frac{3(2)}{(2)^2 + 4} = \frac{6}{8} = \frac{3}{4}\%$

or .0075%

63. (a) $R(t) = t^2(t - 18) + 96t + 1000,$

$0 < t < 8$

$= t^3 - 18t + 96t + 1000$

$R'(t) = 3t^2 - 36t + 96$

Set $R'(t) = 0$.

$3t^2 - 36t + 96 = 0$

$t^2 - 12t + 32 = 0$

$(t - 8)(t - 4) = 0$

$t = 8$ or $t = 4$

8 is not in the domain of R(t).

$R'' = 6t - 36$

$R''(4) = -12 < 0$ implies that R(t) is maximized at t = 4 hours.

(b) $R(4) = 16(-14) + 96(4) + 1000$

$= -224 + 384 + 1000$

$= 1160$

The maximum population is 1160 million.

Section 12.3

1. $x + y = 100$, $P = xy$

(a) $y = 100 - x$

(b) $P = xy = x(100 - x)$

(c) $P' = 100 - 2x$

$100 - 2x = 0$

$2(50 - x) = 0$

$x = 50$

(d) If $x = 50$, $y = 100 - 50 = 50$.

(e) $P''(x) = -2 > 0$ for all x, so P has a maximum when x = 50.

$P(50) = 50(50) = 2500$.

3. Let x and y be the numbers.

$x + y = 200$

$y = 200 - x$

The function to be maintained is

$f(x) = x^2 + y^2$.

$f(x) = x^2 + (200 - x)^2$

$= 40,000 - 400x + 2x^2$

$f'(x) = -400 + 4x$

$-400 + 4x = 0$

$x = 100$

$f''(x) = 4 > 0$ for all x.

If $x = 100$, $y = 200 - 100 = 100$.

f(x) is minimum at $x = 100$, $y = 100$.

5. $x + y = 150$, $x \geq 0$, $y \geq 0$, and x^2y is maximized.

$y = 150 - x$

$f(x) = x^2(150 - x)$

$= 150x^2 - x^3$

$f'(x) = 300x - 3x^2$

$= 3x(100 - x) = 0$ when

$x = 0$ or $x = 100$.

$f''(x) = 300 - 6x$

$f''(0) = 300 > 0$ *Minimum*

$f''(100) = 300 - 600$

$= -300 < 0$ *Maximum*

x^2y is maximized when

$x = 100$, $y = 150 - x = 50$.

7. $x - y = 10$, $x \geq 0$, $y \geq 0$, and xy is minimized.

$x - 10 = y$

$f(x) = x(x - 10)$

$= x^2 - 10x$

$f'(x) = 2x - 10$

$2x - 10 = 0$

$x = 5$

$f''(x) = 2 > 0$ for all x.

f is a minimum for x = 5.

If $x = 5$, then $y = x - 10 = 5 - 10 = -5$.

But since $y \geq 0$, the minimum value for $f(x)$ occurs if $y = 0$, $x = y + 10 = 0 + 10 = 0$. So $x = 10$ and $y = 0$ minimize xy.

9. **(a)** Price in cents for x thousand bars is

$$P(x) = 100 - \frac{x}{10}.$$

Revenue (in cents) is

$$100x \cdot p(x) = 1000x\left(100 - \frac{x}{10}\right).$$

$$R(x) = 100{,}000x - 100x^2$$

(b) $R'(x) = 100{,}000 - 200x = 0$

$$x = 500$$

$R''(x) = -200 < 0$ for all x.

The maximum revenue is attained when $x = 500$.

(c) $R(500) = 100{,}000(500) - 100(500)^2$

$$= 25{,}000{,}000$$

The maximum revenue is 25,000,000 cents (or \$250,000.00).

11. Let x = speed in miles per hour
G = gallons burned per mile.

$$G(x) = \frac{1}{32}\left(\frac{64}{x} + \frac{x}{50}\right)$$

The cost of fuel on the 400-mile trip is \$1.60 per gallon. The total cost is

$$C(x) = 1.60\left(\frac{1}{32}\right)\left(\frac{64}{x} + \frac{x}{50}\right)(400)$$

$$= 20\left(\frac{64}{x} + \frac{x}{50}\right)$$

$$= \frac{1280}{x} + \frac{2}{5}x \text{ dollars.}$$

(a) $C'(x) = \dfrac{-1280}{x^2} + \dfrac{2}{5} = 0$

$$x = \sqrt{3200}$$

$$\approx 56.6 \text{ miles per hour}$$

$C''(x) = \dfrac{-1280(-2)}{x^3} > 0$ for all x.

(Recall x must be positive.)

$C(x)$ is minimum at

$$x = \sqrt{3200}$$

$$\approx 56.6 \text{ miles per hour.}$$

(b) $C(10\sqrt{32}) = \dfrac{1280}{10\sqrt{32}} + \dfrac{2}{5}(10\sqrt{32})$

$$= \frac{128}{\sqrt{32}} + 4\sqrt{32}$$

$$\approx 22.61 + 22.64$$

$$= 45.25$$

The minimum total cost is \$45.25.

13. **(a)** The length of the field in meters is

$$1200 - 2x.$$

(b) $A(x) = x(1200 - 2x)$

$$= 1200x - 2x^2$$

(c) $A'(x) = 1200 - 4x$

$$= 4(300 - x) = 0$$

$$x = 300$$

$A''(x) = -4 < 0$ for all x.

The maximum area is at $x = 300$ meters.

(d) $A(300) = 1200(300 - 2(300)^2$

$$= 360{,}000 - 180{,}000$$

$$= 180{,}000$$

The maximum area is 180,000 square meters.

15. Let x = width of rectangle and y = total length of rectangle.

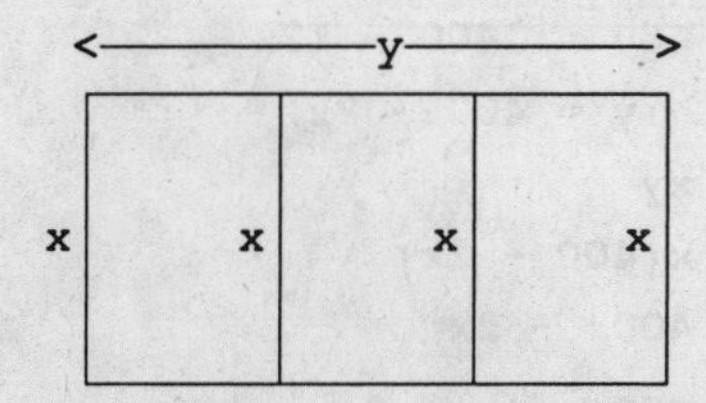

There are 3600 meters of fencing so equation for fencing is

$$3600 = 4x + 2y$$
$$2y = 3600 - 4x$$
$$y = 1800 - 2x$$
$$A = xy$$
$$A(x) = x(1800 - 2x)$$
$$= 1800x - 2x^2$$
$$A'(x) = 1800 - 4x = 0$$
$$x = \frac{1800}{4} = 450$$

$A''(x) = -4 < 0$ for all x.

Area is maximum for x = 450 meters.

$$A(450) = (45)(1800 - 2(450))$$
$$= 450(900)$$
$$= 405,000$$

The maximum area is 405,000 square meters.

17. Let x = length of sides that cost \$6 per foot

y = length of sides that cost \$3 per foot.

An equation for the cost of the fencing is

$$2(y) + 2(6x) = 2400$$
$$6y = 2400 - 12x$$
$$y = 400 - 2x.$$
$$a = xy$$
$$A(x) = x(400 - 2x)$$
$$= 400x - 2x^2$$
$$A'(x) = 400 - 4x = 0$$
$$x = 100 \text{ ft}$$

$A''(x) = -4 < 0$ for all x.

A(x) is maximum when x = 100.

If x = 100, y = 400 - 2(100) = 200.

$$A(100) = 100(400 - 2(100))$$
$$= 100(200) = 20,000$$

The maximum area is 20,000 square feet with 200 feet on the \$3 sides and 100 feet on the \$6 sides.

19. **(a)** Income per pound (in cents):

$$40 - 2x$$

(b) Yield in pounds per tree:

$$100 + 5x$$

(c) Revenue per tree (in cents):

$$R(x) = (100 + 5x)(40 - 2x)$$
$$R(x) = 4000 - 10x^2$$

(d) $R'(x) = -20x$. Since x must be positive, $R'(x)$ is never zero, and the rate of change of revenue is always decreasing.

Pick the peaches now before the revenue falls any more, that is, when x = 0.

(e)

$$R(0) = 4000 - 10(0)^2$$
$$= 4000$$

or the maximum revenue is 4000 cents per tree or \$40 per tree.

21. Profit is 5 dollars per seat for $60 \leq x \leq 80$ seats.

Profit (in dollars) is

$$5 - .05(x - 80)$$

per seat for x > 80 seats.

The number of seats which makes the total profit a maximum will be greater than 80 because after 80 the profit is still increasing, though at a slower rate. (Thus, the function is concave down and its one extrema will be a maximum.)

(a) The total profit for x seats is

$$P(x) = [5 - .05(x - 80)]x$$
$$= [5 - .05x + 4]x$$
$$= [9 - .05x]x$$
$$= 9x - .05x^2.$$
$$P'(x) = 9 - .10x = 0$$
$$9 = .10x$$
$$x = 90$$

(b) $$P(90) = 9(90) - .05\ (90)^2$$
$$= 810 - .05(8100)^2$$
$$= 405$$

The maximum profit is \$405.

23. Let x = a side of the base
h = height of the box.

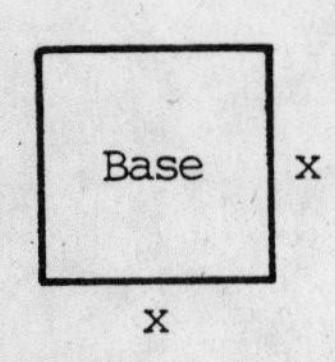

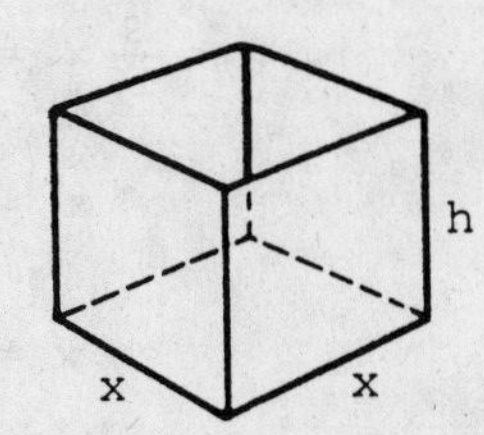

An equation for the volume of the box is

$$32 = x^2h$$
$$h = \frac{32}{x^2}.$$

The box is open at the top so the area of the surface material m(x) in square inches is

$$m(x) = \underset{Base}{x^2} + \underset{4\ sides}{4xh}$$
$$= x^2 + \left(4x\frac{32}{x^2}\right)$$
$$= x^2 + \frac{128}{x}$$
$$m'(x) = 2x - \frac{128}{x^2}$$
$$= \frac{2x^3 - 128}{x^2} = 0$$
$$2x^3 - 128 = 0$$
$$2(x^3 - 64) = 0$$
$$x = 4$$
$$m'(x) = 2 + \frac{256}{x^3} > 0 \text{ since } x > 0.$$

So x = 4 minimizes the volume.

If $x = 4$, $h = \frac{32}{x^2} = \frac{32}{16} = 2.$

The dimensions that will minimize volume are 4 inches by 4 inches by 2 inches.

25. Let x = width

Then 2x = length, and h = height.

An equation for the volume is

$$36 = (2x)(x)h = 2x^2h$$
$$\frac{18}{x^2} = h.$$

The surface area is

$$S(x) = \underset{Base}{(2x)(x)} + \underset{2\ sides}{2xh} + \underset{2\ sides}{2(2x)h}$$
$$= 2x^2 + 6xh$$
$$= 2x^2 + 6x\left(\frac{18}{x^2}\right)$$
$$= 2x^2 + \frac{108}{x}.$$
$$S'(x) = 4x - \frac{108}{x^2}$$
$$= \frac{4x^3 - 108}{x^2} = 0$$
$$4(x^3 - 27) = 0$$
$$x = 3$$
$$S''(x) = 4 + \frac{108(2)}{x^3}$$
$$= 4 + \frac{216}{x^3} > 0 \text{ since } x > 0.$$

So x = 3 minimizes the volume.

If $x = 3$, $h = \frac{18}{x^2} = \frac{18}{9} = 2.$

The dimensions are 3 feet by 6 feet by 2 feet.

27. Let x = speed.
Since the distance is 1000 miles, the cost of the driver at $8 per hour in dollars is

$$\frac{1000}{x}(8).$$

The cost of the fuel is $2 per gallon.

$$G(x) = \frac{1}{200}\left(\frac{800}{x} + x\right) \text{ gallons per mile}$$

$$C(x) = \frac{1}{200}\left(\frac{800}{x} + x\right)(1000)(2) + \frac{1000}{x}(8)$$

$$= 10\left(\frac{800}{x} + x\right) + \frac{8000}{x}$$

$$= \frac{8000}{x} + 10x + \frac{8000}{x}$$

$$= 10x + \frac{16,000}{x}$$

$$C'(x) = 10 - \frac{16,000}{x^2} = 0$$

$$10 = \frac{16,000}{x^2}$$

$$x^2 = 1600$$

$x = 40$ *Discard negative solution*

$$C''(x) = \frac{32,000}{x^3} > 0 \text{ since } x > 0.$$

So cost is minimized if x = 40.

$$C(40) = 10(40) + \frac{16,000}{40}$$

$$= 400 + 400$$

$$= \$800$$

40 miles per hour gives the minimum cost of $800.

29. Distance on shore: 9 - x miles
Cost on shore: $400 per mile
Distance underwater: $\sqrt{x^2 + 36}$
Cost underwater: $500 per mile
Find the distance from A, that is, (9 - x), to minimize cost, C(x).

$$C(x) = (9 - x)400 + (\sqrt{x^2 + 36})500$$

$$= 3600 - 400x + 500(x^2 + 36)^{1/2}$$

$$C'(x) = -400 + 500\left(\frac{1}{2}\right)(x^2 + 36)^{-1/2}(2x)$$

$$= -400 + \frac{500x}{\sqrt{x^2 + 36}}$$

If C'(x) = 0,

$$\frac{500x}{\sqrt{x^2 + 36}} = 400$$

$$\frac{5x}{4} = \sqrt{x^2 + 36}$$

$$\frac{25}{16}x^2 = x^2 + 36$$

$$\left(\frac{25}{16} - 1\right)x^2 = 36$$

$$\frac{9}{16}x^2 = 36$$

$$x^2 = \frac{36 \cdot 16}{9}$$

$$x = \frac{6 \cdot 4}{3}$$

$= 8$ *Discard negative solution*

To yield minimum total cost, the distance is

$$9 - x = 9 - 8$$

$= 1$ mile from point A.

31. Use $x = \sqrt{\frac{kM}{2a}}$, with

x = number of batches per years,
M = 100,00 units produced annually
k = $1, cost to store one unit for one year
a = $500, cost to set up, or fixed cost.

$$x = \sqrt{\frac{kM}{2a}} = \sqrt{\frac{(1)(100,000)}{2(500)}}$$

$$= \sqrt{100} = 10$$

33. $x = \sqrt{\frac{kM}{2a}}$

k = \$3, cost to store for one year
a = \$7, cost to set up, or fixed cost
M = 16,800 units produced annually

$x = \sqrt{\frac{kM}{2a}} = \sqrt{\frac{(3)(16,800)}{(2)(7)}}$

$= \sqrt{3600} = 60$

35. $x = \sqrt{\frac{kM}{2a}}$

M = 100,000 units produced annually
k = \$50, cost to store one unit per year
a = \$1000, cost to set up, or fixed cost

$x = \sqrt{\frac{kM}{2a}} = \sqrt{\frac{(.50)(100,000)}{(2)(1000)}}$

$= \sqrt{25}$

$= 5$

Section 12.4

1. By reading the graph, f(x) is, increasing on (1, ∞) and decreasing on (-∞, 1).

3. By reading the graph, g(x) is increasing on (-∞, -2) and decreasing on (-2, ∞).

5. By reading the graph, h(x) is increasing on (-∞, -4) and (-2, ∞), decreasing on (-4, -2).

7. By reading the graph, f(x) is increasing on (-7, -4) and (-2, ∞), decreasing on (-∞, 7) and (-4, -2).

See the graphs for Exercises 9-25 in the answer section at the back of the textbook.

9. $f(x) = x^2 + 12x - 6$

$f'(x) = 2x + 12 = 0$ when

$2x(x + 6) = 0$

$x = -6.$

Since f(x) is a polynominal function, there are no values where f fails to exist. x = -6 is the only critical point.

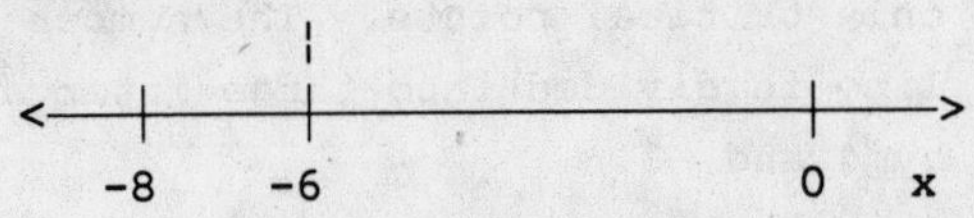

Thus, the number line is divided into two intervals by -6. Testing a point (any point other than x = -6 will do) in each interval, we find the following.

$f'(0) = 12 > 0$, so f(x) is increasing on (-6, ∞).

$f'(-8) = -4 < 0$, so f(x) is decreasing on (-∞, -6).

11. $y = 5 + 9x - 3x^2$

$y' = 9 - 6x = 0$ when

$3(3 - 2x) = 0$

$x = \frac{3}{2}.$

Since y is a polynominal function, there are no values where f fails to exist, and x = 3/2 is the only critical point. There are two intervals.

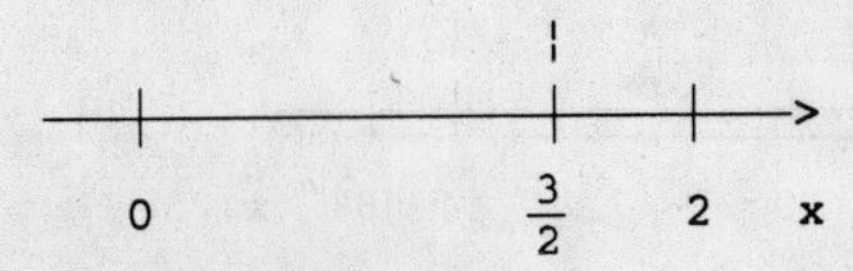

$f'(0) = 9 > 0$, so f(x) is increasing on (-∞, 3/2).

$f'(2) = -3 < 0$, so f(x) is decreasing on (3/2, ∞).

13. $f(x) = 2x^3 - 3x^2 - 72x - 4$

$f(x) = 6x^2 - 6x - 72 = 0$ when

$6(x - 4)(x + 3) = 0$

$x = 4$ or $x = -3$.

Since f(x) is a polynominal function, there are no values where f fails to exist, and $x = 4$ and $x = -3$ are the only critical points. The number line is divided into three intervals by 4 and -3.

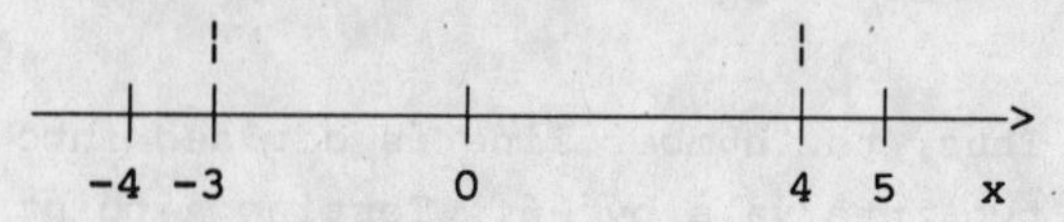

$f'(-4) = 140 > 0$

$f'(0) = -4 < 0$

$f'(5) = 289 > 0$

f(x) is increasing on $(-\infty, -3)$ and $(4, \infty)$.

f(x) is decreasing on $(-3, 4)$.

15. $f(x) = 4x^3 - 15x^2 - 72x + 5$

$f'(x) = 12x^2 - 30x - 72 = 0$ when

$6(2x^2 - 5x - 12) = 0$

$6(2x + 3)(x - 4) = 0$

$x = -\frac{3}{2}$ or $x = 4$.

Since f(x) is a polynominal function, there are no values where f fails to exist, and $x = -3/2$ and $x = 4$ are the only critical points.

2 $\frac{3}{2}$ 0 4 5 x

$f'(-2) = 36 > 0$

$f'(0) = -72 < 0$

$f'(5) = 78 > 0$

f(x) is increasing on $(-\infty, -3/2)$ and $(4, \infty)$.

f(x) is decreasing on $(-3/2, 4)$.

17. $y = -3x + 6$

$y' = -3 < 0$

Since f(x) is a polynominal function, there are no values where f fails to exist, and since the derivative is never zero, there are no critical points. Because the derivative is negative for all x, y(x) is decreasing on $(-\infty, \infty)$.

y is increasing on no interval.

y is decreasing on $(-\infty, \infty)$.

19. $f(x) = \frac{x + 2}{x + 1}$

$$f'(x) = \frac{(x + 1)(1) - (x + 2)1}{(x + 1)^2}$$

$$= \frac{-1}{(x + 1)^2}$$

Since f(x) is a rational function, it fails to exist when $x + 1 = 0$, that is, when $x = -1$. Since the derivative is never zero, there are no other critical points.

By noting that the denominator of the derivative is always positive, we see that $f'(x)$ is negative when $x > -1$ or $x < -1$.

f(x) is increasing on no interval.

f(x) is decreasing on $(-\infty, -1)$ and on $(-1, \infty)$.

21. $y = |x + 4|$ is equivalent to

$$y = \begin{cases} x + 4 & \text{if } x \geq -4 \\ -x - 4 & \text{if } x < -4. \end{cases}$$

The student should test a few values of x to be sure that these functions are equivalent. The function is continuous so it can be thought of as a polynominal over the real numbers.

The derivative is found as follows.

$$y' = \begin{cases} 1 \text{ if } x \ge -4 \\ -1 \text{ if } x < -4. \end{cases}$$

y is increasing for $(-4, \infty)$.
y is decreasing on $(-\infty, -4)$.

23. $f(x) = -\sqrt{x - 1}$ Note f(x) is defined only for $x \ge 1$ and exists for all such points.

$$f'(x) = \frac{-1}{2}(x - 1)^{-1/2}$$

$$= \frac{-1}{2\sqrt{x - 1}} < 0 \text{ for all } x.$$

Since $f'(x)$ is never zero, there are no critical points.
f(x) is increasing on no interval.
f(x) is decreasing on $(1, \infty)$.

25.
$$y = \sqrt{x^2 + 1} = (x^2 + 1)^{1/2}$$

$$y' = \frac{1}{2}(x^2 + 1)^{-1/2}(2x)$$

$$= x(x^2 + 1)^{-1/2}$$

$$= \frac{x}{\sqrt{x^2 + 1}} = 0 \text{ when } x = 0.$$

Since y does not fail to exist for any x, and since $y' = 0$ when $x = 0$, $x = 0$ is the only critical point.

-1 0 1 x

$$y'(1) = \frac{1}{\sqrt{1^2 + 1}} = \frac{1}{\sqrt{2}} > 0$$

$$y'(-1) = \frac{-1}{\sqrt{(-1)^2 + 1}} = \frac{-1}{\sqrt{2}} < 0$$

y is increasing on $(0, \infty)$.
y is decreasing on$(-\infty, 0)$.

27. The function is concave upward on $(2, \infty)$.
The function is concave downward on $(-\infty, 2)$.
The point of inflection is (2, 3).
(The curve changes concavity at that point.)

29. The function is concave upward on $(-\infty, -1)$ and $(8, \infty)$.
The function is concave downward on $(-1, 8)$.
Points of inflection are (-1, 7) and (8, 6).

31. The function is concave upward on $(2, \infty)$.
The function is concave downward on $(-\infty, 2)$.
There is point of inflection.
Although the curve switches concavity at $x = 2$, 2 is not in the domain of the function.

33.
$$f(x) = x^2 + 10x - 9$$
$$f'(x) = 2x + 10$$
$$f''(x) = 2 > 0 \text{ for all } x.$$

Always concave upward
No points of inflection

35.
$$f(x) = -3 + 8x - x^2$$
$$f'(x) = 8 - 2x$$
$$f''(x) = -2 < 0$$

Always concave downward
No points on inflection

37.
$$f(x) = x^3 + 3x^2 - 45x - 3$$
$$f'(x) = 3x^2 + 6x - 45$$
$$f''(x) = 6x + 6$$
$$f''(x) = 6x + 6 > 0 \text{ when}$$
$$6(x + 1) > 0$$
$$x > -1.$$

Concave upward on $(-1, \infty)$

$f''(x) = 6x + 6 < 0$ when

$6(x + 1) < 0$

$x < -1.$

Concave downward on $(-\infty, -1)$

$f''(x) = 6x + 6 = 0$

$6(x + 1) = 0$

$x = -1$

$f(-1) = 44$

Point of inflection at $(-1, 44)$

39. $f(x) = -2x^3 + 9x^2 + 168x - 3$

$f'(x) = -6x^2 + 18x + 168$

$f''(x) = -12x + 18$

$f''(x) = -12x + 18 > 0$ when

$-6(2x - 3) > 0$

$2x - 3 < 0$

$x < \frac{3}{2}.$

Concave upward on $(-\infty, 3/2)$

$f''(x) = -12x + 18 < 0$ when

$-6(2x - 3) < 0$

$2x - 3 > 0$

$x > \frac{3}{2}.$

Concave downward on $(3/2, \infty)$

$f''(x) = -12x + 18 = 0$ when

$-6(2x + 3) = 0$

$2x + 3 = 0$

$x = \frac{3}{2}.$

$f\left(\frac{3}{2}\right) = \frac{525}{2}$

Point of inflection at $(3/2, 525/2)$

41. $f(x) = \frac{3}{x - 5}$

$f'(x) = \frac{-3}{(x - 5)^2}$

$f''(x) = \frac{-3(-2)(x - 5)}{(x - 5)^4}$

$= \frac{6}{(x - 5)^3}$

$f''(x) = \frac{6}{(x - 5)^3} > 0$ when

$(x - 5)^3 > 0$

$x - 5 > 0$

$x > 5.$

Concave upward on $(5, \infty)$

$f''(x) = \frac{6}{(x - 5)^3} < 0$ when

$(x - 5)^3 < 0$

$x - 5 < 0$

$x < 5.$

Concave downward on $(-\infty, 5)$.

$f''(x) = \frac{6}{(x - 5)^3}$ is not equal to 0

for any value of x.

No points of inflection

43. $f(x) = x(x + 5)^2$

$f'(x) = x(2)(x + 5) + (x + 5)^2$

$= (x + 5)(2x + x + 5)$

$= (x + 5)(3x + 5)$

$f''(x) = (x + 5)(3) + 3x + 5$

$= 3x + 15 + 3x + 5$

$= 6x + 20$

$f''(x) = 6x + 20 > 0$ when

$x > -\frac{10}{3}.$

Concave upward on $(-10/3, \infty)$

$f''(x) = 6x + 20 < 0$ when

$x < -\frac{10}{3}.$

Concave downward on $(-\infty, -10/3)$

$f\left(-\frac{10}{3}\right) = -\frac{10}{3}\left(-\frac{10}{3} + 5\right)^2$

$= \frac{-10}{3}\left(\frac{-10 + 15}{3}\right)^2$

$= \frac{-10}{3} \cdot \frac{25}{9}$

$= -\frac{250}{27}$

Point of inflection at

$(-10/3, -250/27)$

See the graphs for Exercises 45-61 in the answer section at the back of the textbook.

45. $f(x) = -x^2 - 10x - 25$

$f'(x) = -2x - 10$

$= -2(x + 5) = 0$

Critical number: $x = -5$

$f''(x) = -2 < 0$ for all x.

Concave downward on $(-\infty, \infty)$

Hence there is a relative maximum at -5.

No points of inflection

47. $f(x) = 3x^3 - 3x^2 + 1$

$f'(x) = 9x^2 - 6x$

$= 3x(3x - 2) = 0$

Critical numbers: $x = 0$ or $x = \frac{2}{3}$

$f''(x) = 18x - 6$

$f''(0) = -6 < 0$

$f''\left(\frac{2}{3}\right) = 6 > 0$

Relative maximum at 0, relative minimum at 2/3

$f''(x) = 18x - 6 = 0$

$x = \frac{1}{3}$

$f''(0) = -6 < 0$

$f''(1) = 12 > 0$

So there is a point of inflection at $x = 1/3$.

49. $f(x) = -2x^3 - 9x^2 + 108x - 10$

$f'(x) = -6x^2 - 18x + 108$

$= -6(x^2 + 3x - 18) = 0$

$(x + 6)(x - 3) = 0$

Critical numbers: $x = -6$ or $x = 3$

$f''(x) = -12x - 18$

$f''(-6) = 54 > 0$

$f''(3) = -54 < 0$

Relative maximum at 3, relative minimum at -6

$f''(x) = -12x - 18 = 0$

$x = -\frac{3}{2}$

$f''(-2) = 6 > 0$

$f''(-1) = -6 < 0$

So there is a point of inflection at $x = -3/2$.

51. $f(x) = 2x^3 + \frac{7}{2}x^2 - 5x + 3$

$f'(x) = 6x^2 + 7x - 5 = 0$

$(2x - 1)(3x + 5) = 0$

Critical numbers: $x = \frac{1}{2}$ or $x = -\frac{5}{3}$

$f''(x) = 12x + 7$

$f''\left(\frac{1}{2}\right) = 13 > 0$

$f''\left(-\frac{5}{3}\right) = -13 < 0$

Relative maximum at $-5/3$, relative minimum at 1/2

$f''(x) = 12x + 7 = 0$

$x = -\frac{7}{12}$

$f''(-1) = -5 < 0$

$f''(0) = 7 > 0$

So there is a point of inflection at $x = -7/12$.

53. $f(x) = (x + 3)^4$

$f'(x) = 4(x + 3)^3(1)$

$= 4(x + 3)^3 = 0$

Critical number: $x = -3$

$f''(x) = 12(x + 3)^2$

$f''(-3) = 12(-3 + 3)^2 = 0$

Second derivative test fails.

Use first derivative test.

$f'(-2) = 4(-2 + 3)^3 = 4(1)^3 = 4 > 0$

$f'(-4) = 4(-4 + 3)^2 = 4(-1)^3 = -4 < 0$

$f(x)$ is increasing on $(-3, \infty)$ and decreasing on $(-\infty, -3)$ So $x = -3$ is at a relative minimum.

$f''(x) = 12(x + 3)^2 > 0$ for all x

So there is no point of inflection.

55. $f(x) = x^4 - 18x^2 + 5$

$f'(x) = 4x^3 - 36x$

$= 4x(x^2 - 9) = 0$

$4x = 0$ or $x^2 - 9 = 0$

$x = 0$ or $x = \pm 3$

$f''(x) = 12x^2 - 36$

$f''(0) = -36 < 0$

$f''(3) = 72 > 0$

$f''(-3) = 72 > 0$

Relative maximum at $x = 0$

Relative minimum at $x = 3$ and $x = -3$

$f''(x) = 12x^2 - 36 = 0$

$x^2 = 3$

$x = \pm\sqrt{3}$

$f''(-2) = 12 > 0$

$f''(-1) = -24 < 0$

So there is a point of inflection at $x = -\sqrt{3}$.

$f''(1) = -24$

$f''(2) = 12$

So there is a point of inflection at $x = \sqrt{3}$.

57. $f(x) = x + \frac{1}{x}$

$f'(x) = 1 - \frac{1}{x^2} = 0$

$\frac{1}{x^2} = 1$

$x = -1$ or $x = 1$, and $f'(x)$ fails to exist at $x = 0$.

$f''(x) = \frac{2}{x^3}$

$f''(-1) = -2 < 0$

$f''(0)$ does not exist.

$f''(1) = 2 > 0$

Relative maximum at -1

Relative minimum at 1

$f''(x) = \frac{2}{x^3}$

No value of x would cause the second derivative to be equal to 0.

So there are no points of inflection.

59. $f(x) = \frac{x^2 + 25}{x}$

$f'(x) = \frac{x(2x) - (x^2 + 25)}{x^2}$

$= \frac{x^2 - 25}{x^2}$

$f'(x) = \frac{(x - 5)(x + 5)}{x^2} = 0$

$x = 5$ or $x = -5$, $f'(x)$ fails to exist at $x = 0$.

$f''(x) = \frac{x^2(2x) - 2x(x^2 - 25)}{x^4}$

$= \frac{50}{x^3}$

$f''(-5) = -\frac{2}{5} < 0$

$f''(0)$ does not exist.

$f''(5) = \frac{2}{5} > 0$

Relative maximum at -5

Relative minimum at 5

$f''(x) = \frac{50}{x^3} \neq 0$ for any x.

So there are no points of inflection.

61. $f(x) = \frac{x - 1}{x + 1}$

$f'(x) = \frac{(x + 1) - (x - 1)}{(x + 1)^2}$

$= \frac{2}{(x + 1)^2}$

$f'(x)$ is never zero.

$f'(x)$ fails to exist for $x = -1$.

$$f''(x) = \frac{(x+1)^2(0) - 2(2)(x+1)^3}{(x+1)^4}$$

$$= \frac{-4(x+1)^3}{(x+1)^4} = \frac{-4}{x+1}$$

$f'(x)$ fails to exist for $x = -1$.

No critical values, no maximum or minimum

$$f''(x) = \frac{-4}{x+1}$$

$f''(x) \neq 0$ for any x.

So there are no points of inflection.

63. $C(x) = x^3 - 2x^2 + 8x + 50$

$C'(x) = 3x^2 - 4x + 8 = 0$

By using the quadratic formula,

$$x = \frac{4 \pm \sqrt{-80}}{6}$$

So $C'(x) \neq 0$.

Hence $C'(x)$ is always positive.

(a) $C(x)$ is never decreasing.

(b) $C(x)$ is increasing everywhere

65. $C(x) = 4.8x - .0004x^2,\ 0 \le x \le 2250$

$R(x) = 8.4x - .002x^2,\ 0 \le x \le 2250$

$P(x) = R(x) - C(x)$

$= 8.4x - .002x^2 - (4.8x - .0004x^{2)}$

$P'(x) = 8.4 - .004x - 4.8 + .0008x$

$= 3.6 - .0032x$

$P'(x) = 0$ when

$$x = \frac{3.6}{.0032}$$

$= 1125.$

0 1125 1200 x

$P'(0) = 3.1 > 0$

$P(1200) = -.24 < 0$

P is increasing on $[0, 1125)$.

67. $P(t) = 2 + 50t - \frac{5}{2}t^2$

$P'(t) = 50 - 5t$

The number of people infected start to decline when $P'(t) < 0$.

$50 - 5t < 0$

$50 < 5t$

$10 < t$

So the number of people infected start to decline after 10 days.

69. $K(x) = \dfrac{4x}{3x^2 + 27}$ for $x \ge 0$

$$K'(x) = \frac{(3x^2 + 27)4 - 4x(6x)}{(3x^2 + 27)^2}$$

$$= \frac{108 - 12x^2}{(3x^2 + 27)^2} = 0$$

$108 - 12x^2 = 0$

$12(9 - x^2) = 0$

$(3 - x)(3 + x) = 0$

$x = 3$ or $x = -3$, but $x \ge 0$.

One critical point.

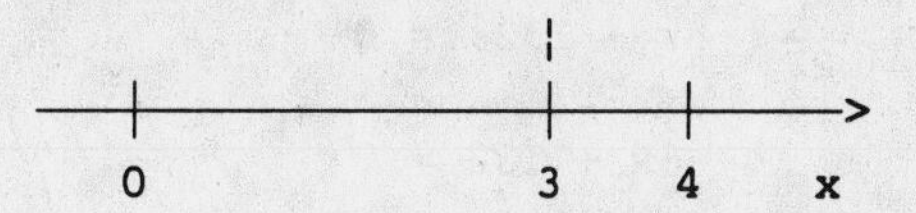

$$K'(0) = \frac{108 - 0}{(0 + 27)^2} = \frac{108}{27^2} > 0$$

$$K'(4) = \frac{108 - 12(4)^2}{(3(4)^2 + 27)^2}$$

$$= \frac{108 - 192}{(48 + 27)^2}$$

$$= \frac{-84}{(48 + 27)^2} < 0$$

(a) $K(x)$ is increasing on $(0, 3)$. (Note: x must be at least 0.)

(b) $K(x)$ is decreasing on $(3, \infty)$.

71. **(a)** Place home videotape recorders between car phones and polaroid cameras.

(b) Place rotary-dial telephones near black-and-white TVs.

(c) Car phones and polaroid cameras are closest to left point of inflection. The rate of growth of sales will now decline.

(d) Black-and-white TVs are closest to right point of inflection. The rate of decline is starting to slow.

73. $R(x) = \frac{4}{27}(-x^3 + 66x^2 + 1050x - 400);$

$0 \le x \le 25$

$R'(x) = \frac{4}{27}(-3x^2 + 132x + 1050)$

$R''(x) = \frac{4}{27}(-6x + 143)$

A point of diminishing returns occurs at a point of inflection, or where $R''(x) = 0$

$\frac{4}{27}(-6x + 132) = 0$

$-6x + 132 = 0$

$6x = 132$

$x = 22$

Test $R''(x)$ to determine whether concavity changes at $x = 22$.

$R''(20) = \frac{4}{27}(-6 \cdot 20 + 132)$

$= \frac{16}{9} > 0$

$R''(24) = \frac{4}{27}(-6 \cdot 24 + 132)$

$= -\frac{16}{9} < 0$

$R(x)$ is concave up on (0, 22) and concave down on (22, 25).

$R(22) = \frac{4}{27}[-22^3 + 66(22^2)$

$+ 1050(22) - 400]$

≈ 6517.9

The point of inflection is (22, 6517.9).

Chapter 12 Review Exercises

1. $f(x) = x^2 - 5x + 3$

$f'(x) = 2x - 5 = 0$

$x = \frac{5}{2}$

$f''(x) = 2 > 0$ for all x.

Then $f(x)$ is a minimum at $x = 5/2$.

$f(x)$ is increasing on $(5/2, \infty)$ and decreasing on $(-\infty, 5/2)$.

3. $f(x) = -x^3 - 5x^2 + 8x - 6$

$f'(x) = -3x^2 - 10x + 8$

$3x^2 + 10x - 8 = 0$

$(3x - 2)(x + 4) = 0$

$x = \frac{2}{3}$ or $x = -4$

$f''(x) = -6x - 10$

$f''\left(\frac{2}{3}\right) = -14 < 0$

$f(x)$ is a maximum at $x = 2/3$.

$f''(-4) = 14 > 0$

$f(x)$ is a minimum at $x = -4$.

$f(x)$ is increasing on $(-4, 2/3)$; decreasing on $(-\infty, -4)$ and on $(2/3, \infty)$.

5. $f(x) = \frac{6}{x - 4}$

$f'(x) = \frac{-6}{(x - 4)^2} < 0$ for all x, but not defined for $x = 4$.

$f(x)$ is never increasing and is decreasing on $(-\infty, 4)$ and $(4, \infty)$.

7. $f(x) = -x^2 + 4x - 8$

$f'(x) = -2x + 4 = 0$

$x = 2$

Critical number: $x = 2$

$f''(x) = -2 < 0$ for all x, so f(2) is a relative maximum.

$f(2) = -2^2 + 4(2) - 8$

$= -4$

Relative maximum of -4 at 2

9. $f(x) = 2x^2 - 8x + 1$

$f'(x) = 4x - 8 = 0$

$x = 2$

Critical number: 2

$f''(x) = 4 > 0$ for all x, f(2) is relative minimum.

$f(2) = 2(2)^2 - 8(2) + 1$

$= -7$

Relative minimum of -7 at 2

11. $f(x) = 2x^3 + 3x^2 - 36x + 20$

$f'(x) = 6x^2 + 6x - 36 = 0$

$6(x^2 + x - 6) = 0$

$(x + 3)(x - 2) = 0$

$x = -3$ or $x = 2$

Critical numbers: -3, 2

$f''(x) = 12x + 6$

$f''(-3) = -30 < 0$ so maximum at $x = -3$.

$f''(2) = 30 > 0$ so minimum at $x = 2$.

$f(-3) = 101$

Relative maximum of 101 at -3

$f(2) = -24$

Relative minimum of -24 at 2

13. $f(x) = -2x^3 - \frac{1}{2}x^2 + x - 3$

$f'(x) = -6x^2 - x + 1 = 0$

$6x^2 + x - 1 = 0$

$(3x - 1)(2x + 1) = 0$

$x = \frac{1}{3}$ or $x = -\frac{1}{2}$ *Critical numbers*

$f''(x) = -12x - 1$

$f''\left(\frac{1}{3}\right) = -5 < 0$

Relative maximum at $x = 1/3$

$f''\left(-\frac{1}{2}\right) = 5 > 0$

Relative minimum at $x = -1/2$

$f\left(\frac{1}{3}\right) = -2\left(\frac{1}{3}\right)^3 - \frac{1}{2}\left(\frac{1}{3}\right)^2 + \frac{1}{3} - 3$

$= \frac{-151}{54}$

$f\left(-\frac{1}{2}\right) = -2\left(\frac{-1}{2}\right)^3 - \frac{1}{2}\left(\frac{-1}{2}\right)^2 - \frac{1}{2} - 3$

$= \frac{-27}{8}$

Relative maximum of -151/54 at 1/3

Relative minimum of -27/8 at -1/2

15. $f(x) = x^4 - \frac{4}{3}x^3 - 4x^2 + 1$

$f'(x) = 4x^3 - 4x^2 - 8x = 0$

$4x(x^2 - x - 2) = 0$

$4x(x - 2)(x + 1) = 0$

$x = 0$, $x = 2$, or $x = -1$

$f''(x) = 12x^2 - 8x - 8$

$f''(0) = -8 < 0$

Relative maximum at $x = 0$

$f''(2) = 8 > 0$

Relative minimum at $x = 2$

$f''(-1) = 12 > 0$

Relative minimum at $x = -1$

$f(0) = 1$

Relative maximum 1 at $x = 0$

$f(2) = -\frac{29}{3}$

Relative minimum of -29/3 at $x = 2$

$f(-1) = -\frac{2}{3}$

Relative maximum of 1 at 0

Relative minimum of -2/3 at -1 and -29/3 at 2

17. $f(x) = \frac{x - 1}{2x + 1}$

$$f'(x) = \frac{(2x + 1) - (x - 1)(2)}{(2x + 1)^2}$$

$$= \frac{3}{(2x + 1)^2}$$

f' is never zero.

$f(x)$ has no relative extrema.

19. $y = x \cdot e^x$

Find the derivative and set it equal to 0.

$$y' = xe^x + e^x = 0$$

$$e^x(x + 1) = 0$$

$$x = -1$$

To determine the minimum or maximum, check the second derivative.

$$y'' = xe^x + e^x + e^x = xe^x + 2e^x$$

At $x = -1$, $y'' = -1e^{-1} + 2e^{-1}$

$$= -\frac{1}{e} + \frac{2}{e} = \frac{1}{e} > 0$$

y is concave up, minimum.

$$y = -1e^{-1}$$

$$= -\frac{1}{e} \approx -.368$$

Relative minimum of $-e^{-1} \approx -.368$ at -1

21. $y = \frac{e^x}{x - 1}$

Find the first derivative and set it equal to zero.

$$y' = \frac{(x - 1)e^x - e^x}{(x - 1)^2} = 0$$

Quotient rule

$$\frac{e^x(x - 1 - 1)}{(x - 1)^2} = 0$$

$$\frac{e^x(x - 2)}{(x - 1)^2} = 0$$

$$x - 2 = 0$$

$$x = 0$$

To determine minimum or maximum, check second derivative.

$$y'' = \frac{(x - 1)^2[e^x + (x - 2)e^x]}{(x - 1)^4} - \frac{e^x(x - 2)(2)(x - 1)}{(x - 1)^4}$$

At $x = 2$,

$$y'' = \frac{1^2(e^2) - 0}{1^4} = e^2 > 0$$

$$y'' = \frac{1^2(e^2) - 0}{1^4} = e^2 > 0.$$

y is concave up, minimum at $x = 2$.

$$y = \frac{e^2}{1} = e^2 \approx 7.39$$

Relative minimum at $e^2 \approx 7.39$ at 2

23. $f(x) = -x^2 + 5x + 1;\ [1, 4]$

$f'(x) = -2x + 5 = 0$ when $x = \frac{5}{2}$

$f(1) = 5$ *Absolute minimum*

$f\left(\frac{5}{2}\right) = \frac{29}{4}$ *Absolute maximum*

$f(4) = 5$ *Absolute minimum*

Absolute maximum of 29/4 at 5/2

Absolute minimum of 5 at 1 and 4

25. $f(x) = x^3 + 2x^2 - 15x + 3;\ [-4, 2]$

$f'(x) = 3x^2 + 4x - 15 = 0$ when

$$(3x - 5)(x + 3) = 0$$

$$x = \frac{5}{3} \text{ or } x = -3$$

$f(-4) = 31$

$f(-3) = 39$ *Absolute maximum*

$f\left(\frac{5}{3}\right) = -\frac{319}{27}$ *Absolute minimum*

$f(2) = -11$

Absolute maximum of 39 at -3

Absolute minimum of $-319/27$ at 5/3

27.

$$f(x) = 3x^4 - 5x^2 - 11x$$
$$f'(x) = 12x^3 - 10x - 11$$
$$f''(x) = 36x^2 - 10$$
$$f''(1) = 36(1)^2 - 10 = 26$$
$$f''(-3) = 36(-3)^2 - 10 = 314$$

29.

$$f(x) = \frac{4 - 3x}{x + 1}$$
$$f'(x) = \frac{-3(x + 1) - (1)(4 - 3x)}{(x + 1)^2}$$
$$= \frac{-7}{(x + 1)^2} = -7(x + 1)^{-2}$$
$$f''(x) = 14(x + 1)^{-3} = \frac{14}{(x + 1)^3}$$
$$f''(1) = \frac{14}{(1 + 1)^3} = \frac{7}{4}$$
$$f''(-3) = \frac{14}{(-3 + 1)^3} = -\frac{7}{4}$$

31.

$$f(x) = e^{-x^2}$$
$$f'(x) = (-2x)e^{-x^2}$$
$$f''(x) = (-2)e^{-x^2} + (-2x)(-2x)e^{-x^2}$$
$$= -2e^{-x^2} + 4x^2e^{-x^2}$$
$$= 2e^{-x^2}(-1 + 2x^2)$$
$$= 2e^{-x^2}(2x^2 - 1)$$
$$f''(1) = 2e^{-1}(+2 - 1) = 2e^{-1}$$
$$f''(-3) = 2e^{-9}(17) = 34e^{-9}$$

See the graphs for Exercises 33-39 in the answer section at the back of textbook.

33.

$$f(x) = -4x^3 - x^2 + 4x + 5$$
$$f'(x) = -12x^2 - 2x + 4$$
$$f''(x) = -24x - 2 = 0 \text{ when}$$
$$-2(12x + 1) = 0$$
$$x = -\frac{1}{12}.$$

If $f''(x) > 0$,
$$-24x - 2 > 0$$
$$-2(12x + 1) > 0$$
$$12x + 1 < 0.$$
Then $x < -\frac{1}{12}$.

Concave upward on $(-\infty, -1/12)$

If $f''(x) < 0$,
$$f''(x) = -12x - 2 < 0$$
$$-2(12x + 1) < 0$$
$$12x + 1 > 0$$
$$x > -\frac{1}{12}.$$

Concave downward on $(-1/12, \infty)$

$$f\left(-\frac{1}{12}\right) = -4\left(-\frac{1}{12}\right)^3 - \left(-\frac{1}{12}\right)^2 + 4\left(-\frac{1}{12}\right) + 5$$
$$= \frac{1007}{216}$$

Points of inflection at $(-1/12, 1007/216)$

35.

$$f(x) = x^4 + 2x^2$$
$$f'(x) = 4x^3 + 4x$$
$$f''(x) = 12x^2 + 4$$
$$= 4(3x^2 + 2) > 0 \text{ for all } x.$$

Concave upward on $(-\infty, \infty)$

Since $3x^2 + 2$ is never zero, $f''(0)$ is never zero. There are no inflection points.

37.

$$f(x) = \frac{x^2 - 4}{x}$$
$$f'(x) = \frac{x(2x) - (x^2 - 4)}{x^2}$$
$$= \frac{x^2 + 4}{x^2}$$
$$f''(x) = \frac{x^2(2x) - (x^2 + 4)2x}{x^4}$$
$$= \frac{-8x}{x^4} = \frac{-8}{x^3}$$

$f''(x)$ is never zero. There are no inflection points.

$$f''(x) = \frac{-8}{x^3} > 0 \text{ when}$$

$$x < 0.$$

Concave upward on $(-\infty, 0)$

$$f''(x) = \frac{-8}{x^3} < 0 \text{ when}$$

$$x > 0.$$

Concave downward on $(0, \infty)$ There are no points of inflection.

39. $f(x) = \dfrac{2x}{3 - x}$

$$f'(x) = \frac{(3 - x)2 - (2x)(-1)}{(3 - x)^2}$$

$$= \frac{6}{(3 - x)^2}$$

$$f''(x) = \frac{(3 - x)^2(0) - 6(3 - x)(-2)}{(3 - x)^4}$$

$$= \frac{12}{(3 - x)^3}$$

$f''(x)$ is never zero.

There are no inflection points.

$$f''(x) = \frac{12}{(3 - x)^3} > 0 \text{ when}$$

$$(3 - x)^3 > 0$$

$$3 - x > 0$$

$$3 > x.$$

Concave upward on $(-\infty, 3)$

$$f''(x) = \frac{12}{(3 - x)^3} < 0 \text{ when}$$

$$(3 - x)^3 < 0$$

$$3 - x < 0$$

$$3 < x.$$

Concave downward on $(3, \infty)$

There are no points of inflection.

41. Let x and y be the numbers.

$$x + y = 25$$

$$y = 25 - x$$

$$f(x) = xy$$

$$f(x) = x(25 - x)$$

$$= 25x - x^2$$

$$f'(x) = 25 - 2x$$

At a maximum, $f' = 0$ and $f'' < 0$.

$$25 - 2x = 0$$

$$2x = 25$$

$$x = \frac{25}{2}$$

$$f'' = -2 < 0$$

So $f(x)$ is maximum at $x = 25/2$ or 12.5.

$$y = 25 - \frac{25}{2}$$

$$= \frac{25}{2} \text{ or } 12.5$$

The numbers are 25/2 and 25/2, or 12.5 and 12.5.

43. $P(x) = 40x - x^2$;

x = price in hundred of dollars

(a) $P'(x) = 40 - 2x$

If $P'(x) = 0$,

$$2x = 40$$

$$x = 20.$$

$$P''(x) = -2 < 0$$

So the maximum profit occurs at 20(100) or $2000.

(b) Maximum profit is found by

$$P(20) = 40(20) - (20)^2$$

$$= 400.$$

The maximum profit is $400.

45. Let x = length and width of a side of base and

h = height.

Volume = 32 cubic meters with a square base and no top. Find height, length, and width for minimum surface area.

$$\text{Volume} = x^2h$$
$$x^2h = 32$$
$$h = \frac{32}{x^2}$$

$$\text{Surface area} = x^2 + 4xh$$
$$A = x^2 + 4x\left(\frac{32}{x^2}\right)$$
$$= x^2 + 128x^{-1}$$
$$A' = 2x - 128x^{-2}$$

If $A' = 0$,

$$\frac{2x^3 - 128}{x^2} = 0$$
$$x^3 = 64$$
$$x = 4.$$
$$A''(x) = 2 + 2(128)x^{-3}$$
$$= 8 > 0$$

So the minimum is at $x = 4$ where

$$h = \frac{32}{4^2}$$
$$= 2.$$

The dimensions are 2 meters by 4 meters by 4 meters.

47. Let x = width of play area
y = length of play area.

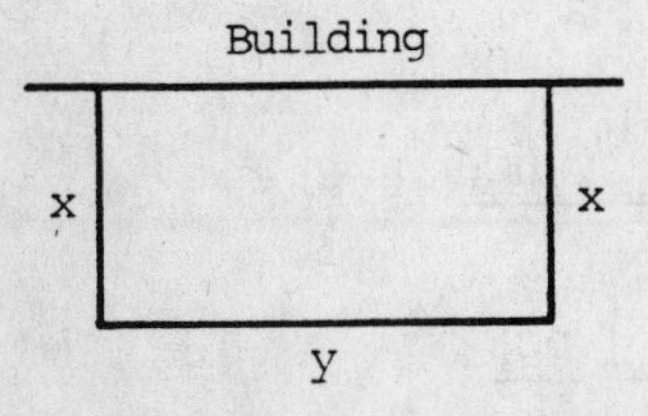

An equation describing the amount of fencing is

$$900 = 2x + y$$
$$y = 900 - 2x.$$
$$A = xy$$
$$A(x) = x(900 - 2x)$$
$$= 900x - 2x^2$$

If $A'(x) = 900 - 4x = 0$,

$$x = 225.$$

Then $y = 900 - 2(225) = 450$.

$A''(x) = -4 < 0$ so
$x = 225$ and
$y = 450$

Dimensions for maximum area are 225 meters by 450 meters.

49. Use $x = \sqrt{\frac{kM}{2a}}$, with

x = number of batches per year
M = 240,000 units produced annually
k = \$2, cost to store one unit for one year
a = \$15 cost to set up, or fixed cost

$$x = \sqrt{\frac{kM}{2a}} = \sqrt{\frac{2(240{,}000)}{2(15)}}$$
$$\approx 126$$

Since the number of batches is a whole number, the answer is 126 batches.

51. Let x = width of play area
y = length of play area

An equation describing the amount of fencing is

$$900 = 2x + 2y$$
$$900 - 2x = 2y$$
$$450 - x = y.$$
$$A = xy$$
$$A = x(450 - x)$$
$$= 450x - x^2$$

If $A'(x) = 450 - 2x = 0$

$$x = 225.$$

Then $y = 450 - 225 = 225$.

$A''(x) = -2 < 0$ so $x = 225$ and $y = 225$.

The dimensions for maximum area are 225 meters by 225 meters

Case 14 Exercises

1. $$\frac{dZ}{dV} = \frac{-V^3 - V^2\frac{L}{16Dp} + \frac{RV_0^3}{2F_0F_cF_p}}{\left(Dp + \frac{L}{24V}\right)^2}$$

Let $L = 1536$, $R = \$86{,}400$, $V_0 = 20$, $D_p = 8$, $F_0 = 50$, $F_c = 250$. Note that $V_m = 10$, so that the interval of concern is $[V_m, V_0] = [10, 20]$.

$$\frac{dZ}{dV} = \frac{-V^3 - V^2\left(\frac{1536}{16(8)}\right) + \frac{86{,}400(20^3)}{5(50)(250)(8)}}{\left(8 + \frac{1536}{24V}\right)^2}$$

$$= \frac{-V^3 - 12V^2 + 3456}{\left(8 + \frac{64}{V}\right)^2} = 0$$

Since the denominator is not 0 unless $V = 8$, which is not in the interval of concern,

$-V^3 - 12V^2 + 3456 = 0$

If 12 is a solution of this equation, 12 is a critical value of the function Z.

$-12^3 - 12(12^2) + 3456$

$= -1728 - 1728 + 3456$

$= 0$

So 12 is a critical value.

Since 12 is a solution of $\frac{dZ}{dV} = 0$, the other solutions in [10, 20] are solutions of

$$\frac{-V^3 - 12V^2 + 3456}{V - 12} = 0$$

$V^2 + 24V + 228 = 0$ *Use synthetic division*

Solve by the quadratic formula.

$$V = \frac{-24 \pm \sqrt{576 - 1152}}{2} = \frac{-24 \pm \sqrt{-576}}{2}$$

There are no real solutions, so there are no other critical values of Z in the interval. Test the endpoints $V_m = 10$ and $V_0 = 20$ as well as $V = 12$ in Z to find the optimum sailing speed.

If $V = 10$, $Z \approx 4305.56$.

If $V = 12$, $Z \approx 4400$.

If $V = 20$, $Z \approx 3142$.

Since the profit Z is a maximum when the cruising speed is V = 12 knots in the interval [10, 20], 12 knots is the optimum cruising speed.

2. C does not appear because it is a constant that appears in every term when the derivative is set equal to 0.

Case 15 Exercises

1. $$Z(m) = \frac{C_1}{m} + DtC_2 + DC_3\left(\frac{m-1}{2}\right)$$

$$= C_1m^{-1} + DtC_2 + \frac{DC_3m}{2} - \frac{DC_3}{2}$$

$$Z'(m) = -C_1m^{-2} + \frac{DC_3}{2}$$

$$= \frac{-C_1}{m^2} + \frac{DC_3}{2}$$

2. $$Z'(m) = 0$$

$$\frac{-C_1}{m^2} + \frac{DC_3}{2} = 0$$

$$\frac{-C_1}{m^2} = \frac{-DC_3}{2}$$

$$\frac{C_1}{m^2} = \frac{DC_3}{2}$$

$$2C_1 = DC_3m^2$$

$$\pm\sqrt{\frac{2C_1}{DC_3}} = m$$

3. $m = \sqrt{\dfrac{2(15,000)}{(3)(900)}}$

$m \approx 3.33$

4. $m^+ = 4$ *Whole number larger than 3.33*

$m^- = 3$ *Whole number smaller than 3.33*

5. Let $C_1 = 15,000$, $m^+ = 4$, $m^- = 3$,

$D = 3$, $C_2 = 100$, $C_3 = 900$, $t = 12$.

$$Z(m) = \frac{C_1}{m} + DtC_2 + DC_3\left(\frac{m-1}{2}\right)$$

$$Z(m^+) = Z(4) = \frac{15,000}{4} + (3)(12)(100) + (3)(900)\left(\frac{4-1}{2}\right)$$

$$= 3750 + 3600 + 4050$$

$$= \$11,400$$

$$Z(m^-) = Z(3) = \frac{15,000}{3} + (3)(12)(100) + (3)(900)\left(\frac{3-1}{2}\right)$$

$$= 5000 + 3600 + 2700$$

$$= \$11,300$$

6. Since $Z(m^-)$ is smaller than $Z(m^+)$, the optimum value of Z is \$11,300. This occurs when t = 3 months.

Now $N = mD$

$= (3)(3)$

$= 9.$

So, the number of trainees in a batch is 9.

7. Answers may vary.

CHAPTER 13 INTEGRAL CALCULUS

Section 13.1

1. $\int 4x\,dx = \frac{4x^2}{2} + C$

$= 2x^2 + C$

3. $\int 5t^2\,dt = 5\int t^2\,dt$

$= \frac{5t^3}{3} + C$

5. $\int 6\,dk = 6\int dk$

$= 6k + C$

7. $\int (2z + 3)dz = \frac{2z^2}{2} + 3z + C$

$= z^2 + 3z + C$

9. $\int (x^2 + 6x)dx = \frac{x^3}{3} + \frac{6x^2}{2} + C$

$= \frac{x^3}{3} + 3x^2 + C$

11. $\int (t^2 - 4t + 5)dt$

$= \frac{t^3}{3} - \frac{4t^2}{2} + 5t + C$

$= \frac{t^3}{3} - 2t^2 + 5t + C$

13. $\int (4z^3 + 3z^2 + 2z - 6)dz$

$= \frac{4z^4}{4} + \frac{3z^3}{3} + \frac{2z^2}{2} - 6z + C$

$= z^4 + z^3 + z^2 - 6z + C$

15. $\int 5\sqrt{z}\,dz = 5\int z^{1/2}dz$

$= \frac{5z^{3/2}}{\frac{3}{2}} + C$

$= 5\left(\frac{2}{3}\right)z^{3/2} + C$

$= \frac{10z^{3/2}}{3} + C$

17. $\int (u^{1/2} + u^{3/2})\,du$

$= \frac{u^{3/2}}{\frac{3}{2}} + \frac{u^{5/2}}{\frac{5}{2}} + C$

$= \frac{2u^{3/2}}{3} + \frac{2u^{5/2}}{5} + C$

19. $\int (15x\sqrt{x} + 2\sqrt{x})\,dx$

$= \int (15x(x^{1/2}) + 2x^{1/2})\,dx$

$= \int (15x^{3/2} + 2x^{1/2})\,dx$

$= \frac{15x^{5/2}}{\frac{5}{2}} + \frac{2x^{3/2}}{\frac{3}{2}} + C$

$= 15\left(\frac{2}{5}\right)x^{5/2} + 2\left(\frac{2}{3}\right)x^{3/2} + C$

$= 6x^{5/2} + \frac{4x^{3/2}}{3} + C$

21. $\int (10u^{3/2} - 14u^{5/2})\,du$

$= \frac{10u^{5/2}}{\frac{5}{2}} - \frac{14u^{7/2}}{\frac{7}{2}} + C$

$= 10\left(\frac{2}{5}\right)u^{5/2} - 14\left(\frac{2}{7}\right)u^{7/2} + C$

$= 4u^{5/2} - 4u^{7/2} + C$

23. $\int\left(\frac{1}{z^2}\right)dz = \int z^{-2}\,dz$

$= \frac{z^{-1}}{-1} + C$

$= -\frac{1}{z} + C$

25. $\int\left(\frac{1}{y^3} - \frac{1}{\sqrt{y}}\right)dy$

$= \int(y^{-3} - y^{-1/2})\,dy$

$= \frac{y^{-2}}{-2} - \frac{y^{1/2}}{\frac{1}{2}} + C$

$= \frac{-1}{2y^2} - 2y^{1/2} + C$

27. $\int(-9t^{-2} - 2t^{-1})\,dt$

$= \frac{-9t^{-1}}{-1} - 2\int\frac{dt}{t}$

$= \frac{9}{t} - 2\ln|t| + C$

29. $\int e^{2t}\,dt = \frac{e^{2t}}{2} + C$

31. $\int 3e^{-0.2x}dx = \frac{3(e^{-.2x})}{-.2} + C$

$= -15e^{-.2x} + C$

33. $\int\left(\frac{3}{x} + 4e^{-.5x}\right)dx$

$= 3\ln|x| + \frac{4e^{-.5x}}{-.5} + C$

$= 3\ln|x| - 8e^{-.5x} + C$

35. $\int\frac{1 + 2t^3}{t}dt = \int\left(\frac{1}{t} + \frac{2t^3}{t}\right)dt$

$= \int\left(\frac{1}{t} + 2t^2\right)dt$

$= \ln|t| + \frac{2t^3}{3} + C$

37. $\int\left(e^{2u} + \frac{u}{4}\right)du = \int\left(e^{2u} + \frac{1}{4}u\right)du$

$= \frac{1}{2}e^{2u} + \frac{1}{4}\frac{u^2}{2} + C$

$= \frac{e^{2u}}{2} + \frac{u^2}{8} + C$

39. $\int(x + 1)^2dx = \int(x^2 + 2x + 1)\,dx$

$= \frac{x^3}{3} + \frac{2x^2}{2} + x + C$

$= \frac{x^3}{3} + x^2 + x + C$

41. $\int\frac{\sqrt{x} + 1}{\sqrt[3]{x}}dx = \int\left(\frac{\sqrt{x}}{\sqrt[3]{x}} + \frac{1}{\sqrt[3]{x}}\right)dx$

$= \int(x^{(1/2-1/3)} + x^{-1/3})dx$

$= \int(x^{1/6} + x^{-1/3})dx$

$= \frac{x^{7/6}}{\frac{7}{6}} + \frac{x^{2/3}}{\frac{2}{3}} + C$

$= \frac{6x^{7/6}}{7} + \frac{3x^{2/3}}{2} + C$

43. $C'(x) = 4x - 5$; fixed cost is \$8

$C(x) - \int(4x - 5)\,dx$

$= \frac{4x^2}{2} - 5x + k$

$C(x) = 2x^2 - 5x + k$

$C(0) = 2(0)^2 - 5(0) + k = k$

Since $C(0) = 8$, $k = 8$.

$C(x) = 2x^2 - 5x + 8$

45. $C'(x) = .2x^2 + 5x$; fixed cost is $10.

$$C(x) = \int (.2x^2 + 5x)dx$$

$$= \frac{.2x^3}{3} + \frac{5x^2}{2} + k$$

$$C(x) = \frac{.2x^3}{3} + \frac{5x^2}{2} + k$$

$$C(0) = \frac{.2(0)^3}{3} + \frac{5(0)^3}{2} + k = k$$

Since $C(0) = 10$, $k = 10$.

$$C(x) = \frac{.2x^3}{3} + \frac{5x^2}{2} + 10$$

47. $C'(x) = x^{1/2}$; 16 units cost $40

$$C(x) = \int x^{1/2}\, dx$$

$$= \frac{x^{3/2}}{\frac{3}{2}} + k$$

$$C(x) = \frac{2x^{3/2}}{3} + k$$

$$C(16) = \frac{2(16)^{3/2}}{3} + k$$

$$= \frac{2(64)}{3} + k$$

Since $C(16) = 40$,

$$\frac{128}{3} + k = 40$$

$$k = \frac{120 - 128}{3}$$

$$= -\frac{8}{3}.$$

$$C(x) = \frac{2x^{3/2}}{3} - \frac{8}{3}$$

49. $C'(x) = x^2 - 2x + 3$; 3 units cost $15

$$C(x) = \int (x^2 - 2x + 3)dx$$

$$= \frac{x^3}{3} - \frac{2x^2}{2} + 3x + k$$

$$C(x) = \frac{x^3}{3} - x^2 + 3x + k$$

$$C(3) = \frac{3^3}{3} - 3^2 + 3(3) + k$$

Since $C(3) = 15$,

$$15 = 9 - 9 + 9 + k$$

$$k = 6.$$

$$C(x) = \frac{x^3}{3} - x^2 + 3x + 6$$

51. $C'(x) = \frac{1}{x} + 2x$; 7 units cost $58.40

$$C(x) = \int \left(\frac{1}{x} + 2x\right)dx$$

$$= \ln|x| + \frac{2x^2}{2} + k$$

$$C(x) = \ln|x| + x^2 + k$$

$$C(7) = \ln|7| + 7^2 + k$$

Since $C(7) = 58.40$,

$$58.40 = \ln 7 + 49 + k$$

$$k = 7.45.$$

$$C(x) = \ln|x| + x^2 + 7.45$$

53. $P'(x) = 2x + 20$, profit is -50 when 0 hamburgers sold.

$$P(x) = \int (-2x + 20)dx$$

$$= -\frac{2x^2}{2} + 20x + k$$

$$P(x) = -x^2 + 20x + k$$

$$P(0) = 0^2 + 20(0) + k$$

Since $P(0) = -50$,

$$k = -50.$$

$$P(x) = -x^2 + 20x - 50$$

55. $P'(t) = 10t - \frac{15}{\sqrt{t}}$

$$= 10t - 15t^{-1/2}$$

$$P(t) = \int (10t - 15t^{-1/2})dt$$

$$= \frac{10t^2}{2} - \frac{15t^{1/2}}{\frac{1}{2}} + k$$

$$= 5t^2 - 30t^{1/2} + k$$

$$P(0) = 5(0)^2 - 30(0)^{1/2} + k$$
$$= k$$

Since $P(0) = 0$, $0 = k$.

$P(t) = 5t^2 - 30t^{1/2}$ tons

57. $F'(x) = x^{2/3}$

$$F(x) = \int x^{2/3}dx$$
$$= \frac{x^{5/3}}{\frac{5}{3}} + C$$
$$F(x) = \frac{3}{5}x^{5/3} + C$$

Since (1, 3/5) is on the curve, $y = 3/5$ when $x = 1$.

So $F(1) = \frac{3}{5}$

$$\frac{3}{5} = \frac{3}{5}(1)^{5/3} + C$$
$$\frac{3}{5} = \frac{3}{5} + C$$
$$0 = C.$$

Thus, $F(x) = \frac{3}{5}x^{5/3}$.

Section 13.2

1. $\int 4(2x + 3)^4\, dx = 2\int 2(2x + 3)^4\, dx$

Rewrite 4 = 2 · 2

Let $u = 2x + 3$

$du = 2dx$.

$$= 2\int u^4 du$$
$$= \frac{2 \cdot u^5}{5} + C$$
$$= \frac{2(2x + 3)^5}{5} + C$$

3. $\int \frac{4}{(y - 2)^3}\, dy = \int 4(y-2)^{-3}\, dy$

$$= 4\int (y - 2)^{-3}\, dy$$

Let $u = y - 2$

$du = dy$.

$$= 4\int u^{-3}du$$
$$= \frac{4u^{-2}}{-2} + C$$
$$= -2(y - 2)^{-2} + C$$

5. $\int \frac{2dm}{(2m + 1)^3} = \int 2(2m + 1)^{-3}\, dm$

Let $u = 2m + 1$

$du = 2\, dm$.

$$= \int u^{-3}du$$
$$= \frac{u^{-2}}{-2} + C$$
$$= \frac{-(2m + 1)^{-2}}{2} + C$$

7. $\int \frac{2x + 2}{(x^2 + 2x - 4)^4}\, dx$

$$= \int (2x + 2)(x^2 + 2x - 4)^{-4}\, dx$$

Let $u = x^2 + 2x - 4$

$du = (2x + 2)dx$

$$= \int u^{-4}\, du$$
$$= \frac{u^{-3}}{-3} + C$$
$$= \frac{-(x^2 + 2x - 4)^{-3}}{3} + C$$

9. $\int z\sqrt{z^2 - 5}\,dz = \int z(z^2 - 5)^{1/2}\,dz$

$= \frac{1}{2}\int 2z(z^2 - 5)^{1/2}\,dz$

Let $u = z^2 - 5$
$du = 2z\,dz.$

$= \frac{1}{2}\int u^{1/2}\,du$

$= \frac{1}{2} \cdot \frac{u^{3/2}}{\frac{3}{2}} + C$

$= \frac{1}{2}\left(\frac{2}{3}\right)u^{3/2} + C$

$= \frac{(z^2 - 5)^{3/2}}{3} + C$

11. $\int(-4e^{2p})\,dp = -2\int 2e^{2p}\,dp$

Rewrite -4 as -2 · 2

Let $u = 2p$
$du = 2\,dp.$

$= -2\int ef^u du$

$= -2e^u + C$

$= -2e^{2p} + C$

13. $\int 3x^2e^{2x^3}dx = \frac{1}{2}\int 2 \cdot 3x^2e^{2x^3}\,dx$

Multiply by $\frac{1}{2} \cdot 2$

Let $u = 2x^3$
$du = 6x^2\,dx.$

$= \frac{1}{2}\int e^u\,du$

$= \frac{1}{2}e^u + C$

$= \frac{e^{2x^3}}{2} + C$

15. $\int(1 - t)e^{2t-t^2}\,dt$

$= \frac{1}{2}\int 2(1 - t)e^{2t-t^2}\,dt$

Multiply by $\frac{1}{2} \cdot 2$

Let $u = 2t - t^2$
$du = (2 - 2t)dt$
$= 2(1 - t)dt.$

$= \frac{1}{2}\int e^u\,du$

$= \frac{e^u}{2} + C$

$= \frac{e^{2t-t^2}}{2} + C$

17. $\int \frac{e^{1/z}}{z^2}\,dz = -\int e^{1/z} \cdot \frac{-1}{z^2}\,dz$

Let $u = \frac{1}{z}$

$du = \frac{-1}{z^2}\,dz.$

$= -\int e^u\,du$

$= -e^u + C$

$= -e^{1/z} + C$

19. $\int \frac{-8}{1 + 3x}\,dx = -8\int \frac{1}{1 + 3x}\,dx$

$= -8\left(\frac{1}{3}\right)\int \frac{3}{1 + 3x}\,dx$

Let $u = 1 + 3x$
$du = 3\,dx.$

$= \frac{-8}{3}\int \frac{du}{u}$

$= \frac{-8}{3}\ln|u| + C$

$= \frac{-8\ln|1 + 3x|}{3} + C$

21. $\int \frac{dt}{2t+1} = \frac{1}{2} \int \frac{2\,dt}{2t+1}$

Let $u = 2t + 1$

$du = 2\,dt.$

$= \frac{1}{2} \int \frac{du}{u}$

$= \frac{1}{2} \ln |u| + C$

$= \frac{\ln |2t+1|}{2} + C$

23. $\int \frac{vdv}{(3v^2+2)^4} = \frac{1}{6} \int \frac{6vdv}{(3v^2+2)^4}$

Let $u = 3v^2 + 2$

$du = 6\,vdv.$

$= \frac{1}{6} \int \frac{du}{u^4}$

$= \frac{1}{6} \int u^{-4}\,du$

$= \left(\frac{1}{6}\right) \frac{u^{-3}}{-3} + C$

$= -\frac{1}{18}(3v^2+2)^{-3} + C$

$= \frac{-(3v^2+2)^{-3}}{18} + C$

25. $\int \frac{x-1}{(2x^2-4x)^2}\,dx = \frac{1}{4} \int \frac{4(x-1)\,dx}{(2x^2-4x)^2}$

Let $u = 2x^2 - 4x$

$du = (4x - 4)dx$

$= 4(x - 1)dx.$

$= \frac{1}{4} \int \frac{du}{u^2}$

$= \frac{1}{4} \int u^{-2}\,du$

$= \frac{1}{4}\left(\frac{u^{-1}}{-1}\right) + C$

$= \frac{-(2x^2-4x)^{-1}}{4} + C$

27. $\int \left(\frac{1}{r} + r\right)\left(1 - \frac{1}{r^2}\right) dr = \int u\,du$

Let $u = \frac{1}{r} + r$

$du = \left(\frac{1}{r^2} + 1\right)dr$

$= \left(1 - \frac{1}{r^2}\right)dr.$

$= \frac{u^2}{2} + C$

$= \frac{1}{2}\left(\frac{1}{r} + r\right)^2 + C$

$= \frac{\left(\frac{1}{r} + r\right)^2}{2} + C$

29. $\int \frac{x^2+1}{(x^3+3x)^{2/3}}\,dx$

$= \frac{1}{3} \int \frac{3(x^2+1)\,dx}{(x^3+3x)^{2/3}}$

Let $u = x^3 + 3x$

$du = (3x^2 + 3)dx$

$= 3(x^2 + 1)dx.$

$= \frac{1}{3} \int \frac{du}{u^{2/3}}$

$= \frac{1}{3} \int u^{-2/3}\,du$

$= \frac{1}{3}\left(\frac{u^{1/3}}{\frac{1}{3}}\right) + C$

$= u^{1/3} + C$

$= (x^3 + 3x)^{1/3} + C$

31. $\int p(p+1)^5\,dp$

Let $u = p + 1$

$du = dp$; also, $p = u - 1.$

$$= \int (u - 1)u^5\, du$$

$$= \int (u^6 - u^5\, du)$$

$$= \frac{u^7}{u} - \frac{u^6}{6} + C$$

$$= \frac{(p + 1)^7}{7} - \frac{(p + 1)^6}{6} + C$$

33. $\int t\sqrt{5t - 1}\, dt$

$$= \frac{1}{5}\int 5t(5t - 1)^{1/2}\, dt$$

Let $u = 5t - 1$

$du = 5\, dt$

$t = \dfrac{u + 1}{5}.$

$$= \frac{1}{5}\int \left(\frac{u + 1}{5}\right)u^{1/2}\, du$$

$$= \frac{1}{25}\int (u^{2/3} + u^{1/2})\, du$$

$$= \frac{1}{25}\left[\frac{u^{5/2}}{\frac{5}{2}} + \frac{u^{3/2}}{\frac{3}{2}}\right] + C$$

$$= \frac{1}{25}\left[\frac{2}{5}(5t - 1)^{5/2} + \frac{2}{3}(5t - 1)^{3/2}\right] + C$$

$$= \frac{2(5t - 1)^{5/2}}{125} + \frac{2(5t - 1)^{3/2}}{75} + C$$

35. $\int \dfrac{u}{\sqrt{u - 1}}\, du$

$$= \int u(u - 1)^{-1/2}\, du$$

Let $w = u - 1$

$dw = du$

$u = w + 1.$

$$= \int (w + 1)w^{-1/2}\, dw$$

$$= \int (w^{1/2} + w^{-1/2})\, dw$$

$$= \frac{w^{3/2}}{\frac{3}{2}} + \frac{w^{1/2}}{\frac{1}{2}} + C$$

$$= \frac{2(u - 1)^{3/2}}{3} + 2(u - 1)^{1/2} + C$$

37. $\int 2x(x^2 + 1)^3 dx$

Let $u = x^2 + 1$

$du = 2x\, dx.$

$$= \int u^3 du$$

$$= \frac{u^4}{4} + C$$

$$= \frac{(x^2 + 1)^4}{4} + C$$

39. $\int (\sqrt{x^2 + 12x})\ (x + 6)dx$

$$= \int (x^2 + 12x)^{1/2}(x + 6)dx$$

Let $x^2 + 12x = u$

$2x + 12 = du$

$x + 6 = \dfrac{du}{2}.$

$$= \frac{1}{2}\int u^{1/2}\, du$$

$$= \frac{1}{2}\left(\frac{2}{3}\right)u^{3/2} + C$$

$$= \frac{(x^2 + 12x)^{3/2}}{3} + C$$

41. $\int \dfrac{t}{t^2 + 2}\, dt$

Let $t^2 + 2 = u$

$2tdt = du$

$tdt = \dfrac{du}{2}.$

$$= \frac{1}{2}\int \frac{du}{u}$$

$$= \frac{1}{2}\ \ln |u|$$

$$= \frac{\ln (t^2 + 2)}{2} + C$$

43. $\int ze^{2z^2}dz$

Let $2z^2 = u$

$4zdz = du$

$zdz = \frac{du}{4}.$

$= \frac{1}{4}\int e^u\, du$

$= \frac{1}{4}e^u + C$

$= \frac{e^{2z^2}}{4} + C$

45. $R'(x) = 2x(x^2 + 50)^3$

$R(x) = \int 2x(x^2 + 50)^3\, du$

Let $u = x^2 + 50$

$du = 2xdx.$

$= \int u^3du$

$= \frac{u^4}{4} + C$

$R(x) = \frac{1}{4}(x^2 + 50)^4 + C$

Since $R = 0$ when $x = 0$ and

$R(0) = \frac{1}{4}(50)^4 + C = 0,$

then $C = -\frac{(50)^4}{4} = -1562500.$

$R(x) = \frac{1}{4}(x^2 + 50)^4 = -1{,}562{,}500$

$R(3) = \frac{1}{4}(3^2 + 50)^4 - 1{,}562{,}500$

$= \frac{(59)^4}{4} - 1{,}562{,}500$

$= 3{,}029{,}340.25 - 1{,}562{,}500$

$\approx \$1{,}466{,}840$

47. $p'(x) = xe^{-x^2}$

$p(x) = \int xe^{-x^2}dx$

Let $u = -x^2$

$du = -2xdx$

$\frac{-du}{2} = xdx.$

$p(x) = \int \frac{-du}{2}(e^u)$

$= -\frac{1}{2}\int e^u\, du$

$= -\frac{1}{2}e^u + C$

$= -\frac{1}{2}e^{-x^2} + C$

If $x = 0$, $p(x) = 0$.

$0 = -\frac{1}{2}e^0 + C$

$0 = -\frac{1}{2} + C$

$\frac{1}{2} = C$

So $p(x) = -\frac{1}{2}e^{-x^2} + \frac{1}{2}.$

$p(3) = -\frac{1}{2}e^{-9} + \frac{1}{2}$

$= -.00006 + .5$

$\approx .5$ million dollars.

Section 13.3

1. $\sum_{i=1}^{3} 3i = 3(1) + 3(2) + 3(3) = 18$

3. $\sum_{i=1}^{5} (2i + 7)$

$= [2(1) + 7] + [2(2) + 7] + [2(3) + 7]$
$+ [2(4) + 7] + [2(5) + 7]$

$= 9 + 11 + 13 + 15 + 17$

$= 65$

5. $\sum_{i=1}^{4} x_i = x_1 + x_2 + x_3 + x_4$

$= -5 + 8 + 7 + 10$

$= 20$

7. $\sum_{i=1}^{3} f(x_i) = \sum_{i=1}^{3} (x_i - 3)$

$= (x_1 - 3) + (x_2 - 3) + (x_3 - 3)$
$= (4 - 3) + (6 - 3) + (7 - 3)$
$= 1 + 3 + 4$
$= 8$

9. $\sum_{i=1}^{4} f(x_i)\Delta x$

$= f(x_1)\Delta x + f(x_2)\Delta x + f(x_3)\Delta x$
$\quad + f(x_4)\Delta x$
$= f(0)(2) + f(2)(2) + f(4)(2)$
$\quad + f(6)(2)$
$= [2(0) + 1](2) + [2(2) + 1](2)$
$\quad + [2(4) + 1]\ (2) + [2(6) + 1](2)$
$= 2 + 5(2) + 9(2) + 13(2)$
$= 56$

11. $f(x) = 3x + 2$ from $x = 1$ to $x = 5$

2 rectangles:

$\Delta x = \frac{5 - 1}{2} = 2$ *Divide by 2 the number of rectangles*

Let $x_1 = 1$ and $x_2 = 1 + \Delta x$.
Then find $f(x_i)$ as follows.

i	x_i	$f(x_i)$
1	1	5
2	3	11

Find the sum of the areas of two rectangles with $f(x_i)$ as height and Δx as width.

$\sum_{i=1}^{2} f(x_i)\Delta x = f(x_1)\Delta + f(x_2)\Delta x$

$= 5(2) + 11(2)$
$= 32$

4 rectangles:

$\Delta x = \frac{5 - 1}{4} = 1$ *Divide by 4*

$x_1 = 1$, $x_2 = 1 + \Delta x$, and so on.

i	x_i	$f(x_i)$
1	1	5
2	2	8
3	3	11
4	4	14

Find the sum of four rectangles with $f(x_i)$ as height and Δx as width.

$\sum_{i=1}^{4} f(x_i)\Delta x$

$= 5(1) + 8(1) + 11(1) + 14(1)$
$= 38$

13. $f(x) = x + 5$ from $x = 2$ to $x = 4$

2 rectangles:

$\Delta x = \frac{4 - 2}{2} = 1$

$x_1 = 2$, $x_2 = 2 + \Delta x$

i	x_i	$f(x_i)$
1	2	7
2	3	8

$\sum_{i=1}^{2} f(x_i)\Delta x = 7(1) + 8(1)$

$= 15$

4 rectangles:

$\Delta x = \frac{4 - 2}{4} = \frac{1}{2} = .5$

$x_1 = 2$, $x_2 = 2 + \Delta x$, and so on.

i	x_i	$f(x_i)$	
1	2	7	*These entries may*
2	2.5	7.5	*also be found in*
3	3	8	*fractional form*
4	3.5	8.5	

$\sum_{i=1}^{4} f(x_i)\Delta x$

$= 7(.5) + 7.5(.5) + 8(.5) + 8.5(.5)$
$= 15.5$

As an improper fraction, this area is 31/2.

15. $f(x) = x^2$ from $x = 1$ to 5

2 rectangles:

$\Delta x = \frac{5 - 1}{2} = 2$

i	x_i	$f(x_i)$
1	1	1
2	3	9

$\sum_{i=1}^{2} f(x_i)\Delta x = 1(2) + 9(2)$

$= 20$

4 rectangles:

$\Delta x = \frac{5 - 1}{4} = 1$

i	x_i	$f(x_i)$
1	1	1
2	2	4
3	3	9
4	4	16

$\sum_{i=1}^{4} f(x_i)\Delta x$

$= 1(1) + 4(1) + 9(1) + 16(1)$

$= 30$

17. $f(x) = x^2 + 2$ from $x = -2$ to $x = 2$

2 rectangles:

$\Delta x = \frac{2 - (-2)}{2} = 2$

i	x_i	$f(x_i)$
1	-2	6
2	0	2

$\sum_{i=1}^{2} f(x_i)\Delta x = 6(2) + 2(2)$

$= 16$

4 rectangles:

$\Delta x = \frac{2 - (-2)}{4} = 1$

i	x_i	$f(x_i)$
1	-2	6
2	-1	3
3	0	2
4	1	3

$\sum_{i=1}^{4} f(x_i)\Delta x = 6 + 3 + 2 + 3$

$= 14$

19. $f(x) = e^x - 1$ from $x = 0$ to $x = 4$

2 rectangles:

$\Delta x = \frac{4 - 0}{2} = 2$

i	x_i	$f(x_i)$
1	0	0
2	2	6.39

$\sum_{i=1}^{2} f(x_i)\Delta x = 0(2) + 6.39(2)$

$= 12.8$

4 rectangles:

$\Delta x = \frac{4 - 0}{4} = 1$

i	x_i	$f(x_i)$
1	0	0
2	1	1.72
3	2	6.39
4	3	19.09

$\sum_{i=1}^{4} f(x_i)\Delta x = 0 + 1.72 + 6.39 + 10.09$

$= 27.2$

21. $f(x) = \frac{1}{x}$ from $x = 1$ to 5

2 rectangles:

$\Delta x = \frac{5 - 1}{2} = 2$

i	x_i	$f(x_i)$
1	1	1
2	3	.333

$$\sum_{i=1}^{2} f(x_i)\Delta x = 1(2) + .333(2)$$

$$= 2.67$$

4 rectangles:

$$\Delta x = \frac{5-1}{4} = 1$$

i	x_i	$f(x_i)$
1	1	1
2	2	.5
3	3	.333
4	4	.25

$$\sum_{i=1}^{4} f(x_i)\Delta x = 1 + .5 + .333 + .25$$

$$= 2.08$$

23. $f(x) = \frac{x}{2}$ between $x = 0$ and $x = 4$

4 rectangles:

(a) $\Delta x = \frac{4-0}{4} = 1$

i	x_i	$f(x_i)$
1	0	0
2	1	.5
3	2	1
4	3	1.5

$$\sum_{i=1}^{4} f(x_I)\Delta x = 0 + .5 + 1 + 1.5$$

$$= 3$$

(b) 8 rectangles:

$$\Delta x = \frac{4-0}{8} = .5$$

i	x_i	$f(x_i)$
1	0	0
2	.5	.25
3	1	.5
4	1.5	.75
5	2	1
6	2.5	1.25
7	3	1.5
8	3.5	1.75

$$\sum_{i=1}^{8} f(x_i)\Delta x$$

$$= 0(.5) + .25(.5) + .5(.5) + .75(.5) + 1(.5) + 1.25(.5) + 1.5(.5) + 1.75$$

$$= 3.5$$

(c) $\int_0^4 f(x)\,dx = \int_0^4 \frac{x}{2}\,dx$

$$= \frac{1}{2}(\text{base})(\text{height})$$

$$= \frac{1}{2}(4)(2)$$

$$= 4$$

25. Refer to the graph. Use $f(x_i)$ given on the graph as left sides of rectangles with width $\Delta x = 1$. Note that the number of cars sold is in thousands.

$$\sum_{i=1}^{12} f(x_i)\Delta x$$

$$\approx 550(1) + 822(1) + 963(1) + 870(1) + 772(1) + 737(1) + 757(1) + 1020(1) + 969(1) + 1027(1) + 1008(1) + 651(1)$$

$$= 10,146$$

Total sales were about 10,000,000 cars.

For Exercises 27-31, readings on the graphs and answers may vary.

27. Read values of the function on the graph at every hour from 0 to 7. These are left sides of rectangles with $\Delta x = 1$ as width.

$$\sum_{i=1}^{8} f(x_i)\Delta x$$

$$\approx 0(1) + 1(1) + 2(1) + 3(1) + 3.4(1) + 3.6(1) + 3.2(1) + 2.4(1)$$

$$\approx 18.6$$

Total concentration is about 19 units.

29. Read values of the function on the graph for every 5 seconds from t = 0 to t = 25. These are speeds in miles per hour, so multiply these values by 5280/3600 to get the speeds in feet per second. Then to estimate the distance traveled, find the sum of areas of rectangles with heights the sides of the rectangles and with widths of 5 seconds. The last rectangle has width of 3 seconds. The distributive property may be used to find the sum as follows.

$$\frac{5280}{3600}\sum_{i=0}^{5} f(x_i)\Delta x = \frac{5280}{3600}[0(5) + 38(5) + 62(5) + 76(5) + 86(5) + 96(3)]$$

$$= \frac{5280}{3600}(1598)$$

$$\approx 2000 \text{ feet}$$

31. Read values of the graph with snow depth index of 1.0 for every month until April. These are the left sides of rectangles with width $\Delta x = 1$.

$$\sum_{i=0}^{7} f(x_i)\Delta x$$

$$\approx 0(1) + 6(1) + 4(1) + 9(1) + 8(1) + 4(1) + 3(.5)$$

$$= 32.5$$

Read values of the given graph for every month until April. These are the left sides of rectangles with width $\Delta x + 1$.

$$\sum_{i=0}^{7} f(x_i)\Delta x$$

$$\approx 0(1) + 4(1) + 20(1) + 37(1) + 36(1) + 24(1) + 14(.5)$$

$$= 128$$

Snow depth index = ratio = $\frac{128}{32.5} \approx 4$

33. $f(x) = x \ln x$; [1, 5]

$$\Delta x = \frac{5 - 1}{8} = .5$$

i	x_i	$f(x_i)$
1	1	0
2	1.5	.6082
3	2	1.3863
4	2.5	2.2907
5	3	3.2958
6	3.5	4.3847
7	4	5.5452
8	4.5	6.7683
		24.2792 Total

$$\sum_{i=1}^{8} f(x_i)\Delta x = \Delta x\sum_{i=1}^{8} f(x_i) \quad \textit{Distributive property}$$

$$= .5(24.2792)$$

$$\approx 12.14$$

35. $f(x) = \frac{\ln x}{x}$; [1, 5]

$$\Delta x = \frac{5 - 1}{8} = .5$$

i	x_i	$f(x_i)$
1	1	0
2	1.5	.2703
3	2	.3466
4	2.5	.3665
5	3	.3662
6	3.5	.3579
7	4	.3466
8	4.5	.3342
		2.3883 Total

$$\sum_{i=1}^{8} f(x_i)\Delta x = \Delta x\sum_{i=1}^{8} f(x_i) \quad \textit{Distributive property}$$

$$= .5(2.3883)$$

$$\approx 1.19$$

Section 13.4

1. $\int_{-2}^{4} (-dp) = -1\int_{-2}^{4} dp$

$$= -1 \cdot p\Big]_{-2}^{4}$$

$$= -1[4 - (-2)]$$

$$= -6$$

3. $\int_{-1}^{2} (3t - 1)dt = 3\int_{-1}^{2} tdt - \int_{-1}^{2} dt$

$= \frac{3}{2}t^2\Big]_{-1}^{2} - t\Big]_{-1}^{2}$

$= \frac{3}{2}[2^2 - (-1)]^2 - [2 - (-1)]$

$= \frac{3}{2}(4 - 1) - (2 + 1)$

$= \frac{9}{2} - 3 = \frac{9 - 6}{2}$

$= \frac{3}{2}$

5. $\int_{0}^{2} (5x^2 - 4x + 2)dx$

$= 5\int_{0}^{2} x^2dx - 4\int_{0}^{2} xdx + 2\int_{0}^{2} dx$

$= \frac{5}{3}x^3\Big]_{0}^{2} - 2x^2\Big]_{0}^{2} + 2x\Big]_{0}^{2}$

$= \frac{5}{3}(2^3 - 0^3) - 2(2^2 - 0^2) + 2(2 - 0)$

$= \frac{5}{3}(8) - 2(4) + 2(2)$

$= \frac{40 - 24 + 12}{3}$

$= \frac{28}{3} \approx 9.33$

7. $\int_{4}^{9} 3\sqrt{u}\, du$

$= \int_{4}^{9} 3u^{1/2}\, du = 3\int_{4}^{9} u^{1/2}\, du$

$= 3\frac{u^{3/2}}{\frac{3}{2}}\Big]_{4}^{9} = 2u^{3/2}\Big]_{4}^{9}$

$= 2(9^{3/2} - 4^{3/2}) = 2(27 - 8)$

$= 2(19) = 38$

9. $\int_{0}^{1} 2(t^{1/2} - t)dt$

$= 2\int_{0}^{1} t^{1/2} - 2\int_{0}^{1} tdt$

$= 2\left(\frac{t^{3/2}}{\frac{3}{2}} - \frac{t^2}{2}\right)\Big]_{0}^{1}$

$= 2\left(\frac{2t^{3/2}}{3} - \frac{t^2}{2}\right)\Big]_{0}^{1}$

$= \left(\frac{2}{3} - \frac{1}{2}\right) - 2(0)$

$= 2\left(\frac{4 - 3}{6}\right)$

$= \frac{1}{3}$

11. $\int_{1}^{4} (5y\sqrt{y} + \sqrt{3y})dy$

$= \int_{1}^{4} (5y^{3/2} + 3y^{1/2})dy$

$= 5\left(\frac{y^{5/2}}{\frac{5}{2}}\right)\Big]_{1}^{4} + 3\left(\frac{y^{3/2}}{\frac{3}{2}}\right)\Big]_{1}^{4}$

$= 2y^{5/2}\Big]_{1}^{4} + 2y^{3/2}\Big]_{1}^{4}$

$= 2(4^{5/2} - 1) + 2(4^{3/2} - 1)$

$= 2(32 - 1) + 2(8 - 1)$

$= 62 + 14$

$= 76$

13. $\int_{1}^{3} \frac{2}{x^2dx} = \int_{1}^{3} 2x^{-2}\, dx = 2\int_{1}^{3} x^{-2}\, dx$

$= 2\left(\frac{x^{-1}}{-1}\right)\Big]_{1}^{3} = 2\left(-\frac{1}{x}\right)\Big]_{1}^{3}$

$= -2\left(\frac{1}{3} - 1\right) = -2\left(-\frac{2}{3}\right) = \frac{4}{3}$

15. $\int_1^5 (5n^{-1} + n^{-3})dn$

$= \int_1^5 5n^{-1}\, dn + \int_1^5 n^{-3}\, dn$

$= 5\int_1^5 \frac{1}{n}\, dn + \int_1^5 n^{-3}\, dn$

$= 5 \ln n\Big]_1^5 + \frac{n^{-2}}{-2}\Big]_1^5$

$= 5(\ln 5 - \ln 1) - \frac{1}{2}\left(\frac{1}{n^2}\right)\Big]_1^5$

$= 5 \ln 5 - \frac{1}{2}\left(\frac{1}{25} - 1\right)$

$= 5 \ln 5 - \frac{1}{2}\left(-\frac{24}{25}\right)$

$= 5 \ln 5 + \frac{12}{25}$

≈ 8.527

17. $\int_2^3 \left(2e^{-.1A} + \frac{3}{A}\right)dA$

$= 2\int_2^3 e^{-.1A}\, dA + 3\int_2^3 \frac{1}{A}\, dA$

$= 2\frac{e^{-.1A}}{-.1}\Big]_2^3 + 3 \ln |A|\Big]_2^3$

$= -20e^{-.1A}\Big]_2^3 + 3 \ln |A|\Big]_2^3$

$= 20e^{-.2} - 20e^{-.3} + 3 \ln 3 - 3 \ln 2$

≈ 2.775

19. $\int_1^2 \left(e^{5u} - \frac{1}{u^2}\right)dw = \int_1^2 e^{5u}\, du - \int_1^2 \frac{1}{u^2}\, du$

$= \frac{e^{5u}}{5}\Big]_1^2 + \frac{1}{u}\Big]_1^2$

$= \frac{e^{10}}{5} - \frac{e^5}{5} + \frac{1}{2} - 1$

$= \frac{e^{10}}{5} - \frac{e^5}{5} - \frac{1}{2}$

≈ 4375.1

21. $\int_{-1}^0 (2y - 3)^2\, dy$

Let $2y - 3 = u$

$dy = \frac{du}{2}$.

When $y = 0$, $u = -3$

$y = -1$, $u = -5$.

$\int_{-1}^0 (2y - 3)^2\, dy = \frac{1}{2}\int_{-5}^{-3} u^2\, du$

$= \frac{u^3}{6}\Big]_{-5}^{-3}$

$= \frac{1}{6}[(-3)^3 - (-5)^3]$

$= \frac{1}{6}(-27 + 125)$

$= \frac{98}{6}$

$= \frac{49}{3} \approx 16.33$

23. $\int_1^{64} \frac{\sqrt{z} - 2}{\sqrt[3]{z}}dz$

$= \int_1^{64} \left(\frac{z^{1/2}}{z^{1/3}} - 2z^{-1/3}\right)dz$

$= \int_1^{64} z^{1/6}\, dz - 2\int_1^{64} z^{-1/3}\, dz$

$= \frac{z^{7/6}}{\frac{7}{6}}\Big]_1^{64} - 2\frac{z^{2/3}}{\frac{2}{3}}\Big]_1^{64}$

$= \frac{6z^{7/6}}{7}\Big]_1^{64} - 3z^{2/3}\Big]_1^{64}$

$= \frac{6(64)^{7/6}}{7} - \frac{6(1)^{7/6}}{7} - 3(64^{2/3} - 1^{2/3})$

$= \frac{6(128)}{7} - \frac{6}{7} - 3(16 - 1)$

$= \frac{768 - 6 - 315}{7}$

$= \frac{447}{7}$

≈ 63.857

25. $f(x) = 2x + 3$; [8, 10]

Sketch the graph of the function and note that the graph does not cross x-axis in the given interval [8, 10].

$$\int_8^{10} (2x + 3)dx = \left(\frac{2x^2}{2} + 3x\right)\Bigg]_8^{10}$$

$$= (10^2 + 30) - (8^2 + 24)$$

$$= 42$$

27. $f(x) = 2 - 2x^2$; [0, 5]

Sketch the graph of the parabola and note that it crosses the x-axis. Find the points where the graph crosses the x-axis by solving

$$2 - 2x^2 = 0$$

$$2x^2 = 2$$

$$x^2 = 1$$

$$x = \pm 1.$$

The only solution in the interval [0, 5] is 1.

The total area is

$$\int_0^1 (2 - 2x^2)dx + \left|\int_1^5 (2 - 2x^2)dx\right|$$

$$= \left(2x - \frac{2x^3}{3}\right)\Bigg]_0^1 + \left|\left(2x - \frac{2x^3}{3}\right)\Bigg]_1^5\right|$$

$$= 2 - \frac{2}{3} + \left|10 - \frac{2(5^3)}{3} - 2 + \frac{2}{3}\right|$$

$$= \frac{4}{3} + \left|\frac{-224}{3}\right|$$

$$= \frac{228}{3}$$

$$= 76.$$

29. $f(x) = x^2 + 4x - 5$; [-1, 3]

Draw the graph of the parabola and note that it crosses the x-axis.

Solve $x^2 + 4x - 5 = 0$

$$(x + 5)(x - 1) = 0$$

$$x = -5 \text{ or } x = 1.$$

The graph crosses the x-axis at 1 in the interval [-1, 3].

The total area is

$$\left|\int_{-1}^1 (x^2 + 4x - 5)dx\right| + \int_1^3 (x^2 + 4x - 5)dx$$

$$= \left|\left(\frac{x^3}{3} + \frac{4x^2}{2} - 5x\right)\Bigg]_{-1}^1\right|$$

$$+ \left(\frac{x^3}{3} + \frac{4x^2}{2} - 5x\right)\Bigg]_1^3$$

$$= \left|\frac{1}{3} + 2 - 5 - \left(-\frac{1}{3} + 2 + 5\right)\right|$$

$$+ (9 + 18 - 15) - \left(\frac{1}{3} + 2 - 5\right)$$

$$= \left|\frac{2}{3} - 10\right| + 12 - \frac{1}{3} + 3$$

$$= \frac{28}{3} + 15 - \frac{1}{3}$$

$$= 24$$

31. $f(x) = x^3$; [-1, 3]

The solution of

$$x^3 = 0$$

$$x = 0$$

indicates that the graph crosses the x-axis at 0 in the given interval [-1, 3].

The total area is

$$\left|\int_{-1}^0 x^3dx\right| + \int_0^3 x^3dx$$

$$= \left|\frac{x^4}{4}\Bigg]_{-1}^0\right| + \frac{x^4}{4}\Bigg]_0^3$$

$$= \left|\left(0 - \frac{1}{4}\right)\right| + \left(\frac{3^4}{4} - 0\right)$$

$$= \frac{1}{4} + \frac{81}{4} = \frac{82}{4}$$

$$= \frac{41}{2}$$

33. $f(x) = e^x - 1$; $[-1, 2]$

Solve

$$e^x - 1 = 0$$
$$e^x = 1$$
$$x \ln e = \ln 1$$
$$x = 0.$$

The graph crosses the x-axis at 0 in the given interval [-1, 2]. The total area is

$$\left|\int_{-1}^{0} (e^x - 1)\,dx\right| + \int_{0}^{2} (e^x - 1)\,dx$$

$$= \left|(e^x - x)\Big]_{-1}^{0}\right| + (e^x - x)\Big]_{0}^{2}$$

$$= |(1 - 0) - (e^{-1} + 1)|$$
$$+ (e^2 - 2) - (1 - 0)$$
$$= |1 - e^{-1} - 1| + e^2 - 2 - 1$$
$$= \frac{1}{e} + e^2 - 3$$
$$\approx 4.757$$

35. $f(x) = \frac{1}{x}$; $[1, e]$

$\frac{1}{x} = 0$ has no solution so the graph does not cross the x-axis in the given interval [1, e].

$$\int_1^e \frac{1}{x}\,dx = \ln x\Big|_1^e$$
$$= \ln e - \ln 1$$
$$= 1$$

37. $\int_0^1 2x(x^2 + 1)^3\,dx$

Let $u = x^2 + 1$

$du = 2x\,dx.$

If $x = 0$, $u = 1$.

If $x = 1$, $u = 2$.

$$\int_0^1 2x(x^2 + 1)^3\,dx = \int_1^2 u^3\,du$$
$$= \frac{u^4}{4}\Big]_1^2$$
$$= \frac{1}{4}(2^4 - 1^4)$$
$$= \frac{1}{4}(16 - 1)$$
$$= \frac{15}{4}$$

39. $\int_0^4 \sqrt{x^2 + 12x}\,(x + 6)\,dx$

Let $u = x^2 + 12x$

$du = (2x + 12)\,dx = 2(x + 6)dx.$

So $(x + 6)dx = \frac{du}{2}$.

If $x = 0$, $u = 0$.

If $x = 4$, $u = 16 + 48 = 64$.

$$\int_0^4 \sqrt{x^2 + 12x}\,(x + 6)\,dx = \int_0^{64} \sqrt{u}\,\frac{du}{2}$$
$$= \frac{1}{2}\int_0^{64} u^{1/2}du$$
$$= \frac{1}{2}\frac{u^{3/2}}{\frac{3}{2}}\Bigg]_0^{64}$$
$$= \frac{1}{3}(64^{3/2} - 0^{3/2})$$
$$= \frac{1}{3}(512)$$
$$= \frac{512}{3} \approx 170.67$$

41. $\int_{-1}^{e-2} \frac{t}{t^2 + 2}\,dt$

Let $u = t^2 + 2$

$du = 2tdt$ or $tdt = \frac{du}{2}$.

If $t = -1$, $u = 3$.

If $t = e - 2$, $u = (e - 2)^2 + 2$

$= e^2 - 4e + 4 + 2$

$= e^2 - 4e + 6.$

$$\int_{-1}^{e-2} \frac{t}{t^2+2}\,dt = \frac{1}{2}\int_{3}^{e^2-4e+6} \frac{du}{u}$$

$$= \frac{1}{2}(\ln|u|)\Big]_3^{e^2-4e+6}$$

$$= \frac{1}{2}[\ln(e^2 - 4e + 6) - \ln 3]$$

$$= \ln\left(\frac{e^2 - 4e + 6}{3}\right)^{1/2}$$

$$\approx -.0880$$

43. $\int_0^1 ze^{2z^2d}\,dz$

Let $u = 2z^2$

$du = 4z\,dz$ or $z\,dz = \frac{du}{4}$.

If $z = 0$, $u = 0$

If $z = 1$, $u = 2$.

$$\int_0^1 ze^{2x^2}\,dz = \frac{1}{4}\int_0^2 e^u\,du$$

$$= \frac{1}{4}(e^u)\Big]_0^2$$

$$= \frac{1}{4}(e^2 - e^0)$$

$$= \frac{1}{4}(e^2 - 1)$$

$$\approx 1.597$$

45. $\int_0^2 (4 - x^2)\,dx + \left|\int_2^3 (4 - x^2)\,dx\right|$

$$= \left(4x - \frac{x^3}{3}\right)\Big]_0^2 + \left|\left(4x - \frac{x^3}{3}\right)\Big]_2^3\right|$$

$$= \left(8 - \frac{8}{3} - 0\right) + \left|(12 - 9) - \left(8 - \frac{8}{3}\right)\right|$$

$$= \frac{16}{3} + \frac{7}{3}$$

$$= \frac{23}{3} \approx 7.67$$

47. $y = e^2 - e^x$; $[1, 3]$

From the graph, we see that the total area is

$$\int_1^2 (e^2 - e^x)\,dx + \left|\int_2^3 (e^2 - e^x)\,dx\right|$$

$$= (e^2x - e^x)\Big]_1^2 + \left|(e^2x - e^x)\Big]_2^3\right|$$

$$= (2e^2 - e^2) - (e^2 - e)$$
$$+ \left|(3e^2 - e^3) - (2e^2 - e^2)\right|$$

$$= e^2 - e^2 + e$$
$$+ \left|3e^2 - e^3 - 2e^2 + e^2\right|$$

$$= e + \left|2e^2 - e^3\right|$$

$$= e + e^3 - 2e^2 \qquad e^3 > 2e^2$$

$$\approx 8.026.$$

49. $P(t) = 15t^{3/2} + 10$

Pollution in 25 months:

$$\int_0^{25} (15t^{3/2} + 10)\,dt$$

$$= \left(\frac{15t^{5/2}}{\frac{5}{2}} + 10t\right)\Bigg]_0^{25}$$

$$= (6t^{5/2} + 10t)\Big]_0^{25}$$

$$= [6(25)^{5/2} + 10(25)] - (0 + 0)$$
$$= 6(3125) + 250$$
$$= 19{,}000 \text{ units}$$

51. $E'(x) = 4x + 2$ is the rate of expenditure per day.

(a) The total expenditure in hundreds of dollars in 10 days is

$$\int_0^{10} (4x + 2)\,dx$$

$$= \left(\frac{4x^2}{2} + 2x\right)\Big]_0^{10}$$

$$= 2(100) + 20 - 0$$
$$= 220.$$

Therefore, since $220(100) = 22{,}000$, the total expenditure is \$22,000.

(b) From the tenth to the twentieth day:

$$\int_{10}^{20} (4x + 2)\,dx$$

$$= \left(\frac{4x^2}{2} + 2x\right)\Big]_{10}^{20}$$

$$= [2(400) + 40] - [2(200) + 20]$$
$$= 620$$

That is, \$62,000 is spent.

(c) If no more than \$5000, or 50(100), is spent,

$$\int_0^a (4x + 2)\,dx = 50$$

$$= (2x^2 + 2x)\Big]_0^a$$

$$= 2a^2 + 2a.$$

Solve $50 = 2a^2 + 2a$ by the quadratic formula.

$$2a^2 + 2a - 50 = 0$$
$$a^2 + a - 25 = 0$$

$$a = \frac{-1 \pm \sqrt{1 - 4(-25)}}{2}$$

Since the number of days must be positive,

$$a = \frac{-1 + \sqrt{101}}{2}$$

$$\approx 4.5 \text{ days.}$$

53. $R'(x) = 200e^x$

$$R(x) = 200 \int_0^{2.5} e^x dx$$

$$= 200(e^x)\Big]_0^{2.5}$$

$$= 200(e^{2.5} - e^0)$$
$$= 200(e^{2.5} - 1)$$
$$= 200(12.1825 - 1)$$
$$\approx 2236.5$$

55. $R'(t) = \frac{5}{t} + \frac{2}{t^2}$

(a) Total reaction from t = 1 to t = 12 is

$$\int_1^{12} \left(\frac{5}{t} + 2t^{-2}\right)dt$$

$$= (5 \ln t - 2t^{-1})\Big]_1^{12}$$

$$= \left(5 \ln 12 - \frac{1}{6}\right) - (5 \ln 1 - 2)$$

$$\approx 14.26.$$

(b) Total reaction from t = 12 to t = 24 is

$$\int_{12}^{24} \left(\frac{5}{t} + 2t^{-2}\right)dt$$

$$= \left(5 \ln t - \frac{2}{t}\right)\Big]_{12}^{24}$$

$$= \left(5 \ln 24 - \frac{1}{12}\right) - \left(5 \ln 12 - \frac{1}{6}\right)$$

$$\approx 3.55.$$

57. $c'(t) = 1.2e^{.04t}$

(a) $\int_0^{10} 1.2e^{.04t}dt$

(b) $= \frac{1.2e^{.04t}}{.04}\Big]_0^{10}$

$$= 30e^{.04t}\Big]_0^{10}$$

$$= 30e^{.4} - 30$$
$$\approx 14.75 \text{ billion}$$

(c) $\int_0^T 1.2e^{.04t}dt = 30e^{.04t}\Big]_0^T$

$$= 30e^{.04T} - 30$$

Solve

$$20 = 30e^{.04T} - 30.$$
$$50 = 30e^{.04T}$$
$$\frac{5}{3} = e^{.04T}$$
$$\ln \frac{5}{3} = .04T \ln e$$
$$T = \frac{\ln \frac{5}{3}}{.04}$$
$$\approx 12.8 \text{ years}$$

(d) $\int_0^T 1.2e^{.02t}dt = 60e^{.02t}\Big]_0^T$

$$= 60e^{.02T} - 60$$

Solve

$$20 = 60e^{.02T} - 60$$
$$80 = 60e^{.02T}$$
$$\frac{4}{3} = e^{.02T}$$
$$\ln \frac{4}{3} = .02T \ln e$$
$$T = \frac{\ln \frac{4}{3}}{.02}$$
$$\approx 14.4 \text{ years}$$

59. $C'(t) = 72e^{.014t}$

$$\int_0^T 72e^{.014t}dt = 72\frac{e^{.014t}}{.014}\Big]_0^T$$
$$= \frac{72e^{.014T}}{.014} - \frac{72e^0}{.014}$$
$$= \frac{72}{.014}(e^{.014T} - 1)$$

Section 13.5

1. $M'(x) = 60(1 + x^2)$

$$M(x) = \int_0^2 60(1 + x^2)dx$$
$$= 60 \int_0^2 (1 + x^2)dx$$
$$= 60\left(x + \frac{x^3}{3}\right)\Big]_0^2 = 60\left(2 + \frac{2^3}{3} - 0\right)$$
$$= 60\left(2 + \frac{8}{3}\right) = 120 + 160$$
$$= \$280$$

Monthly addition for maintenance:

$$\frac{280}{24} = \$11.67$$

3. Rate of savings:

$s(t) = 1000(t + 2)$

<u>During the first year</u>

Total savings:

$$\int_0^1 1000(t + 2)dt = 1000 \int_0^1 (t + 2)dt$$
$$= 1000\left(-\frac{t^2}{2} + 2t\right)\Big]_0^1$$
$$= 1000\left(\frac{1}{2} + 2 - 0\right)$$
$$= 1000\left(\frac{5}{2}\right)$$
$$= \$2500$$

<u>During the first 6 years</u>

Total savings:

$$\int_0^6 1000(t + 2)dt = 100 \int_0^6 (t + 2)dt$$
$$= 1000\left(\frac{t^2}{2} + 2t\right)\Big]_0^6$$
$$= 1000\left(\frac{36}{2} + 12 - 0\right)$$
$$= 1000(30)$$
$$= \$30,000$$

5. Rate of production:

$P(x) = 1000e^{.2x}$

In the first 4 years, total production:

$$\int_0^4 1000e^{.2x}dx = 1000\int_0^4 e^{.2x}dx$$

$$= 1000\left(\frac{e^{.2x}}{.2}\right)\Big]_0^4$$

$$= \frac{1000}{.2}(e^{.8} - e^0)$$

$$= 5000(2.226 - 1)$$

$$= 6127.7$$

so a production of 20,000 units in the first 4 years will not be met.

7. Rate of savings: $S(x) = 294 - x^2$

Cost: $C(x) = x^2 + \frac{1}{2}x$

(a) Rate of savings = Cost

$$294 - x^2 = x^2 + \frac{1}{2}x$$

$$0 = 2x^2 + \frac{1}{2}x - 294$$

$$0 = 4x^2 + x - 588$$

$$0 = (4x + 49)(x - 12)$$

$4x + 49 = 0$ or $x - 12 = 0$

$x = \frac{-49}{4}$ or $x - 12 = 0$

$x = 12$

Only 12 is a meaningful solution, so it will be profitable to use this machine for 12 years.

(b) Total net savings:

$$= \int_0^1 (294 - x^2)dx - \int_0^1 (x^2 + \frac{1}{2}x)dx$$

$$= \left(294 - \frac{x^3}{3}\right)\Big]_0^1 - \left(\frac{x^3}{3} + \frac{x^2}{4}\right)\Big]_0^1$$

$$= \left(294 - \frac{1}{3} - 0\right) - \left(\frac{1}{3} + \frac{1}{4}\right)$$

$$= 294 - \frac{1}{3} - \frac{1}{3} - \frac{1}{4}$$

$$\approx \$293$$

(c) Total net savings over the entire period:

$$\int_0^{12} (294 - x^2)dx - \int_0^{12} (x^2 + \frac{1}{2}x)dx$$

$$= \left(294x - \frac{x^3}{3}\right)\Big]_0^{12} - \left(\frac{x^3}{3} + \frac{x^2}{4}\right)\Big]_0^{12}$$

$$= \left[(294)(12) - \frac{12^3}{3} - 0\right]$$

$$- \left(\frac{12^3}{3} + \frac{12^2}{4} - 0\right)$$

$$= 2952 - 612$$

$$= \$2340$$

9. **(a)** The job should last until

$E(x) = I(x)$

$4x + 2 = 100 - x$

$5x = 98$

$x = 19.6,$

or for 19.6 days.

(b) $\int_0^{19.6} (100 - x)dx$

$$= \left(100x - \frac{x^2}{2}\right)\Big]_0^{19.6}$$

$$= 100(19.6) - \frac{(19.6)^2}{2}$$

$$= 1767.92$$

The total income is 1767.92(100), or \$176,792, since rate are in hundreds of dollars.

(c) $\int_0^{19.6} (4x + 2)dx$

$$= \left(\frac{4x^2}{2} + 2x\right)\Big]_0^{19.6}$$

= 17.50 cents or \$.175

= 807.52

The total expenditure is 807.52(100), or \$80,752.

(d) $\int_0^{19.6} [(100 - x) - (4x + 2)]dx$

$= \int_0^{19.6} (-5x + 98)dx$

$= \left(\frac{-5x^2}{2} + 98x\right)\Big]_0^{19.6}$

$= \frac{-5(19.6)^2}{2} + 98(19.6)$

$= 960.40$

The maximum profit is 960.40(100), or $96,040.

11. Producer's surplus:

$\int_0^3 [p(3) - p(x)]dx$

$= \int_0^3 [118 - (100 + 3x + x^2)]dx$

$= \int_0^3 (18 - 3x - x^2)dx$

$= \left(18x - \frac{3x^2}{2} - \frac{x^3}{3}\right)\Big]_0^3$

$= (18)(3) - \frac{3}{2}(3)^2 - \frac{1}{3}(3)^3$

$= \$31.50$

13. Consumer's surplus:

$\int_0^3 [p(x) - p(3)]dx$

$= \int_0^3 ([-(x + 4)^2 + 66] - 17)dx$

$= \int_0^3 (-x^2 - 8x + 33)dx$

$= \left(\frac{-x^3}{3} - 4x^2 + 33x\right)\Big]_0^3$

$= -\frac{1}{3}(3)^3 - 4(3)^2 + 33(3)$

$= \$54$

15. At equilibrium,
supply = demand = 20 units.
$P(20) = 20^2 + 2(20) + 50 = 490$
Producer's surplus is represented by the area between $y = 490$ and $P(x) = x^2 + 2x + 50$.

$\int_0^{20} 490\,dx - \int_0^{20} (x^2 + 2x + 50)dx$

$= 490x\Big]_0^{20} - \left(\frac{x^3}{3} + x^2 + 50x\right)\Big]_0^{20}$

$= 9800 - \left[\frac{20^3}{3} + 20^2 + 50(20)\right]$

$\approx \$5733.33$

17. Supply = Demand

$x^2 + \frac{11}{4}x = 150 - x^2$

$2x^2 + \frac{11}{4}x - 150 = 0$

$8x^2 + 11x - 600 = 0$

$(x - 8)(8x + 75) = 0$

$x - 8 = 0$ or $8x + 75 = 0$

$x = 8$ or $x = \frac{-75}{8}$

Since the number of units would not be negative, the equilibrium point occurs when $x = 8$.
When $x = 8$, equilibrium demand is $150 - (8)^2 = 86$.
So consumer's surplus is

$\int_0^8 [(150 - x^2) - 86]dx$

$= \int_0^8 (64 - x^2)dx$

$= \left(64x - \frac{x^3}{3}\right)\Big]_0^8$

$= (64)(8) - \frac{1}{3}(8)^3$

$= \frac{1024}{3}$

$\approx \$341.33$

When x = 8, equilibrium supply is

$(8)^2 + \frac{11}{4}(8) = 86.$

So producer's surplus is

$$\int_0^8 [86 - (x^2 + \frac{11}{4}x)]\,dx$$

$$= \int_0^8 (86 - x^2 - \frac{11}{4}x)\,dx$$

$$= \left(86x - \frac{x^3}{3} - \frac{11x^2}{8}\right)\Big]_0^8$$

$$= (86)(8) - \frac{1}{3}(8)^3 - \frac{11}{8}(8)^2$$

$$= 688 - \frac{512}{3} - 88$$

$$= \frac{1288}{3}$$

$\approx$ \$429.33

19. Total increase in costs:

$$\int_0^4 .45t^{3/2} = \frac{.45t^{5/2}}{\frac{5}{2}}\Big]_0^4$$

$$= .18t^{5/2}\Big]_0^4$$

$= .18(4^{5/2} - 0^{5/2})$

$= .18(32)$

$= 5.76$

Since cost is measured in millions of dollars, total increase in costs is \$5.76 millions.

21. **(a)** Since x thousand = 100,000, x = 100,000/1000 = 100.

Total repair costs for 100,000 miles:

$$\int_0^{100} .05x^{3/2}dx = \frac{.05x^{5/2}}{\frac{5}{2}}\Big]_0^{100} = .02x^{5/2}\Big]_0^{100}$$

$= .02(100)^{5/2}$

= \$2000

(b) x = 400,000/1000 = 400

Total repair costs for 400,000 miles:

$$\int_0^{400} .05x^{3/2}dx = \frac{.05x^{5/2}}{\frac{5}{2}}\Big]_0^{400} = .02x^{5/2}\Big]_0^{400}$$

$= .02(400)^{5/2}$

= \$64,000

23. Number of pollutants in lake in 4 years:

$$\int_0^4 4000\, e^{-t/10}\, dt = \frac{4000\, e^{-t/10}}{\frac{-1}{10}}\Big]_0^4$$

$$= 40{,}000e^{-t/10}\Big]_0^4$$

$= -40{,}000(e^{-4/10} - e^0)$

$= 40{,}000(.6703 - 1)$

= 13,187

Since total pollution level is above 4850 when all fish die, the factory cannot operate for 4 years.

25. Total profit in the first 3 years:

$$\int_0^3 (6t^2 + 4t + 5)\,dt = \frac{6t^3}{3} + \frac{4t^2}{2} + 5t\Big]_0^3$$

$= 2(3)^3 + 2(3)^3 + 5(3)$

= 54 + 18 + 15

= 87

Since profit is in hundreds of dollars, total profit is \$8700.

27. Total number of barrels leaked on first day:

$$\int_0^{24} (20t + 50)\,dt = (10t^2 + 50t)\Big]_0^{24}$$

$= 10(24)^2 + 50(24)$

= 6960

29. (a) Total number of hours required to produce 5 items:

$$\int_0^5 (20 - 2x)\,dx = (20x - x^2)\Big]_0^5$$

$= (20)(5) - (5)^2$

$= 100 - 25$

$= 75$ hours

(b) Total number of hours required to produce 10 items:

$$\int_0^{10} (20 - 2x)\,dx = (20x - x^2)\Big|_0^{10}$$

$= 20(10) - (10)^2$

$= 200 - 100$

$= 100$ hours

31. Number of feet the tree will grow in the third year:

$$\int_2^3 (.2 + 4t^{-4})\,dt = \left(.2t + \frac{4t^{-3}}{-3}\right)\Big]_2^3$$

$$= \left(.2t - \frac{4}{3t^3}\right)\Big]_2^3$$

$$= \left[(.2)(3) - \frac{4}{3(3)^3}\right] - \left[(.2)(2) - \frac{4}{3(2)^3}\right]$$

$= (.6 - .0494) - (.4 - .1667)$

$= .5506 - .2333$

$= .3173$

$\approx .32$ feet

33. $I(x) = .9x^2 + .1x$

(a) $I(.1) = .9(.1)^2 + .1(.1) = .019$

The lower 10% of the population earn 1.9% of the total income.

(b) $I(.4) = .9(.4)^4 + .1(.4) = .184$

The lower 40% of the population earn 18.4% of the total income.

(c) $I(.6) = .9(.6)^2 + .1(.6) = .384$

The lower 60% of the population earn 38.4% of the total income.

(d) $I(1.9) = .9(.9)^2 + .1(.9) = .819$

The lower 90% of the population earn 81.9% of the total income.

Section 13.6

1. $$\int \frac{-4}{\sqrt{x^2 + 36}}\,dx$$

$$= -4\int \frac{dx}{\sqrt{x^2 + 36}}$$

$$= -4 \ln\left|\frac{x + \sqrt{x^2 + 36}}{6}\right| + C$$

or $-4 \ln|x + \sqrt{x^2 + 36}| + C$

3. $$\int \frac{6}{x^2 - 9}\,dx$$

$$= 6\int \frac{1}{x^2 - 9}\,dx$$

Use entry 8 from table with $a = 3$. Note $x^2 > 3^2$.

$$= 6\left(\frac{1}{2(3)} \ln\left|\frac{x - 3}{x + 3}\right|\right) + C$$

$$= \ln\left|\frac{x - 3}{x + 3}\right| + C, \; x^2 > 9$$

5. $$\int \frac{-4}{x\sqrt{9 - x^2}}\,dx$$

$$= -4\int \frac{dx}{x\sqrt{9 - x^2}}$$

Use entry 9 from table with $a = 3$. Note $0 < x < 3$.

$$= \frac{4}{3} \ln\left|\frac{3 + \sqrt{9 - x^2}}{x}\right| + C,$$

$0 < x < 3$

7. $\int \frac{-2x}{3x + 1}\,dx$

$= -2\int \frac{x}{3x + 1}\,dx$

Use entry 11 from table with a = 3, b = 1.

$= -2\left(\frac{x}{3} - \frac{1}{9}\ln|3x + 1|\right) + C$

$= -\frac{2x}{3} + \frac{2}{9}\ln|3x + 1| + C$

9. $\int \frac{2}{3x(3x - 5)}\,dx$

$= \frac{2}{3}\int \frac{1}{x(3x - 5)}\,dx$

Use entry 13 from table with a = 3, b = 5.

$= \frac{2}{3}\left(-\frac{1}{5}\ln\left|\frac{x}{3x - 5}\right|\right) + C$

$= -\frac{2}{15}\ln\left|\frac{x}{3x - 5}\right| + C$

11. $\int \frac{4}{4x^2 - 1}\,dx$

$= 4\int \frac{dx}{4x^2 - 1}$

$= 2\int \frac{2\,dx}{4x^2 - 1}$

Let $u = 2x$

$du = 2\,dx.$

$= 2\int \frac{du}{u^2 - 1}$

Use entry 8 from table with a = 1, $u^2 > 1$.

$= 2 \cdot \frac{1}{2}\ln\left|\frac{u - 1}{u + 1}\right| + C,\ u^2 > 1,$

Substitute 2x for u. Since $u^2 > 1$, $4x^2 > 1$, or $x^2 > 1/4$.

$= \ln\left|\frac{2x - 2}{2x + 1}\right| + C,\ x^2 > \frac{1}{4}$

13. $\int \frac{3}{x\sqrt{1 - 9x^2}}\,dx$

$= 3\int \frac{3\,dx}{3x\sqrt{1 - 9x^2}}$

Let $u = 3x$

$du = 3\,dx.$

$= 3\int \frac{du}{u\sqrt{1 - u^2}}$

Use entry 9 from table with a = 1. Note $0 < u < 1$.

$= -3\ln\left|\frac{1 + \sqrt{1 - u^2}}{u}\right| + C$

$= -3\ln\left|\frac{1 + \sqrt{1 - 9x^2}}{3x}\right| + C,$

$0 < 3x < 1$, or $0 < x < \frac{1}{3}$

15. $\int \frac{4x}{2x + 3}\,dx$

$= 4\int \frac{x}{2x + 3}\,dx$

Use entry 11 from table with a = 2, b = 3.

$= 4\left(\frac{x}{2} - \frac{3}{4}\ln|2x + 3|\right) + C$

$= 2x - 3\ln|2x + 3| + C$

17. $\int \frac{-x}{(5x - 1)^2}\,dx$

$= -\int \frac{x\,dx}{(5x - 1)^2}$

Use entry 12 from table with a = 5, b = -1.

$= \left(\frac{-1}{25(5x - 1)} + \frac{1}{25}\ln|5x - 1|\right) + C$

$= \frac{1}{25(5x - 1)} - \frac{\ln|5x - 1|}{25} + C$

19. $\int x^4 \ln |x| dx$

Use entry 16, with n = 4.

$$= x^5\left(\frac{\ln|x|}{5} - \frac{1}{25}\right) + C$$

21. $\int \frac{\ln|x|}{x^2} dx = \int x^{-2} \ln|x| dx$

Use entry 16 from table with n = -2.

$$= x^{-1}\left(\frac{\ln|x|}{-1} - \frac{1}{1}\right) + C$$

$$= \frac{1}{x}(-\ln|x| - 1) + C$$

23. $\int xe^{-2x} dx$

Use entry 17 from table, with n = 1 and a = -2.

$$= \frac{xe^{-2x}}{-2} - \frac{1}{-2}\int x^0e^{-2x} dx$$

$$= -\frac{1}{2}xe^{-2x} + \frac{1}{2}\int e^{-2x} dx$$

$$= -\frac{1}{2}xe^{-2x} + \frac{1}{2}\left(\frac{e^{-2x}}{-2}\right) + C$$

$$= -\frac{1}{2}xe^{-2x} - \frac{1}{4}e^{-2x} + C$$

$$= -\frac{xe^{-2x}}{2} - \frac{e^{-2x}}{4} + C$$

25. Total revenue:

$$\int_0^{20} 10\sqrt{x^2 + 4}\, dx$$

Use entry 15 with a = 2.

$$= 10\left(\frac{x}{2}\sqrt{x^2 + 4} + \frac{4}{2}\ln\left|x + \sqrt{x^2 + 4}\right|\right)\Bigg]_0^{20}$$

$$= 10\left(\frac{20}{2}\sqrt{404} + 2\ln|20 + \sqrt{404}| - 2\ln 2\right)$$

$= 10(200.998 + 7.383 - 1.386)$

$\approx \$2070$

27. Total growth:

$$\int_0^3 30xe^{2x} dx$$

Use entry 17 with n = 1 and a = 2.

$$= 30\left(\frac{xe^{2x}}{2}\Bigg]_0^3 - \frac{1}{2}\int_0^3 e^{2x} dx\right)$$

$$= 15\left(3e^6 - \frac{e^{2x}}{2}\Bigg]_0^3\right)$$

$$= 15\left[3e^6 - \frac{1}{2}(e^6 - e^0)\right]$$

$$= 45e^6 - \frac{15}{2}(e^6 - 1)$$

$= 18154.30 - 3018.22$

$\approx$ 15,100 microbes

Section 13.7

1. $\frac{dy}{dx} = 8x - 7$

$dy = (8x - 7)dx$

$\int dy = \int (8x - 7)dx$

$y = \frac{8x^2}{2} - 7x + C$

$y = 4x^2 - 7x + C$

3. $\frac{dy}{dx} = -2x + 3x^2$

$dy = (-2x + 3x^2)dx$

$\int dy = \int(-2x + 3x^2)dx$

$y = \frac{-2x^2}{2} + \frac{3x^3}{3} + C$

$y = -x^2 + x^3 + C$

5. $\frac{dy}{dx} = e^{4x}$

$dy = e^{4x}\,dx$

$\int dy = \int e^{4x}\,dx$

$y = \frac{e^{4x}}{4} + C$

$y = \frac{1}{4}e^{4x} + C$

7. $\frac{dy}{dx} = x\sqrt{1 - x^2}$

$dy = x\sqrt{1 - x^2}\,dx$

$\int dy = \int x\sqrt{1 - x^2}\,dx$

Let $u = 1 - x^2$

$du = -2x\,dx$

$\frac{du}{-2} = x\,dx.$

$y = -\frac{1}{2}\int \sqrt{u}\,du$

$y = -\frac{1}{2}\,\frac{u^{3/2}}{\frac{3}{2}} + C$

$y = -\frac{1}{3}(1 - x^2)^{3/2} + C$

$y = \frac{-(1 - x^2)^{3/2}}{3} + C$

9. $y\frac{dy}{dx} = x$

$y\,dy = x\,dx$

$\frac{y^2}{2} = \frac{x^2}{2} + C$

$y^2 = x^2 + C$

11. $\frac{dy}{dx} = 2xy$

$\frac{dy}{y} = 2x\,dx$

$\int \frac{dy}{y} = \int 2x\,dx$

$\ln y = x^2 + C$

$y = e^{x^2+C} = e^{x^2}e^{C}$

$y = ke^{x^2}$

13. $\frac{dy}{dx} = 3x^2y = 2xy$

$\frac{dy}{dx} = y(3x^2 - 2x)$

$\frac{dy}{y} = (3x^2 - 2x)dx$

$\int \frac{dy}{y} = \int (3x^2 - 2x)dx$

$\ln|y| = \frac{3x^3}{3} - \frac{2x^2}{2} + C$

$\ln|y| = x^3 - x^2 + C$

$y = e^{x^3-x^2+C}$

$y = e^{x^3-x^2}e^{C}$

$y = ke^{x^3-x^2}$

15. $\frac{dy}{dx} = \frac{y}{x}$

$\frac{dy}{y} = \frac{dx}{x}$

$\int \frac{dy}{y} = \int \frac{dx}{x}$

$\ln|y| = \ln|x| + C$

$y = e^{\ln|x|+C}$

$y = e^{\ln x}e^{C}$

$y = Mx$

17. $\frac{dy}{dx} = y - 5$

$\frac{dy}{y - 5} = dx$

$\ln|y - 5| = x + C$

$y - 5 = e^{x+C} = e^{x}e^{C}$

$y = 5 + e^{x}e^{C}$

$y = 5 + Me^{x}$

19. $\frac{dy}{dx} = y^2e^x$

$\frac{dy}{y^2} = e^x dx$

$\int y^{-2}dy = \int e^x dx$

$-y^{-1} = e^x + C$

$-\frac{1}{y} = e^x + C$

$y = \frac{-1}{e^x + C}$

21. $\frac{dy}{dx} = \frac{x^2}{y}$; $y = 3$ when $x = 0$

$y\,dy = x^2dx$

$\int y\,dy = \int x^2dx$

$\frac{y^2}{2} = \frac{x^3}{3} + C$

Let $x = 0$ and $y = 3$.

$\frac{9}{2} = C$

$\frac{y^2}{2} = \frac{x^3}{3} + \frac{9}{2}$

$y^2 = \frac{2}{3}x^3 + 9$

23. $(2x + 3)y = \frac{dy}{dx}$; $y = 1$ when $x = 0$

$(2x + 3)dx = \frac{dy}{y}$

$\int (2x + 3)dx = \int \frac{dy}{y}$

$x^2 + 3x = \ln|y| + C$

If $x = 0$ and $y = 1$, $0 = \ln|1| + C$

$0 = C.$

$x^2 + 3x = \ln|y|$

$e^{x^2+3x} = y$

or $y = e^{x^2+3x}$

25. $\frac{dy}{dx} = \frac{y^2}{x}$; $y = 5$ when $x = e$

$\frac{dy}{y^2} = \frac{dx}{x}$

$\int \frac{dy}{y^2} = \int \frac{dx}{x}$

$\int y^{-2}dy = \int \frac{dx}{x}$

$-y^{-1} = \ln|x| + C$

$-\frac{1}{y} = \ln|x| + C$

If $x = e$ and $y = 5$,

$-\frac{1}{5} = \ln|e| + C$

$-\frac{1}{5} = 1 + C$

$\frac{-6}{5} = C.$

So $-\frac{1}{y} = \ln|x| - \frac{6}{5}$

$-\frac{1}{y} = \frac{5\ln|x| - 6}{5}$

$y = \frac{-5}{5\ln|x| - 6}.$

27. $\frac{dy}{dx} = \frac{2x + 1}{y - 3}$; $y = 4$ when $x = 0$

$(y - 3)dy = (2x + 1)dx$

$\int (y - 3)dy = \int (2x + 1)dx$

$\frac{y^2}{2} - 3y = x^2 + x + C$

If $x = 0$ and $y = 4$,

$\frac{16}{2} - 12 = C$

$-4 = C.$

So $\frac{y^2}{2} - 3y = x^2 + x - 4.$

29. $\frac{dy}{dx} = \frac{50}{y}\sqrt{x}$

$\int y\,dy = \int 50x^{-1/2}\,dx$

$\frac{y^2}{2} = \frac{50x^{1/2}}{\frac{1}{2}} + C$

$= 100x^{1/2}$ s+ C

$y^2 = 200x^{1/2} + k$

Since y = 150 when x = 4,

$150^2 = 200\sqrt{4} + k$

k = 22,100.

So $y^2 = 200\sqrt{x} + 22,100$.

(a) If x = 100,

$y^2 = 200(10) + 22,100$

$y \approx \$155.24$.

(b) If x = 400,

$y^2 = 200(20) + 22,100$

$y \approx \$161.55$.

(c) If x = 625,

$y^2 = 200(25) + 22,100$

$y \approx \$164.62$.

31. $\frac{dy}{dt} = -.06y$ *See Example 3*

$\int \frac{dy}{y} = \int -.06\,dt$

$\ln|y| = -.06t + C$

$e^{\ln|y|} = e^{-.06t+C}$

$e^{\ln|y|} = e^{-.06t} \cdot e^C$

$|y| = e^{-.06t} \cdot e^C$

$y = Me^{-.06t}$

Let y = 1 when t = 0.

Solve for M:

$1 = Me^0$

M = 1.

So $y = e^{-.06t}$.

If y = .50,

$.50 = e^{-.06t}$

$\ln .5 = -.06t \ln e$

$t = \frac{-\ln .5}{.06}$

≈ 11.6 years.

33. Let y = the number of firms that will be bankrupt.

$\frac{dy}{dt} = .06(1500 - y)$

$\int \frac{dy}{1500-y} = .06\int dt$

$-\ln|1500 - y| = .06t + C$

$\ln|1500 - y| = -.06t - C$

$|1500 - y| = e^{-.06t-C}$

$1500 - y = ke^{-.06t}$

$y = 1500 - ke^{-.06t}$

Since t = 0 when y = 100,

100 = 1500 - k(1)

= 1400.

$y = 1500 - 1400e^{-.06t}$

When t = 2 years,

$y = 1500 - 1400e^{-.06(2)}$

y = 258.31.

About 260 firms will be bankrupt in 2 years.

35. $\frac{dy}{dt} = -.03y$

$\int \frac{dy}{y} = -.03\int dt$

$\ln|y| = -.03t + C$

$e^{\ln|y|} = e^{-.03t+C}$

$y = Me^{-.03t}$

Since y = 6 when t = 0,

$6 = Me^0$

M = 6.

So $y = 6e^{-.03t}$.

If t = 10,

$y = 6e^{-.03(10)}$

≈ 4.4 cc.

37. $\frac{dy}{dx} = .01(5000 - y)$

$\int \frac{dy}{5000 - y} = \int .01\, dx$

$-\ln|5000 - y| = .01x + C$

$\ln|5000 - y| = -.01x + C$

$|5000 - y| = e^{-.01x-C}$

$|5000 - y| = e^{-.01x} \cdot e^{-C}$

$5000 - y = Me^{-.01x}$

$y = 500 - Me^{-.01x}$

Since $y = 150$ when $x = 0$,

$150 = 5000 - Me^0$

$M = 4850.$

So $y = 5000 - 4850e^{-.01x}$.

If $x = 5$,

$y = 5000 - 4850e^{-.01(5)}$

$\approx 387.$

39. (a) $\frac{dy}{dt} = ky$

$\int \frac{dy}{y} = \int k\, dt$

$\ln|y| = kt + C$

$e^{\ln|y|} = e^{kt+C}$

$y = \pm(e^{kt})(e^C)$

$y = Me^{kt}$

If $y = 1$ when $t = 0$ and $y = 5$ when $t = 2$, we have the following system of equations.

$1 = Me^{k(0)}$ *(1)*

$5 = Me^{2k}$ *(2)*

$1 = M(1)$ *From equation 1*

$M = 1$

$5 = (1)e^{2k}$ *Substitute into equation 2*

$e^{2k} = 5$

$2k \ln e = \ln 5$

$k = \frac{\ln 5}{2}$

$\approx .8$

(b) If $k = .8$ and $M = 1$,

$y = e^{.8t}$.

When $t = 3$,

$y = e^{.8(3)}$

$= {}^{2.4}$

$\approx 11.$

(c) When $t = 5$,

$y = e^{.8(5)}$

$= e^4$

$\approx 55.$

(d) When $t = 10$,

$y = e^{.8(10)}$

$= e^8$

$\approx 2981.$

41. (a) $\frac{dy}{dt} = -.05y$

(b) $\int \frac{dy}{y} = -\int .05\, dt$

$\ln|y| = -.05t + C$

$e^{\ln|y|} = e^{-.05t+C}$

$y = \pm e^{-0.5t} \cdot e^C$

$y = Me^{-.05t}$

(c) Since $y = 90$ when $t = 0$,

$90 = me^0$

$M = 90.$

So $y = 90e^{-.05t}$.

(d) At $t = 10$,

$y = 90e^{-.05(10)}$

≈ 55 grams.

Chapter 13 Review Exercises

1. $\int (2x^3 - x)dx = \int 2x^3dx - \int xdx$

$= \frac{2x^4}{4} - \frac{x^2}{2} + C$

$= \frac{x^4}{2} - \frac{x^2}{2} + C$

3. $\int (6x^2 - 2x + 5)dx$

$= \int 6x^2dx - \int 2x\ dx + \int 5dx$

$= \frac{6x^3}{3} - \frac{2x^2}{2} + \frac{5x}{1} + C$

$= 2x^3 - x^2 + 5x + C$

5. $\int (x^{3/2} - x + x^{2/5})dx$

$= \int x^{3/2}dx - \int x\ dx + \int x^{2/5}dx$

$= \frac{x^{5/2}}{5/2} - \frac{x^2}{2} + \frac{x^{7/5}}{7/5} + C$

$= \frac{2x^{5/2}}{5} - \frac{x^2}{2} + \frac{5x^{7/5}}{7} + C$

7. $\int (\sqrt{x})^3dx = \int (x^{1/2})^3dx = \int x^{3/2}dx$

$= \frac{x^{5/2}}{5/2} + C$

$= \frac{2x^{5/2}}{5} + C$

9. $\int \left(\frac{1}{x^3}\right)dx = \int x^{-3}dx$

$= \frac{x^{-2}}{-2} + C$

$= \frac{-x^{-2}}{2} + C$

11. $\int \left(\frac{-2}{\sqrt{x}} - \frac{3}{x}\right)dx = \int \left(-\frac{2}{x^{1/2}} - \frac{3}{x}\right)dx$

$= \int -2x^{-1/2}\ dx - \int \frac{3}{x}dx$

$= \frac{-2x^{1/2}}{1/2} - 3\ \ln\ |x| + C$

$= -4x^{1/2} - 3\ \ln\ |x| + C$

13. $\int e^{-5x}\ dx = \frac{e^{-5x}}{-5} + C$

$= \frac{-e^{-5x}}{5} + C$

15. $\int \frac{-5}{x}dx = -5\int \frac{dx}{x}$

$= -5\ \ln\ |x| + C$

17. $\int 2x\sqrt{x^2 - 3}\ dx = \int 2x(x^2 - 3)^{1/2}\ dx$

Let $u = x^2 - 3$

$du = 2x\ dx.$

$= \int u^{1/2}\ du$

$= \frac{u^{3/2}}{3/2} + C$

$= \frac{2(x^2 - 3)^{3/2}}{3} + C$

19. $\int \frac{x^2\ dx}{(x^3 + 5)^4} = \frac{1}{3}\int \frac{3x^2\ dx}{(x^3 + 5)^4}$

Let $u = x^3 + 5$

$du = 3x^2\ dx.$

$= \frac{1}{3}\int \frac{du}{u^4}$

$= \frac{1}{3}\int u^{-4}\ du$

$= \frac{1}{3}\left(\frac{u^{-3}}{-3}\right) + C$

$= \frac{-(x^3 + 5)^{-3}}{9} + C$

$= -\frac{1}{9(x^3 + 5)^3} + C$

21. $\int \frac{4x - 5}{2x^2 - 5x}\, dx$

Let $u = 2x^2 - 5x$

$du = (4x - 5)dx.$

$= \int \frac{du}{u}$

$= \ln |u| + C$

$= \ln |2x^2 - 5x| + C$

23. $\int \frac{x^3}{e^{3x^4}} dx = \int x^3 e^{-3x^4}$

$= -\frac{1}{12}\int -12x^3 e^{-3x^4}\, dx$

Let $u = -3x^4$

$du = -12x^3\, dx.$

$= -\frac{1}{12}\int e^u\, du$

$= -\frac{1}{12}e^u + C$

$= \frac{-e^{-3x^4}}{12} + C$

25. $\int -2e^{-5x}\, dx = \frac{-2}{-5}\int -5e^{-5x}\, dx$

Let $u = 5x$

$du = -5dx.$

$= \frac{2}{5}\int e^u\, du$

$= \frac{2}{5}e^u + C$

$= \frac{2e^{-5x}}{5} + C$

27. $C'(x) = 6x + 2;$ fixed cost is $75

$C(x) = \int (6x + 2)dx = \int 6x\, dx + \int 2dx$

$= \frac{6x^2}{2} + 2x + k$

$= 3x^2 + 2x + k$

When $x = 0$ and $C(x) = \$75$,

$75 = 3(0)^2 + 2(0) + k$

$75 = k.$

So $C(x) = 3x^2 + 2x + 75.$

29. $C'(x) = x^{3/4} + 2x;$ 16 units cost $375

$C(x) = \int (x^{3/4} + 2x)dx = \int x^{3/4}dx + \int 2x\, dx$

$= \frac{x^{7/4}}{7/4} + \frac{2x^2}{2} + k$

$= \frac{4x^{7/4}}{7} + x^2 + k$

$C(16) = 375$

$375 = \frac{4}{7}(16)^{7/4} + (16)^2 + k$

$375 = \frac{512}{7} + 256 + k$

$375 = \frac{512}{7} + 256 + k$

$\frac{321}{7} = k$

So $C(20) = \frac{4}{7}x^{7/4} + x^2 + \frac{321}{7}.$

31. Divide the time axis into 8 intervals of length 2 and one interval (16-20) of length 4.

Area $= (2)(5) + (2)(3) + (2)(2) + (2)(1.9) + (2)(1.3) + (2)(1) + (2)(.8) + (2)(.6) + (4)(.2)$

≈ 32 *Answers may vary*

33. $\int_1^2 (3x^2 + 5)dx = \left(\frac{3x^3}{3} + 5x\right)\Big]_1^2$

$= (2^3 + 10) - (1 + 5)$

$= 18 - 6$

$= 12$

35. $\int_1^5 (3x^{-2} + x^{-3})dx$

$= \left(\frac{3x^{-1}}{-1} + \frac{x^{-2}}{-2}\right)\Big]_1^5$

$= \left(-\frac{3}{5} - \frac{1}{50}\right) - \left(-3 - \frac{1}{2}\right)$

$= \frac{-31}{50} + \frac{7}{2}$

$= \frac{-31 + 175}{50}$

$= \frac{144}{50}$

$= \frac{72}{25} \approx 2.88$

37. $\int_1^3 2x^{-1}\,dx = 2\int_1^3 \frac{dx}{x}$

$= 2\ln|x|\Big]_1^3$

$= 2\ln 3 - 2\ln 1$

$= 2\ln 3$ or $\ln 9$

≈ 2.1972

39. $\int_0^4 2e^x\,dx = 2e^x\Big]_0^4$

$= 2e^4 - 2e^0$

$= 2e^4 - 2$

≈ 107.1963

41. $S' = \sqrt{x} + 2$

$S(x) = \int_0^9 (x^{1/2} + 2)\,dx$

$= \left(\frac{x^{3/2}}{3/2} + 2x\right)\Big]_0^9$

$= \frac{2}{3}(9)^{3/2} + 18$

$= 36$

Total sales are 36(1000), or 36,000 units.

43. $f(x) = 5 - x^2$, $g(x) = x^2 - 3$

Points of intersection:

$5 - x^2 = x^2 - 3$

$2x^2 - 8 = 0$

$2(x^2 - 4) = 0$

$x = \pm 2$

Since $f(x) \geq g(x)$ in $[-2, 2]$, the area between the graphs is

$\int_{-2}^2 [f(x) - g(x)]dx$

$= \int_{-2}^2 [5 - x^2) - (x^2 - 3)]dx$

$= \int_{-2}^2 (-2x^2 + 8)dx$

$= \left(\frac{-2x^3}{3} + 8x\right)\Big]_{-2}^2$

$= -\frac{2}{3}(8) + 16 + \frac{2}{3}(-8) - 8(-2)$

$= \frac{-32}{3} + 32$

$= \frac{64}{3} \approx 21.33.$

45. $\frac{dy}{dx} = -.2x$

$y = \int_0^{10} -.2x\,dx$

$y = \frac{-.2x^2}{2}\Big]_0^{10} = -.1x^2\Big]_0^{10}$

$y = -.1(100 - 0) = -10$

If $y = 100$ when $x = 0$ and $y = -10$ at 10, then $y = 90$ at $x = 10$.

47. $\int \frac{1}{\sqrt{x^2 - 64}}\,dx$

Use entry 6 from table with $a = 8$.

$= \ln\left|\frac{x + \sqrt{x^2 - 64}}{8}\right| + C$

49. $\int \frac{-2}{x(3x + 4)}\,dx = -2\int \frac{dx}{x(3x + 4)}$

Use entry 13 from table with $a = 3$ and $b = 4$.

$= -2\left(\frac{1}{4}\ln\left|\frac{x}{3x + 4}\right|\right) + C$

$= \frac{-1}{2}\ln\left|\frac{x}{3x + 4}\right| + C$

51. $\int 3x^2 \ln |x|\, dx$

Use entry 16 from table with n = 2.

$$= 3 \int x^2 \ln |x|\, dx$$

$$= 3(x^3)\left(\frac{\ln |x|}{3} - \frac{1}{9}\right) + C$$

$$= 3x^3\left(\frac{\ln |x|}{3} - \frac{1}{9}\right) + C$$

$$= 3x^3\left(\frac{3 \ln |x| - 1}{9}\right) + C$$

$$= x^3\left(\frac{3 \ln |x| - 1}{3}\right) + C$$

53. $\int \frac{12}{x^2 - 9} dx = 12\int \frac{dx}{x^2 - 9}$

Use entry 8 from table with a = 3.

$$= 12\left(\frac{1}{6} \ln \left|\frac{x - 3}{x + 3}\right|\right) + C$$

$$= 2 \ln \left|\frac{x - 3}{x + 3}\right| + C$$

55. $\frac{dy}{dx} = 2x^3 + 6x$

$$dy = 2x^3 + 6x)dx$$

$$\int dy = \int (2x^3 + 6x)dx$$

$$y = \frac{2x^4}{4} + \frac{6x^2}{2} + C$$

$$y = \frac{x^4}{2} + 3x^2 + C$$

57. $\frac{dy}{dx} = 4e^x$

$$dy = 4e^x dx$$

$$\int dy = \int 4e^x dy$$

$$y = 4e^x + C$$

59. $\frac{dy}{dx} = x^2 - 5x;\quad y = 1 \text{ when } x = 0$

$$dy = (x^2 - 5x)dx$$

$$\int dy = \int (x^2 - 5x)dx$$

$$y = \frac{x^3}{3} - \frac{5x^2}{2} + C$$

Let x = 0 and y = 1.

$$1 = C$$

So $y = \frac{x^3}{3} - \frac{5}{2}x^2 + 1.$

61. $\frac{dy}{dx} = 5(e^{-x} - 1);\ y = 17 \text{ when } x = 0.$

$$dy = 5(e^{-x} - 1)dx$$

$$\int dy = \int 5(e^{-x} - 1)dx$$

$$y = 5\left[\frac{e^{-x}}{-1} - x\right] + C$$

$$y = -5e^{-x} - 5x + C$$

Let x = 0 and y = 17.

$$17 = -5e^0 - 5(0) + C$$

$$17 = -5 + C$$

$$22 = C$$

$$y = -5e^{-x} - 5x + 22$$

63. $\frac{dy}{dx} = \frac{3x + 1}{y}$

$$ydy = (3x + 1)dx$$

$$\int ydy = \int (3x + 1)dx$$

$$\frac{y^2}{2} = \frac{3x^2}{2} + x + C$$

$$y^2 = 3x^2 + 2x + C$$

65. $\frac{dy}{dx} = \frac{2y + 1}{x}$

$\frac{dy}{2y + 1} = \frac{dx}{x}$

$\int \frac{dy}{2y + 1} = \int \frac{dx}{x}$

$\frac{1}{2} \int \frac{2dy}{2y + 1} = \int \frac{dx}{x}$

$\frac{1}{2} \ln |2y + 1| = \ln |x| + \ln C$

$\ln |2y + 1|^{1/2} = \ln C|x|$

$(2y + 1)^{1/2} = Cx$

$2y + 1 = Cx^2$

$2y = Cx^2 - 1$

$y = \frac{Cx^2 - 1}{2}$

67. $(5 - 2x)y = \frac{dy}{dx}$; $y = 2$ when $x = 0$

$(5 - 2x)dx = \frac{dy}{y}$

$\int (5 - 2x)dx = \int \frac{dy}{y}$

$5x - x^2 = \ln y + C$

Let $x = 0$ and $y = 2$.

$0 = \ln 2 + C$

$C = -\ln 2$

$5x - x^2 = \ln y - \ln 2 = \ln \frac{y}{2}$

$e^{5x-x^2} = \frac{y}{2}$

$2e^{5x-x^2} = y$

$y = 2e^{5x-x^2}$

69. $\frac{dy}{dx} = \frac{1 - 2x}{y + 3}$; $y = 16$ when $x = 0$

$(y + 3)dy = (1 - 2x)dx$

$\int (y + 3)dy = \int (1 - 2x)dx$

$\frac{y^2}{2} + 3y = (x - x^2) + C$

Let $x = 0$ and $y = 16$.

$\frac{(16)^2}{2} + 3(16) = C$

$176 = C$

$\frac{y^2}{2} + 3y = (x - x^2) + 176$

$y^2 + 6y = 2(x - x^2) + 352$

$y^2 + 6y = 2x - 2x^2 + 352$

71. (a) $\frac{dy}{dx} = 5e^{.2x}$

$dy = 5e^{.2x}\, dx$

$y = \frac{5}{.2}e^{.2x} + C$

$y = 25e^{.2x} + C$

When $x = 0$, $y = 0$.

$0 = 25e^0 + C$

$c = -25$ $\qquad e^0 = 1$

$y = 25e^{.2x} - 25$

When $x = 6$,

$y = 25e^{1.2} - 25$

$\approx 58.$

Sales are $5800.

(b) When $x = 12$,

$y = 25e^{2.4} - 25$

$\approx 251.$

Sales are $25,100.

73. $\frac{dA}{dt} = rA - D$; $t = 0$, $A = \$10,000$;

$r = .05$; $D = \$1000$

$\frac{dA}{dt} = .05A - 1000$

$\int \frac{dA}{.05A - 1000} = \int dt$

$\frac{1}{.05} \ln |.05A - 1000| = t + C_2$

$\ln |.05A - 1000| = .05t + C_1$

$.05A - 1000 = Ce^{.05t}$

$A = 20Ce^{.05t} + 20,000$

Since $t = 0$ when $A = 10{,}000$

$$10{,}000 = 20C(1) + 20{,}000$$
$$C = -500.$$
$$A = 20(-500)e^{.05t} + 20{,}000$$
$$A = 20{,}000 - 10{,}000e^{.05t}$$

Find t for A = 0.

$$0 = 20{,}000 - 10{,}000e^{.05t}$$
$$e^{.05t} = \frac{20{,}000}{10{,}000} = 2$$
$$\ln e^{.05t} = \ln 2$$
$$t = 20 \ln 2 = 13.86$$

It will take about 13.9 years.

75. $$\frac{dy}{dt} = ky;\ k = .1,\ t = 0,\ y = 120$$
$$\frac{dy}{y} = k\,dt$$
$$\ln |y| = kt + C_1$$
$$|y| = e^{kt+C_1}$$
$$y = Me^{kt}$$
$$y = Me^{.1t}$$
$$120 = Me^0$$
$$M = 120$$
$$y = 120e^{.1t}$$

Let t = 6 and find y.

$$y = 120e^{.6}$$
$$\approx 219$$

After 6 weeks about 219 are present.

Case 16 Exercises

1. $2{,}000{,}000 \div 19{,}600 \approx 102$ years

2. $$2{,}000{,}000 = \frac{19{,}600}{.02}(e^{.02t_1} - 1)$$
$$\frac{2{,}000{,}000(.02)}{19{,}600} = e^{.02t_1} - 1$$
$$2.0408 + 1 = e^{.02t_1}$$
$$3.0408 = e^{.02t_1}$$
$$\ln 3.0408 = .02t_1$$
$$t_1 = \frac{\ln 3.0408}{.02}$$
$$t_1 = 55.6$$

or about 55.6 years.

3. $$15{,}000{,}000 = \frac{63{,}000}{.06}(e^{.06t_1} - 1)$$
$$\frac{15{,}000{,}000(.06)}{63{,}000} + 1 = e^{.06t_1}$$
$$15.286 \approx e^{.06t_1}$$
$$\ln 15.286 \approx .06t_1 \ln e$$
$$t_1 = \frac{\ln 15.286}{.06}$$
$$\approx 45.4$$

or about 45.4 years.

4. $$2{,}000{,}000 = \frac{2200}{.04}(e^{.04t_1} - 1)$$
$$\frac{2{,}000{,}000(.04)}{2200} + 1 = e^{.04t_1}$$
$$37.36 = e^{.04t_1}$$
$$\ln 37.36 = .04t_1$$
$$t_1 = 90.5$$

or about 90 years

Case 17 Exercises

1. $$r = 50\left(1 - \frac{48}{24} + \frac{48}{24}e^{-24/48}\right)$$
$$= 50(1 - 2 + 2e^{-.5})$$
$$= 10.65$$

2. $$r = 1000\left(1 - \frac{60}{24} + \frac{60}{24}e^{-24/60}\right)$$
$$= 1000(1 - 2.5 + 2.5e^{-.4})$$
$$\approx 175.8$$

3. $$r = 1200\left(1 - \frac{30}{30} + \frac{30}{30}e^{-1}\right)$$
$$= 1200(e^{-1})$$
$$\approx 441.46$$

CHAPTER 14 MULTIVARIATE CALCULUS

Section 14.1

1. $f(x, y) = 4x + 5y + 3$

(a) $f(2, -1) = 4(2) + 5(-1) + 3 = 6$

(b) $f(-4, 1) = 4(-4) + 5(1) + 3 = -8$

(c) $f(-2, -3) = 4(-2) + 5(-3) + 3 = -20$

(d) $f(0, 8) = 4(0) + 5(8) + 3 = 43$

3. $h(x, y) = \sqrt{x^2 + 2y^2}$

(a) $h(5, 3) = \sqrt{25 + 2(9)} = \sqrt{43}$

(b) $h(2, 4) = \sqrt{4 + 32} = 6$

(c) $h(-1, -3) = \sqrt{1 + 18} = \sqrt{19}$

(d) $h(-3, -1) = \sqrt{9 + 2} = \sqrt{11}$

See the graphs for Exercises 5-13 in the answer section at the back of the textbook. In each exercise the graph of the plane is obtained by finding the x, y, and z intercepts.

5. $x + y + z = 6$
x-intercept: Let $y = 0$ and $z = 0$, then $x = 6$. y-intercept: Let $x = 0$ and $z = 0$, then $y = 6$. z-intercept: Let $x = 0$ and $y = 0$, then $z = 6$. Plot the points (6, 0, 0), (0, 6, 0), and (0, 0, 6). The plane through these points is the graph of the function. Show the region in the first octant, which is a triangular surface.

7. $2x + 3y + 4z = 12$
Let $y = 0$ and $z = 0$, then $x = 6$. Let $x = 0$ and $z = 0$, then $y = 4$. Let $x = 0$ and $y = 0$, then $z = 3$.
Plot the points (6, 0, 0), (0, 4, 0) and (0, 0, 3), and show the part of the plane through them in the first octant.

9. $3x + 2y + z = 18$
Let $y = 0$ and $z = 0$, then $x = 6$. Let $x = 0$ and $z = 0$, then $y = 9$. Let $x = 0$ and $y = 0$, then $z = 18$.
Plot the points (6, 0, 0), (0, 9, 0) and (0, 0, 18).

11. $x + y = 4$
Let $y = 0$, then $x = 4$. Let $x = 0$, then $y = 4$.
Since there is no z in the equation there can be no z-intercept. A plane that has no z-intercept is parallel to the z-axes. Plot the points (4, 0, 0) and (0, 4, 0) and draw the plane parallel to the z-axes.

13. $x = 2$
Since there is no y nor z in the equation, the graph is a plane parallel to the y- and z-axes through the point (2, 0, 0).

15. $m = \frac{2.5(T - F)}{w^{.6T}}$

(a) $m = \frac{2.5(38 - 6)}{(32)^{.67}} \approx 7.85$

(b) $m = \frac{2.5(40 - 20)}{(43)^{.67}} \approx 4.02$

17. $P(x, y) = 100\left[\frac{3}{5}x^{-2/5} + \frac{2}{5}y^{-2/5}\right]^{-5}$

(a) $P(32,1) = 100\left[\frac{3}{5}(32)^{-2/5} + \frac{2}{5}(1)^{-2/5}\right]^{-5}$

$= 100\left[\frac{3}{5}\left(\frac{1}{4}\right) + \frac{2}{5}(1)\right]^{-5}$

$= 100\left(\frac{11}{20}\right)^{-5}$

$= 100\left(\frac{20}{11}\right)^{-5}$

$\approx 1986.95.$

(b) $P(1, 32) = 100\left[\frac{3}{5}(1)^{-2/5} + \frac{2}{5}(32)^{-2/5}\right]^{-5}$

$= 100\left[\frac{3}{5}(1) + \frac{2}{5}\left(\frac{1}{4}\right)\right]^{-5}$

$= 100\left(\frac{7}{10}\right)^{-5}$

$= 100\left(\frac{10}{7}\right)^{-5}$

≈ 595

(c) 32 work hours means that $x = 32$. 243 units of capital means that $y = 243$.

$P(32, 243) = 100\left[\frac{3}{5}(32)^{-2/5} + \frac{2}{5}(243)^{-2/5}\right]^{-5}$

$= 100\left[\frac{3}{5}\left(\frac{1}{4}\right) + \frac{2}{5}\left(\frac{1}{9}\right)\right]^{-5}$

$= 100\left(\frac{7}{36}\right)^{-5}$

$= 100\left(\frac{36}{7}\right)^{5}$

$\approx 359{,}767.81$

19. $z = x^{.4}y^{.6}$ where $z = 500$

$500 = x^{2/5}y^{3/5}$

$\frac{500}{x^{2/5}} = y^{3/5}$

$\left(\frac{500}{x^{2/5}}\right)^{5/3} = (y^{3/5})^{5/3}$

$y = \frac{(500)^{5/3}}{x^{2/3}}$

$y \approx \frac{31{,}498}{x^{2/3}}$

See the graph in the answer section at the back of the textbook.

21. $z = x^2 + y^2$

Find the traces.

xy-trace: $0 = x^2 + y^2$

xz-trace: $z = x^2$

yz-trace: $z = y^2$

These traces match with the graph in (c), the correct graph.

23. $x^2 - y^2 = z$

xy-trace: $x^2 - y^2 = 0$

yz-trace: $y^2 = -z$

xz-trace: $x^2 = z$

These traces match with the graph of (e), the correct graph.

25. $\frac{x^2}{16} + \frac{y^2}{25} + \frac{z^2}{4} = 1$

xy-trace: $\frac{x^2}{16} + \frac{y^2}{25} = 1$

yz-trace: $\frac{y^2}{25} + \frac{z^2}{4} = 1$

xz-trace: $\frac{x^2}{16} + \frac{z^2}{4} = 1$

These traces match with the graph of (b), the correct graph.

27. $z = 3 - x^2 - y^2$

xy-trace: $3 = x^2 + y^2$

xz-trace: $z = 3 - x^2$

yz-trace: $z = 3 - y^2$

<u>Level curve</u>

$z = 1$: $-2 = -x^2 - y^2$

$2 = x^2 + y^2$

$z = 2$: $-1 = -x^2 - y^2$

$1 = x^2 + y^2$

$z = 3$: $0 = -x^2 - y^2$

$0 = x^2 + y^2$

At $z = 1$, draw a circle of radius $\sqrt{2}$.
At $z = 2$, draw a circle of radius 1.
At $z = 3$, draw a point.
See the graph in the answer section at the back of this textbook.

29. **(a)** If $f(x, y) = 9x^2 - 3y^2$, then

$$\frac{f(x + h, y) - f(x, y)}{h}$$

$$= \frac{[9(x + h)^2 - 3y^2] - (9x^2 - 3y^2)}{h}$$

$$= \frac{9x^2 + 18xh + 9h^2 - 3y^2 - 9x^2 + 3y^2}{h}$$

$$= \frac{h(18x + 9h)}{h}$$

$$= 18x + 9h.$$

(b) If $f(x, y) = 9x^2 - 3y^2$, then

$$\frac{f(x, y + h) - f(x, y)}{h}$$

$$= \frac{[9x^2 - 3(y + h)^2] - (9x^2 - 3y^2)}{h}$$

$$= \frac{9x^2 - 3y^2 - 6yh - 3h^2 - 9x^2 + 3y^2}{h}$$

$$= \frac{-h(6y + 3h}{h}$$

$$= -(6y + 3h)$$
$$= -6y - 3h.$$

Section 14.2

1. $z = f(x, y) = 12x^2 - 8xy + 3y^2$

(a) $\frac{\partial z}{\partial x} = 24x - 8y$

(b) $\frac{\partial z}{\partial y} = -8x + 6y$

(c) $\left(\frac{\partial f}{\partial x}\right)(2, 3) = 24(2) - 8(3) = 24$

(d) $f_y(1, -2) = -8(1) + 6(-2)$
$= -20$

3. $f(x, y) = -2xy + 6y^3 + 2$

$f_x = -2y$

$f_y = -2x + 18y^2$

$f_x(2, -1) = -2(-1) = 2$

$f_y(-4, 3) = -2(-4) + 18(3)^2$
$= 8 + 18(9)$
$= 170$

5. $f(x, y) = 3x^3y^2$

$f_x = 9x^2y^2$

$f_y = 6x^3y$

$f_x(2, -1) = 9(4)(1) = 36$

$f_y(-4, 3) = 6(-64)3 = -1152$

7. $f(x, y) = e^{x+y}$

$f_x = e^{x+y}$

$f_y = e^{x+y}$

$f_x(2, -1) = e^{2-1} = e$

$f_y(-4, 3) = e^{-4+3} = e^{-1} = \frac{1}{e}$

9. $f(x, y) = -5e^{3x-4y}$

$f_x = -15e^{3x-4y}$

$f_y = 20e^{3x-4y}$

$f_x(2, -1) = -15e^{10}$

$f_y(-4, 3) = 20e^{-12-12} = 20e^{-24}$ or $\frac{20}{e^{24}}$

11. $f(x, y) = \dfrac{x^2 + y^3}{x^3 - y^2}$

$f_x = \dfrac{2x(x^3 - y^2) - 3x^2(x^2 + y^3)}{(x^3 - y^2)^2}$

$= \dfrac{2x^4 - 2xy^2 - 3x^4 - 3x^2y^3}{(x^3 - y^2)^2}$

$= \dfrac{-x^4 - 2xy^2 - 3x^2y^3}{(x^3 - y^2)^2}$

$f_y = \dfrac{3y^2(x^3 - y^2) - (-2y)\ (x^2 + y^3)}{(x^3 - y^2)^2}$

$= \dfrac{3x^3y^2 - 3y^4 + 2x^2y + 2y^4}{(x^3 - y^2)^2}$

$= \dfrac{3x^3y^2 - y^4 + 2x^2y}{(x^3 - y^2)^2}$

$f_x(2, -1)$

$= \dfrac{-2^4 - 2(2)(-1)^2 - 3(2^2)(-1)^3}{[2^3 - (-1)^2]^2}$

$= -\dfrac{8}{49}$

$f_y(-4, 3)$

$= \dfrac{3(-4)^3(3)^2 - 3^4 + 2(-4)^2(3)}{[(-4)^3 - 3^2]^2}$

$= -\dfrac{1713}{5329}$

13. $f(x, y) = \ln(1 + 3x^2y^3)$

$f_x = \dfrac{1}{1 + 3x2y^3}.6xy^3$

$= \dfrac{6xy^3}{1 + 3x^2y^3}$

$f_y = \dfrac{9x^2y^2}{1 + 3x^2y^3}$

$f_x(2, -1) = \dfrac{6(2)(-1)}{1 + 3(4)(-1)}$

$= \dfrac{-12}{1 - 12} = \dfrac{12}{11}$

$f_y(-4, 3) = \dfrac{9(16)(9)}{1 + 3(16)(27)}$

$= \dfrac{1296}{1297}$

15. $f(x, y) = xe^{x^2y}$

$f_x = e^{x^2y}.1 + x(2xy)(e^{x^2y})$

$= e^{x^2y}(1 + 2x^2y)$

$f_y = x^3e^{x^2y}$

$f_x(2, -1) = e^{-4}(1 - 8) = -7e^{-4}$ or $\dfrac{-7}{e^4}$

$f_y(-4, 3) = -64e^{48}$

17. $f(x, y) = 6x^3y - 9y^2 + 2x$

$f_x = 18x^2y + 2$

$f_y = 6x^3 - 18y$

$f_{xx} = 36xy$

$f_{yy} = -18$

$f_{xy} = 18x^2 = f_{yx}$

19. $R(x, y) = 4x^2 - 5xy^3 + 12y^2x^2$

$R_x = 8x - 5y^3 + 24y^2x$

$R_y = -15xy^2 + 24yx^2$

$R_{xx} = 8 + 24y^2$

$R_{yy} = -30xy + 24x^2$

$R_{xy} = -15y^2 + 48xy = R_{yx}$

21. $r(x, y) = \dfrac{4x}{x + y}$

$r_x = \dfrac{4(x + y) - 4x}{(x + y)^2}$

$= 4y(x + y)^{-2}$

$r_y = \dfrac{-4x}{(x + y)^2}$

$r_{xx} = -8y(x + y)^{-3} = \dfrac{-8y}{(x + y)^3}$

$r_{yy} = 8x(x + y)^{-3} = \dfrac{8x}{(x + y)^3}$

$r_{xy} = 4(x - y)(x + y)^{-3}$

$= \dfrac{4x - 4y}{(x + y)^3} = r_{yx}$

23.
$$z = 4xe^y$$
$$z_x = 4e^y$$
$$z_y = 4xe^y$$
$$z_{xx} = 0$$
$$z_{yy} = 4xe^y$$
$$z_{xy} = 4e^y = z_{yx}$$

25.
$$r = \ln |x + y|$$
$$r_x = \frac{1}{x + y}$$
$$r_y = \frac{1}{x + y}$$
$$r_{xx} = \frac{-1}{(x + y)^2}$$
$$r_{yy} = \frac{-1}{(x + y)^2}$$
$$r_{xy} = \frac{-1}{(x + y)^2} = r_{yx}$$

27.
$$z = x \ln |xy|$$
$$z_x = \ln xy + 1$$
$$z_y = \frac{x}{y}$$
$$z_{xx} = \frac{1}{x}$$
$$z_{yy} = -xy^{-2} = \frac{-x}{y^2}$$
$$z_{xy} = \frac{1}{y} = z_{yx}$$

29. $f(x, y, z) = x^2 + yz + z^4$
$$f_x = 2x$$
$$f_y = z$$
$$f_z = y + 4x^3$$
$$f_{yz} = 1$$

31. $f(x, y, z) = \dfrac{6x - 5y}{4z + 5}$
$$f_x = \frac{6}{4z + 5}$$
$$f_y = \frac{-5}{4z + 5}$$
$$f_z = \frac{-4(6x - 5y)}{(4z + 5)^2}$$
$$f_{yz} = \frac{20}{(4z + 5)^2}$$

33. $f(x, y, z) = \ln |x^2 - 5xz^2 + y^4|$
$$f_x = \frac{2x - 5z^2}{x^2 - 5xz^2 + y^4}$$
$$f_y = \frac{4y^3}{x^2 - 5xz^2 + y^4}$$
$$f_z = \frac{-10xz}{x^2 - 5xz^2 + y^4}$$
$$f_{yz} = \frac{4y^3(10zx)}{(x^2 - 5xz^2 + y^4)^2}$$
$$= \frac{40xy^3z}{(x^2 - 5xz^2 + y^4)^2}$$

35. $M(x, y) - 40x^2 + 30y^2 - 10xy + 30$

(a)
$$M_y = 60y - 10x$$
$$M_y(4, 2) = 120 - 40 = 80$$

(b)
$$M_x = 80x - 10y$$
$$M_x(3, 6) = 240 - 60 = 180$$

(c)
$$\left(\frac{\partial M}{\partial x}\right)(2, 5) = 80(2) - 10(5)$$
$$= 110$$

(d)
$$\left(\frac{\partial M}{\partial y}\right)(6, 7) = 60(7) - 10(6)$$
$$= 360$$

37. $f(p, i) = 132p - 2pi - .01p^2$

(a) $f(9400, 8)$

$= 132(9400) - 2(9400)(8) - .01(9400)^2$

$= \$206{,}800$

(b) $f_p = 132 - 2i - .02p$, which represents the rate of change in weekly sales revenue per unit change in price when the interest rate remains constant.

$f_i = -2p$, which represents the rate of change in weekly sales revenue per unit change in interest rate when the list price remains constant.

(c) $p = 9400$ remains constant and i changes by 1 unit from 8 to 9.

$f_i(p, i) = f_i(9400, 8)$
$= -2(9400)$
$= -18{,}800$

Therefore, sales revenue declines by $18,800.

39. $f(x, y) = \left[\frac{1}{3}x^{-1/3} + \frac{2}{3}y^{-1/3}\right]^{-3}$

(a) $f(27, 64)$

$= \left[\frac{1}{3}(27)^{-1/3} + \frac{2}{3}(64)^{-1/3}\right]^{-3}$

$= \left[\frac{1}{3}\left(\frac{1}{3}\right) + \frac{2}{3}\left(\frac{1}{4}\right)\right]^{-3}$

$= \left(\frac{1}{9} + \frac{1}{6}\right)^{-3}$

$= \left(\frac{5}{18}\right)^{-3}$

$= \left(\frac{18}{5}\right)^{3}$

$= 46.656$

(b) $f_x = -3\left[\frac{1}{3}x^{-1/3} + \frac{2}{3}y^{-1/3}\right]\left(-\frac{1}{9}x^{-4/3}\right)$

$f_x(27, 64)$

$= -3\left[\frac{1}{3}(27)^{-1/3} + \frac{2}{3}(64)^{-1/3}\right]^{-4}\cdot$

$\left(-\frac{1}{9}\right)(27)^{-4/3}$

$= -3\left(\frac{5}{18}\right)^{-4}\left(-\frac{1}{9}\right)\left(\frac{1}{81}\right)$

$= \frac{432}{625}$

$f_x(27, 64) = .6912$, which represents the rate at which production is changing when labor changes by 1 unit from 27 to 28 and capital remains constant.

$f_y = -3\left[\frac{1}{3}x^{-1/3} + \frac{2}{3}y^{-1/3}\right]^{-4}\left(-\frac{2}{9}y^{-4/3}\right)$

$f_y(27, 64)$

$= -3\left[\frac{1}{3}(27)^{-1/3} + \frac{2}{3}(64)^{-1/3}\right]^{-4}\cdot$

$\left(-\frac{2}{9}\right)(64)^{-4/3}$

$= -3\left(\frac{5}{18}\right)^{-4}\left(-\frac{2}{9}\right)\left(\frac{1}{256}\right)$

$= \frac{2187}{5000}$

$= .4374$, which represents the rate at which production is changing when capital changes by 1 unit from 64 to 65 and labor remains constant.

(c) If labor increases by 1 unit then production would increase by .6912(100) or about 69 units. (See part (b) of this solution.)

41. $z = x^{.7}y^{.3}$, where x is labor, y is capitol.

Marginal productivity of labor is

$\frac{\partial z}{\partial x} = .7x^{-.3}y^{.3}$ or $\frac{.7y^{.3}}{x^{.3}}$.

Marginal productivity of capital is

$\frac{\partial z}{\partial y} = .3x^{.7}y^{-.7}$ or $\frac{.3x^{.7}}{y^{.7}}$.

43. $M(x, y) = 2xy + 10xy^2 + 30y^2 + 20$

(a) $\frac{\partial M}{\partial x} = 2y + 10y^2$

$\left(\frac{\partial M}{\partial x}\right)(20, 4) = 168$

(b) $\frac{\partial M}{\partial y} = 2x + 20yx + 60y$

$\left(\frac{\partial M}{\partial y}\right)(24, 10) = 5448$

(c) $M_x(17, 3) = 2(3) + 10(9)$

$= 96$

(d) $M_y(21, 8)$

$= 2(21) + 20(21)(8) + 60(8)$

$= 3882$

45. $A(W, H) = .202W^{.425}H^{.725}$

(a) $\frac{\partial A}{\partial W} = .08585W^{-.575}H^{.725}$

(b) $\left(\frac{\partial A}{\partial W}\right)(72, 1.8)$

$= .08585(72)^{-.575}(1.8)^{.725}$

$= .0112$

(c) $\frac{\partial A}{\partial H} = .14645W^{.425}H^{-.275}$

(d) $\left(\frac{\partial A}{\partial H}\right)(70, 1.6)$

$= .14645(70)^{.425}(1.6)^{-.275}$

$\approx .783$

Section 14.3

1. $f(x, y) = xy + x - y$

$f_x = y + 1$

$f_y = x - 1$

If $f_x = 0$, $y = -1$

If $f_y = 0$, $x = 1$.

Therefore, $(1, -1)$ is the critical point.

$f_{xx} = 0$

$f_{yy} = 0$

$f_{xy} = 1$

For $(1, -1)$,

$M = 0 \cdot 0 - 1^2 = -1 < 0$.

A saddle point is at $(1, -1)$.

3. $f(x, y) = x^2 - 2xy + 2y^2 + x - 5$

$f_x = 2x - 2y + 1$

$f_y = -2x + 4y$

Solve the system $f_x = 0$, $f_y = 0$.

$$\begin{aligned} 2x - 2y + 1 &= 0 \\ \underline{-2x + 4y \quad} &\underline{= 0} \\ 2y + 1 &= 0 \\ y &= -\tfrac{1}{2} \end{aligned}$$

$$-2x + 4\left(-\frac{1}{2}\right) = 0$$

$-2x = 2$

$x = -1$

Therefore, $(-1, -1/2)$ is the critical point.

$f_{xx} = 2$

$f_{yy} = 4$

$f_{xy} = -2$

For $\left(-1, -\frac{1}{2}\right)$,

$M = 2 \cdot 4 - (-2)^2$

$= 4 > 0$.

Since $f_{xx} = 2 > 0$, then a relative minimum is at $\left(-1, -\frac{1}{2}\right)$.

5. $f(x, y) = x^2 - xy + y^2 + 2x + 2y + 6$

$f_x = 2x - y + 2$

$f_y = -x + 2y + 2$

Solve the system $f_x = 0$, $f_y = 0$.

$2x - y + 2 = 0$

$-x + 2y + 2 = 0$

$2x - y + 2 = 0$

$\underline{-2x + 4y + 4 = 0}$

$3y + 6 = 0$

$y = -2$

$-x + 2(-2) + 2 = 0$

$x = -2$

$(-2, -2)$ is the critical point.

$f_{xx} = 2$

$f_{yy} = 2$

$f_{xy} = -1$

For $(-2, -2)$,

$M = (2)(2) - (-1)^2 = 3 > 0.$

Since $f_{xx} > 0$, a relative minimum is at $(-2, -2)$.

7. $f(x, y) = x^2 + 3xy + 3y^2 - 6x + 3y$

$f_x = 2x + 3y - 6$

$f_y = 3x + 6y + 3$

Solve the system $f_x = 0$, $f_y = 0$.

$2x + 3y - 6 = 0$

$3x + 6y + 3 = 0$

$-4x - 6y + 12 = 0$

$\underline{3x + 6y + 3 = 0}$

$-x + 15 = 0$

$x = 15$

$3(15) + 6y + 3 = 0$

$y = -8$

$(15, -8)$ is the critical point.

$f_{xx} = 2$

$f_{yy} = 6$

$f_{xy} = 3$

For $(15, -8)$,

$M = 2 \cdot 6 - 9 = 3 > 0.$

Since $f_{xx} > 0$, then a relative minimum is at $(15, -8)$.

9. $f(x, y) = 4xy - 10x^2 - 4y^2 + 8x + 8y + 9$

$f_x = 4y - 20x + 8$

$f_y = 4x - 8y + 8$

$4y - 20x + 8 = 0$

$4x - 8y + 8 = 0$

$4y - 20x + 8 = 0$

$\underline{-4y + 2x + 4 = 0}$

$-18x + 12 = 0$

$x = \frac{2}{3}$

$4y - 20\left(\frac{2}{3}\right) + 8 = 0$

$y = \frac{4}{3}$

The critical point is $\left(\frac{2}{3}, \frac{4}{3}\right)$.

$f_{xx} = -20$

$f_{yy} = -8$

$f_{xy} = 4$

For $\left(\frac{2}{3}, \frac{4}{3}\right)$,

$M = (-20)(-8) - 16 = 144 > 0.$

Since $f_{xx} < 0$, then a relative minimum is at $\left(\frac{2}{3}, \frac{4}{3}\right)$.

11. $f(x, y) = x^2 + xy - 2x - 2y + 2$

$f_x = 2x + y - 2$

$f_y = x - 2$

$2x + y - 2 = 0$

$x - 2 = 0$

$x = 2$

$2(2) + y - 2 = 0$

$y = -2$

The critical point is $(2, -2)$.

$f_{xx} = 2$

$f_{yy} = 0$

$f_{xy} = 1$

For $(2, -2)$

$M = 2 \cdot 0 - 1^2 = -1 < 0$

A saddle point is at $(2, -2)$.

13. $f(x, y) = x^2 - y^2 - 2x + 4y - 7$

$f_x = 2x - 2$

$f_y = -2y + 4$

If $2x - 2 = 0$, $x = 1$.

If $-2y + 4 = 0$, $y = 2$.

The critical point is (1, 2).

$f_{xx} = 2$

$f_{yy} = -2$

$f_{xy} = 0$

For (1, 2)

$M = -4 - 0 < 0$

A saddle point is at (1, 2).

15. $f(x, y) = 2x^2 + 3y^2 - 12xy + 4$

$f_x = 6x^2 - 12y$

$f_y = 6y - 12x$

$6x^2 - 12y = 0$

$6y - 12x = 0$

Solve the second equation for y:

$y = 2x$

Substitute for y in the first equation.

$6x^2 - 12(2x) = 0$

$6x(x - 4) = 0$

$x = 0$ or $x = 4$

Then, $y = 0$ or $y = 8$.

The critical points are (0, 0) and (4, 8).

$f_{xx} = 12x$

$f_{yy} = 6$

$f_{xy} = -12$

For (0, 0)

$M = 12(0)6 - (-12)^2$

$= -144 < 0.$

A saddle point is at (0, 0).

For (4, 8),

$M = 12(4)6 - (-12)^2$

$= -144 > 0.$

Since $f_{xx} = 12(4) = 48 > 0$, a relative minimum is at (4, 8).

17. $f(x, y) = x^2 + 4y^3 - 6xy - 1$

$f_x = 2x - 6y$

$f_y = 12y^2 - 6x$

$2x - 6y = 0$

$x = 3y$

Substitute for x in $12y^2 - 6x = 0$.

$12y^2 - 6(3y) = 0$

$6y(2y - 3) = 0$

$y = 0$ or $y = \frac{3}{2}$

Then, $x = 0$ or $x = \frac{9}{2}$.

The critical points are (0, 0) and (9/2, 3/2).

$f_{xx} = 2$

$f_{yy} = 24y$

$f_{xy} = -6$

For (0, 0),

$M = 2 \cdot 24(0) - (-6)^2$

$= -36 < 0.$

A saddle point is at (0, 0).

For (9/2, 3/2),

$M = 2 \cdot 24\left(\frac{3}{2}\right) - (-6)^2$

$= 36 > 0.$

Since $f_{xx} > 0$, a relative minimum is at (9/2, 3/2).

19. $f(x, y) = e^{xy}$

$f_x = ye^{xy}$

$f_y = xe^{xy}$

$ye^{xy} = 0$

$xe^{xy} = 0$

$x = y = 0$

The critical point is (0, 0).

$f_{xx} = y^2e^{xy}$

$f_{yy} = x^2e^{xy}$

$f_{xy} = e^{xy} + xye^{xy}$

For (0, 0),

$M = 0 \cdot 0 - (e^0)^2 = -1 < 0$

A saddle point is at (0, 0).

21. $f(x, y) = 3xy - x^3 - y^3 + \frac{1}{8}$

$f_x = 3y - 3x^2$

$f_y = 3x - 3y^2$

Solve the system $f_x = 0$, $f_y = 0$.

$3y - 3x^2 = 0$

$3x - 3y^2 = 0$

Solve the first equation for y.

$3y = 3x^2$

$y = x^2$

Substitute into the second equation and solve for x.

$3x - 3(x^2)^2 = 0$

$3x - 3x^4 = 0$

$3x(1 - x^3) = 0$

$x = 0$ or $x = 1$

Then, $y = 0$ or $y = 1$.

The critical points are (0, 0) and (1, 1).

$f_{xx} = -6x$

$f_{yy} = -6y$

$f_{xy} = 3$

For (0, 0):

$M = f_{xx}(0, 0) \cdot f_{yy}(0, 0) - [f_{xy}(0, 0)]^2$

$= 0 \cdot 0 - (3)^2$

$= -9 < 0$

So (0, 0) is a saddle point.

For (1, 1):

$M = f_{xx}(1, 1) \cdot f_{yy}(1, 1) - [f_{xy}(1, 1)]^2$

$= (-6)(-6) - [3]^2$

$= 36 - 9$

$= 27 > 0$

$f_{xx} = -6(1) = -6 < 0$

A relative maximum exists at (1, 1).

23. $f(x, y) = x^4 - 2x^2 + y^2 + \frac{17}{16}$

$f_x = 4x^3 - 4x$

$f_y = 2y$

Solve the system $f_x = 0$, $f_y = 0$.

$4x^3 - 4x = 0 \quad (1)$

$2x = 0 \quad (2)$

Equation (1) gives

$4x(x^2 - 1) = 0$

$x = 0$ or $x = 1$ or $x = -1$.

Equation (2) gives

$y = 0$.

The critical points are (0, 0) (1, 0) and (-1, 0).

$f_{xx} = 12x^2 - 4$

$f_{yy} = 2$

$f_{xy} = 0$

For (0, 0):

$M = (-4)(2) - 0^2 = -8 < 0$

A saddle point is at (0, 0).

For (1, 0):

$M = (8)(2) - 0^2 = 16 > 0$

$f_{xx} = 8 > 0$

So (1, 0) is a relative minimum.

For (-1, 0)

$M = (8)(2) - 0^2 = 16 > 0$

$f_{xx} = 8 > 0$

So (-1, 0) is a relative minimum.

25. $f(x, y) = x^4 - y^4 - 2x^2 + 2y^2 + \frac{1}{16}$

$f_x = 4x^3 - 4x$

$f_y = -4y^3 + 4y$

Solve $f_x = 0$, $f_y = 0$.

$4x^3 - 4x = 0 \quad (1)$

$-4y^3 + 4y = 0 \quad (2)$

$4x(x^2 - 1) = 0 \quad (1)$

$4x(x + 1)(x - 1) = 0$

$x = 0$ or $x = -1$ or $x = 1$

$-4y^3 + 4y = 0 \quad (2)$

$-4y(y^2 \ 1) = 0$

$-4y(y + 1)(y - 1) = 0$

$-4y = 0$ or $y + 1 = 0$ or $y - 1 = 0$

$y = 0$ or $y = 1$ or $y = -1$.

The critical points are (0, 0) (0, -1), (0, 1), (-1, 0), (-1, -1), (-1, 1), (1, 0) (1, -1), and (1, 1).

$f_{xx} = 12x^2 - 4$

$f_{yy} = -12y^2 + 4$

$f_{xy} = 0$

For (0, 0):

$M = (-4)(4) - 0 = -16 < 0$

Saddle point

For (0, -1):

$M = (-4)(-8) - 0 = 32 > 0$

and $f_{xx} = -4$ *Relative maximum*

For (0, 1):

$M = (-4)(-8) - 0 = 32 > 0$

and $f_{xx} = -4$ *Relative maximum*

For (-1, 0):

$M = (8)(4) - 0 = 32 > 0$

and $f_{xx} = 12 - 4 = 8$

Relative maximum

For (-1, -1):

$M = (8)(-8) - 0 = -64 < 0$

Saddle point

For (-1, 1):

$M = (8)(-8) - 0 = -64 < 0$

Saddle point

For (1, 0):

$M = (8)(4) - 0 = 32 > 0$

and $f_{xx} = 8$

Relative maximum

For (1, -1):

$M = (8)(-8) - 0 = -64 < 0$

Saddle point

For (1, 1):

$M = (8)(-8) - 0 = -64 < 0$

Saddle point

So saddle points are at (0, 0), (-1, -1), (-1, 1), (1, -1), and (1, 1).

Relative minima are at (1, 0) and (-1, 0).

Relative maxima are at (0, 1) and (0, -1).

27. $P(x, y) = 1000 + 24x - x^2 + 80y - y^2$

$P_x = 24 - 2x$

$P_y = 80 - 2y$

$24 - 2x = 0$

$80 - 2y = 0$

$12 = x$

$40 = y$ *Critical point (12, 40)*

$P_{xx} = -2$

$P_{yy} = -2$

$P_{xy} = 0$

For (12, 40),

$M = -2(02) - 0^2 = 4 > 0$

Since $P_{xx} < 0$, $x = 12$ and $y = 40$ maximize the profit.

Therefore,

$P(12, 40) = 1000 + 24(12) - (12)^2 + 80(40) - (40)^2 = 2744$

is the maximum profit.

29. $R(x, y) = 800 - 2x^3 + 12xy - y^2$

$R_x = -6x^2 + 12y$

$R_y = 12x - 2y$

$-6x^2 + 12y = 0$ *(1)*

$12x - 2y = 0$ *(2)*

Solve equation 2 for y and substitute in equation 1.

$12x = 2y$

$6x = y$

$-6x^2 + 12(6x) = 0$

$-6x(x - 12) = 0$

$x = 0$ or $x = 12$

Then $y = 0$ or $y = 72$.

$R_{xx} = -12x$

$R_{yy} = -2$

$R_{xy} = 12$

For (0, 0):

$M = -144 < 0$

A saddle point is at (0, 0).

For (12, 72):

$M = -12(12)(-2) - 12^2 = 144 > 0$

Since $R_{xx} < 0$, a relative maximum is at (12, 72).

So, 12 lb of Super chicken feed and 72 lb of Brand A feed produce the maximum number of chickens.

31. $C(x, y) = 2x^2 + 3y^2 - 2xy$
$+ 2x - 126y + 3800$

$C_x = 4x - 2y + 2$
$C_y = 6y - 2x - 126$
$0 = 4x - 2y + 2$
$0 = 6y - 2x - 126$
$0 = 2x - y + 1$
$\underline{0 = -2x + 6y - 126}$
$0 = 5y - 125$
$y = 25$
$0 = 2x - 25 + 1$
$12 = x$ *Critical point (12, 25)*
$C_{xx} = 4$
$C_{yy} = 6$
$C_{xy} = -2$

For (12, 25):
$M = 4 \cdot 6 - 4 = 20 > 0$

Since $C_{xx} > 0$, 12 units of electrical tape and 25 units of packing tape should be produced to yield a minimum cost.

$C(12, 25) = 2(12)^2 + 3(25)^2 - 2(12 \cdot 25)$
$+ 2(12) - 126(25) + 3800$
$= 2237$

The minimum cost is 2237.

33. The volume is

$V = xyz - 27$, so $z = \frac{27}{xy}$.

The surface area is

$S = 2xy + 2yz + 2xz$

$= 2xy + 2y\left(\frac{27}{xy}\right) + 2x\left(\frac{27}{xy}\right)$

$= 2xy + \frac{54}{x} + \frac{54}{y}$.

$S_x = 2y - \frac{54}{x^2}$

$S_y = 2x - \frac{54}{y^2}$

Let $S_x = 0$ and $S_y = 0$

$2y - \frac{54}{x^2} = 0$ *(1)*

$2x - \frac{54}{y^2} = 0$ *(2)*

Solve equation 1 for y.

$2y = \frac{54}{x^2}$

$y = \frac{27}{x^2}$

Substitute into equation 2.

$$2x - \frac{54}{\left(\frac{27}{x^2}\right)^2} = 0$$

$$x = \frac{27}{\frac{(27)^2}{x^4}}$$

$27x = x^4$
$0 = x^4 - 27x$
$0 = x(x^3 - 27)$
$x = 0$ or $x^3 - 27 = 0$
$x^3 = 27$
$x = 3$

So $y = \frac{27}{x^2} = \frac{27}{3^2} = \frac{27}{9} = 3$.

Thus, the critical point is (3, 3).

$S_{xx} = \frac{108}{x^3}$

$S_{yy} = \frac{100}{y^3}$

$S_{xy} = 2$.
$M = (-4)(-4) - 2 = 14 > 0$

$S_{xx} = \frac{108}{27} = 4 > 0$

So a minimum surface area will occur, when $x = 3$, $y = 3$, and $z = 27/3(3) = 3$.
The dimensions will be 3 meters by 3 meters by 3 meters.

Chapter 14 Review Exercises

1. $f(x, y) = -4x^2 + 6xy - 3$
$f(-1, 2) = -4(-1)^2 = 6(-1)(2) - 3$
$= -19$
$f(6, -3) = -4(6)^2 + 6(6)(-3) - 3$
$= -4(36) + (-108) - 3$
$= -255$

3. $f(x, y) = \frac{x - 3y}{x + 4y}$

$f(-1, 2) = \frac{-1 - 6}{-1 + 8} = \frac{-7}{7} = -1$

$f(6, -3) = \frac{6 - 3(-3)}{6 + 4(-3)} = \frac{6 + 9}{6 - 12}$

$= \frac{15}{-6} = -\frac{5}{2}$

See the graphs for Exercises 5-9 in the answer section at the back of the textbook.

5. The plane $x + y + z = 4$ intersects the axes at (4, 0, 0), (0, 4, 0), and (0, 0, 4).

7. The plane $5x + 2y = 10$ intersects the x- and y-axes at (2, 0, 0) and (0, 5, 0). Note there is no z-intercept since $x = y = 0$ is not a solution of the equation of the plane.

9. The plane $x = 3$ intersects the x-axis at (3, 0, 0). It is parallel to the y- and z-axes since there is no y nor z in the equation.

11. $C(x, y) = 2x^2 + 4y^2 - 3xy + \sqrt{x}$

(a) $C(10, 5)$

$= 2(10)^2 + 4(5)^2 - 3(10)(5) + \sqrt{10}$

$= 200 + 100 - 150 + \sqrt{10}$

$= \$(150 + \sqrt{10})$

(b) $C(15, 10)$

$= 2(15)^2 + 4(10)^2 - 3(15)(10) + \sqrt{15}$

$= 450 + 400 - 450 + \sqrt{15}$

$= \$(400 + \sqrt{15})$

(c) $C(20, 20)$

$= 2(20)^2 + 4(20)^2 - 3(20)(20) + \sqrt{20}$

$= 800 + 1600 - 1200 + \sqrt{20}$

$= \$(1200 + 2\sqrt{5})$

13. $z = f(x, y) = \frac{x + y^2}{x - y^2}$

(a) $\frac{\partial z}{\partial y} = \frac{(x - y^2)\cdot 2y - (x + y^2)(-2y)}{(x - y^2)^2}$

$= \frac{4xy}{(x - y^2)^2}$

(b) $\frac{\partial z}{\partial x} = \frac{(x - y^2)\cdot 1 - (x + y^2)\cdot 1}{(x - y^2)^2}$

$= \frac{-2y^2}{(x - y^2)^2}$

$= -2y^2(x - y^2)^{-2}$

$\left(\frac{\partial z}{\partial x}\right)(0, 2) = \frac{-8}{(-4)^2} = -\frac{1}{2}$

(c) $f_{xx} = 4y^2(x - y^2)^{-3}$

$= \frac{4y^2}{(x - y^2)^3}$

$f_{xx}(-1, 0) = \frac{0}{1} = 0$

15. $f(x, y) = 6x^5y - 8xy^9$

$f_x = 30x^4y - 8y^9$, $f_y = 6x^5 - 72xy^8$

17. $f(x, y) = \dfrac{2x + 5y^2}{3x^2 + y^2}$

$f_x = \dfrac{(3x^2 + y^2)\cdot 2 - (2x + 5y^2)\cdot 6x}{(3x^2 + y^2)^2}$

$= \dfrac{2y^2 - 6x^2 - 30xy^2}{(3x^2 + y^2)^2}$

$f_y = \dfrac{(3x^2 + y^2)\cdot 10y}{(3x^2 + y^2)^2}$

$- \dfrac{(2x + 5y^2)\cdot 2y}{(3x^2 + y^2)^2}$

$= \dfrac{30x^2y - 4xy}{(3x^2 + y^2)^2}$

19. $f(x, y) = (y - 2)e^{x+2y}$

$f_x = (y - 2)^2e^{x+2y}$

$f_y = e^{x+2y}\cdot 2(y - 2)$

$+ (y - 2)^2\cdot 2e^{x+2y}$

$= 2(y - 2)[1 + (y - 2)]e^{x+2y}$

$= 2(y - 2)(y - 1)e^{x+2y}$

21. $f(x, y) = \ln |2 - x^2y^3|$

$f_x = \dfrac{1}{2 - x^2y^3}\cdot(-2xy^3)$

$= \dfrac{-2xy^3}{2 - x^2y^3}$

$f_y = \dfrac{1}{2 - x^2y^3}\cdot(-3x^2y^3)$

$= \dfrac{-3x^2y^2}{2 - x^2y^3}$

23. $f(x; y) = -6xy^4 + x^2y$

$f_x = -6y^4 + 2xy$

$f_{xx} = 2y$

$f_{xy} = -24y^3 + 2x$

25. $f(x, y) = \dfrac{3x + y}{x - 1}$

$f_x = \dfrac{(x - 1)\cdot 3 - (3x + y)\cdot 1}{(x - 1)^2}$

$= \dfrac{-3 - y}{(x - 1)^2} = (-3 - y)(x - 1)^{-2}$

$f_{xx} = -2(-3 - y)(x - 1)^{-3}$

$= \dfrac{2(3 + y)}{(x - 1)^3}$

$f_{xy} = \dfrac{-1}{(x - 1)^2}$

27. $f(x, y) = ye^{x^2}$, $f_x = 2xye^{x^2}$

$f_{xx} = 2xy\cdot 2xe^{x^2} + e^{x^2}\cdot 2y$

$= 2ye^{x^2}(2x^2 + 1)$

$f_{xy} = 2xe^{x^2}$

29. $f(x, y) = \ln(1 - 3xy^2)$

$f_x = \dfrac{1}{1 + 3xy^2}\cdot 3y^2$

$= \dfrac{3y^2}{1 + 3xy^2}$

$= 3y^2(1 + 3xy^2)^{-1}$

$f_{xx} = 3y^2\cdot(-3y^2)(1 + 3xy^2)^{-2}$

$= \dfrac{-9y^4}{(1 + 3xy^2)^2}$

$f_{xy} = \dfrac{(1 + 3xy^2)\cdot 6y - 3y^2(6xy)}{(1 + 3xy^2)^2}$

$= \dfrac{6y}{(1 + 3xy^2)^2}$

31. $z = x^2 + y^2 + 9x - 8y + 1$

$z_x = 2x + 9,\ z_y = 2y - 8$

$2x + 9 = 0$

$x = -\frac{9}{2}$

$2y - 8 = 0$

$y = 4$

$z_{xx} = 2,\ z_{yy} = 2,\ z_{xy} = 0$

$M = 2(2) - (0)^2 = 4 > 0$ and $z_{xx} > 0$.

Relative minimum at $\left(-\frac{9}{2}, 4\right)$.

33. $z = x^3 - 8y^2 + 6xy + 4$

$z_x = 3x^2 + 6y,\ z_y = -16y + 6x$

Set $z_x = 0$ and solve for y.

$3x^2 + 6y = 0$

$x^2 + 2y = 0$

$y = -\frac{x^2}{2}$

Solve $z_y = 0$.

$-16y + 6x = 0$

$-8y + 3x = 0$

Substitute for y in the last equation.

$-8\left(-\frac{x^2}{2}\right) + 3x = 0$

$4x^2 + 3x = 0$

$x(4x + 3) = 0$

$x = 0$ or $x = -\frac{3}{4}$

$y = 0$ or $y = -\frac{9}{32}$

$z_{xx} = 6x,\ z_{yy} = -16,\ z_{xy} = 6$

$M = 6x(-16) - (6)^2 = -96x - 36$

At (0, 0), $M = -36 < 0$;

Saddle point at (0, 0)

At $\left(-\frac{3}{4}, -\frac{9}{32}\right)$, $M = 36 > 0$ and

$z_{xx} = -\frac{9}{32} < 0$

Relative minimum at $\left(-\frac{3}{4}, -\frac{9}{32}\right)$

35. $f(x, y) = 3x^2 + 2xy + 2y^2 - 3x + 2y - 9$

$f_x = 6x + 2y - 3,$

$f_y = 2x + 4y + 2$

Solve the system $f_x = 0$ and $f_y = 0$

$6x + 2y - 3 = 0$

$2x + 4y + 2 = 0$

$$\begin{array}{r} -12x - 4y + 6 = 0 \\ \underline{2x + 4y + 2 = 0} \\ -10x \qquad + 8 = 0 \end{array}$$

$x = \frac{4}{5}$

$2\left(\frac{4}{5}\right) + 4y + 2 = 0$

$4y = -\frac{18}{5}$

$y = -\frac{9}{10}$

$f_{xx} = 6,\ f_{yy} = 4,\ f_{xy} = 2$

$M = 6(4) - (2)^2 = 2 > 0$ and $f_{xx} > 0$.

Relative minimum at $\left(\frac{4}{5}, -\frac{9}{10}\right)$

37. $f(x, y) = 7x^2 + y^2 - 3x + 6y - 5xy$

$f_x = 14x - 3 - 5y$

$f_y = 2y + 6 - 5x$

Solve $f_x = 0$ and $f_y = 0$.

$14x - 5y - 3 = 0$

$-5x + 2y + 6 = 0$

$$\begin{array}{r} 28x - 10y - 6 = 0 \\ \underline{-25x + 10y + 30 = 0} \\ 3x \qquad + 24 = 0 \end{array}$$

$x = -8$

$-5(-8) + 2y + 6 = 0$

$2y = -46$

$y = -23$

$f_{xx} = 14,\ f_{yy} = 2,\ f_{xy} = -5$

$M = 14(2) - (-5)^2 = 3 > 0$ and $f_{xx} > 0$.

Relative minimum at (-8, -23).

39. $z = x^{(.6)}y^{(.4)}$

(a) The marginal productivity of labor is

$$\frac{\partial z}{\partial x} = .6x^{-(.4)}y^{(.4)}$$

$$= .6\left(\frac{y}{x}\right)^{.4}.$$

(b) The marginal productivity of capital is

$$\frac{\partial z}{\partial y} = .4x^{(.6)}y^{-(.6)}$$

$$= .4\left(\frac{x}{y}\right)^{.6}.$$